CONTENTS

CONTENTS

SCHAUM'S OUTLINE OF

THEORY AND PROBLEMS

OF

ELECTRONIC CIRCUITS

•

BY

EDWIN C. LOWENBERG, Ph.D.
Professor of Electrical Engineering
University of Nebraska

•

SCHAUM PUBLISHING CO.
257 Park Avenue South, New York 10010

PRINTED IN THE UNITED STATES OF AMERICA

Preface

Electronics may be broadly divided into two categories: the application of the physical properties of materials in the development of electronic control devices, and the utilization of electronic control devices in circuit applications. The second category is often identified as electronic circuit theory.

This book is designed for a first course in electronic circuits for engineers. It should also be valuable to non-engineers who are taking formal courses or are learning electronic instrumentation by self-study. It is based on wide experience in teaching electronics, covering the range from courses for life and behavioral scientists to those for electrical engineers.

A brief review of basic electric circuits is followed by the introduction of one port electronic control devices. Diodes are examples; ideal diodes are models. Circuits containing nonlinear resistances are analyzed graphically. Resistance concepts are generalized as required for representing electronic control devices. A number of applications involving diodes are included to illustrate the methods of analysis. No attempt has been made to include all of the common circuits; they appear in numerous handbooks.

Two port networks are introduced in Chapter 4. Two port electronic control devices for which the input port is an open circuit, such as vacuum tubes, are discussed in Chapter 5. Graphical analysis techniques for finding operating points are developed for these circuits. Simple amplifier circuits are analyzed graphically and analytically.

Chapter 6 continues the discussion by introducing electronic control devices for which the input port is not an open circuit; transistors are the most common examples. The most general two port models for electronic control devices are now used. A discussion of linear amplifiers, including frequency response problems and a representation of a complete amplifier circuit using two port network models, follows logically in Chapter 7.

Chapter 8 is an introduction to more advanced topics and thereby provides a transition to advanced-level books. The discussion of amplifier topics illustrates how the methods of analysis are applied in electronic system analysis.

This book emphasizes fundamentals which will not change with time. The methods of analysis will continue to be used in the future for new circuits and circuits with new devices. Two port representations are stressed because they are useful for modeling an entire linear amplifier, whether it is a conventional one or an integrated circuit module of the future.

I am indebted to many colleagues for their helpful suggestions, and to all the students who have helped in many ways. The staff of the Schaum Publishing Company has been most cooperative and has greatly contributed to the manuscript. Special thanks are extended to Mrs. Sally Curtis and Mrs. Marge Alles who typed most of the manuscript.

EDWIN C. LOWENBERG

University of Nebraska
August, 1967

Chapter 1

Introduction

1.1 HISTORICAL INTRODUCTION

The area of electronics probably began with the discovery by Hertz (1888) that electromagnetic energy could be propagated and detected. Electronics and electronic circuits entered a period of rapid evolution with the development of a vacuum tube diode by Fleming (1903) followed by the discovery of a crystal detector by Pickard (1906), and later the vacuum tube triode by De Forest (1907). The impact of the many important discoveries and inventions on the area of electronics is a matter of record. Electronics has traditionally included the problems of developing and manufacturing electronic control devices, and their applications in electronic circuits and systems.

The characteristics of electronic control devices have often been limiting factors in the achievement of the desired performance of electronic systems. Some problems were solved by developing new types of electronic control devices, and others by developing new circuits. For example, vacuum tube tetrodes and pentodes solved some of the problems encountered when triodes were used in tuned amplifier circuits of radio receivers. Some of the performance characteristics of linear amplifiers were improved by incorporating negative (inverse) feedback.

The invention of the transistor by Bardeen and Brattain (1948), followed by the development of theories of *P-N* junctions by Shockley (1949) and others, was the beginning of another revolution in electronics. Electronic engineers who were on intimate terms with vacuum tubes found themselves confronted with new concepts. In addition to new physical principles, more general circuit representations were needed for representing transistors as circuit elements. The first equivalent circuits for transistors were developed to somewhat resemble the familiar ones employed for representing vacuum tubes.

As the family of solid state electronic devices began to enlarge, it became apparent that there are two general types of problems which are closely related. One of these is largely concerned with the physical principles and their applications in the development and improvement of electronic control devices. This is an area which sometimes changes rapidly, and then it is difficult to keep abreast of the state of the art. The other type of problem is more concerned with the development of circuit models and concepts, and with circuit applications for electronic control devices.

The fact that vacuum tubes, transistors, tunnel diodes, and other electronic control devices all have similar circuit representations in system applications, suggests the need for developing a general approach. This led to unified treatments of tubes and transistors. This book will continue this trend by concentrating on electronic control devices of the resistive class as compared to those of the reactive class. When the parameters of the control devices are resistive (in a general sense, as will be explained), the procedures for developing circuit representations and circuit applications have much in common regardless of the type of device. It is of course necessary to develop the general concepts using specific examples. Hence vacuum tubes and transistors will be employed in the development of general concepts. The introduction of a revolutionary new device should not be cause for concern so long as it belongs to the resistive class of electronic control devices. The more classical material discussed under electronic circuits may eventually be (and probably should be) included in the sequence of electrical circuits courses.

In this book it is assumed that the reader has completed introductory courses in calculus and electrical circuits. It is anticipated, however, that some non-engineering readers may not have this background. In such case, some of the derivations may be omitted and the emphasis placed on the circuit models and concepts.

1.2 CIRCUIT MODELS

The three linear, passive circuit models shown in Fig. 1-1 should be familiar to the reader. As a matter of convenience, in this book the polarity of a voltage will be indicated by an arrow as shown in Fig. 1-1. The arrow points in the direction of a voltage rise; therefore the direction opposite to the arrow is a voltage drop. Electrical potential difference is a scalar quantity and the arrow is employed to indicate polarity.

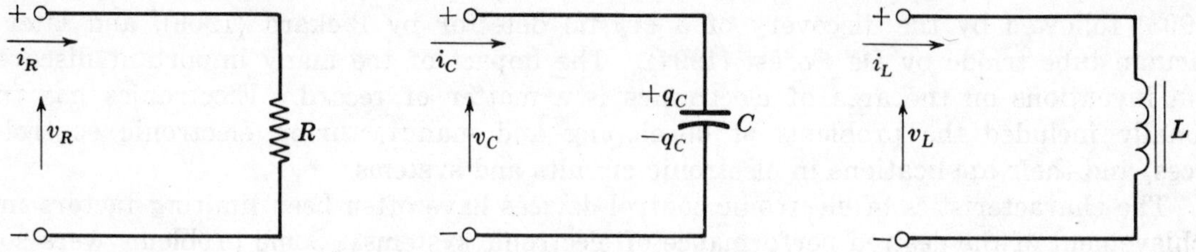

Resistor $\quad v_R = Ri_R$

Capacitor $\quad v_C = \dfrac{q_C}{C} = \dfrac{1}{C}\displaystyle\int i_C \, dt$

Inductor $\quad v_L = L\dfrac{di_L}{dt}$

Fig. 1-1. Passive Circuit Models

The direction of a current is also denoted by an arrow. It is probably appropriate to emphasize that the total instantaneous current is a scalar quantity. It has become customary to show a direction for a current because it is necessary to specify the direction of the current density, which is a vector. In this book the arrow shown with a current is the direction of the current density (which is the direction of the drift velocity of positive charges). It is convenient to say that this is the direction of the current. Every effort will be made to avoid using the term "current flow" because this is confusing. It is electrical charge that flows, and electrical current is the rate of flow of charge.

The two active element circuit models are the constant current generator and the constant voltage generator shown in Fig. 1-2. The constant voltage generator is symbolized

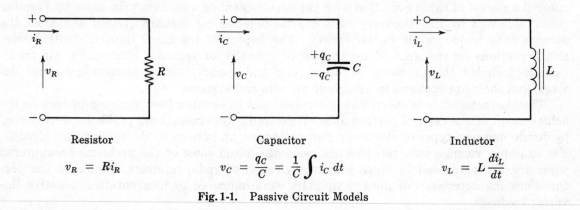

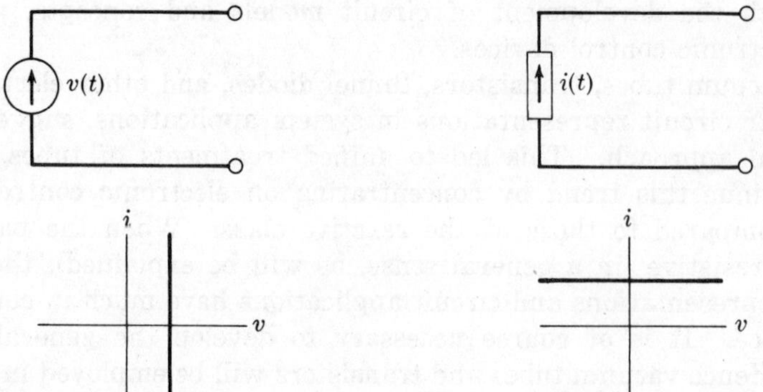

Constant voltage

Constant current

Fig. 1-2. Active Circuit Models

by a circle and an arrow showing the polarity of the generated voltage. The constant current generator is denoted by a rectangle with an arrow indicating the direction of the generated current. In neither model does the word constant mean independent of time. In the case of the voltage generator, the generated voltage is independent of the generator current; and in the current generator, the generated current is independent of the generator terminal voltage, as shown in Fig. 1-2. These graphs are for a given instant of time. A constant voltage generator which is also time independent is often symbolized as a battery, as shown in Fig. 1-3. The battery is a special case of the voltage generator of Fig. 1-2 in which $v(t) = V$ (a constant).

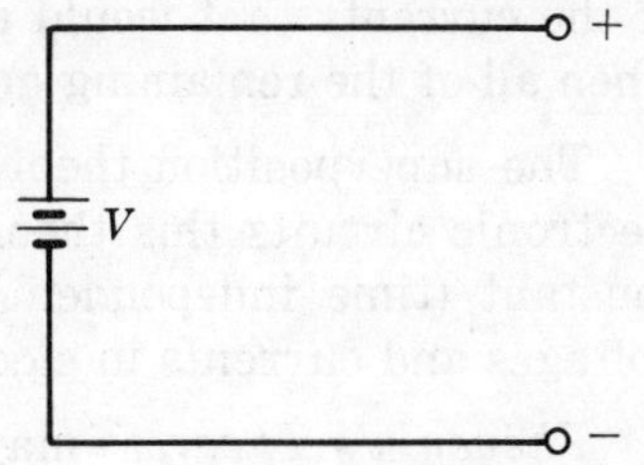

Fig. 1-3. Time Independent Constant Voltage Generator

Practical circuit elements are usually represented by using one or more of the linear models. The simplest representation which is adequate for the particular problem is usually used.

1.3 KIRCHHOFF'S LAWS

Kirchhoff's Voltage Law states that the algebraic sum of the instantaneous voltage drops around any closed circuit path is zero. In this book a voltage drop is written with a plus sign regardless of whether it is across an active or a passive element. A voltage rise is written with a minus sign.

Loop equations will usually be written by starting at some point of a loop and then writing the algebraic sum of the voltage drops, going in the reference direction for the loop current. The algebraic sum of the voltage drops is set equal to zero. Mistakes in sign are more likely to occur when the generator voltages of a loop are combined by inspection.

When a set of simultaneous equations is solved, Cramer's rule and determinants will usually be used. This method is most convenient, especially when there are more than three equations. The equations may then be easily written in matrix form when this is desirable.

Kirchhoff's Current Law states that the algebraic sum of the instantaneous currents away from any junction is zero. The current away from a junction is written with a plus sign and the current toward a junction is written with a minus sign.

Nodal equations will also be written by setting the algebraic sum of the currents equal to zero.

In general when there are many circuit elements connected in series, voltage generators are used to represent sources and loop equations are written. When there are many circuit elements in parallel, current generators are employed to represent sources and nodal equations are written. In some circuits it is convenient to use both voltage and current generators; then the resulting set of equations contains both loop and nodal equations.

Equations will usually be written in terms of instantaneous values of voltages and currents. When a circuit containing only linear passive elements is excited by a sinusoidal source, then the steady-state voltages and currents are sinusoidal and the frequency is that of the source. The voltages and currents are then specified by their magnitudes and phase angles with respect to a reference (usually a source).

1.4 NETWORK THEOREMS

The following network theorems are used quite frequently in electronic circuit problems.

The Superposition Theorem states that: Given a network containing only linear elements and more than one generator, the current in any branch is equal to the algebraic sum of the currents that would exist in the branch if each generator is considered individually when all of the remaining generators are replaced by their internal impedances.

The superposition theorem is often applied to circuits having a number of loops. In electronic circuits this theorem is also used for single loops containing both varying and constant (time independent) sources. This is the basis for considering only the signal voltages and currents in electronic circuits.

Thévenin's Theorem may be stated as follows: Given a two terminal network containing linear elements including any number of generators, the terminal electrical characteristics of the network are the same as a series combination of a voltage generator and an impedance provided that the equivalent circuit elements satisfy the following:

(*a*) The instantaneous value of the equivalent voltage generator is the same as the instantaneous open-circuit terminal voltage of the network.

(*b*) The impedance of the equivalent circuit is the impedance of the network when each generator has been replaced by its equivalent internal impedance.

Norton's Theorem is the dual of Thévenin's theorem and may be stated as follows: Given a two terminal network containing linear elements including any number of generators, the terminal electrical characteristics of the network are the same as a parallel combination of a current generator and an admittance provided that the equivalent circuit elements satisfy the following:

(*a*) The instantaneous value of the equivalent current generator is the same as the instantaneous short-circuit current in the terminals of the network.

(*b*) The admittance is the admittance of the network when each generator is replaced by its internal impedance.

It should be understood that the terminal characteristics of a network and its equivalent are the same. In other words, no experiment involving the terminals of the network can distinguish it from its Thévenin or Norton equivalents.

1.5 VOLTAGE AND CURRENT REPRESENTATIONS

The voltages and currents of a network may be specified in the time domain by writing the instantaneous values as functions of time. These representations are convenient to use in electronic circuit problems. Electronic circuit problems which are solved graphically on a point by point basis require that the instantaneous values of voltages and currents be used.

When capacitors or inductors are included in a circuit, the time domain equations are either differential or integral equations. This makes some of the simpler problems unnecessarily complicated. In beginning electrical circuits courses, frequency domain representations are usually introduced. The interest is usually in the steady-state solution to problems involving linear circuits excited by sinusoidal driving functions. In such problems phasor representations are most convenient because it is necessary to determine only the amplitude of each voltage and current and the phase angle of each with respect to a reference phasor.

When phasor representations are used and a single frequency is involved, the impedance or admittance of a circuit element or a two terminal circuit is readily defined. Then the impedance is the ratio of the phasor voltage to the phasor current. Impedance is dimensionally resistance and may be complex. The real part of the impedance is the resistance and the imaginary part is the reactance. The impedance is not a phasor. The admittance of a two terminal circuit is the reciprocal of the impedance; it is the ratio of the phasor current to the phasor voltage. The real part of the admittance is the conductance and the imaginary part is the susceptance.

Frequency domain representations are easily generalized by introducing Laplace transforms. General solutions (including transient solutions) may be found, including solutions for other than sinusoidal driving functions. The concepts of impedance and admittance are generalized by defining them in terms of the ratios of the Laplace transforms of the voltages and currents. However, since this book is designed for a first course in electronic circuits, it is assumed that Laplace transformation techniques may not be familiar to the reader.

To avoid differential and integral equations, impedance and admittance concepts are employed in writing circuit equations in the time domain. This results in a hybrid notation. However, there are no inconsistencies with the frequency domain definitions of impedance and admittance. In fact, this approach seems to be less troublesome to beginning students than phasor representations.

Considering the special case involving steady-state solutions for sinusoidal excitation, the currents are all sinusoidal. If the current is $i(t) = I \sin \omega t$, the voltage drop across an inductor is

$$v_L = L\,di/dt = L\omega I \cos \omega t = L\omega I \sin(\omega t + \pi/2) \tag{1.1}$$

When the notation is understood, the voltage across the inductor is

$$v_L = Z_L i = j\omega L I \sin \omega t \tag{1.2}$$

where the rotational operator j indicates that the voltage is $\pi/2$ radians ahead of the current. The voltage v_L given by equation (*1.2*) is the same as that given by (*1.1*).

When the current in a capacitor is $i(t) = I \sin \omega t$, the steady-state voltage drop is

$$v_C = Z_C i = -j(I/\omega C) \sin \omega t \quad \text{or} \quad v_C = (I/\omega C) \sin(\omega t - \pi/2) \tag{1.3}$$

When the time domain equations are written in this way, it is understood that only the steady-state response to sinusoidal excitation is being considered. This allows the impedance and admittance concepts as developed in beginning courses to be applied in time domain representations.

Solved Problems

1.1. Given the circuit of Fig. 1-4, find the current in the generator V_3 if $V_1 = 4$ v, $V_2 = 1$ v, $V_3 = 6$ v.

The three loop currents are identified as shown in Fig. 1-4. The three loop equations are written starting at the common node as follows:

$$+V_2 + 2I_1 - V_1 + 6(I_1 - I_2) = 0$$

$$6(I_2 - I_1) + 3I_2 + 7(I_2 - I_3) = 0$$

$$7(I_3 - I_2) + 4I_3 - V_3 = 0$$

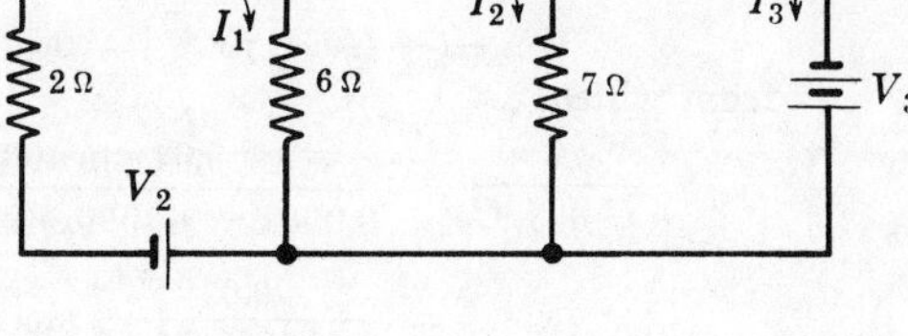

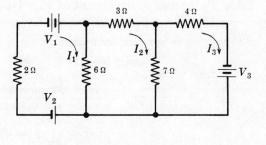

Fig. 1-4

Substituting numerical values for the generators, the equations may be written with the dependent variables forming columns to the left and the independent variables in a column to the right.

$$\begin{aligned} 8I_1 - 6I_2 \qquad\qquad &= 3 \\ -6I_1 + 16I_2 - 7I_3 &= 0 \\ -7I_2 + 11I_3 &= 6 \end{aligned}$$

Solving for I_3, using determinants and Cramer's rule,

$$I_3 = \frac{\begin{vmatrix} 8 & -6 & 3 \\ -6 & 16 & 0 \\ 0 & -7 & 6 \end{vmatrix}}{\begin{vmatrix} 8 & -6 & 0 \\ -6 & 16 & -7 \\ 0 & -7 & 11 \end{vmatrix}} = \frac{894 - 216}{1408 - 392 - 396} = \frac{678}{620} = 1.094 \text{ amps}$$

1.2. Given the circuit of Fig. 1-5, determine the voltage drop V_{AO}.

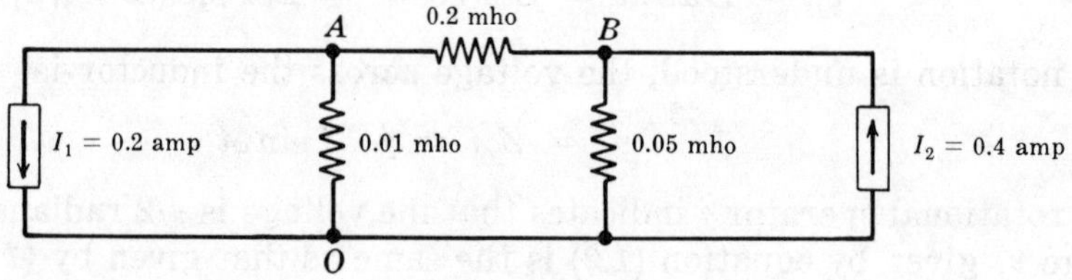

Fig. 1-5

The equations for nodes A and B are

$$+0.2 + 0.01V_{AO} + 0.2(V_{AO} - V_{BO}) = 0$$
$$+0.2(V_{BO} - V_{AO}) + 0.05V_{BO} - 0.4 = 0$$

which simplify to

$$0.21V_{AO} - 0.2V_{BO} = -0.2$$
$$-0.2V_{AO} + 0.25V_{BO} = 0.4$$

Solving for V_{AO},

$$V_{AO} = \frac{\begin{vmatrix} -0.2 & -0.2 \\ 0.4 & 0.25 \end{vmatrix}}{\begin{vmatrix} 0.21 & -0.2 \\ -0.2 & 0.25 \end{vmatrix}} = \frac{-0.05 + 0.08}{0.0525 - 0.04} = 2.4 \text{ volts}$$

1.3. Referring to the circuit of Fig. 1-6, calculate the steady-state voltage drop v assuming that $i = 0.01 \sin 4000\pi t$ ampere.

The equation for node A is

$$-i + Gv + j\omega Cv = 0$$

from which

$$v = \frac{i}{G + j\omega C} = \frac{0.01 \sin 4000\pi t}{0.0001 + j(4000\pi)(0.01 \times 10^{-6})}$$
$$= \frac{0.01}{0.000161/\underline{51.6^\circ}} \sin 4000\pi t$$
$$= 62.1 \sin(4000\pi t - 0.9) \text{ volts}$$

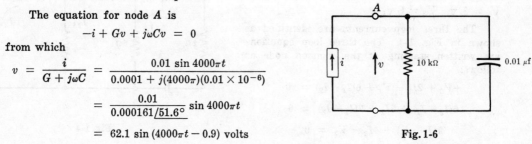

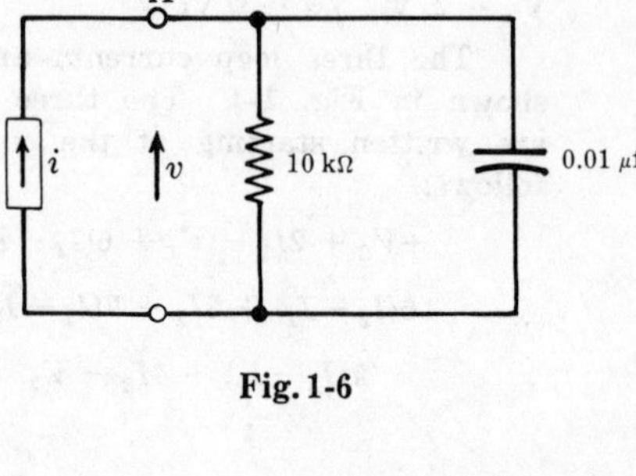

Fig. 1-6

1.4. For the network of Fig. 1-7 find (*a*) the Thévenin equivalent circuit and (*b*) the Norton equivalent circuit.

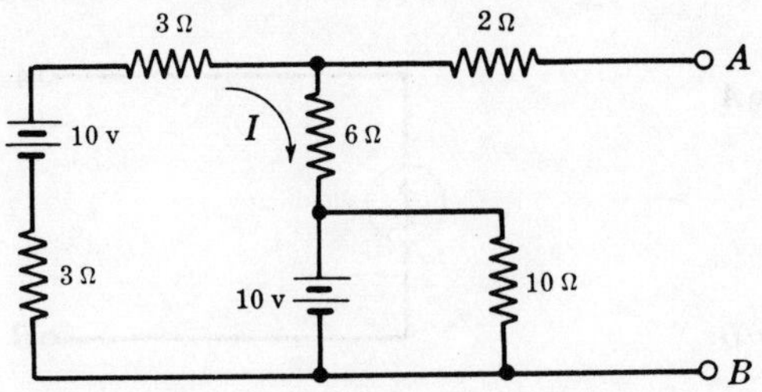

Fig. 1-7

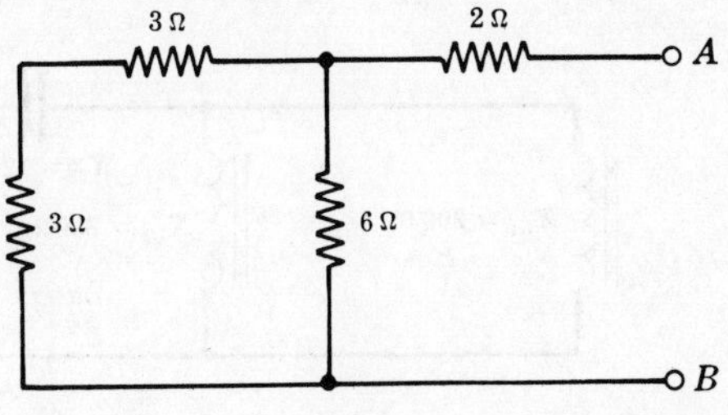

Fig. 1-8

(*a*) The equation for the closed loop is

$$3I - 10 + 3I + 6I + 10 = 0 \quad \text{or} \quad I = 0$$

The open circuit voltage $V_{AB} = 2(0) + 6(0) + 10 = 10$ volts. The voltage drops across the 2 Ω and 6 Ω resistors are zero.

When each battery is replaced by zero resistance, the circuit of Fig. 1-7 is drawn as in Fig. 1-8. The resistance of the branch containing the two 3 Ω resistors is 6 Ω. The parallel combination of the two branches containing 6 Ω resistors is 3 Ω. Hence $R_{AB} = 2 + 3 = 5$ Ω. The Thévenin equivalent circuit is shown in Fig. 1-9.

Thévenin's theorem could be applied to the parallel combination of the 10 volt battery and the 10 Ω resistor. When this is done, the combination is replaced by a 10 volt battery. When the 10 volt battery is ideal, the terminal voltage remains 10 volts regardless of the magnitude of the current.

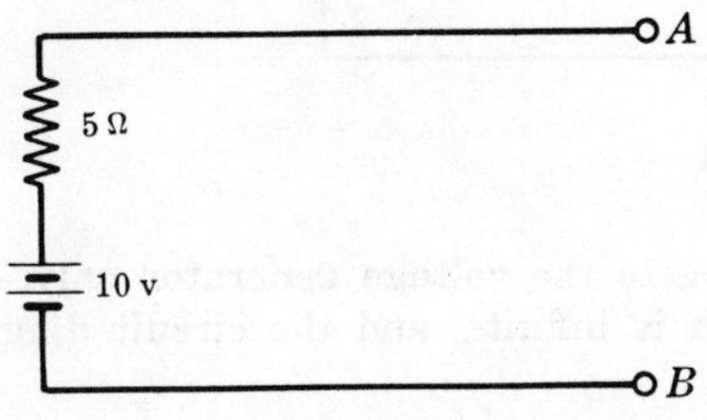

Fig. 1-9

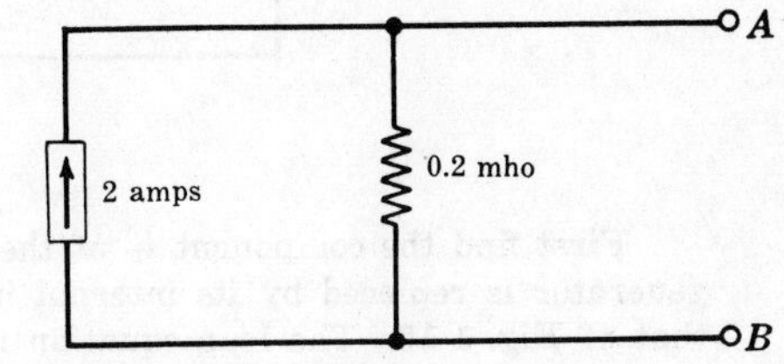

Fig. 1-10

(*b*) The Norton equivalent circuit may be found from the Thévenin equivalent. The short circuit current, found by shorting the terminals *A* and *B* together, is $I_{AB} = 10/5 = 2$ amps. The Norton equivalent circuit is shown in Fig. 1-10.

1.5. In the circuit of Fig. 1-11, the reactance of each impedance is specified for the given generator frequency. Find the Thévenin and Norton equivalent circuits for the network.

The equation for the closed loop is

$$-100 \sin \omega t + j200i + j300i = 0$$

from which

$$i = \frac{100}{j500} \sin \omega t = 0.2 \sin (\omega t - \pi/2) \text{ amp}$$

The open circuit voltage v_{AB} is

$$v_{AB} = -j120(0) + j300[0.2 \sin (\omega t - \pi/2)]$$
$$= 60 \sin \omega t \text{ volts}$$

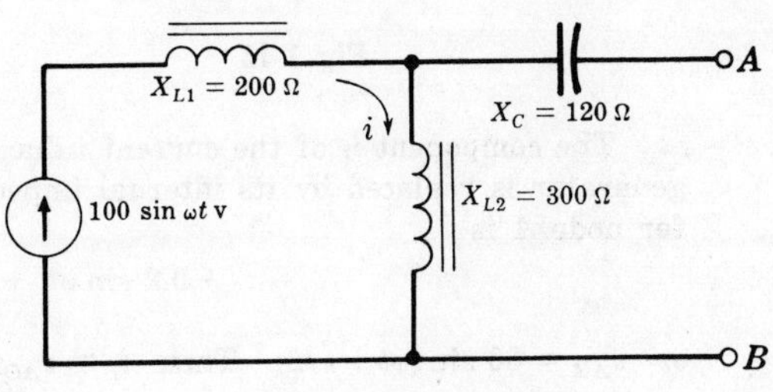

Fig. 1-11

The network with the generator replaced by its internal impedance is drawn in Fig. 1-12. The impedance of the parallel combination of the two inductors is $Z = \dfrac{(j200)(j300)}{j200 + j300} = j120\ \Omega$. The impedance $Z_{AB} = -j120 + j120 = 0\ \Omega$. The Thévenin equivalent circuit is shown in Fig. 1-13.

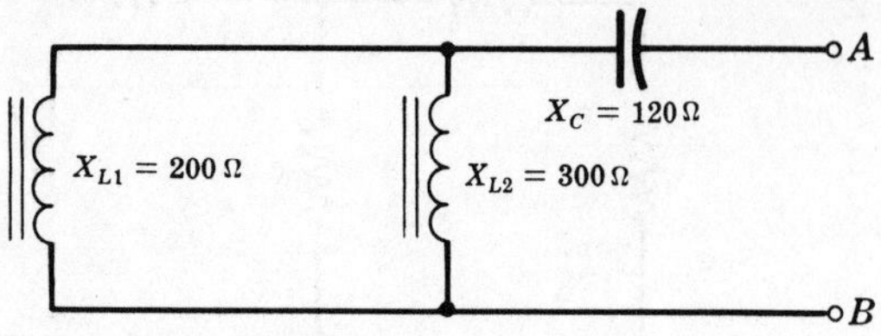

Fig. 1-12

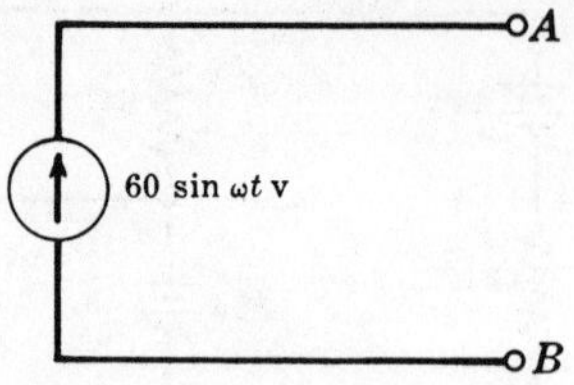

Fig. 1-13

The network of Fig. 1-11 cannot be represented by a Norton equivalent circuit because the effective internal admittance is infinite. In any practical problem there would be a finite real component of the impedance of the inductors of Fig. 1-11, and a Norton equivalent would then be meaningful.

1.6. Given the circuit of Fig. 1-14. The reactances of the impedances are specified for the generator frequency. The voltage and current generators are both operating at the same frequency and have the phase relationships shown. Find the current i using superposition.

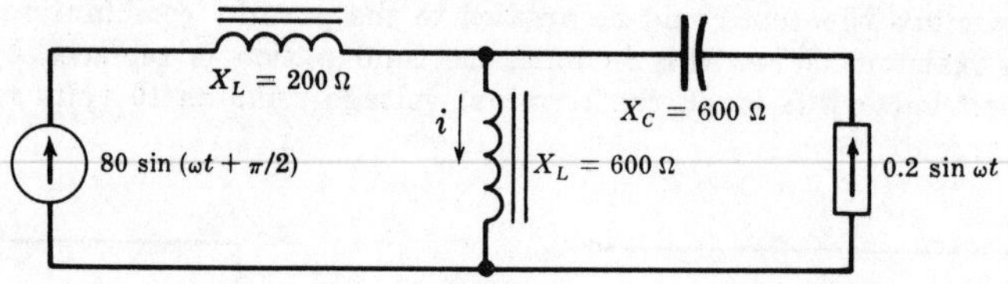

Fig. 1-14

First find the component i_V of the current i owing to the voltage generator only. The current generator is replaced by its internal impedance which is infinite, and the circuit diagram becomes that of Fig. 1-15. The loop equation is

$$-80 \sin(\omega t + \pi/2) + j200 i_V + j600 i_V = 0$$

from which $i_V = \dfrac{80 \sin(\omega t + \pi/2)}{j800} = 0.1 \sin \omega t$ amp.

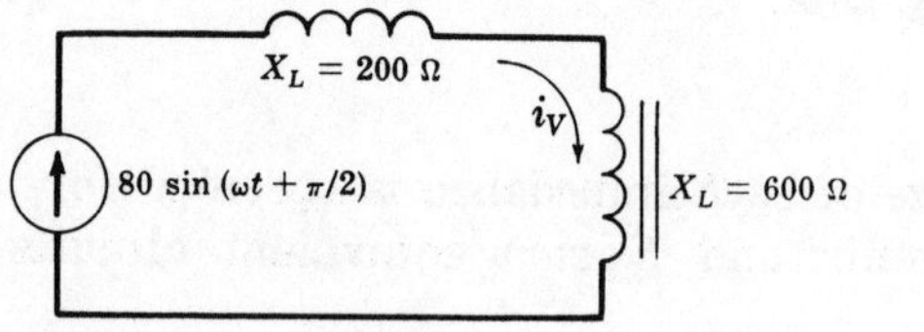

Fig. 1-15

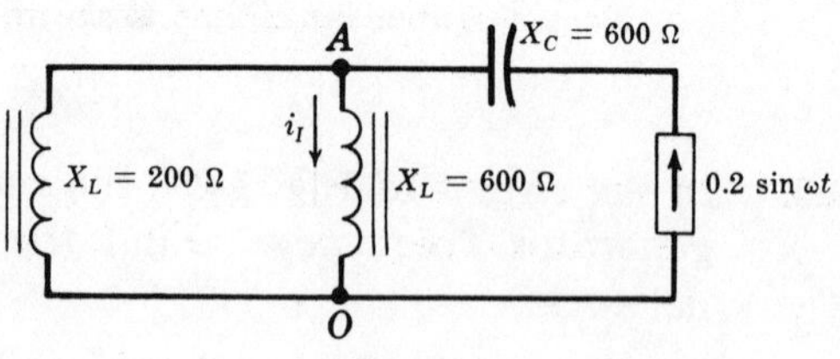

Fig. 1-16

The component i_I of the current i due to the current generator only is found next. The voltage generator is replaced by its internal impedance which is zero as shown in Fig. 1-16. The equation for node A is

$$-0.2 \sin \omega t + \frac{1}{j600} v_{AO} + \frac{1}{j200} v_{AO} = 0$$

or $v_{AO} = 30 \sin(\omega t + \pi/2)$. Then $i_I = v_{AO}/j600 = 0.05 \sin \omega t$ amp.

The current $i = i_V + i_I = 0.15 \sin \omega t$ amp.

1.7. **For the circuit of Fig. 1-17, find the instantaneous voltage v_{AB} using superposition.**

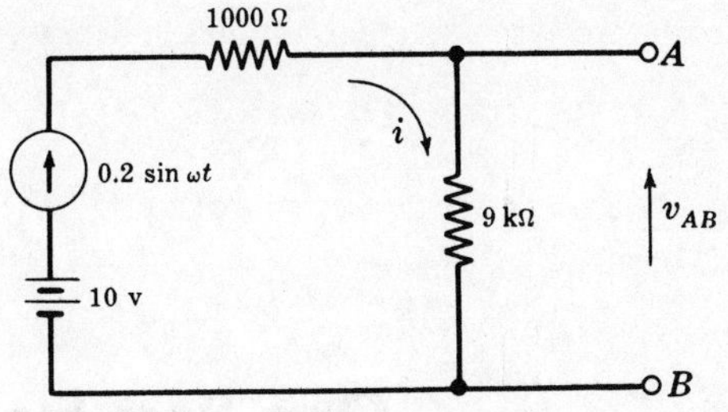

Fig. 1-17

First replace the varying generator by its equivalent internal impedance, yielding the circuit of Fig. 1-18. Since this network contains only a constant source, the current is constant and will be denoted by an upper case letter. The loop equation is

$$-10 + 1000I_1 + 9000I_1 = 0 \quad \text{or} \quad I_1 = 1 \text{ ma}$$

The voltage across terminals A and B due to the battery is a constant and is $V_{AB} = 9000I_1 = 9$ volts.

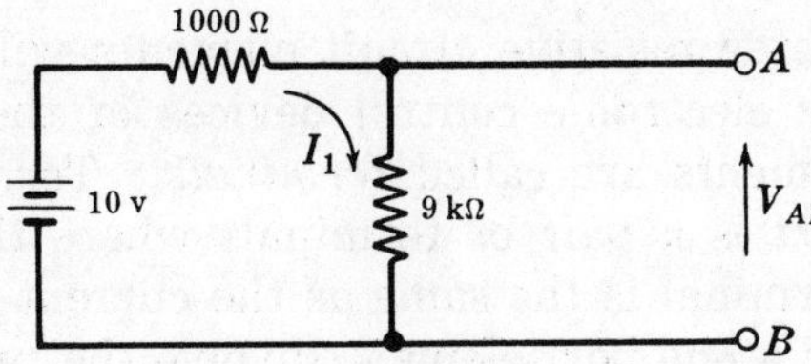

Fig. 1-18

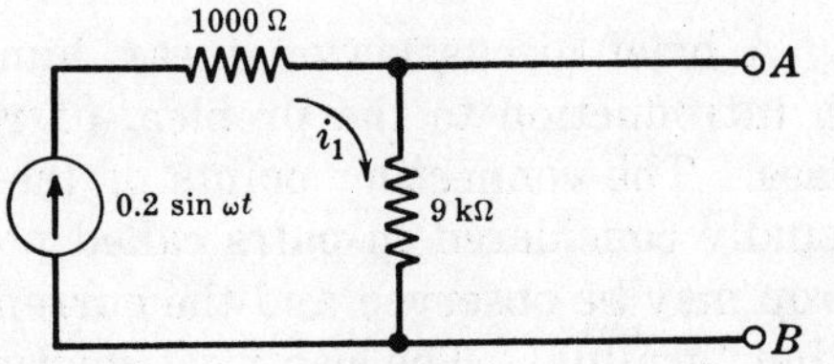

Fig. 1-19

Next replace the battery by its equivalent internal impedance, resulting in the circuit of Fig. 1-19. The current in this network is varying and denoted by a lower case letter. The loop equation is

$$-0.2 \sin \omega t + 1000i_1 + 9000i_1 = 0 \quad \text{or} \quad i_1 = 20 \times 10^{-6} \sin \omega t \text{ amp}$$

The voltage across terminals A and B due to the varying generator is $v_{ab} = 9000i_1 = 0.18 \sin \omega t$ volts.

The total or instantaneous voltage drop across terminals A and B of Fig. 1-17 is

$$v_{AB} = V_{AB} + v_{ab} = 9 + 0.18 \sin \omega t \text{ volts}$$

This illustrates how superposition is used in linear electronic circuits to separate instantaneous values into average values and instantaneous values of varying components only.

Chapter 2

One Port Electronic Control Devices

2.1 LINEAR RESISTORS

A brief discussion of linear, lumped-parameter resistive circuit elements will serve as an introduction to the problem of representing electronic control devices of the resistive class. The connecting points of the circuit elements are called *terminals.* Terminals are usually considered in pairs called ports. A *port* is a pair of terminals where the voltage drop may be observed and the current *in* one terminal is the same as the current *out* of the other terminal. The electrical characteristics of a one port element involve the relationship between the instantaneous current in the element and the instantaneous voltage drop across the terminals of the element. When the characteristics are independent of the direction of the current in the element, it is a *bilateral* element. If the characteristics are different for each direction of current, the element is *unilateral.*

A *linear resistor* is a bilateral resistive element which has a constant resistance. The linear resistor is an idealization or a model useful for representing practical resistances as circuit elements. In practical situations it may be necessary to include stray capacitance, inductance in leads, thermal noise sources, power dissipation, temperature dependence, and other considerations which are not included in the linear resistor model. The static characteristics of a linear resistor are indicated in Fig. 2-1. Included are the reference directions for the voltage drop across the resistor terminals and for the current. The characteristics are usually plotted with current as the dependent variable and voltage drop as the independent variable; hence the slope of the straight line is the conductance.

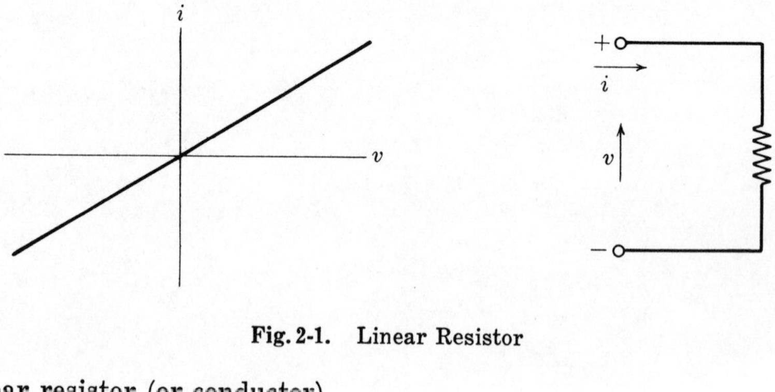

Fig. 2-1. Linear Resistor

For the linear resistor (or conductor),

$$G = i/v \quad \text{or} \quad i = Gv = v/R \tag{2.1}$$

$$R = v/i \quad \text{or} \quad v = Ri = i/G \tag{2.2}$$

Notation conventions are introduced by showing the constants R and G with upper case letters, and using lower case letters for i and v because these quantities may vary with time. To be consistent, the instantaneous power dissipated in the resistor is written with a lower case p:

$$p = vi = v^2/R = Gv^2 = i^2R = i^2/G \tag{2.3}$$

The average power is denoted by an upper case P and is

$$P = \frac{1}{T}\int_0^T p\,dt = \frac{1}{T}\int_0^T vi\,dt \tag{2.4}$$

where T is the period over which p is averaged.

2.2 IDEAL DIODE

A diode is a two terminal (one port) unilateral element which is considered resistive. A diode model is the *ideal diode* which has zero *forward resistance* (low resistance polarity) and infinite reverse resistance as shown in Fig. 2-2. The diagram also includes the symbol for the ideal diode and the reference direction (forward) for the current and the voltage drop. Note that the arrow in the symbol points in the direction of the forward current. The forward and reverse polarities are established in terms of the polarity of the diode voltage which results in low or high resistance, respectively. The ideal diode characteristics have a discontinuity at the origin. Although the ideal diode does not exist, it is a useful model for representing practical diodes. The same symbol may be used for the ideal diode or for a practical diode, so it is necessary to determine which is intended from the context. The terminals of the ideal diode are identified as the plate (or anode) and the cathode so that they correspond to the terminals of practical diodes.

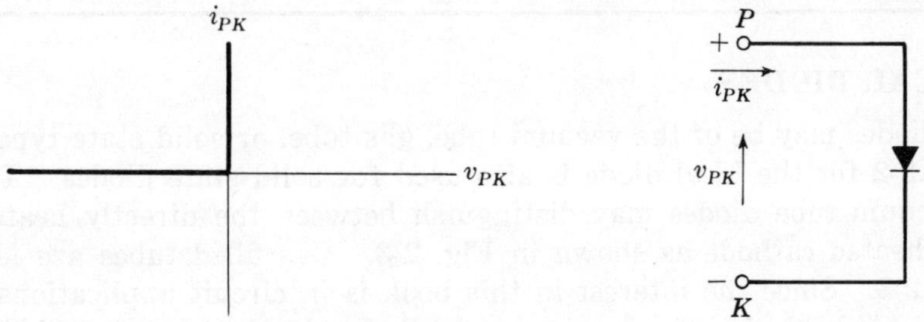

Fig. 2-2. Ideal Diode

The ideal diode does not dissipate any power. In the forward direction, the current is not zero but the voltage is zero. In the reverse direction, the voltage drop is not zero but the current is zero. Thus the instantaneous power is always zero.

2.3 DIODE NOTATION

The first problem in representing electronic control devices as circuit elements is that of notation. In general, one terminal of the device is chosen as a reference (or common) terminal. The reference voltage drops are from the non-reference terminals to the reference terminal. The reference currents are in the non-reference terminals. Double subscripts are used to identify voltage drops and currents. The first subscript identifies the non-reference terminal and the second identifies the reference terminal. For example, v_{PK} is the instantaneous voltage drop from plate to cathode. When the reference terminal is clearly understood, it is usually omitted. This is usually the case for diodes and vacuum tubes where the cathode is the reference terminal.

The problem of notation is rather troublesome because of many inconsistencies. In this book the notation will be consistent with IRE Standards on Letter Symbols for Semiconductor Devices, 1956. Using this notation it is only necessary to know the significance of the case of the letter and the case of the subscripts. In the older vacuum tube notation it was necessary to remember the actual subscripts used, in addition to the case of the letters.

The notation for diodes is given in Table 2.1, with the cathode understood to be the reference terminal. Power supplies are denoted by repeating the subscripts, as shown in the table.

TABLE 2.1

Symbol	Description
v_P	Instantaneous plate to cathode voltage drop
V_P	Average value of v_P
v_p	Instantaneous value of the varying component of v_P
V_p	Rms or effective value of v_p
i_P	Instantaneous current into the plate
I_P	Average value of i_P
i_p	Instantaneous value of the varying component of i_P
I_p	Rms or effective value of i_p
v_{PP}	Instantaneous plate power supply voltage
V_{PP}	Average value of v_{PP}
v_{pp}	Instantaneous value of the varying component of v_{PP}
V_{pp}	Rms or effective value of v_{pp}

2.4 PRACTICAL DIODES

Practical diodes may be of the vacuum tube, gas tube, or solid state types. The symbol shown in Fig. 2-2 for the ideal diode is also used for solid state diodes. The symbols for thermionic vacuum tube diodes may distinguish between the directly heated cathode and the indirectly heated cathode as shown in Fig. 2-3. Gas filled tubes are identified by the dot as in Fig. 2-3. Since the interest in this book is in circuit applications, there is little need to consider physical differences among diodes in detail but rather to emphasize applications.

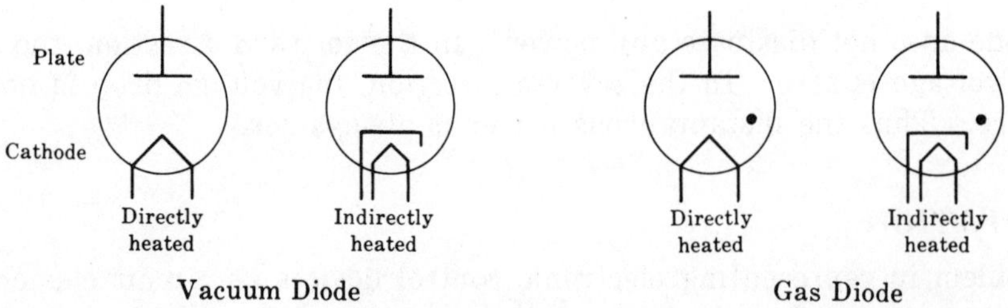

Fig. 2-3. Diode Symbols

A thermionic vacuum tube diode includes a plate (anode) and a cathode in a sealed envelope which is evacuated. The cathode is heated to a temperature sufficiently high to cause emission of electrons from its surface. The Richardson-Dushman equation for the magnitude of the current density provides a useful model for discussing electron emission:

$$|J| = A_0 T^2 \epsilon^{-(b_0/T)} \text{ amp/m}^2 \tag{2.5}$$

where $|J|$ = magnitude of the cathode current density in amperes per square meter;

A_0 = $4\pi mek^2/h^3$, a constant which involves the electron mass, electron charge, Boltzmann's constant and Planck's constant (in MKS units);

T = cathode temperature in degrees Kelvin;

$b_0 = e\phi_w/k = 11{,}600\phi_w$ where ϕ_w is the work function of the cathode surface material in electron volts.

In this model the cathode emission current (cathode surface area is constant) is a function of temperature only for given values of the constants. The electric field external to the cathode surface, caused by free electrons near the cathode surface and external voltages applied between the plate and cathode, are neglected in this model. Electrons emitted from the surface of the cathode either recombine with the cathode or travel to the plate.

The terminal characteristics of the vacuum diode are of most interest in this book. The current-voltage relationship (often called the volt-ampere characteristic) may be determined experimentally for given cathode temperatures. Curves are sketched for several different temperatures in Fig. 2-4. The region $v_P > 0$, where the three curves are for all practical purposes coincident, is the space charge limited region. This is the normal region of operation for most vacuum tubes. In this region the terminal characteristics are relatively independent of temperature so long as the temperature is above some minimum value. Hence the heater voltage is not critical in most applications. At higher values of v_P a point is reached at which all the electrons emitted from the cathode are drawn to the plate. The plate current saturates, but for second order effects on the emission current of the free electrons between the plate and cathode and the electric field resulting from the applied plate voltage. If it were not for these effects, the plate current curves would be parallel to the v_P axis in the temperature limited region.

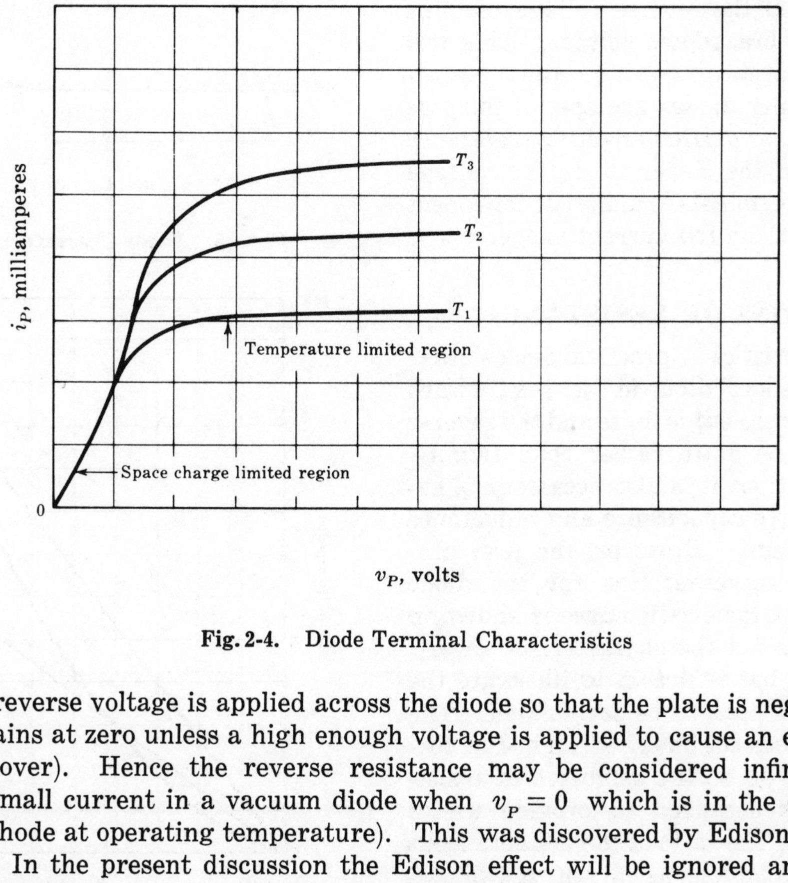

Fig. 2-4. Diode Terminal Characteristics

When a reverse voltage is applied across the diode so that the plate is negative, the plate current remains at zero unless a high enough voltage is applied to cause an electrical breakdown (flash-over). Hence the reverse resistance may be considered infinite. There is, however, a small current in a vacuum diode when $v_P = 0$ which is in the order of microamperes (cathode at operating temperature). This was discovered by Edison experimentally about 1883. In the present discussion the Edison effect will be ignored and diode curves drawn as though they pass exactly through the origin.

The gas diode also has a plate and a cathode but they are in a gas filled envelope. The physical principles of operation become more complicated because ionization phenomena are also involved. In addition to free electrons which may be emitted from a hot cathode or result from ionization, there are positive ions which contribute to the total current. Most gas diodes are designed for particular applications. A discussion of the principles of operation of gas diodes is beyond the scope of this book.

Solid state diodes include a number of types designed for particular applications. Copper-cuprous oxide rectifiers, selenium rectifiers and magnesium-copper sulfide rectifiers have been used in power rectifier applications. The present interest is in semiconductor diodes of the types employed in electronic circuits. Point-contact semiconductor diodes were developed in the early days of radio. The *P-N* junction diodes have more recently been developed as a result of the research which led to the discovery of the transistor. A simple model for the *P-N* junction diode is

$$i_P = I_{CO}[e^{(ev_P/kT)} - 1] \tag{2.6}$$

where i_P = diode current, I_{CO} = inverse saturation current (assumed constant for given materials), e = electron charge, v_P = diode voltage drop, k = Boltzmann's constant, T = temperature in degrees Kelvin. A *P-N* junction diode characteristic curve based on this model is shown in Fig. 2-5. (See Problem 2.1, page 24.)

The *P-N* junction model given by equation (*2.6*) assumes that the reverse voltage remains above the Zener breakdown voltage. This will be true in applications involving ordinary junction diodes. Zener diodes are special purpose devices designed for particular voltage reference applications. In the Zener diode the voltage drop across the terminals is almost independent of current over a limited current range.

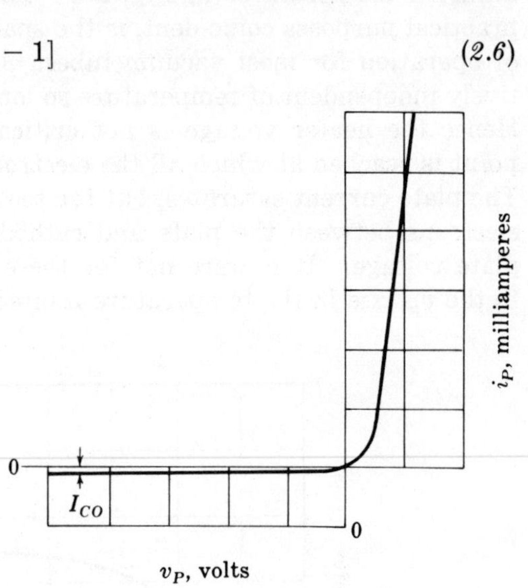

Fig. 2-5. Diode Characteristic Curve

2.5 DEFINITIONS OF RESISTANCE

The characteristics of practical diodes differ from those of the ideal diode in that the forward resistance is not zero but is finite and the reverse resistance may be finite rather than infinite. In many applications it is also necessary to examine the effective capacitance and inductance of the diode circuit. However, the first consideration is a representation for the diode resistance. A characteristic curve is shown in Fig. 2-6 which is not the characteristic of any particular diode but is drawn to illustrate the definitions which need to be formulated. The practical diode does not have a straight-line forward characteristic, so the definition of resistance must be generalized to provide useful representations. The ratio of the instantaneous voltage to the instantaneous current is still useful, so this is called the *instantaneous plate resistance* and is the lumped resistance between

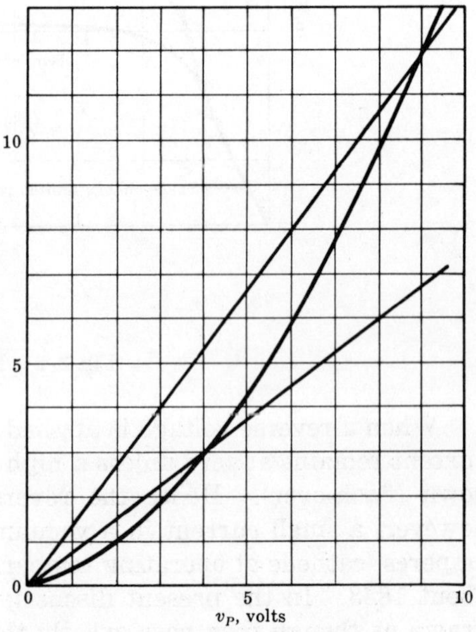

Fig. 2-6. Diode Volt-ampere Characteristic Curve

the plate and the cathode. The instantaneous plate resistance is by definition

$$r_P = v_P/i_P \tag{2.7}$$

Using a lower case letter and an upper case subscript for the instantaneous resistance is consistent with the adopted notation. The instantaneous plate conductance is by definition

$$g_P = i_P/v_P = 1/r_P \tag{2.8}$$

The characteristic curve of Fig. 2-6 indicates that g_P is not constant but varies with v_P. Consider two specific values of v_P. When $v_P = 4$ volts, $g_P = 0.003/4 = 0.00075$ mho $=$ 750 μmho; this is the slope of the straight line through the origin and the point on the characteristic curve where $v_P = 4$ volts. When $v_P = 9$ volts, $g_P = 0.012/9 = 0.001333$ mho $=$ 1333 μmho.

In many applications the interest is in small variations about some average value. This special case involves the "small signal" problem and is important because the definitions and concepts developed for simple diode circuits are easily extended to vacuum tube and transistor circuits. For this reason more emphasis will be placed on these concepts and circuits than might otherwise be justifiable. To include the average values and small variations about the average values, the following definitions are used:

Instantaneous plate resistance	$r_P = v_P/i_P$
Instantaneous plate conductance	$g_P = i_P/v_P = 1/r_P$
Static plate resistance	$R_P = V_P/I_P$
Static plate conductance	$G_P = I_P/V_P = 1/R_P$
Differential plate resistance	$r_p = \partial v_P/\partial i_P$
Differential plate conductance	$g_p = \partial i_P/\partial v_P = 1/r_p$

The average values specify the *operating point* or quiescent point. By definition, the *static plate resistance* is the instantaneous plate resistance at the operating point. The *differential plate conductance* at the operating point is the slope of the characteristic curve at the operating point. The differential resistance is the reciprocal of the slope of the characteristic curve at the operating point. These definitions will be illustrated in later sections.

2.6 DIODES IN SERIES

Consider a simple circuit consisting of two nonlinear resistive elements (practical diodes in this case) in series as shown in Fig. 2-7. Characteristic curves for two diodes have been constructed to best illustrate the method of solution, as shown in Fig. 2-8. The loop equation for the series circuit in terms of instantaneous values is

$$-V_{PP} + v_{P1} + v_{P2} = 0 \qquad \text{or} \qquad v_{P1} + v_{P2} = V_{PP} \tag{2.9}$$

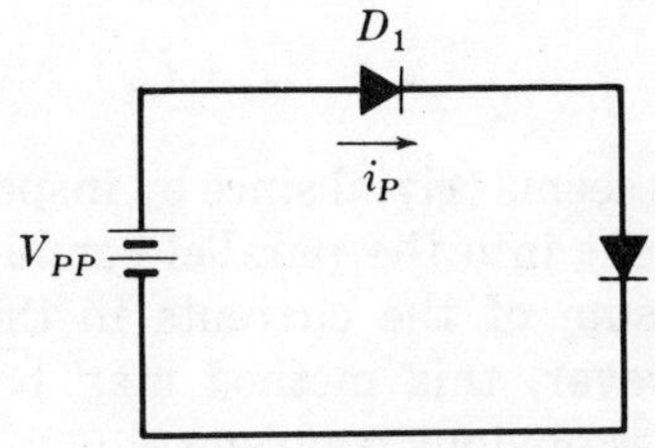

Fig. 2-7. Diodes in Series

Since v_{P1} is not linearly related to i_{P1}, it is not correct to write $v_{P1} = ki_{P1}$ where k is a constant.

To solve this nonlinear equation, it is convenient to use graphical techniques. Equation (*2.9*) shows that the instantaneous voltage drop across the series combination of diodes is equal to the source voltage V_{PP}. Because the two diodes are in series, the current is the same in each. To solve graphically, a characteristic curve for the series combination of diodes is constructed.

First assume an arbitrary value of current i_P. Then add the voltage v_{P1} for this value of i_P to the voltage v_{P2}, thus obtaining one point on the characteristic curve for the series combination as shown in Fig. 2-8 below. Repeat this procedure, choosing a sufficient number and distribution of values of i_P to define the characteristic curve for the series combination indicated as $D_1 + D_2$. The specific value of V_{PP} may then be located on the v_P axis. The current i_P is the intersection of the line $v_P = V_{PP}$ and the $D_1 + D_2$ curve; knowing i_P, v_{P1} and v_{P2} may be found by noting the intersection of the i_P line and the characteristic curves for each of the diodes. In the example shown in Fig. 2-8, i_P is about 5.4 ma, v_{P1} is about 3.8 v, and v_{P2} is about 5.2 v. Because these values are taken from the graph, there is a limit to the accuracy. A check may be made by adding v_{P1} and v_{P2} to see if their sum is V_{PP}.

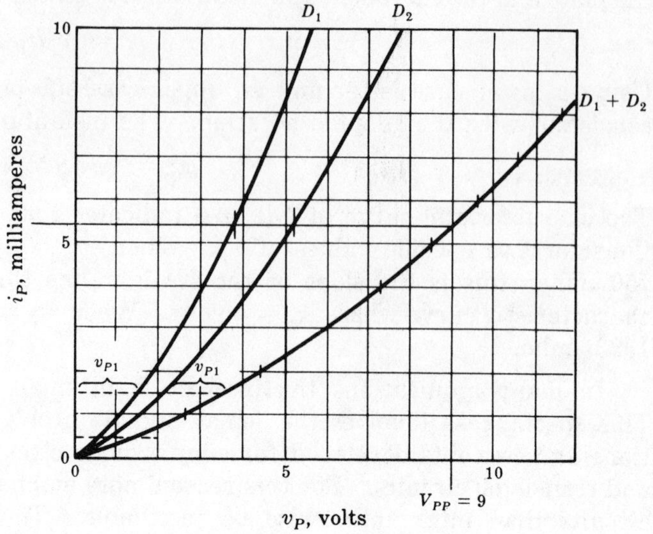

Fig. 2-8. Series Combination of Two Diodes

2.7 DIODES IN PARALLEL

A simple circuit with two diodes D_1 and D_2 connected in parallel is shown in Fig. 2-9, and the diode characteristic curves are displayed in Fig. 2-10.

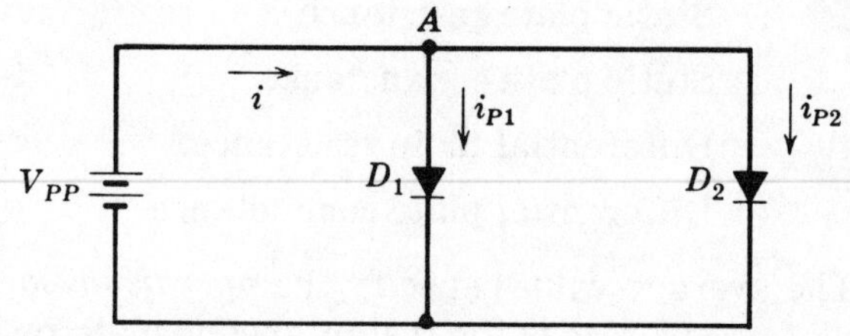

Fig. 2-9. Diodes in Parallel

In general, when circuit elements are in parallel, nodal equations are most convenient. The equation for node A of Fig. 2-9 is

$$-i + i_{P1} + i_{P2} = 0$$

or

$$i = i_{P1} + i_{P2} \tag{2.10}$$

This seems trivial since by inspection the current into the parallel combination is the sum of the currents in the diodes. However, this method may be applied to less obvious problems.

Since the diodes are in parallel, the voltage across each is the same. Choose an arbitrary voltage v_P and then add the two currents i_{P1} and i_{P2} for this value of v_P, thus giving one point on the characteristic curve for the parallel combination. Repeat this procedure using a sufficient number and distribution of values of v_P to define the characteristic curve for the parallel combination denoted D_1D_2. Then locate V_{PP} on the v_P

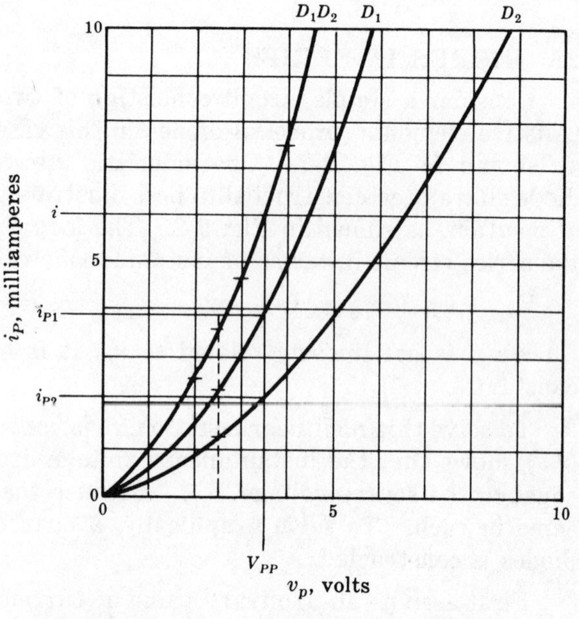

Fig. 2-10. Parallel Combination of Two Diodes

axis and draw a vertical line through V_{PP}. The current in the parallel combination will be the intersection of the line V_{PP} and the D_1D_2 curve. The current in each diode is the intersection of the V_{PP} line and the characteristic curve for the particular diode. In the indicated example, $V_{PP} = 3.5$ volts and the values taken from the graph are $i = 6$ ma, $i_{P1} = 3.8$ ma, $i_{P2} = 2.2$ ma. The two diode currents may be added to check that the result is the battery current.

2.8 DIODES IN SERIES-PARALLEL COMBINATIONS

The graphical methods of solving the equations representing nonlinear resistive circuits can be applied to any combination of resistive elements. Usually the interest is only in the forward characteristics of the diodes; hence the examples have included only these, but the methods apply equally well to the reverse directions.

As an example, consider the series-parallel combination of three diodes as shown in Fig. 2-11. The problem is to find the voltage drop across each diode and the current in each diode. The characteristic curves for the diodes are given in Fig. 2-12. In solving graphically, the diodes are combined in parallel and series combinations following the same order that would be used in circuits with linear resistors. The order is as follows:

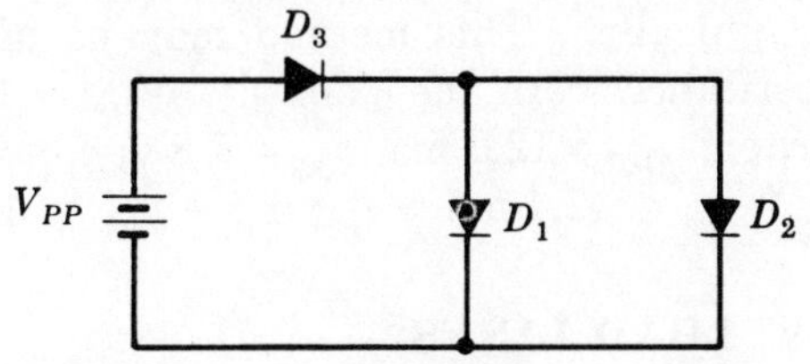

Fig. 2-11. Series-parallel Combination of Diodes

1. Determine the characteristic curve for the parallel combination of D_1 and D_2 as outlined in Section 2.7.

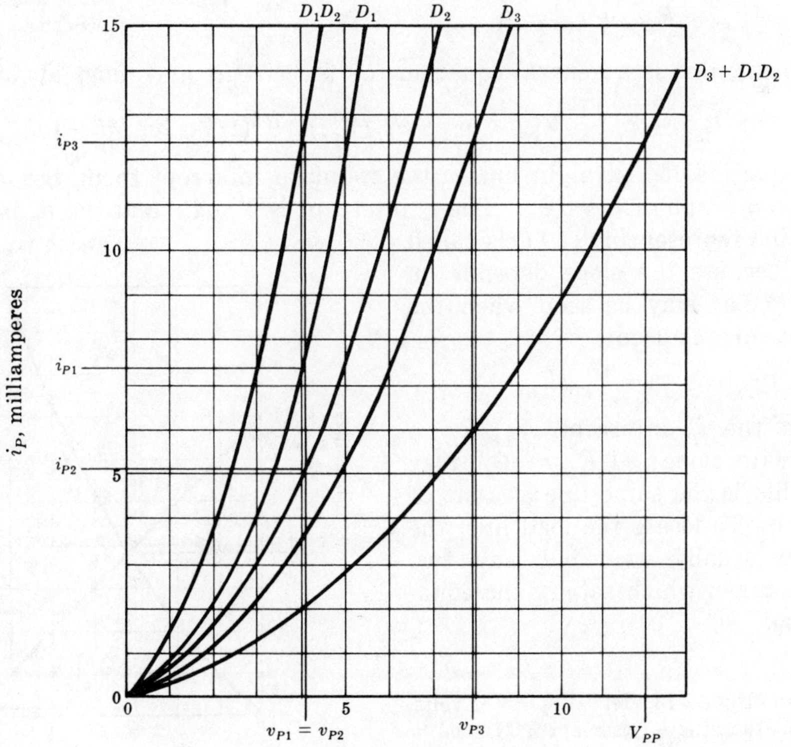

Fig. 2-12. Series-parallel Combination of Diodes

2. Determine the characteristic curve for the series combination of D_3 and D_1D_2 (parallel combination of D_1 and D_2) as explained in Section 2.6. This gives the characteristic curve for the series-parallel combination of the three diodes.
3. Locate V_{PP} on the v_P axis and draw the line $v_P = V_{PP}$. The current i_{P3} is the intersection of this line and the curve $D_3 + D_1D_2$.
4. The intersection of the line $i_P = i_{P3}$ and the D_3 curve gives the value of v_{P3}.
5. The current i_{P3} is in the parallel combination D_1D_2, hence the voltage drop across this parallel combination is the intersection of the curve D_1D_2 and the line i_{P3}. The voltage $v_{P1} = v_{P2}$.
6. The current i_{P1} is the intersection of the line v_{P1} and the curve D_1.
7. The current i_{P2} is the intersection of the line v_{P2} and the curve D_2.

The voltage drop across and the current in each diode has now been determined graphically. This method may be used for any series-parallel combination of nonlinear resistances. In the example, $V_{PP} = 12$ volts and the following values are taken from the graph: $i_{P3} = 12.3$ ma, $v_{P3} = 7.8$ v; $i_{P2} = 5.1$ ma, $v_{P2} = 4.1$ v; $i_{P1} = 7.3$ ma, $v_{P1} = 4.1$ v. Check: $i_{P1} + i_{P2} = 12.4 \text{ ma} = i_{P3}$; $v_{P1} + v_{P3} = 11.9 \text{ v} = V_{PP}$.

2.9 LOAD LINES

Circuits involving only one nonlinear and one linear resistive element in series can be solved by a simple method using the *load line*.

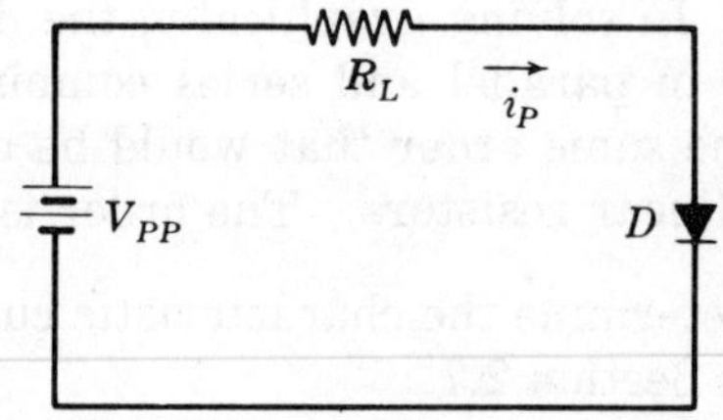

Fig. 2-13. Diode and Linear Resistor in Series

Consider the series circuit of Fig. 2-13. The loop equation in terms of instantaneous values is

$$-V_{PP} + R_L i_P + v_P = 0$$

Since R_L is linear, this is a general equation for the circuit and must always be satisfied. Rewriting,

$$R_L i_P + v_P = V_{PP} \quad \text{or} \quad (R_L/V_{PP})i_P + (1/V_{PP})v_P = +1 \tag{2.11}$$

which is the equation of a straight line in the standard intercept form, i.e. it will plot as a straight line on a graph of i_P vs. v_P. The i_P intercept is V_{PP}/R_L and the v_P intercept is V_{PP}. The straight line representing (*2.11*) is called the *load line* because the slope depends on the value of R_L, as may be seen when the slope intercept form is used:

$$i_P = -(1/R_L)v_P + V_{PP}/R_L$$

Thus through the i_P intercept, V_{PP}/R_L, a straight line with slope $-1/R_L = -G_L$ may be drawn. This is the same line as that of equation (*2.11*). To locate the load line, the intercepts are usually used but any two points may be taken which satisfy the equation of the line.

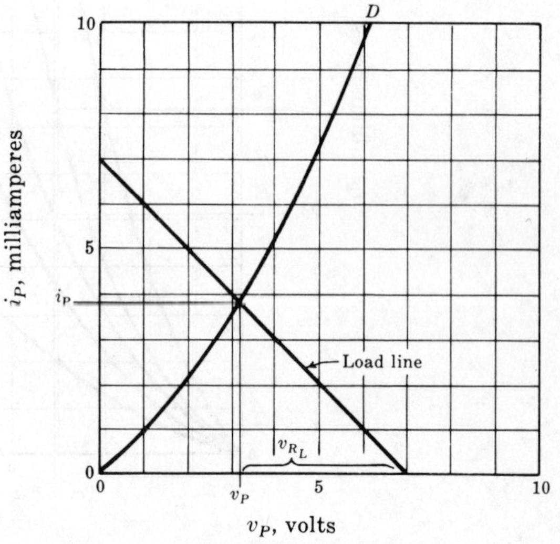

Fig. 2-14. Load Line

Example 1.

Referring to Fig. 2-13, let $V_{PP} = 7$ volts, $R_L = 1000$ ohms; then the i_P intercept is $7/1000 = 0.007 \text{ amp} = 7$ ma. Draw the load line on the graph of Fig. 2-14 which is the characteristic curve for the diode. The intersection of the load line and the characteristic curve for the diode gives the current i_P in the series circuit and the voltage v_P. The voltage drop across R_L is the difference between v_P and V_{PP} on the graph. For the given values, we find $i_P = 3.8$ ma, $v_P = 3.2$ v. Note that the magnitude of the voltage drop across the linear resistor R_L is proportional to the current.

2.10 DIODE CIRCUITS WITH VARYING GENERATORS

Only circuits with constant voltage sources have been studied, hence the next extension will involve voltage sources which vary with time. Circuits consisting of combinations of linear and nonlinear resistive elements may be solved graphically on a point by point basis for any voltage source, employing the methods previously outlined. For circuits with more than two resistive elements, this may become very tedious. The most common type of circuit involves a linear resistor in series with a nonlinear electronic control device, so this combination will be emphasized. The circuits which will be examined in this section are not necessarily practical but serve very well to introduce the concepts needed for studying vacuum tube and transistor circuits.

The circuit of Fig. 2-15 includes a varying voltage source which may be represented as a constant voltage source in series with a sinusoidal voltage source. The loop equation for the circuit in terms of instantaneous values is

$$-V_{PP} - v_{pp} + R_L i_P + v_P = 0 \qquad (2.12)$$

This equation is general but nonlinear, so again the solution may be found graphically by plotting the characteristic curve for the series combination of R_L and D, as shown in Fig. 2-16. The instantaneous voltage drop v_P and the instantaneous current i_P may be determined point by point.

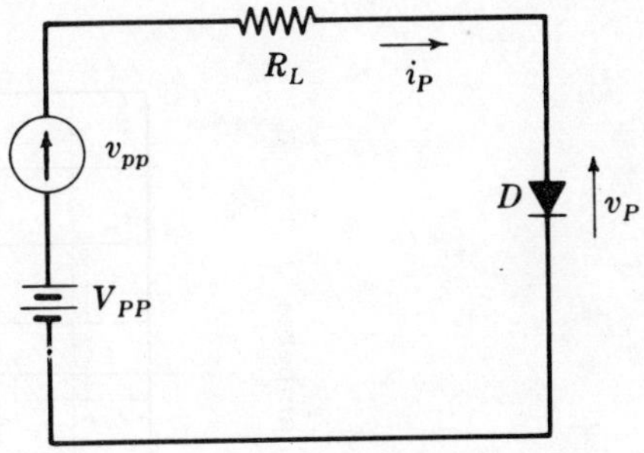

Fig. 2-15. Diode Circuit with Varying Voltage Source

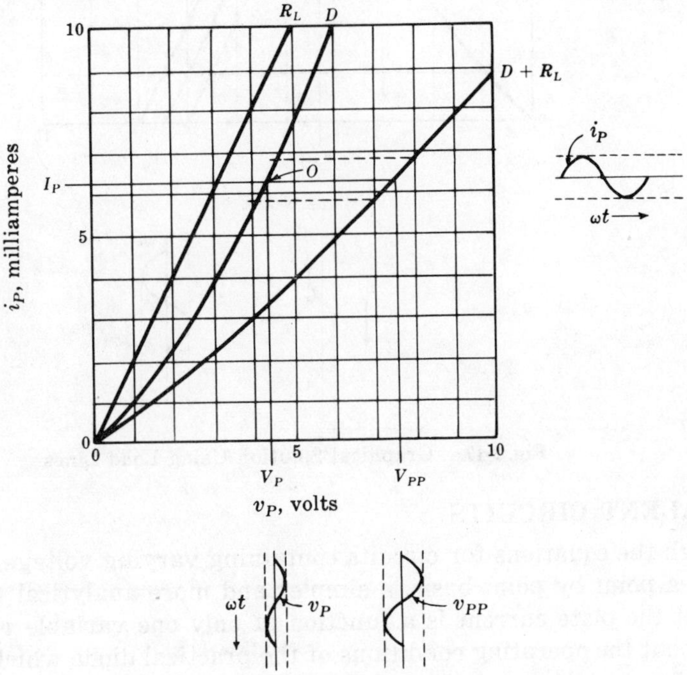

Fig. 2-16. Graphical Solution

Example 2.

Referring to Fig. 2-15, let $V_{PP} = 7.5$ volts, $v_{pp} = 0.5 \sin \omega t$ volts, $R_L = 500$ ohms. Plot the characteristic line for R_L and determine the characteristic curve $D + R_L$. Assume the diode characteristic curve is approximately a straight line for the actual range of v_P for the values chosen. Then the current i_P will have a constant component I_P (average value) and a varying component i_p which is sinusoidal, as sketched in Fig. 2-16 above. The operating point O is found by determining I_P and V_P. The intersection of the line V_{PP} and the $D + R_L$ curve gives I_P. The intersection of the line I_P and the D curve gives V_P. The instantaneous values could be found point by point by taking various values of v_{PP}. In the example, the maximum and minimum values of v_{PP} were taken to determine the maximum and minimum values of i_P and v_P. The sinusoidal curves were then sketched, indicating instantaneous values for one complete cycle of v_{PP}. For the values used in the example,

$$V_P = 4.3 \text{ v} \qquad v_{P(\max)} = 4.5 \text{ v} \qquad v_{P(\min)} = 4.1 \text{ v}$$
$$I_P = 6.3 \text{ ma} \qquad i_{P(\max)} = 6.8 \text{ ma} \qquad i_{P(\min)} = 5.8 \text{ ma}$$

In most problems of this type, load lines would be used. In this case a line with slope $-G_L$ is drawn through the various values of v_{PP}. The instantaneous values of i_P and v_P are the intersections of each of the load lines and the curve D. This is illustrated in Fig. 2-17 using the same values as in Fig. 2-16.

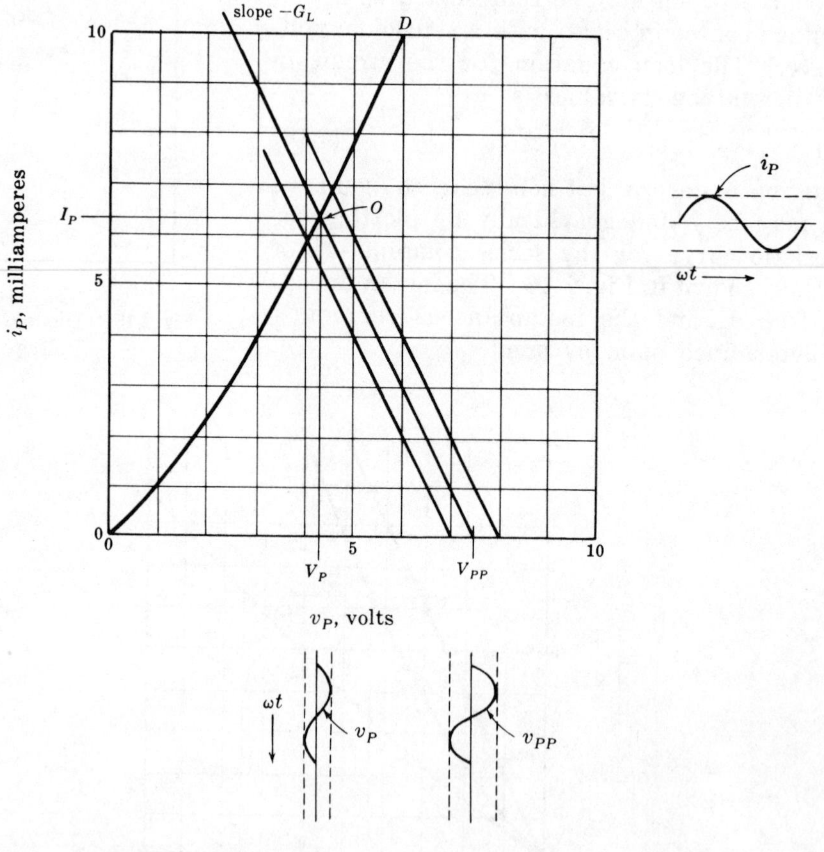

Fig. 2-17. Graphical Solution Using Load Lines

2.11 EQUIVALENT CIRCUITS

Even though the equations for circuits containing varying voltage sources can be solved graphically on a point by point basis, a simpler and more analytical method is desired. It is assumed that the plate current is a function of only one variable v_P. This implies some assumptions about the operating conditions of the practical diode which will not be discussed at this point. Assume that

$$i_P = f(v_P) \tag{2.13}$$

A more analytical approach to the problem of representing the diode as an electrical circuit element includes the use of Taylor's series to expand the function about an operating point.

$$i_P = f(V_P) + (v_P - V_P)\frac{\partial i_P}{\partial v_P} + \frac{(v_P - V_P)^2}{2!}\frac{\partial^2 i_P}{\partial v_P^2} + \cdots \tag{2.14}$$

If the second and higher derivatives of i_P with respect to v_P are zero, then operation is on the linear part of the operating range and thus

$$i_P = f(V_P) + (v_P - V_P)\frac{\partial i_P}{\partial v_P} \tag{2.15}$$

In this case the characteristic curve for the diode is a straight line for the range of v_P, as was previously assumed (see Example 2 in Section 2.10). Note that in equation (*2.15*) the term $f(V_P)$ is I_P (average value) and by definition $\partial i_P/\partial v_P = g_p$, the differential plate conductance. Substituting into (*2.15*),

$$i_P = I_P + (v_P - V_P)g_p \tag{2.16}$$

where g_p is evaluated at the operating point.

When operation is limited to a straight line portion of the diode characteristic curve where g_p is constant $(\partial^2 i_P/\partial v_P^2 = \partial g_p/\partial v_P = 0)$, superposition may be applied to write an instantaneous value as the sum of the average value and the instantaneous value of the varying component. Thus

$$v_P = V_P + v_p \quad \text{and} \quad i_P = I_P + i_p \tag{2.17}$$

Substituting (*2.17*) into (*2.16*) and solving for g_p,

$$g_p = i_p/v_p \quad \text{or} \quad r_p = v_p/i_p$$

The equations developed provide a basis for representing the diode as a circuit element. Even though the ranges of the voltages and currents are restricted to the linear range for the diode, in terms of instantaneous values the diode must be represented as a nonlinear resistance (r_P is not constant). However, when superposition is applied and only the average values are considered, the diode may be represented as a constant resistance R_P (static plate resistance); then only the first term of the Taylor's series expansion is involved and this is found graphically. Considering only the varying components, the diode is represented as a constant resistance r_p (differential plate resistance) and the value of r_p may be found graphically.

Since equation (*2.13*) has only one dependent variable, a geometrical representation involves a curve as was shown earlier.

Consider the diode and linear resistor in series with a varying source, as in Fig. 2-18. An equivalent circuit in terms of instantaneous values includes a nonlinear resistance representation for the diode. This resistance is defined as the instantaneous plate resistance, and the equivalent circuit is shown in Fig. 2-19.

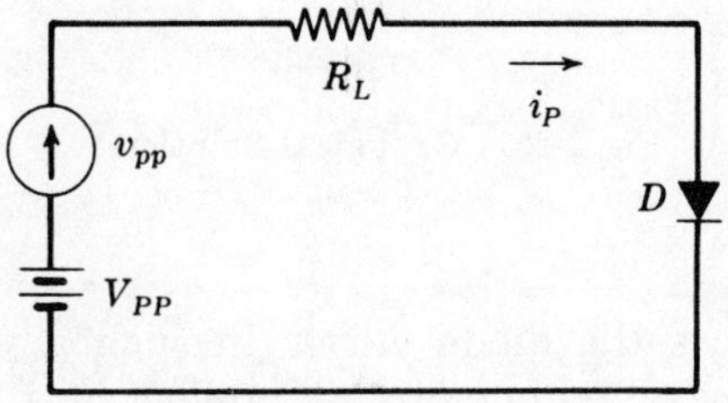

Fig. 2-18. Diode and Resistor in Series

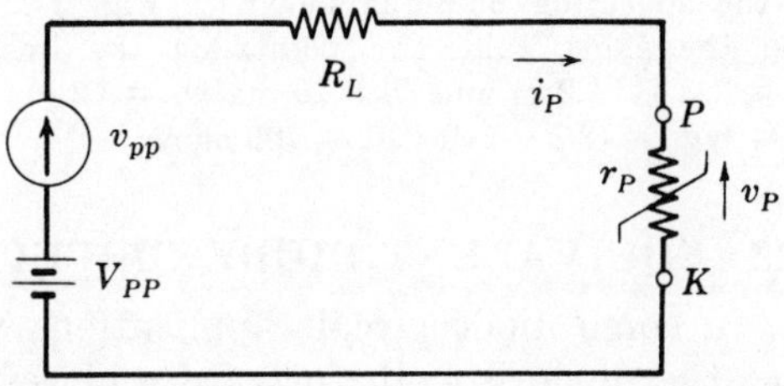

Fig. 2-19. Equivalent Circuit Considering Instantaneous Values

To find an expression for the instantaneous plate resistance r_P, rewrite (*2.16*):

$$i_P = I_P + g_p v_P - g_p V_P \quad \text{or} \quad v_P = (1/g_p)(i_P - I_P) + V_P = r_p i_P - r_p I_P + V_P$$

Then by definition,
$$r_P = v_P/i_P = r_p - (1/i_P)(r_p I_P - V_P) \tag{2.18}$$

Since r_P is a function of i_P even when r_p is constant, r_P is nonlinear.

An equivalent circuit in terms of average values for the circuit of Fig. 2-18 is given in Fig. 2-20. This is sometimes called the "dc equivalent circuit". The values of V_P and I_P are determined graphically, and from these values the static plate resistance R_P can be calculated.

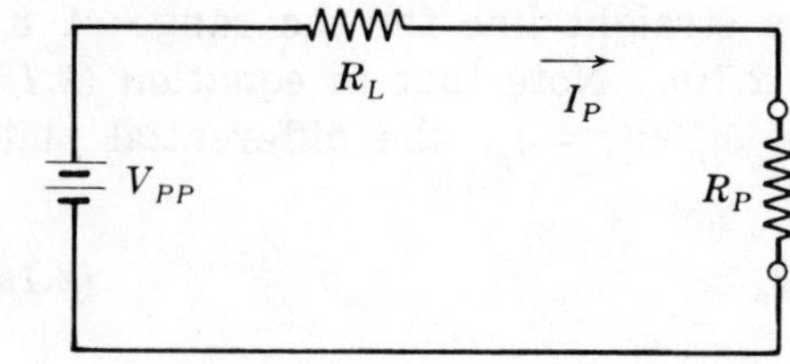

Fig. 2-20. Equivalent Circuit Considering Average Values

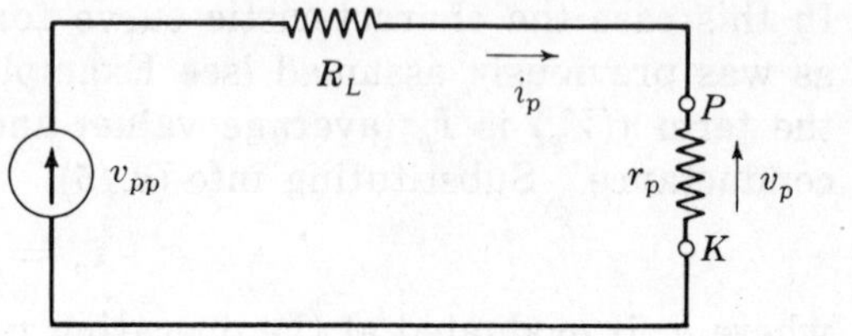

Fig. 2-21. Equivalent Circuit Considering Only Varying Components

An equivalent circuit in terms of varying components only for the circuit of Fig. 2-18 is shown in Fig. 2-21. This is sometimes called the "ac equivalent circuit", and will be most frequently used in later work because the interest will be in the behavior of the circuit to applied signals.

The value of the differential plate resistance at the operating point may be determined graphically by finding the slope of the diode characteristic curve at the operating point. To improve the accuracy of the graphical determination of r_p (or g_p), a line is drawn tangent to the curve at the operating point, and the slope of the tangent line is found by using large increments.

Example 3.

Referring to the circuit of Fig. 2-18, let $V_{PP} =$ 8 volts, $R_L = 1000$ ohms, $v_{pp} = 0.5 \sin \omega t$ volts. The diode characteristic curve is given in Fig. 2-22. Find the static plate resistance R_P for the circuit values, and the differential plate resistance r_p at the operating point.

The load line is drawn through its intercepts V_{PP} and V_{PP}/R_L. The intersection of the load line and the diode characteristic curve is the operating point. From the graph, $V_P = 3.9$ v and $I_P =$ 4.1 ma. Then $R_P = 3.9/0.0041 = 950$ ohms.

To find r_p, draw the line tangent to the curve at the operating point as shown in Fig. 2-22. To find the slope, take two points on the line, say ($i_P = 0$, $v_P = 1.2$ v) and ($i_P = 10$ ma, $v_P = 8.2$ v). Then $r_p = 1/g_p = (8.2 - 1.2)/0.01 = 700$ ohms.

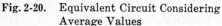

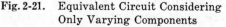

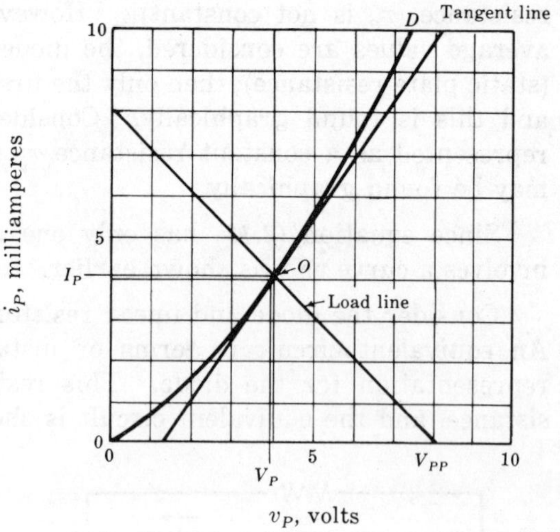

Fig. 2-22. Graphical Solution

2.12 EQUIVALENT DIODE CIRCUITS

In some diode circuit applications the voltage across the diode varies in such a way that the diode is both forward and reverse biased. In switching circuit applications the diode may be switched between a point on the reverse characteristic curve and a point on the forward characteristic curve. When the diode is forward biased it is represented as a resistor whose value is the instantaneous forward resistance. When it is reverse biased it is represented by the instantaneous reverse resistance. Thus the diode appears as a combination of a high resistance (reverse) and a low resistance (forward) connected to a two position switch. In some applications a sufficiently good approximation is to consider the forward resistance of the diode as one constant and the reverse resistance as another constant. The ideal diode combined with two linear resistors can be used to represent the practical diodes for these applications. This representation is shown in Fig. 2-23 and is based on the forward resistance r_{PF} and the reverse resistance r_{PR} being constants.

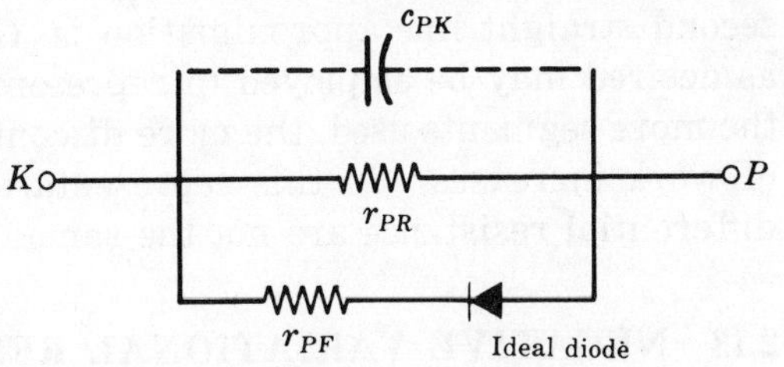

Fig. 2-23. Diode Equivalent Circuit

In most cases the reverse resistance is so large compared to the forward resistance that the parallel combination is just the forward resistance, so that it is not necessary to combine them in parallel. When the instantaneous forward resistance is constant over some range beginning at the origin, then the instantaneous forward resistance, the static resistance and the differential resistance have all the same value; and similarly for the reverse resistance. This simplifies the problem of representing the diode. With this representation it is only necessary to distinguish between the forward and reverse diode polarities. In some applications the effective diode capacitance may be represented as a capacitor between the plate and cathode terminals of the practical diode as shown with dashed lines in Fig. 2-23.

The diode characteristic curves for some semiconductor diodes approach the v_P axis at a point to the right of the origin. An example is shown in Fig. 2-24. To better approximate this type of characteristic curve using circuit models, a battery is added to the equivalent circuit which represents the diode for the forward polarity as shown in Fig. 2-25. The battery voltage is the intercept of the characteristic curve on the v_P axis, or it may be the intercept of a straight line approximation to the curve. The circuit of Fig. 2-25 with the ideal diode approximates the characteristic curve of Fig. 2-24 as a straight line intersecting the v_P axis, as shown in Fig. 2-24 by a dashed line. The slope of the approximating line is G_1. This is drawn to illustrate the use of a straight line approximation.

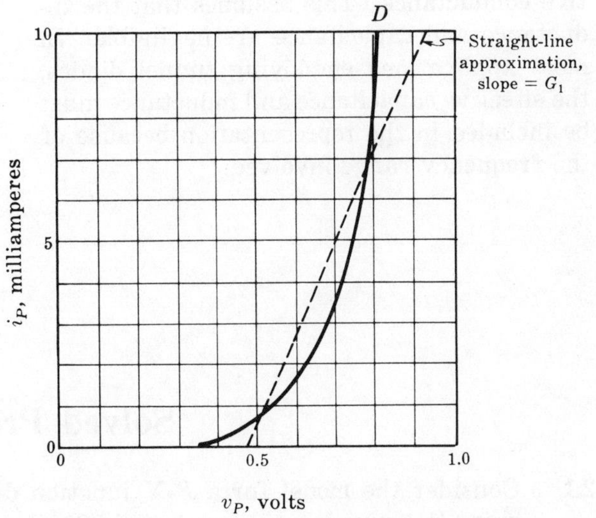

Fig. 2-24. Diode Characteristic Curve

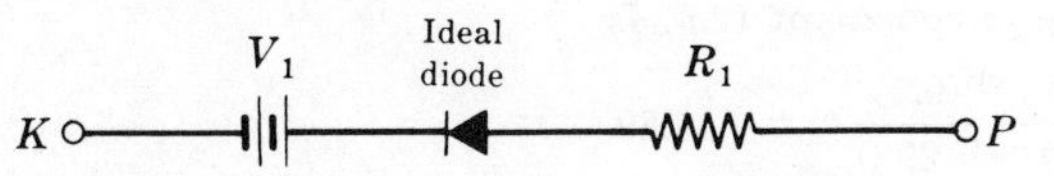

Fig. 2-25. Diode Equivalent Circuit

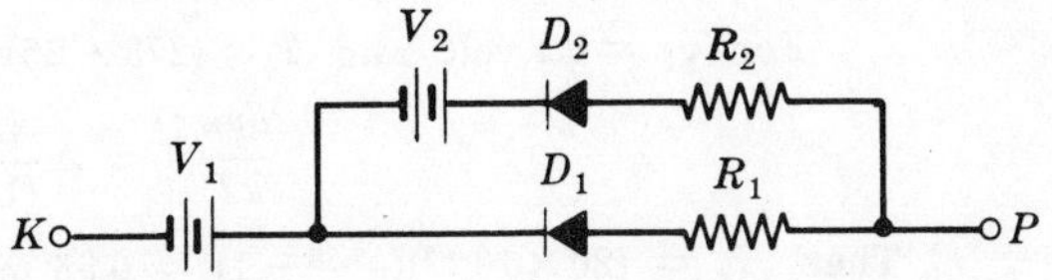

Fig. 2-26. Diode Equivalent Circuit

In a practical problem we must consider the best fit in the range of interest. If a better approximation to the characteristic curve of the diode is desired, more straight line segments may be used as shown in Fig. 2-26 where there are two. Both diodes are reverse biased until the voltage v_P from the plate to the cathode terminals exceeds V_1, and then D_1 is forward biased. The slope of the straight line approximation is $G_1 = 1/R_1$ for $V_1 < v_P < V_2$. When v_P reaches V_2, the diode D_2 is also forward biased, and the slope of the second straight line approximation is $G = 1/R_1 + 1/R_2$. As many straight line segments as desired may be employed to represent the forward characteristic of a diode. However, the more segments used, the more discontinuities in the approximation, and thus the amount of work increases. In this representation, the values of the instantaneous resistance and differential resistance are not the same.

2.13 NEGATIVE VARIATIONAL RESISTANCE

Another example of a diode characteristic curve which will illustrate the definitions of resistance and conductance is the tunnel diode characteristic curve of Fig. 2-27.

When v_P is below V_1, the instantaneous conductance and differential conductance are both positive. At the point $v_P = V_1$, the differential conductance is zero. In the region where v_P is between V_1 and V_2, the differential conductance is negative. Thus in the equivalent circuit representing the tunnel diode for this region, considering varying components only, the tunnel diode is represented as a negative conductance. This assumes that the inductance and capacitance are negligible. In most applications employing tunnel diodes, the effective capacitance and inductance must be included in the representation because of the frequency range involved.

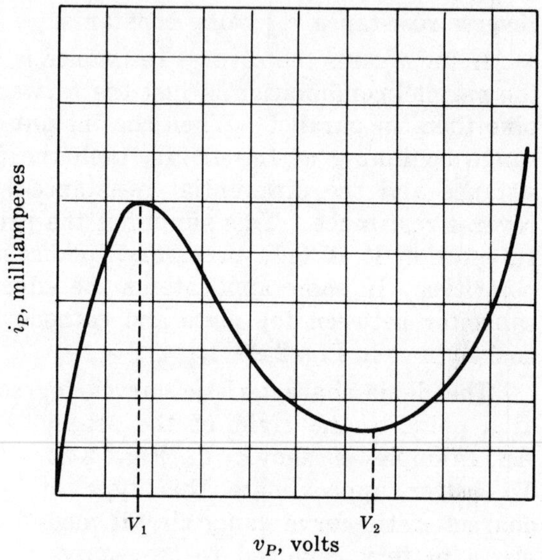

Fig. 2-27. Static Characteristic of a Tunnel Diode

Solved Problems

2.1. Consider the model for a *P-N* junction diode given in equation (*2.6*), page 14. Assume the junction temperature is 25°C and $I_{CO} = 20$ μamp. Calculate the instantaneous plate resistance r_P and differential plate resistance r_p for this diode when $v_P = 0.1$ volt.

For $v_P = 0.1$ volt and $T = (273 + 25)°\text{K}$, the exponent of (*2.6*) is

$$\frac{ev_P}{kT} = \frac{(1.6 \times 10^{-19})(0.1)}{(1.38 \times 10^{-23})(298)} = 3.89$$

Then $i_P = (20 \times 10^{-6})(e^{3.89} - 1) = 0.96$ ma, and $r_P = 0.1/0.00096 = 104.3\ \Omega$.

The differential plate conductance is

$$g_p = \partial i_P/\partial v_P = I_{CO}(e/kT)\, e^{(ev_P/kT)} = (20 \times 10^{-6})(38.9)(49) = 38.1 \times 10^{-3}$$

Then $r_p = 1/g_p = 1/(38.1 \times 10^{-3}) = 27.3\ \Omega$.

2.2. Given the diode characteristic curve of Fig. 2-28. Approximate this curve using two straight line segments and draw the equivalent circuit representing the diode when biased in the forward direction.

First, draw two straight line segments to best approximate the curve. Then the slopes of the two lines are determined by taking large increments to improve accuracy.

The slope of the first line is $G_1 = 0.0069/10 = 690\ \mu\text{mho}$; then

$$R_1 = 1/G_1 = 1/0.00069 = 1450\ \Omega$$

The slope of the second line is

$$G = 0.01/(6.7 - 2.3) = 0.00227 \text{ mho}$$

Since the slope of the second line is the conductance of two conductors in parallel, $G = G_1 + G_2$; then $G_2 = G - G_1 = 0.00227 - 0.00069 = 0.00158$ and

$$R_2 = 1/G_2 = 1/0.00158 = 633\ \Omega$$

The two line segments intersect at $V_2 = 3.3$ volts. In this example the first segment passes through the origin. The equivalent circuit is drawn in Fig. 2-29.

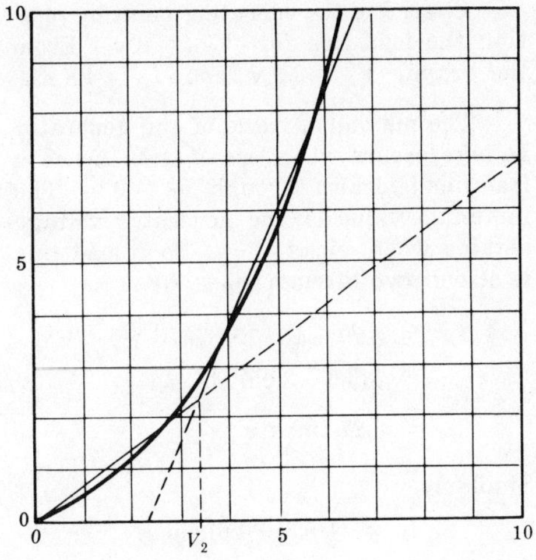

Fig. 2-28

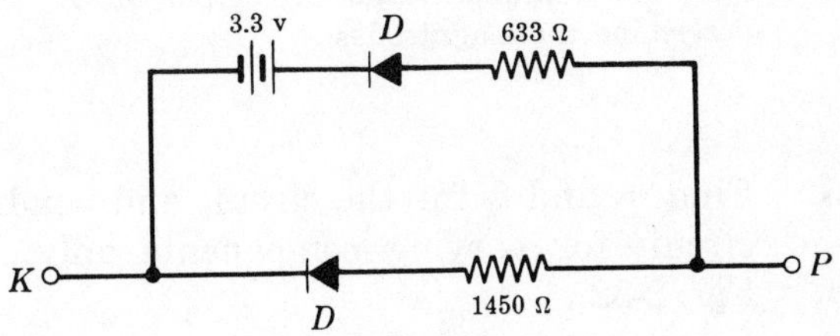

Fig. 2-29

2.3. Given the diode circuit of Fig. 2-18, page 21. Assume linear operation and that $v_P = 4 + 0.2 \sin \omega t$ volts and $i_P = 4 + 0.3 \sin \omega t$ milliamperes. Calculate the instantaneous power dissipated in the diode and then find the average power dissipated. Find the maximum and the minimum values of the instantaneous power.

The instantaneous power dissipated in the diode is

$$p = v_P i_P = (4 + 0.2 \sin \omega t)(4 + 0.3 \sin \omega t)10^{-3} \text{ watts} \qquad (1)$$

Substituting $\sin^2 \omega t = \frac{1}{2} - \frac{1}{2} \cos 2\omega t$,

$$p = (16.03 + 2 \sin \omega t - 0.03 \cos 2\omega t)10^{-3} \text{ watts} \qquad (2)$$

Since the voltage and current are in phase, $p_{\max}$ occurs when $\sin \omega t = 1$ and $p_{\min}$ when $\sin \omega t = -1$. Then using (1),

$$p_{\max} = (4.2)(4.3)10^{-3} = 0.018 \text{ watts}, \qquad p_{\min} = (3.8)(3.7)10^{-3} = 0.014 \text{ watts}$$

For an integer number of cycles the average power is, using (2),

$$P = \frac{\omega}{2\pi}\int_0^{2\pi/\omega} p\,dt = 0.016 \text{ watts}$$

2.4. The diode characteristic curve of the circuit of Fig. 2-30 is given in Fig. 2-31 below. Find the operating point using the load line method and then find v_p and i_p graphically.

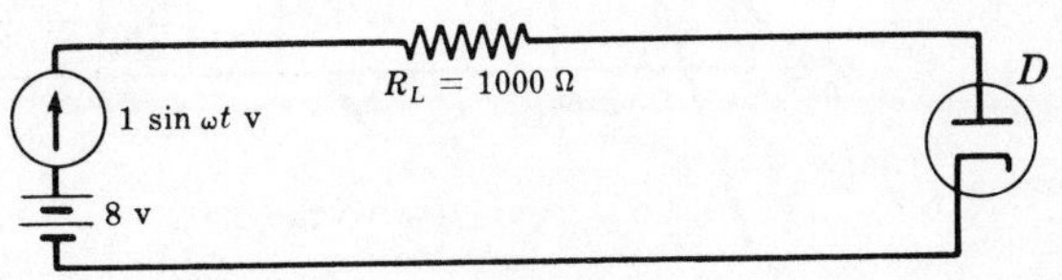

Fig. 2-30

First, find the operating point by plotting the load line for $V_{PP} = 8$ v. From the graph, $V_P = 3.3$ v and $I_P = 4.8$ ma.

The maximum value of the generator voltage occurs when $\sin \omega t = 1$, hence a load line is drawn through $v_P = 9$ v. The minimum value of the generator voltage obtains when $\sin \omega t = -1$, so a load line is also drawn through $v_P = 7$ v.

$$v_p \;=\; \tfrac{1}{2}[v_{P(\max)} - v_{P(\min)}] \sin \omega t$$
$$=\; \tfrac{1}{2}(3.6 - 3.05) \sin \omega t$$
$$=\; 0.27 \sin \omega t \text{ v}$$

Similarly,

$$i_p \;=\; \tfrac{1}{2}(5.5 - 4.0) \sin \omega t$$
$$=\; 0.75 \sin \omega t \text{ ma}$$

Unless a large graph is used, the accuracy is poor; for this reason the maximum and minimum values are employed to determine the magnitudes.

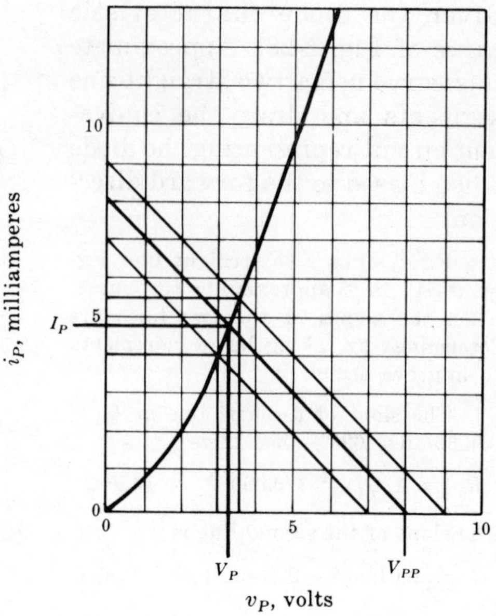

Fig. 2-31

2.5. Find v_p and i_p for the circuit and conditions of Problem 2.4 by using the equivalent circuit for varying components only. The diode characteristic curve is given in Fig. 2-32.

A load line is drawn to locate the operating point. A line is drawn tangent to the characteristic curve at the operating point; the differential plate resistance is the reciprocal of the slope of this line.

$$r_p \;=\; 1/g_p \;=\; (6.2 - 1.6)/(0.013 - 0) \;=\; 354\ \Omega$$

The equivalent circuit considering varying components only, is shown in Fig. 2-33. Then

$$i_p \;=\; (1 \sin \omega t)/(1000 + 354)$$
$$=\; 0.738 \sin \omega t \text{ ma}$$

$$v_p \;=\; r_p i_p \;=\; 354(0.738 \times 10^{-3} \sin \omega t)$$
$$=\; 0.261 \sin \omega t \text{ v}$$

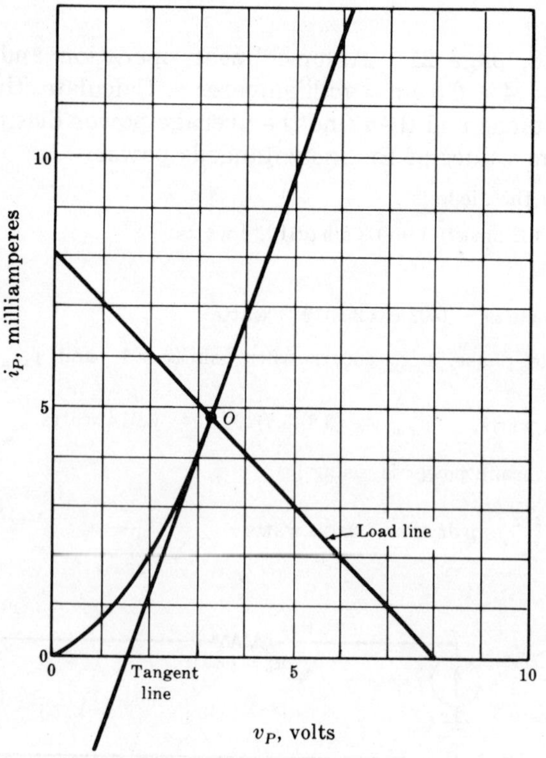

Fig. 2-32

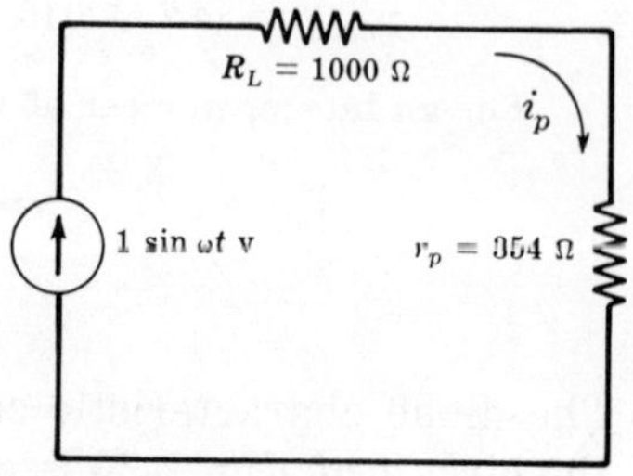

Fig. 2-33. Equivalent Circuit

Supplementary Problems

2.6. For the circuit of Fig. 2-34, find the voltage drop and current in each diode. Diode characteristics are given in Fig. 2-35.

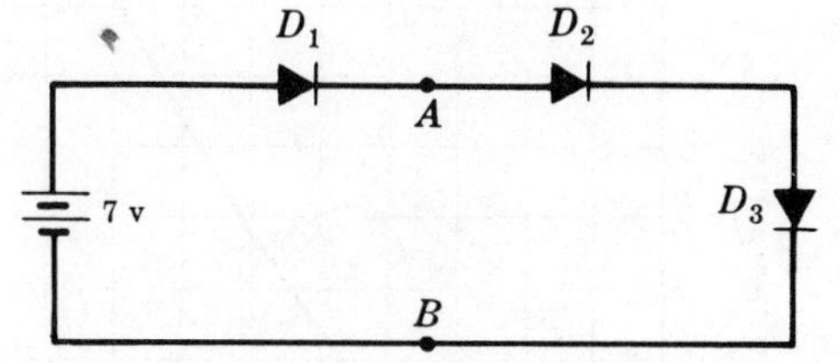

Fig. 2-34

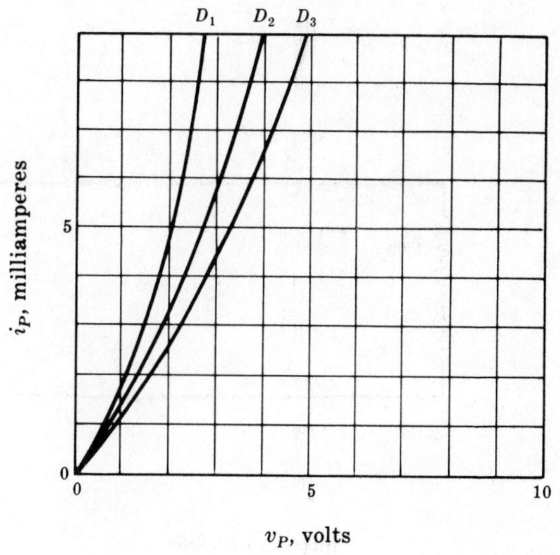

Fig. 2-35

2.7. Repeat Problem 2.6 for the case where there is a linear 1000 ohm resistor connected from point A to point B in Fig. 2-34. (*Hint.* On the graph draw the characteristic curve for the resistor and use the methods outlined in the previous sections. The methods developed for non-linear resistances will also work for linear resistances.)

2.8. Given the circuit of Fig. 2-36 and the characteristic curves of Fig. 2-37. Find the voltage drop across each element and the current in each.

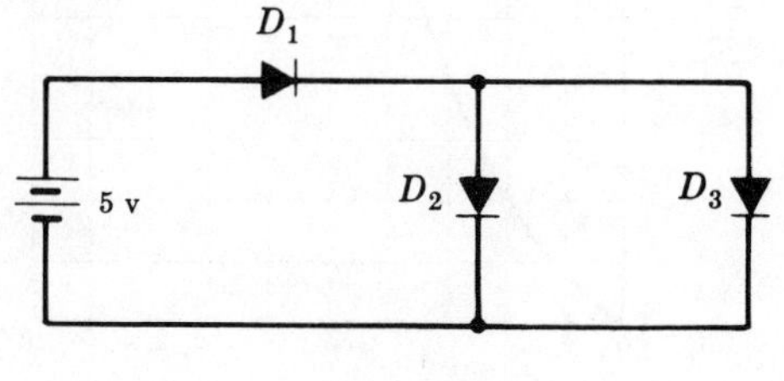

Fig. 2-36

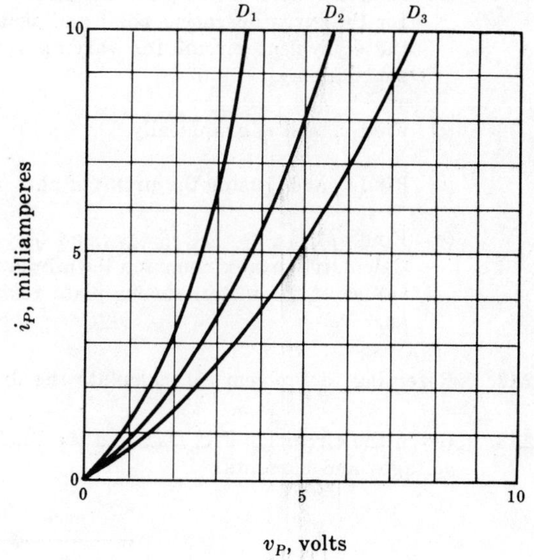

Fig. 2-37

2.9. Referring to Problem 2.8, calculate (*a*) the instantaneous plate resistance for each diode for the voltages determined, (*b*) the instantaneous plate conductance for each diode for the given voltage.

2.10. Given the circuit of Fig. 2-38 and the diode characteristic curve of Fig. 2-39. Find the diode voltage v_P and current i_P. (*Hint.* Since the circuit to the left of terminals A and B has only linear elements, Thévenin's theorem may be applied.)

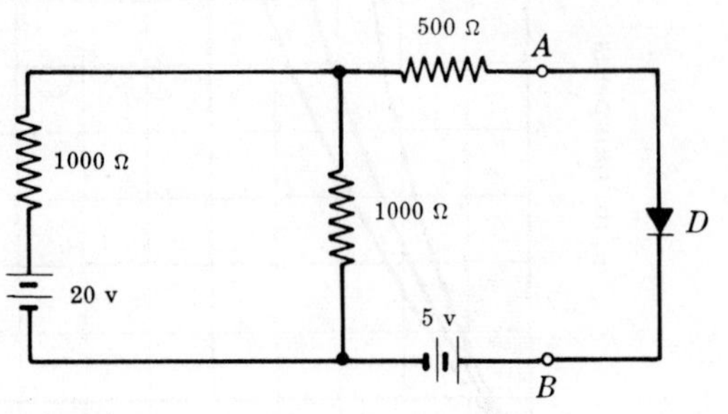

Fig. 2-38

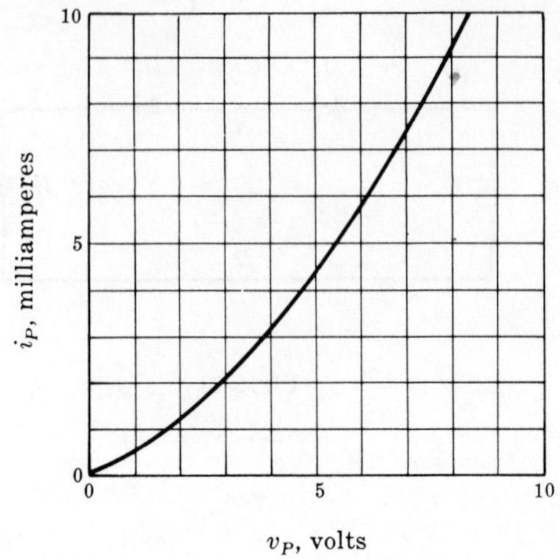

Fig. 2-39

2.11. Refer to the circuit diagram of Fig. 2-18, page 21, and use the values $V_{PP} = 7$ volts, $R_L = 500$ ohms, and $v_{pp} = 2 \sin \omega t$ volts. The diode characteristic curve is given in Fig. 2-40.

(*a*) Find V_P and I_P. Calculate the static plate resistance R_P and draw the equivalent circuit based on average values.

(*b*) Find the differential plate resistance r_p for the given operating point and sketch the equivalent circuit for varying components only.

(*c*) Find v_p and i_p graphically.

(*d*) Find v_p and i_p using the circuit of part (*b*).

(*e*) Find $v_{P(\max)}$, $v_{P(\min)}$, $i_{P(\max)}$, and $i_{P(\min)}$. Calculate the maximum and the minimum values of the instantaneous plate resistance r_P.

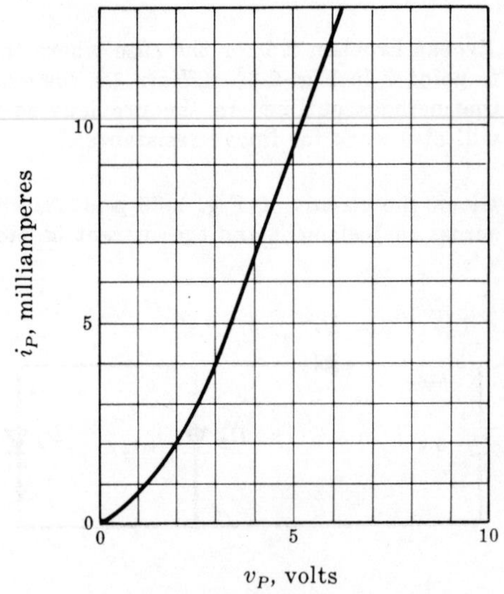

Fig. 2-40

2.12. Referring to Problem 2.4, calculate the differential plate resistance r_p.

2.13. Given the circuit of Fig. 2-41 and the diode characteristic curves of Fig. 2-42 below, find the diode voltages and currents.

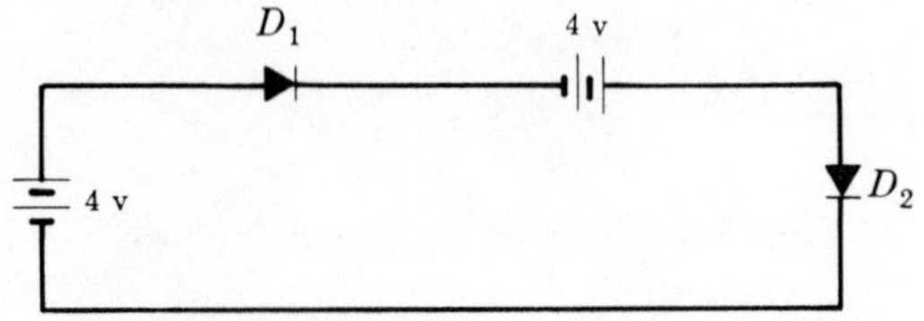

Fig. 2-41

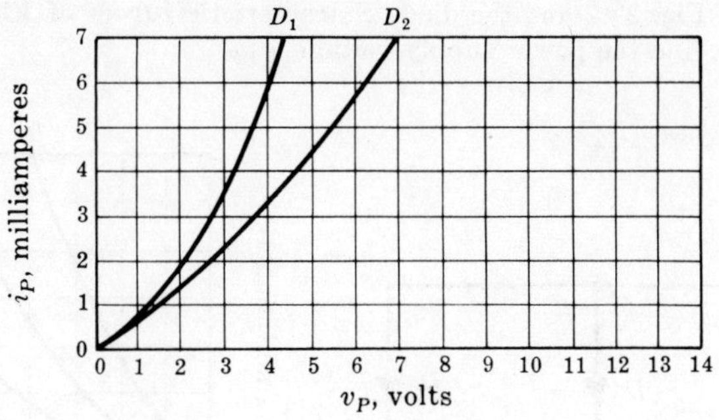

Fig. 2-42

2.14. Given the circuit of Fig. 2-43 and the diode characteristic curve of Fig. 2-44. Determine (*a*) V_P and I_P, (*b*) the differential plate resistance r_p, (*c*) v_p and i_p graphically and by using the equivalent circuit.

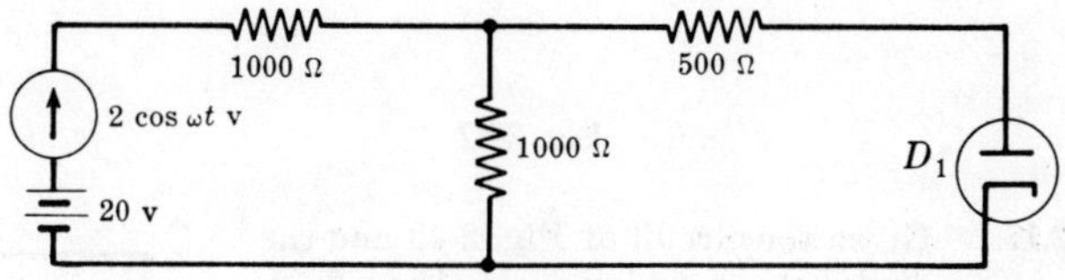

Fig. 2-43

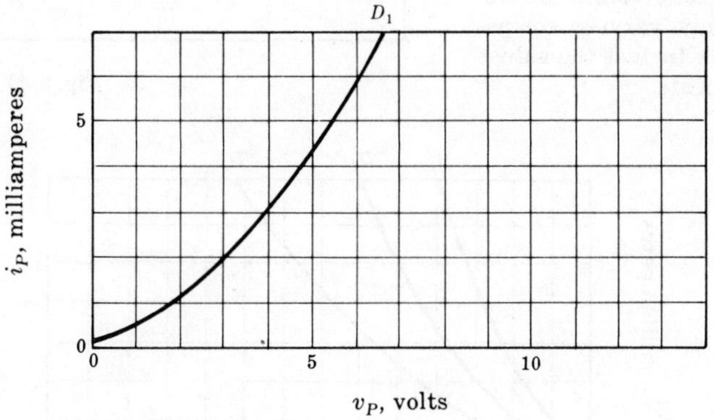

Fig. 2-44

2.15. Given the circuit of Fig. 2-45 and the diode characteristic curve of Fig. 2-46. (*a*) Find the operating point using the load line method and specify V_P and I_P. (*b*) If $v_p = 1 \sin \omega t$ volts, find the generator voltage v_{pp}. Assume linear operation.

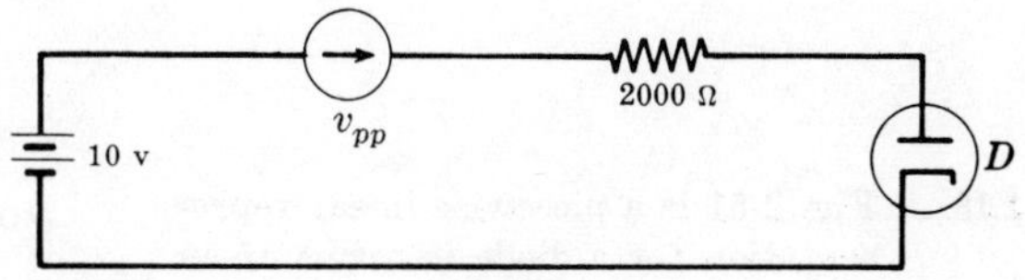

Fig. 2-45

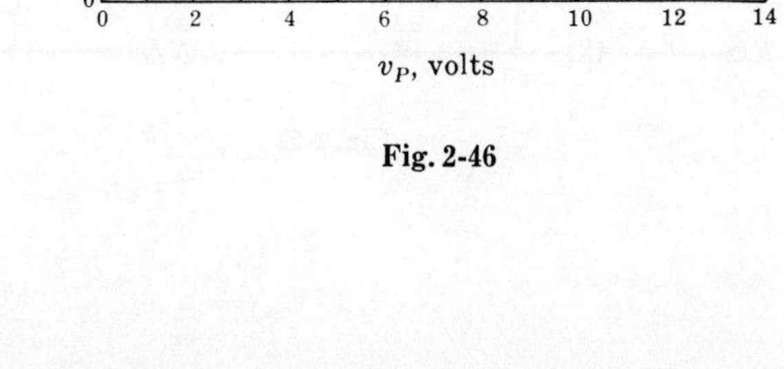

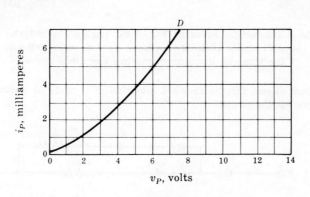

Fig. 2-46

2.16. Given the circuit of Fig. 2-47 and the diode characteristic curves of Fig. 2-48 below. If the current in D_2 is $i_{P2} = 2$ ma, find the power supply voltage V_{PP}.

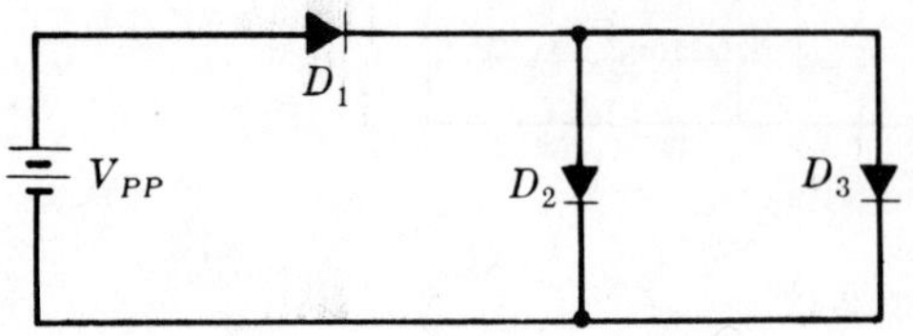

Fig. 2-47

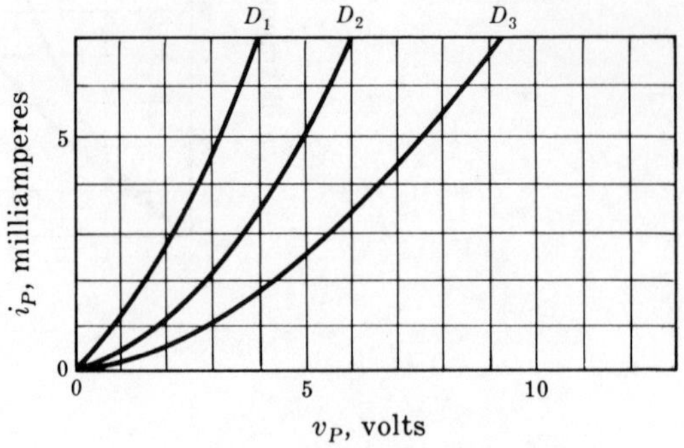

Fig. 2-48

2.17. Given the circuit of Fig. 2-49 and the diode characteristics of Fig. 2-50. Draw the equivalent circuit considering (*a*) only average values (dc circuit) and (*b*) only varying components (ac circuit). Include the values of all circuit elements.

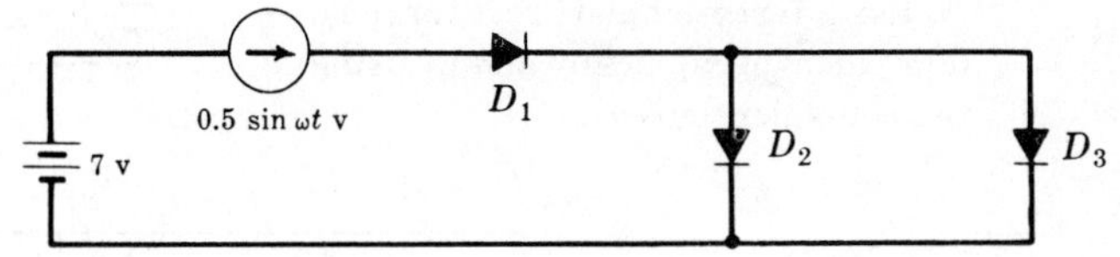

Fig. 2-49

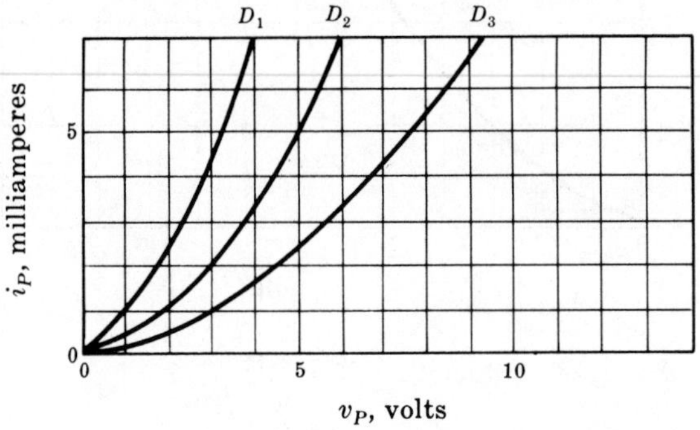

Fig. 2-50

2.18. Fig. 2-51 is a piecewise linear representation for a diode in terms of an ideal diode D. Plot the characteristic curve for this model.

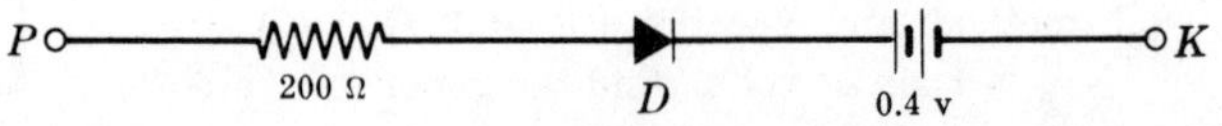

Fig. 2-51

2.19. Refer to the diode characteristic curve of Fig. 2-24, page 23. Find the instantaneous plate conductance g_P and the differential plate conductance g_p for $v_P = 0.8$ volts. Compare these values with the slope G_1 of the straight line approximating the diode curve.

2.20. Given the equivalent circuit for the forward direction of a diode as shown in Fig. 2-52. Plot the characteristic curve for the circuit.

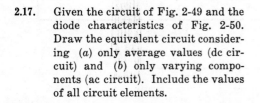

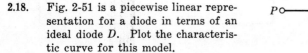

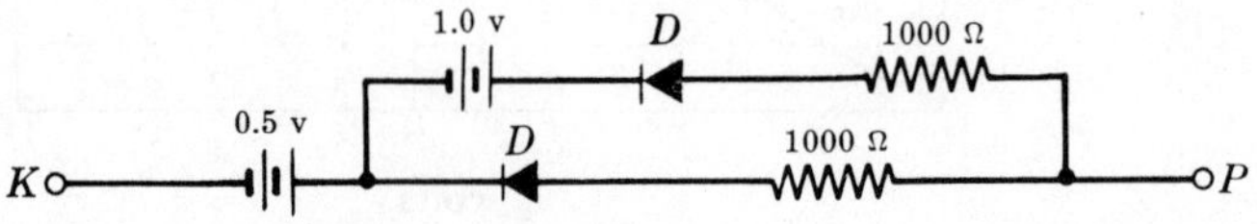

Fig. 2-52

2.21. Draw the characteristic curve for the circuit of Fig. 2-53. A different current scale should be used for the reverse direction.

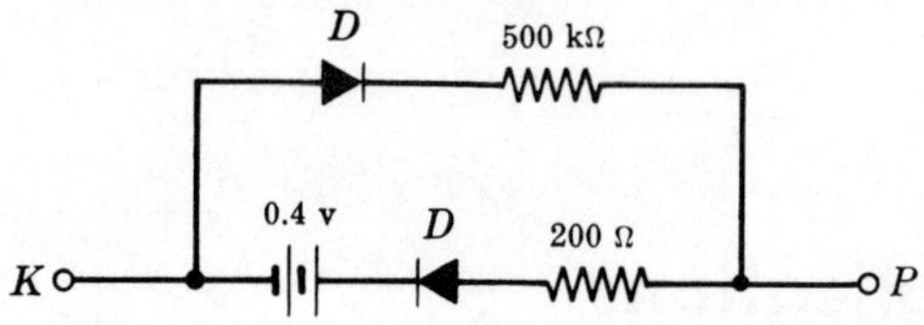

Fig. 2-53

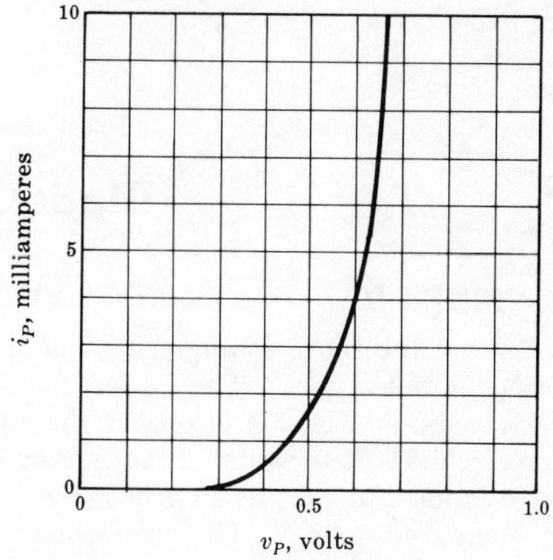

Fig. 2-54

2.22. Given the characteristic curve of Fig. 2-54, diagram an equivalent circuit which approximates the characteristic curve using two line segments. Include the values of the elements of the equivalent circuit.

2.23. Given the circuit of Fig. 2-55 with two diodes which are assumed ideal. What is the potential of the point a with respect to point b? Assume an open circuit.

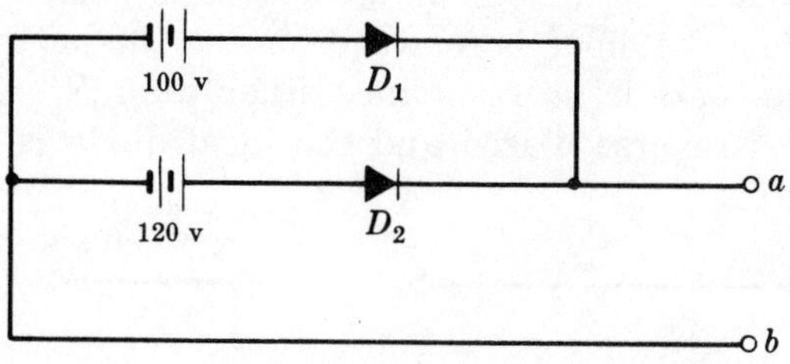

Fig. 2-55

2.24. Given the circuit of Fig. 2-56 and the diode characteristic curve of Fig. 2-57. Find the operating point and specify V_P and I_P.

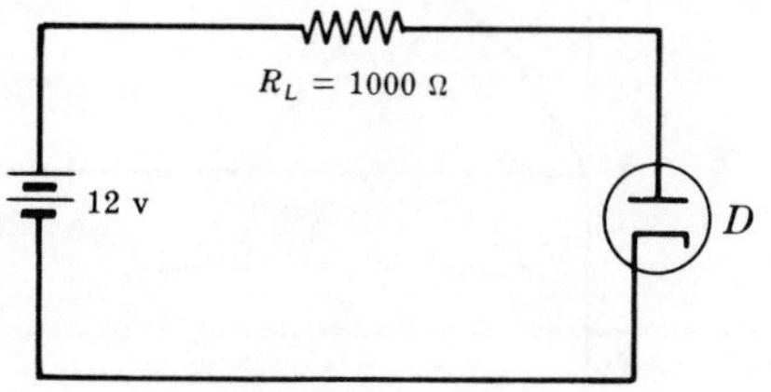

Fig. 2-56

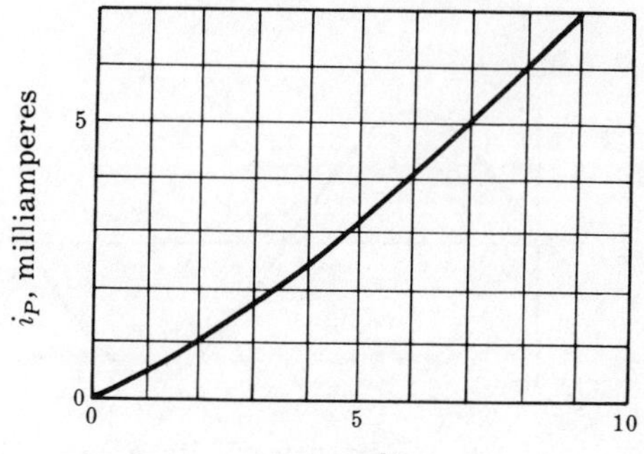

Fig. 2-57

Chapter 3

Diode Applications

3.1 SERIES HALF-WAVE RECTIFIERS

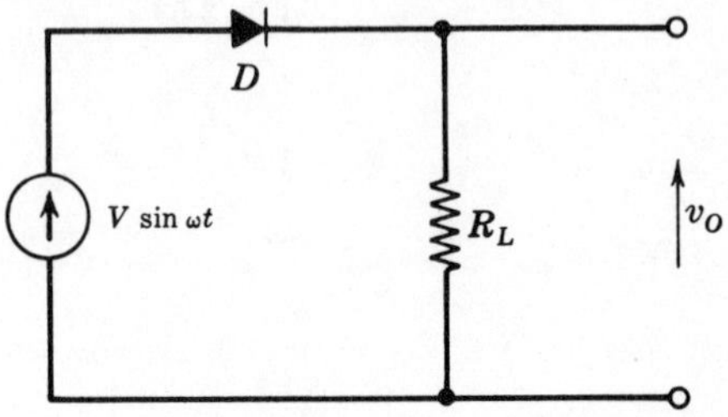

Fig. 3-1. Half-wave Rectifier Circuit

One of the more common uses of diodes is in rectifier applications. The series, half-wave rectifier circuit of Fig. 3-1 is one of the simplest rectifier circuits. It is assumed that either the voltage source is ideal or that its internal resistance is very low compared to R_L. The loop equation may be written in terms of instantaneous values:

$$-V \sin \omega t + v_P + v_O = 0 \qquad (3.1)$$

Since the diode is nonlinear, equation (*3.1*) is nonlinear and may be solved graphically on a point by point basis. However, this circuit can be approximated by two linear circuits, which is usually an adequate representation. One circuit representation is for the case when the diode is forward biased and the other is for the reverse biased case.

If the diode is assumed ideal, then the two circuits are very simple. Fig. 3-2(*a*) shows the equivalent circuit for the forward direction and Fig. 3-2(*b*) for the reverse direction. The diode is forward biased when the voltage source is positive. The output voltage v_O is shown in Fig. 3-3. When the diode is forward biased, the output voltage is the same as the generator voltage since there is no resistance other than R_L. When the generator voltage is negative, the diode is reverse biased and the ideal diode is an infinite resistance; hence the current in R_L is zero.

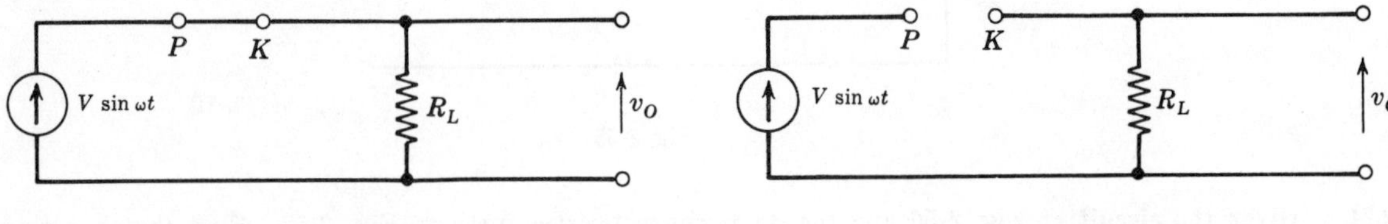

$V \sin \omega t \geq 0$
(*a*) Forward Direction

$V \sin \omega t < 0$
(*b*) Reverse Direction

Fig. 3-2

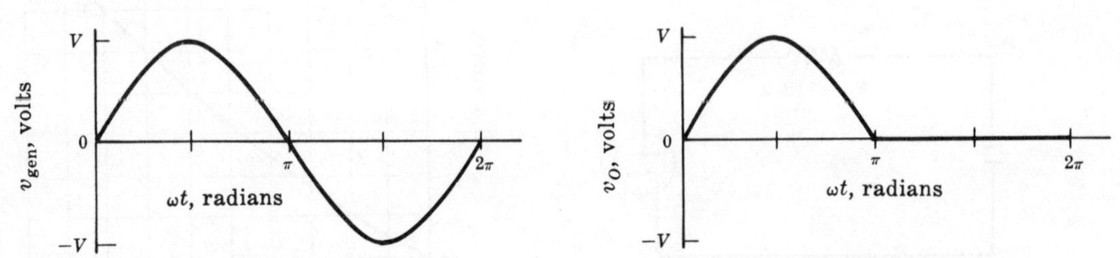

Generator Voltage

$$v_O = \begin{cases} V \sin \omega t & \text{when } V \sin \omega t \geq 0 \\ 0 & \text{when } V \sin \omega t < 0 \end{cases}$$

Fig. 3-3. Output Voltage of the Half-wave Rectifier Circuit

Example 1.

Refer to Fig. 3-1. If $V = 10$ volts and $R_L = 10{,}000$ ohms, find the instantaneous voltage drop across the diode and the instantaneous current in R_L. Assume the generator and diode are ideal.

When the diode is forward biased, $v_P = 0$ since the diode forward resistance is zero. When the diode is reverse biased, the diode resistance is infinite and the current is zero. When the current is zero, v_O is zero. When v_O is zero in equation (*3.1*), $v_P = V \sin \omega t$; this obtains when $V \sin \omega t < 0$. A sketch of v_P is shown in Fig. 3-4.

The current in the circuit is zero when the diode is reverse biased, and is $(V \sin \omega t)/R_L$ when it is forward biased. A sketch of the current in R_L is given in Fig. 3-5.

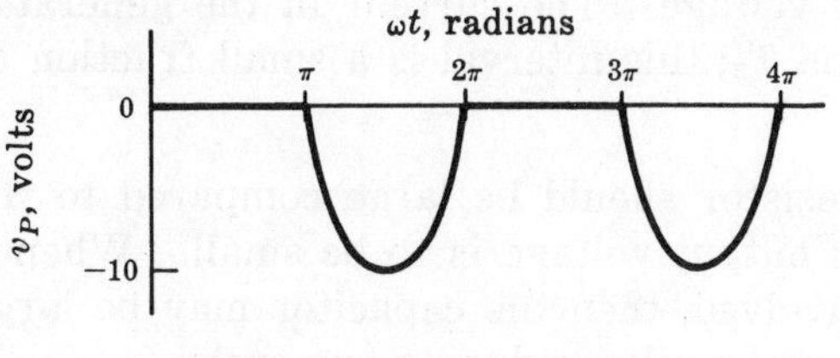

Fig. 3-4. Diode Plate Voltage

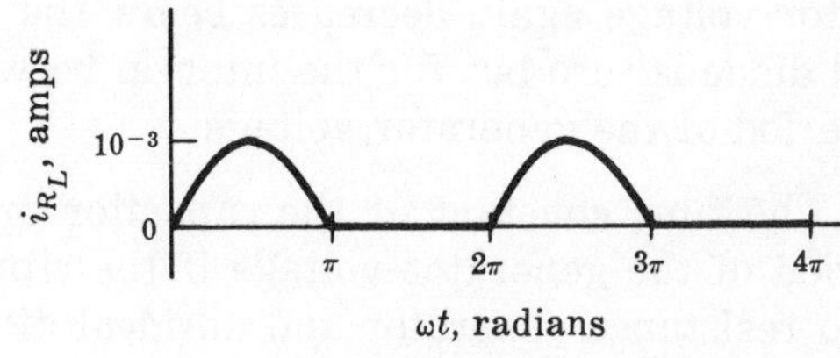

Fig. 3-5. Current in the Load Resistor

3.2 CAPACITOR FILTER

A capacitor may be added to the circuit of Fig. 3-1 to provide some smoothing of the output voltage v_O. To study this circuit, assume a switch is included and that it is closed at $t = 0$. This circuit is shown in Fig. 3-6.

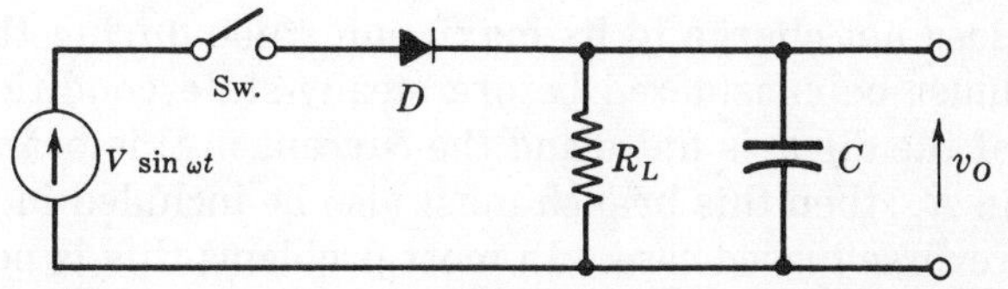

Fig. 3-6. Half-wave Rectifier Circuit with Capacitor

When the switch is closed at $t = 0$, the diode is forward biased. In a circuit containing ideal circuit elements, there is no resistance in series with the voltage source. Thus when the diode is forward biased the capacitor terminal voltage is the same as the generator voltage. The practical situation approaches this when the total series resistance in the circuit is so small that the generator and capacitor voltages are approximately the same. It is assumed that R_L represents all the resistance across the output terminals, including any external load resistance.

When $\omega t = \pi/2$ radians, the voltage across the capacitor is $v_C = V = v_O$. As ωt increases beyond $\pi/2$, the voltage of the generator decreases below the capacitor voltage, causing the diode to be reverse biased. The output voltage is now determined only by the parallel combination of C and R_L; the capacitor has initial voltage V at the instant the diode bias is reversed. The problem involving a charged capacitor C in parallel with a resistor R_L is usually solved in elementary circuits courses, the solution being

$$v_O = v_C = Ve^{-t/R_LC}$$

The diode remains reverse biased until the generator voltage is again equal to the voltage across the capacitor. This will occur during the first quarter of the next cycle at an instant of time T_1 as shown in Fig. 3-7.

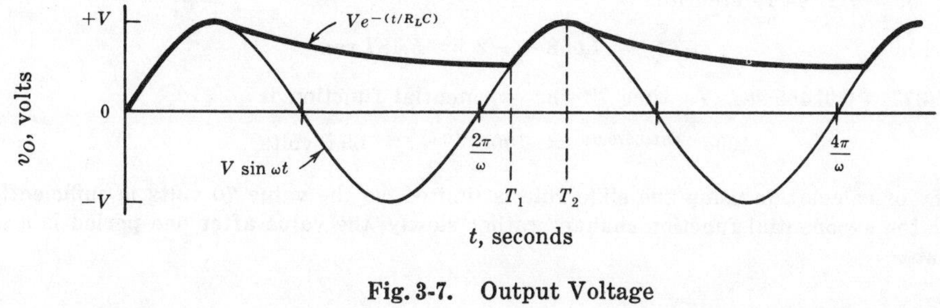

Fig. 3-7. Output Voltage

The time T_1 when the generator voltage is again equal to the capacitor voltage may be determined graphically by plotting the two voltages and locating the point at which they intersect. This is the instant of time T_1 when

$$V \sin \omega T_1 = V e^{-T_1/R_L C} \tag{3.2}$$

The proper time references must be used as shown in Example 2. Equation (*3.2*) is transcendental and can be solved graphically or by some method of successive approximations.

The diode is again forward biased from time T_1 until time T_2 (Fig. 3-7) when the generator voltage again decreases below the capacitor voltage. The current in the generator and diode is zero but for the interval between T_1 and T_2; this interval is a small fraction of a period of the generator voltage.

The time constant of the capacitor and load resistor should be large compared to the period of the generator voltage if the ripple of the output voltage is to be small. When a zero resistance generator and an ideal diode are involved, then the capacitor may be large and it will still charge to the peak value of the generator voltage during one cycle.

When there is appreciable resistance in the generator and diode circuit representing the forward diode polarity, the problem becomes more complicated. In this case the capacitor does not charge to its maximum value during the first quarter cycle, so a number of cycles must be considered before steady-state conditions are attained. If the reverse resistance of the diode is finite and the current in this branch is not negligible compared to the current in R_L, then this branch must also be included in the equivalent circuit representation for the reverse biased case. In most problems this is not done because of the order of accuracy and amount of work entailed.

The varying component of the output voltage is called the *ripple*. Since the ripple waveform is usually nonsinusoidal, the peak-to-peak value is usually measured. Sometimes the rms value is measured, but this involves special problems because the waveform is not sinusoidal. The ripple voltage is sometimes compared to the average output voltage, thus defining a ripple factor.

Example 2.

Referring to the circuit of Fig. 3-6, let $V = 100$ v, $R_L = 1000\ \Omega$, and $C = 40\ \mu$f. Assume the diode and generator are ideal. If $\omega = 377$ radians/sec, find the peak-to-peak value of the ripple voltage and the time the diode is conducting. Compare the conduction time with the period of the generator voltage.

First find the time T_1 when the diode begins conducting again. As a starting point, evaluate the exponential function after one period:

$$V e^{-(2\pi/\omega)/R_L C} = 100 e^{-0.417} = 66 \text{ volts}$$

The output voltage is an exponential function beginning at time $\pi/2\omega$ as shown in Fig. 3-7. The time T_1 is less than one period later than $\pi/2\omega$. The capacitor voltage at T_1 will be greater than 66 because the discharging time is less than one period; hence 70 volts seems a reasonable estimate. At the time T when $100 \sin \omega T = 70$ volts, the angle ωT is 0.775 radian measured from 2π radians. The total angle using $\omega t = 0$ as reference is

$$\omega T + 2\pi = 0.775 + 6.283 = 7.058 \text{ rad}$$

The angle using $\omega t = \pi/2$ as reference is

$$\omega T^* = 7.058 - \pi/2 = 5.487 \text{ rad}$$

and $T^* = 5.487/377 = 0.01455$ sec. At time T^* the exponential function is

$$100 e^{-0.01455/0.04} = 100 e^{-0.363} = 69.5 \text{ volts}$$

The accuracy of calculation using the slide rule is limited, so the value 70 volts is sufficiently close to 69.5 volts. Since the exponential function changes rather slowly, the value after one period is a good basis for a first estimate.

Using $t = 0$ as reference, $T_1 = 7.058/377 = 0.0187$ sec. The time $T_2 = 5\pi/(2 \times 377) = 0.0208$ sec. The conduction time is $T_2 - T_1 = 0.0021$ sec. The period of the generator voltage is $T = 2\pi/377 = 0.0167$ sec. The conduction time is 12.6% of a period.

The maximum output voltage is 100 volts and the minimum is 70 volts; thus the peak-to-peak ripple voltage = 100 − 70 = 30 volts. The ripple is rather high because for this type of circuit the 1000 ohms resistance is rather low, but this value was chosen to better illustrate the method of solution.

3.3 SERIES FULL-WAVE RECTIFIERS

In many power supply applications a power transformer is used with two diodes (or a double diode) in a full-wave rectifier circuit as shown in Fig. 3-8. The secondary winding is center-tapped so that v_1 and v_2 have equal magnitudes but are 180° out of phase with the center tap as reference. When v_1 and v_2 are positive (using the reference polarities of Fig. 3-8), the diode D_1 is forward biased and is conducting; the diode D_2 is reverse biased and is not conducting. When v_1 and v_2 are negative, D_2 is forward biased and D_1 is reverse biased. Thus there is current in R_L through either D_1 or D_2 for all time but the instant when $v_1 = 0 = v_2$.

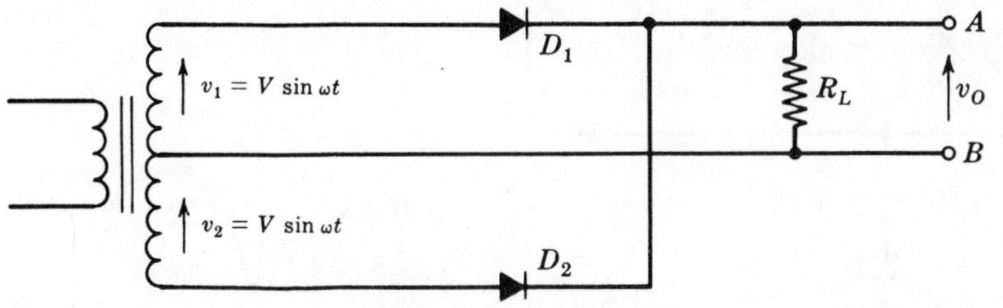

Fig. 3-8. Full-wave Rectifier Circuit

The output voltage v_O for the circuit of Fig. 3-8 is sketched in Fig. 3-9. Terminal A is positive with respect to terminal B. If the diodes are reversed, the polarity of the output is also reversed. In power supply applications a voltage positive with respect to the reference is more common, but the diodes can be connected either way if individual diodes are used. In the case of double-diode vacuum tubes where the cathode is common, the two diode sections cannot be reversed.

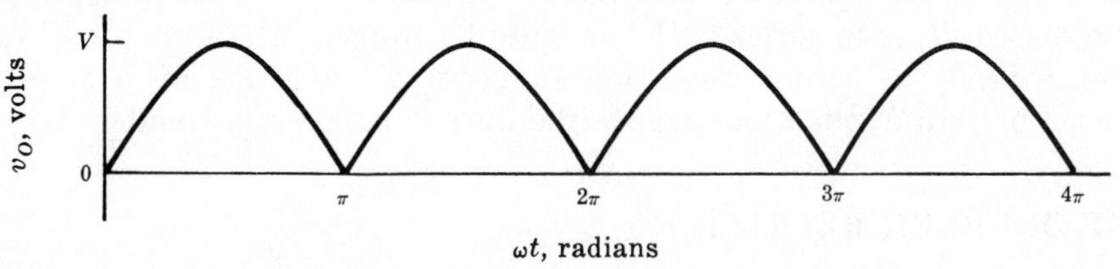

Fig. 3-9. Full-wave Rectifier Output Voltage

The full-wave rectified voltage has a higher average value than a half-wave rectified voltage (see Problems 3.2 and 3.10). In addition, less filtering is required to meet a given ripple specification. The fundamental frequency of the full-wave rectified voltage is twice that of the half-wave rectified voltage. One of the problems involved in the analysis of power supply circuits is that it is difficult to represent the voltage at the output of the rectifier. When there is only a resistive load, there is no problem because the rectified voltage can be represented using Fourier series. However, when there is a filter connected to the rectifier output, the voltage is difficult to represent analytically except for very simple cases. Thus power supply design problems are usually not completely solved analytically even when linear elements are used in the filter.

Another point of interest is that an ideal full-wave rectifier performs an operation on the input voltage which gives the instantaneous magnitude as an output. By applying the instantaneous magnitude to a properly chosen low-pass filter, the average magnitude may be found.

3.4 SHUNT RECTIFIERS

In some applications involving little power, a shunt rectifier is sometimes used. A simple half-wave shunt rectifier circuit is shown in Fig. 3-10. In this case a series resistance R_1 is usually included unless the generator internal resistance is rather high. The shunt resistance R_2 is usually large compared to R_1. When $V \sin \omega t$ is positive, the diode is reverse biased, so the output voltage is easily determined:

$$v_O = [R_2/(R_1+R_2)]V \sin \omega t \quad \text{when} \quad V \sin \omega t \geqq 0 \tag{3.3}$$

$$v_O \cong V \sin \omega t \quad \text{when} \quad R_2 \gg R_1 \tag{3.4}$$

When $V \sin \omega t$ is negative, the diode is forward biased:

$$v_O = 0 \quad \text{when} \quad V \sin \omega t < 0 \tag{3.5}$$

The output voltage v_O is sketched in Fig. 3-11.

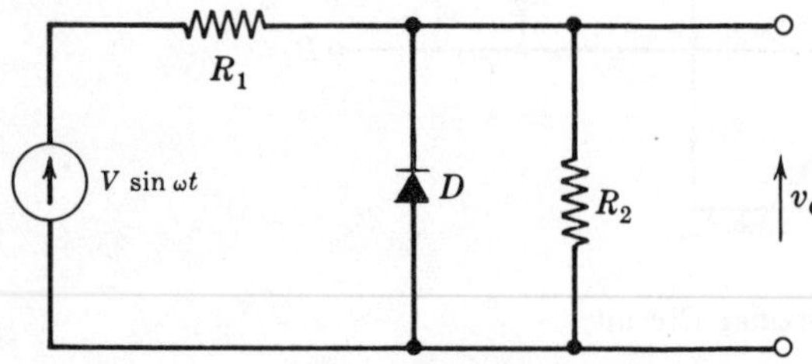

Fig. 3-10. Shunt Rectifier Circuit

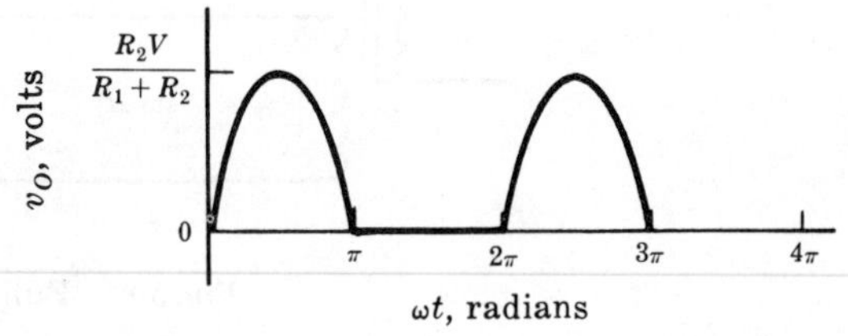

Fig. 3-11. Shunt Rectifier Output Voltage

If a negative voltage is desired, the diode may be reversed. The output voltage will then be the negative of that shown in Fig. 3-11. A filter may be added to the output of the rectifier to provide smoothing. Because of the low efficiency, this type of rectifier is not used when a large amount of power is involved. During the part of the cycle when the diode is forward biased, the generator voltage is applied directly across R_1; but when the diode is reverse biased, R_1 is in series with the output voltage. The first condition suggests a high value of R_1 and the second condition suggests a low value of R_1. Hence shunt rectifiers are used only in applications where the output current is small.

3.5 VOLTAGE-DOUBLER RECTIFIER

The voltage-doubler, half-wave rectifier circuit is employed very frequently in radio and TV sets which use no power transformer. The basic rectifier circuit is shown in Fig. 3-12. Consider first the circuit to the left of the terminals X and Y assuming both switches are open. If Switch 1 is closed at $t = 0$ and the initial voltage of the capacitor

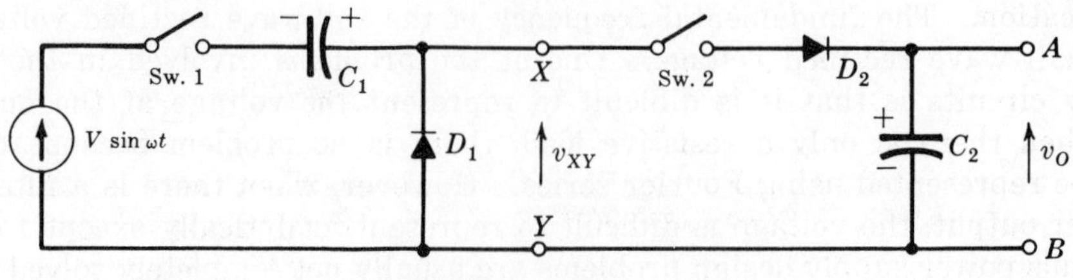

Fig. 3-12. Voltage-doubler Circuit

is zero, the diode is reverse biased until $\omega t = \pi$ radians. When $V \sin \omega t$ is negative, the diode is forward biased and the capacitor C_1 charges until the terminal voltage is V with the polarity shown in Fig. 3-12; this occurs at $\omega t = 3\pi/2$ radians. If nothing is connected to the terminals X and Y, then the diode is never forward biased again (assuming ideal circuit elements).

If the current in the circuit is zero, the capacitor voltage is a constant V and the capacitor appears to be a battery. Then v_{XY} is

$$v_{XY} = +V + V \sin \omega t \tag{3.6}$$

Thus the voltage v_{XY} is $V \sin \omega t$ shifted upward by a constant amount V. This is a sine wave and its peak value is $2V$ with respect to zero. The minimum instantaneous value is zero and the maximum is $2V$ as shown in Fig. 3-13. If Switch 2 is closed at $\omega t = 3\pi/2$ radians, then the circuit to the right of terminals X and Y is connected when v_{XY} is zero. Assuming the initial voltage of the capacitor C_2 is zero, D_2 is forward biased and C_2 begins to charge. When D_1 is reverse biased, there is current only in the series circuit including C_1. Hence the voltage across C_1 decreases as charge flows into C_2.

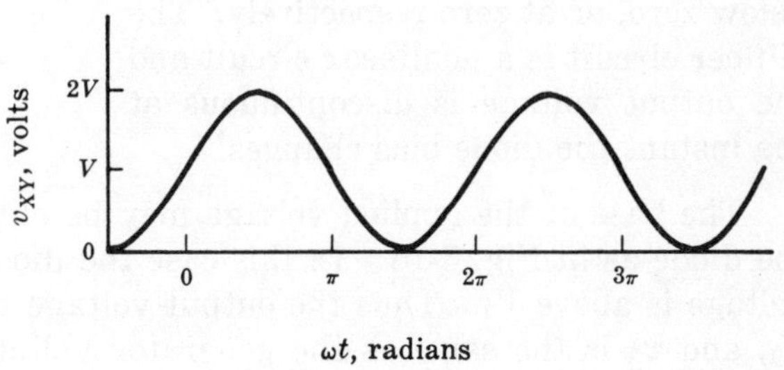

Fig. 3-13. Instantaneous Voltage v_{XY}

When v_{XY} decreases below v_O, then D_2 is reverse biased and the current in C_2 is zero (assuming no external load). When v_{XY} drops to zero, then D_1 is forward biased, so that C_1 charges until the terminal voltage is again V (assuming ideal elements). With no external load, v_O increases during each cycle and finally approaches $2V$. When there is a load, the output voltage v_O never reaches $2V$ but is much higher than it would be if only a single series rectifier were used. The capacitor C_1 is usually rather large and can be an electrolytic type since the voltage across such a capacitor always has the same polarity. A filter may be added at A and B to provide smoothing of the output voltage.

The discussion here has been limited to the case where ideal circuit elements are used, because of the complexity of the problem when this is not the case. To appreciate this, the circuit of Fig. 3-12 should be considered when the generator internal resistance and the diode forward and reverse resistances are included as well as an external load.

3.6 DIODE CLIPPING CIRCUITS

A simple peak *clipping circuit* using a diode is shown in Fig. 3-14. A switch is included so that the operation of the circuit may be more easily studied. If the switch is closed at $t = 0$, the diode remains reverse biased until $V \sin \omega t = V_1$, at which time it is forward biased. Assuming no external load at the output terminals, the output voltage v_O is the same as the generator voltage until the diode is forward biased.

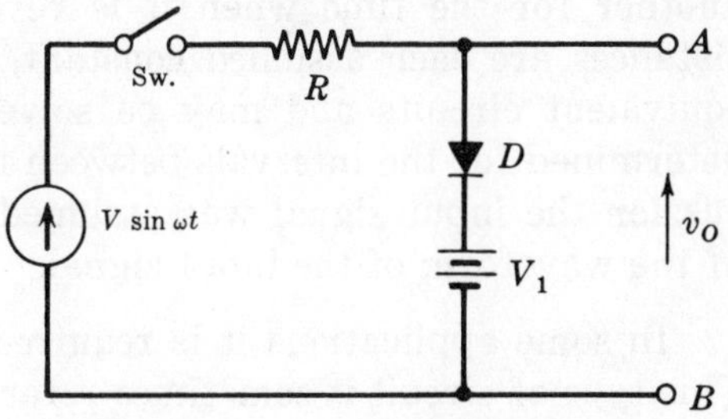

Fig. 3-14. Peak Clipping Circuit

When the diode is forward biased, v_O remains at V_1 because there is no series resistance in the diode branch. When $V \sin \omega t$ drops below V_1, the diode is reverse biased and v_O is again the same as the generator voltage, as shown in Fig. 3-15. The cycle then repeats itself, giving the periodic voltage in Fig. 3-15.

Because of the polarity of the diode in Fig. 3-14, it is the peak of the applied voltage that is clipped. The resistor R is included to limit the current during the conduction time of the diode. The voltage source indicated as V_1 may be either positive, negative, or zero, thus clipping the peak at a point above zero, below zero, or at zero respectively. The clipper circuit is a nonlinear circuit and the output voltage is discontinuous at the instant the diode bias changes.

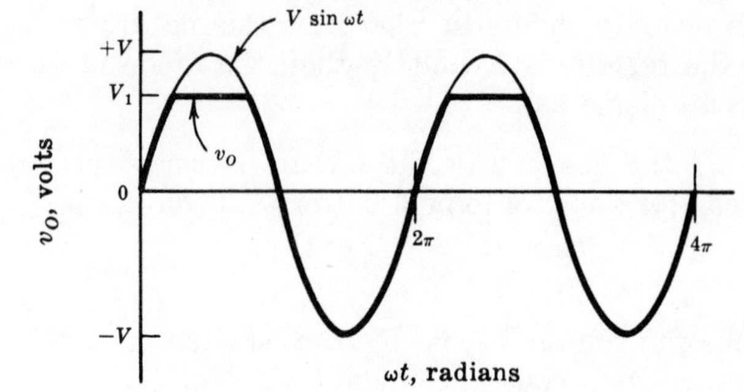

Fig. 3-15. Peak Clipper Output Voltage

The base of the applied voltage may be clipped at some particular value by connecting the diode as in Fig. 3-16. In this case the diode is reverse biased so long as the generator voltage is above V_1. Thus the output voltage v_O is V_1 when the generator voltage is below V_1, and v_O is the same as the generator voltage when it is above V_1. Again, V_1 may be positive, negative, or zero, thus clipping the base above zero, below zero, or at zero respectively.

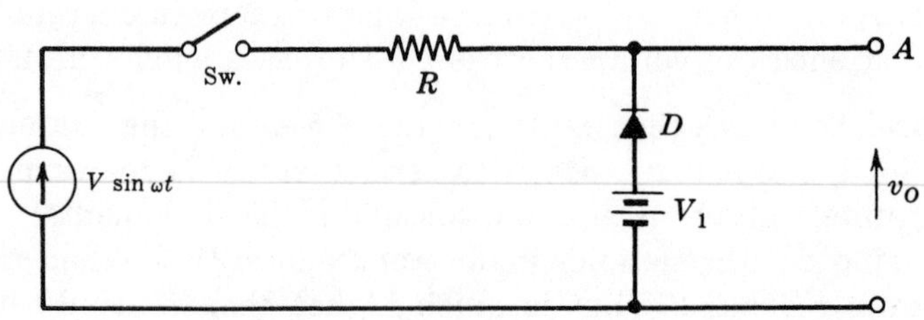

Fig. 3-16. Base Clipping Circuit

If the diode forward resistance is appreciable, the voltage for either the peak or base clipper will not remain at exactly V_1 when the diode is forward biased but will be slightly above or below V_1 depending on the current direction (diode connection). If the reverse resistance is not large compared to R, then the output voltage will not be exactly equal to the generator voltage during the time the diode is reverse biased. In these cases, the appropriate representation is used for the diode and for the source V_1.

An equivalent circuit is drawn for the time when the diode is forward biased, and another for the time when it is reverse biased. If the diode forward and reverse resistances are each assumed constant, the circuit equations are linear for each of the two equivalent circuits and may be solved analytically. The output voltage v_O can then be determined for the intervals between the discontinuities, giving v_O for all time. In this discussion the input signal was assumed sinusoidal, but the approach is the same regardless of the waveform of the input signal.

In some applications it is required that both the peak and base of a signal be clipped. This type of circuit is sometimes referred to as a *slicer*. A simple diode slicer circuit which is a combination base clipper and peak clipper is shown in Fig. 3-17 below. The output v_O is the same as the applied input v_I only when the diodes are both reverse biased, which is the case when $V_2 < v_I < V_1$. When $v_I > V_1$, the output voltage is V_1; when $v_I < V_2$, the output voltage is V_2. This assumes ideal diodes and ideal voltage sources in the diode branches.

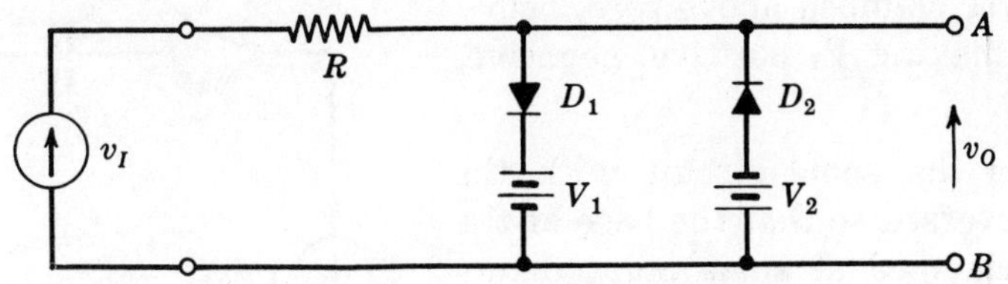

Fig. 3-17. Slicer Circuit

The circuits may be analyzed using methods previously outlined. Now, however, there are added discontinuities in v_O. If the diodes are ideal or if they have constant resistance in the forward and reverse directions, linear circuit representations may be used for other than the points of discontinuity.

3.7 DIODE CLAMPING CIRCUITS

Another application for diodes is in *clamping circuits*. An example of a circuit where the peak is clamped at some value determined by the voltage V_1 is shown in Fig. 3-18. The operation of the circuit is more easily studied by including a switch. Assume the switch is closed at $t = 0$ and that the capacitor has no initial charge. The diode remains reverse biased until the input voltage $V \sin \omega t = V_1$, at which time the diode is forward biased. So long as the diode is forward biased, the output remains at V_1. During this time a current i causes the capacitor to charge.

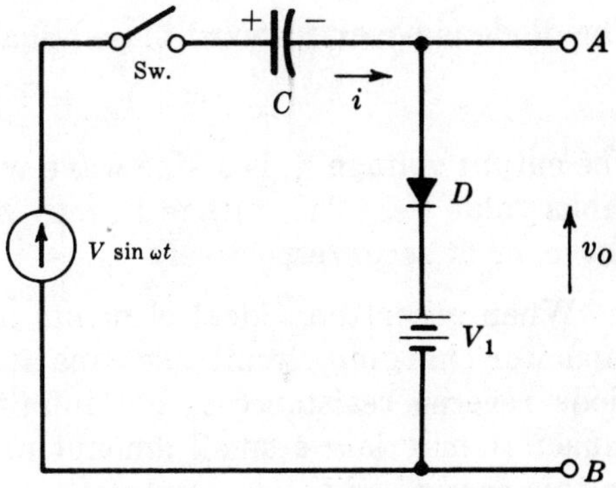

Fig. 3-18. Peak Clamping Circuit

When the generator voltage reaches its maximum value at $\omega t = \pi/2$ radians, the loop equation may be written as

$$-V + v_C + V_1 = 0 \qquad \text{when} \quad \omega t = \pi/2 \tag{3.7}$$

Then $v_C = V - V_1$, with the polarity shown in Fig. 3-18. As ωt increases above $\pi/2$, the diode is reverse biased and is never forward biased again, assuming ideal elements. The capacitor voltage remains constant at $v_C = V - V_1$, so the output voltage is

$$v_O = -v_C + V \sin \omega t = -V + V_1 + V \sin \omega t \tag{3.8}$$

When $\sin \omega t = +1$, $v_O = +V_1$; when $\sin \omega t = -1$, $v_O = V_1 - 2V$. The steady-state output voltage is a sine wave with a peak-to-peak value $2V$ but with a maximum instantaneous value V_1. When $V_1 < V$, the capacitor will charge so that the maximum value of v_O is V_1; thus it may be said that the output peak is clamped at V_1.

When the elements are ideal, there is distortion of the output waveform only during the time the capacitor is charging. The output v_O is sketched in Fig. 3-19 for the circuit of Fig. 3-18 assuming the switch is closed at $t = 0$.

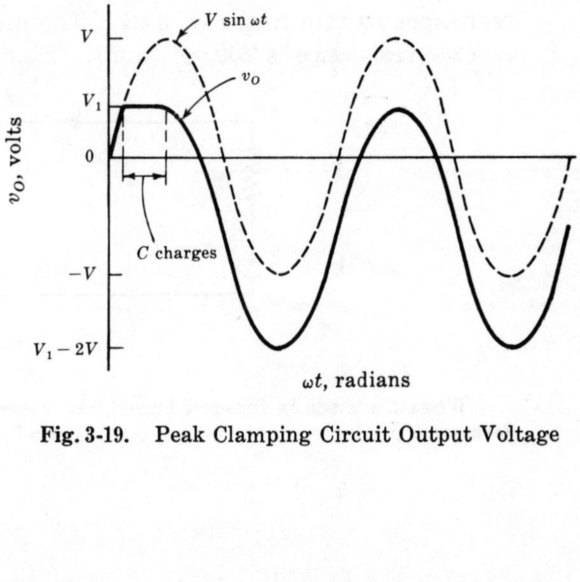

Fig. 3-19. Peak Clamping Circuit Output Voltage

The peak of v_O can be clamped above zero, below zero, or at zero by making V_1 positive, negative, or zero respectively.

Fig. 3-20 shows the same circuit with the diode connections reversed so that the base of the applied voltage is clamped at some point determined by V_1. Assume the capacitor has no initial charge and that the switch is closed at $\omega t = \pi/2$ so that the diode will be reverse biased (assuming $V > V_1$). When $V \sin \omega t = V_1$, the diode is forward biased and the current i is in the direction shown in Fig. 3-20. The loop equation for the instant when $\omega t = 3\pi/2$ radians is

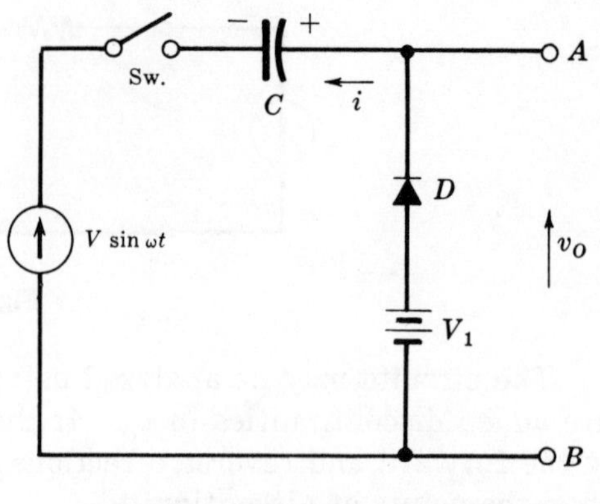

Fig. 3-20. Base Clamping Circuit

$$-V_1 + v_C - V = 0 \quad \text{or} \quad v_C = V_1 + V \tag{3.9}$$

The diode is never forward biased again (assuming ideal elements). Hence

$$v_O = v_C + V \sin \omega t = V_1 + V + V \sin \omega t \tag{3.10}$$

The output voltage v_O is a sine wave with a peak-to-peak value $2V$ and a minimum instantaneous value V_1. The voltage V_1 may be negative, positive, or zero to clamp the base below, above, or at zero, respectively.

When other than ideal elements are involved so that there is series resistance in the capacitor charging circuit, the capacitor does not fully charge during the first cycle. If the diode reverse resistance is not infinite or if there is a terminating resistance, then the capacitor may lose a small amount of charge during the time the diode is reverse biased. In this case steady-state conditions are reached after a number of cycles and there may be a small amount of distortion in the output v_O depending on how constant the capacitor voltage remains.

Solved Problems

3.1. A half-wave rectifier circuit is connected to a generator which has 100 ohms internal resistance as shown in Fig. 3-21. The diode forward resistance is 200 ohms and the reverse resistance is 200,000 ohms. Find the instantaneous output voltage v_O.

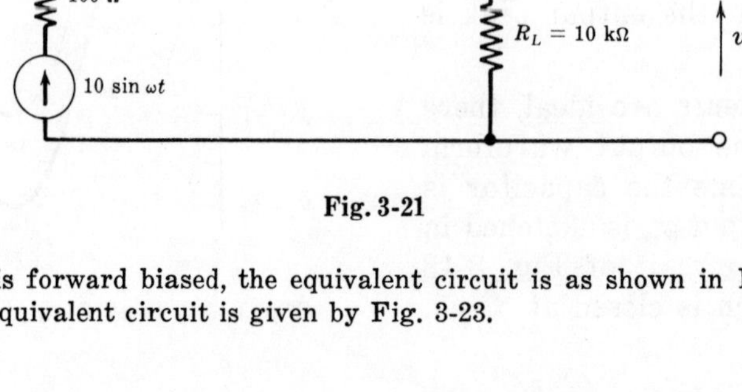

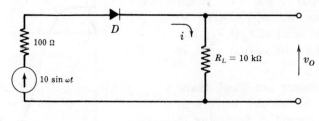

Fig. 3-21

When the diode is forward biased, the equivalent circuit is as shown in Fig. 3-22. When it is reversed biased, the equivalent circuit is given by Fig. 3-23.

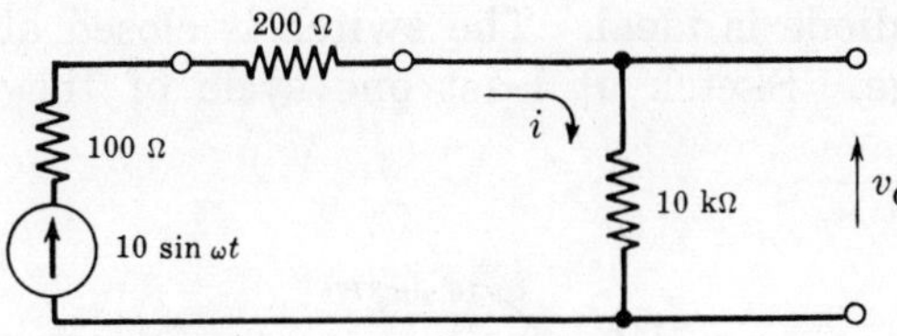

Fig. 3-22

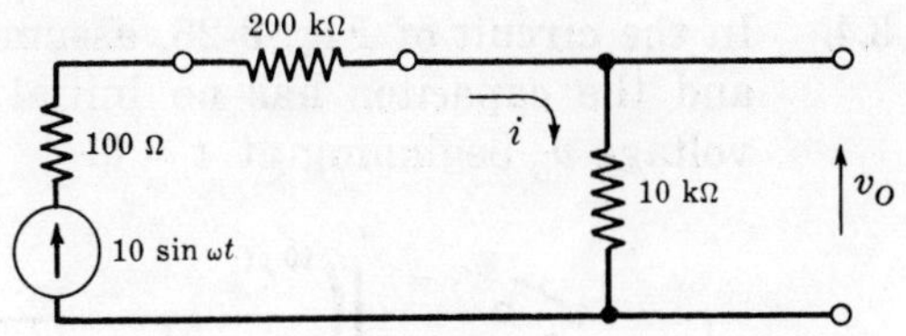

Fig. 3-23

When $10 \sin \omega t \geqq 0$: $i = (10 \sin \omega t)/10{,}300$ amp and $v_O = 10{,}000i = 9.7 \sin \omega t$ volts.

When $10 \sin \omega t < 0$: $i = (10 \sin \omega t)/210{,}000$ amp and $v_O = 0.475 \sin \omega t$ volts.

3.2. Find the average value of the output voltage v_O in terms of V for the circuit of Fig. 3-1, page 32.

$$v_O = \frac{1}{T}\int_0^T v_O(t)\,dt = \frac{\omega}{2\pi}\int_0^{\pi/\omega} V \sin \omega t\,dt + \int_{\pi/\omega}^{2\pi/\omega} 0\,dt = V/\pi$$

3.3. Given the circuit of Fig. 3-16 with $R = 10$ kΩ, $V = 20$ volts, $V_1 = -5$ volts, and the diode resistances $r_{PF} = 500$ Ω and $r_{PR} = 100$ kΩ. Sketch one complete cycle of v_O.

The discontinuity in the output voltage will occur at approximately $V \sin \omega t = V_1 = -5$ volts; thus when $V \sin \omega t \geqq -5$ volts the diode is reverse biased and the equivalent circuit is shown in Fig. 3.24(a). When $V \sin \omega t < -5$ volts, the diode is forward biased and the equivalent circuit is given by Fig. 3-24(b).

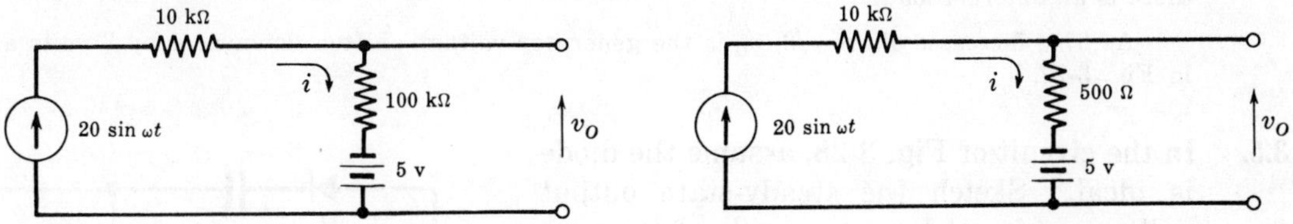

$20 \sin \omega t \geqq -5$
(a)

$20 \sin \omega t < -5$
(b)

Fig. 3-24

The loop equation for the circuit of Fig. 3-24(a) is

$$-20 \sin \omega t + 110{,}000i - 5 = 0 \quad \text{or} \quad i = (5 + 20 \sin \omega t)/110{,}000 \text{ amps}$$

$$v_O = 100{,}000i - 5 = -0.46 + 18.18 \sin \omega t \text{ volts}$$

The loop equation for the circuit of Fig. 3-24(b) is

$$-20 \sin \omega t + 10{,}500i - 5 = 0 \quad \text{or} \quad i = (5 + 20 \sin \omega t)/10{,}500 \text{ amps}$$

$$v_O = 500i - 5 = -4.76 + 0.952 \sin \omega t \text{ volts}$$

In Fig. 3-25 v_O is sketched for the two intervals, using the two equations for v_O.

$v_{O(\max)}$ occurs when $\sin \omega t = +1$: $v_{O(\max)} = -0.46 + 18.18 = 17.72$ volts.

$v_{O(\min)}$ occurs when $\sin \omega t = -1$: $v_{O(\min)} = -4.76 - 0.952 = -5.712$ volts.

The output voltage v_O is zero when

$$-0.46 + 18.18 \sin \omega t = 0$$

or $\sin \omega t = 0.46/18.18 = 0.0253$

Then $\omega t \cong 0.0253$ radians $= 1.45°$ and $\omega t \cong 180° - 1.45° = 178.55°$. The exact points of discontinuity of the output v_O could be found in a similar manner.

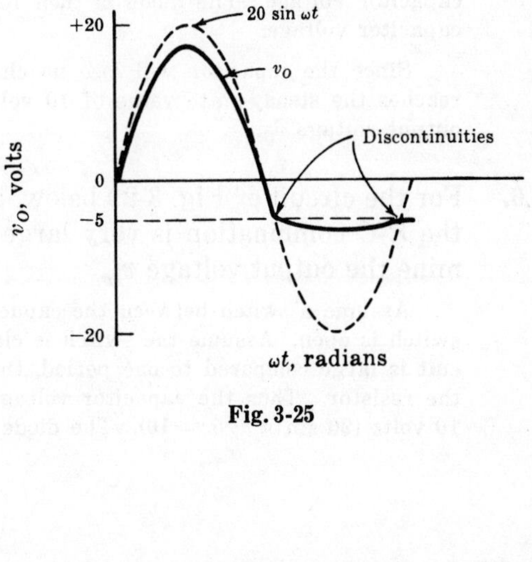

Fig. 3-25

3.4. In the circuit of Fig. 3-26, assume the diode is ideal. The switch is closed at $t = 0$ and the capacitor has no initial charge. Sketch at least one cycle of the output voltage v_o beginning at $t = 0$.

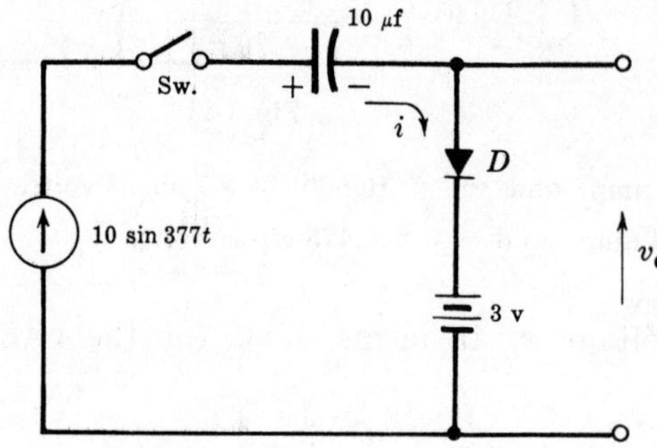

Fig. 3-26

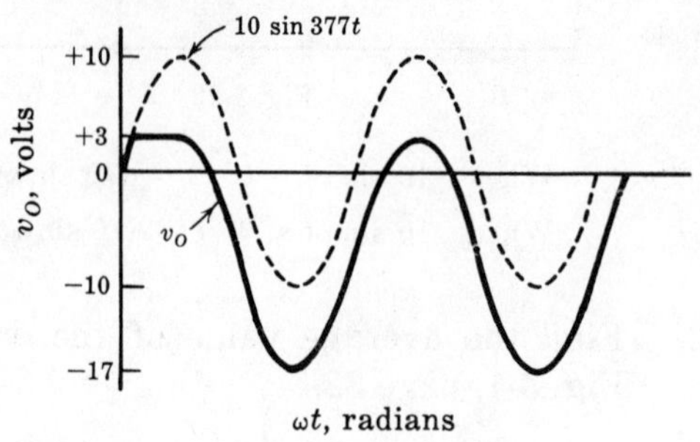

Fig. 3-27

When the switch is closed at $t = 0$, the diode is reverse biased and the output voltage is the same as the generator voltage.

When the generator voltage reaches 3 volts, the diode is forward biased and the output voltage remains at 3 volts until the peak of the generator voltage is reached as shown in Fig. 3-27.

When $377t = \pi/2$, the generator voltage is 10 volts and the capacitor voltage is 7 volts (polarity shown in Fig. 3-26). The capacitor voltage remains constant because the current is zero unless there is an external load.

As $377t$ increases above $\pi/2$, v_O is the generator voltage shifted downward by 7 volts as shown in Fig. 3-27.

3.5. In the circuit of Fig. 3-28, assume the diode is ideal. Sketch the steady-state output voltage v_O for at least one cycle of the generator voltage.

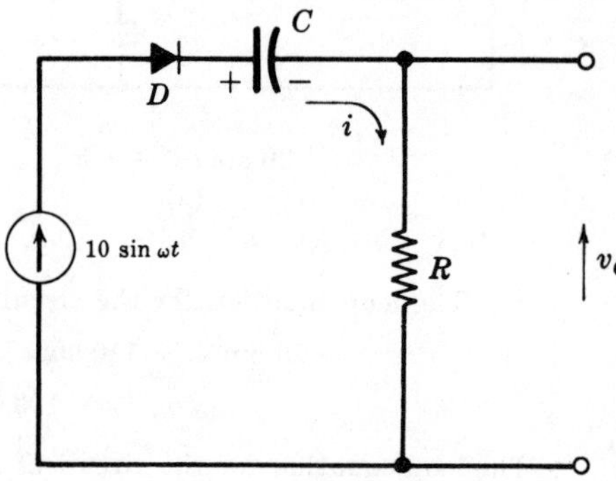

Fig. 3-28

Assume there is a series switch which is closed at $t = 0$ and that the capacitor is initially uncharged. The diode is forward biased, so that the capacitor charges until the generator voltage drops below the capacitor voltage. When this occurs the diode is reverse biased and the current is zero. Since $v_O = Ri$, v_O is zero during this time. If there were no series resistance R, the capacitor would have a terminal voltage of 10 volts when $\omega t = \pi/2$ radians.

The diode is reverse biased and the current is zero until the generator voltage is equal to the capacitor voltage. The diode is then forward biased until the generator voltage drops below the capacitor voltage.

Since the capacitor will lose no charge in this circuit, it charges until the terminal voltage reaches the steady-state value of 10 volts. The steady-state current is zero as is the steady-state output voltage v_O.

3.6. For the circuit of Fig. 3-29 below, assume an ideal diode and that the time constant of the R-C combination is very large compared to one period of the generator. Determine the output voltage v_O.

Assume a switch between the capacitor and generator. The diode is reverse biased when this switch is open. Assume the switch is closed when $\omega t = 0$. Since the time constant of the R-C circuit is large compared to one period, there will be negligible charge flowing in the loop including the resistor. Thus the capacitor voltage remains at zero until the voltage across the resistor is 10 volts ($20 \sin \omega t + 5 = 10$). The diode is then forward biased and the capacitor charges through the diode branch. When $\omega t = \pi/2$ radians, the capacitor voltage is 15 volts (polarity shown in Fig. 3-29). The capacitor can discharge only through the loop including the resistor, but the time constant of this circuit is large. Hence the capacitor voltage remains constant at 15 volts with the polarity shown, and the diode remains reverse biased.

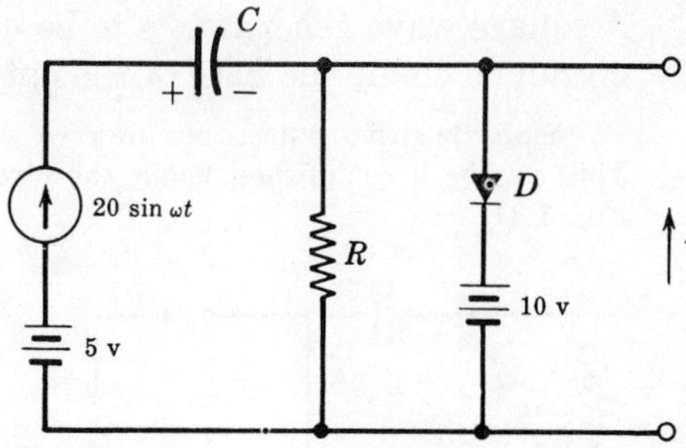

Fig. 3-29

To find v_O, consider the branch which includes the generator:

$$v_O = -15 + 20 \sin \omega t + 5$$
$$= -10 + 20 \sin \omega t \text{ volts}$$

3.7. For the circuit of Fig. 3-30, assume ideal diodes and determine the output voltage v_O.

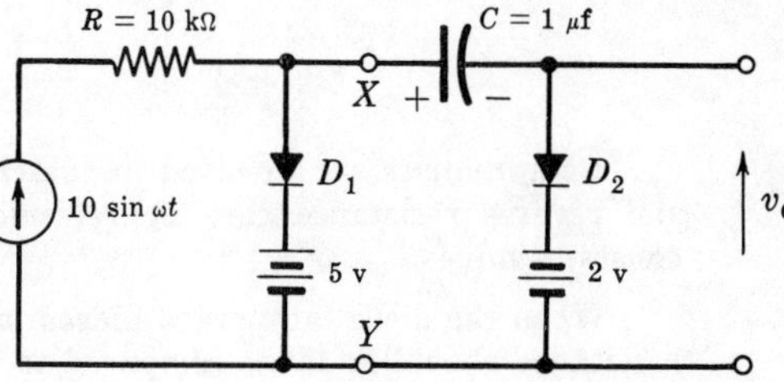

Fig. 3-30

The first consideration is the steady-state capacitor voltage. If a switch is added in the capacitor branch at point X and if this switch is open, the circuit to the left of terminals X-Y is a clipping circuit. The voltage v_{XY} can never exceed 5 volts when D_1 and the 5 volt battery are ideal.

Assume the switch is closed at $t = 0$. Both diodes are reverse biased until the generator voltage is 2 volts; hence all currents are zero until then. When the generator voltage is 2 volts, D_2 is forward biased and the capacitor begins to charge. If v_{XY} reaches 5 volts, then D_1 is forward biased. The voltage v_{XY} can never exceed 5 volts when D_1 and the 5 volt battery are ideal.

A complete solution for the transient period may become very tedious. The circuit is linear between the points of discontinuity of the diode characteristics. For example, in the first period when $2 \leq v_{XY} < 5$ volts, the equivalent circuit is that of Fig. 3-31. Switch 2 is closed when $10 \sin \omega t = 2$.

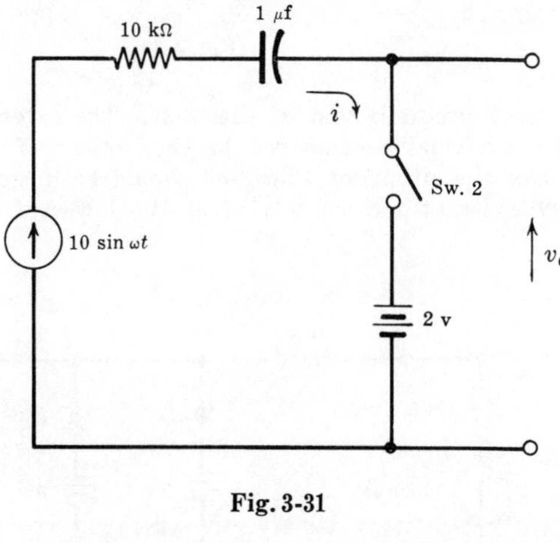

Fig. 3-31

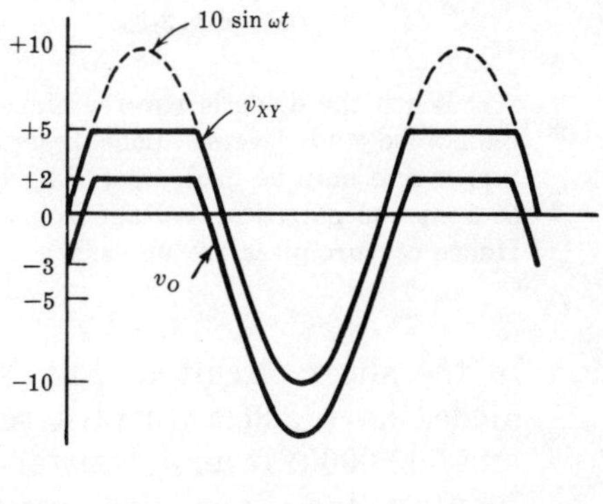

Fig. 3-32

At the next discontinuity, a new equivalent circuit is drawn and the initial charge of the capacitor included in the analysis. This process is repeated until the capacitor voltage reaches the steady-state value.

It is not necessary to consider the entire transient period to obtain the steady-state capacitor voltage. When the current in the capacitor is zero (capacitor fully charged) and v_{XY} is 5 volts, the capacitor voltage must be 3 volts with the polarity shown in Fig. 3-30.

The steady-state output voltage v_O is a clipped sine wave which is shifted downward by 3 volts because of the clamping circuit to the right of X-Y. Considering steady-state conditions, the circuit is a combination of a clipping and clamping circuit. v_O is shown in Fig. 3-32.

3.8. A square wave generator is to be used to generate positive pulses. Design a clamping circuit to clamp the base of the output voltage at zero.

Since positive pulses are desired, the base of the square wave should be clamped at zero. This can be accomplished using the circuit of Fig. 3-33. The generator voltage v_{gen} is shown in Fig. 3-34.

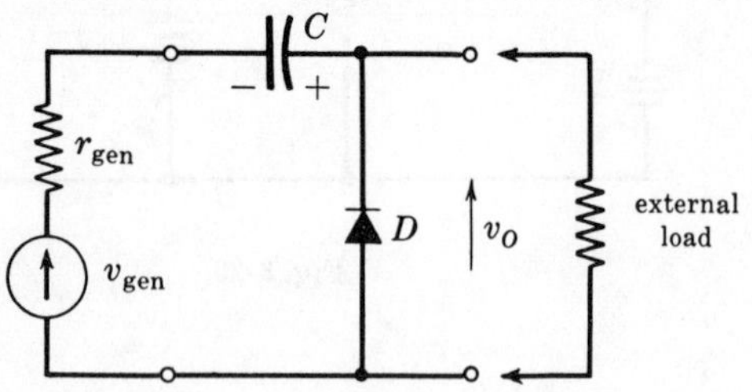

Fig. 3-33

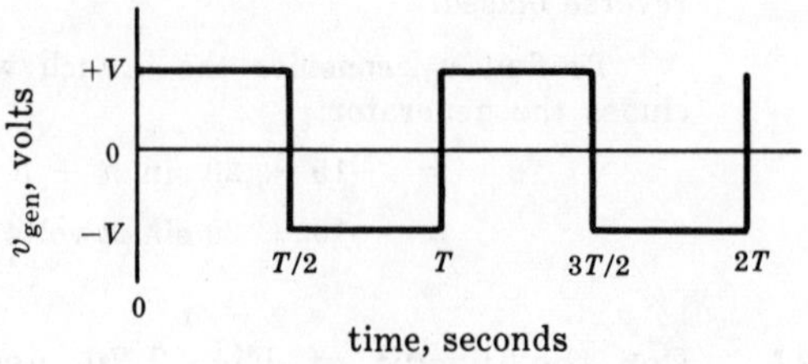

Fig. 3-34

Compromises are involved in the choice of the value of the capacitor C. If the diode forward and reverse resistances are appreciable, they must be considered and may be assumed to have constant values.

When the diode is forward biased, the equivalent circuit is that of Fig. 3-35. The external load resistance should be large compared to the forward diode resistance. The time constant should be small enough so that at the highest frequency, the capacitor voltage reaches V at the instant the generator voltage is $-V$.

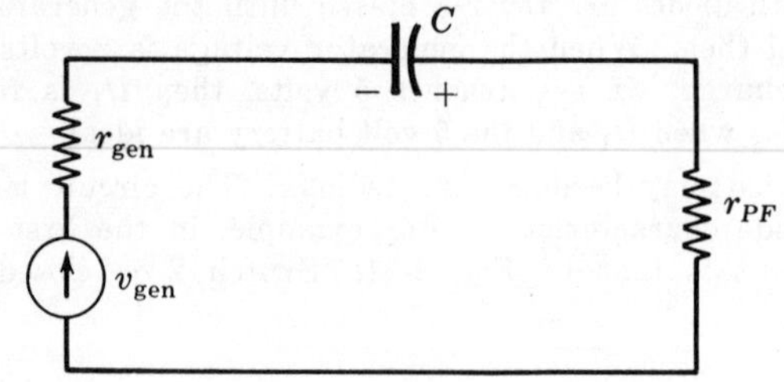

Fig. 3-35

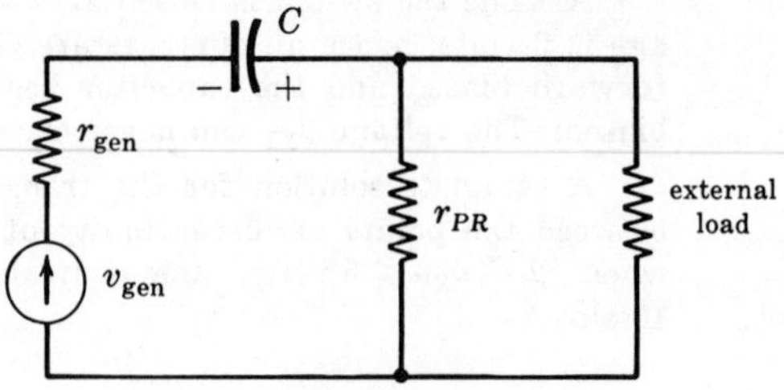

Fig. 3-36

When the diode is reverse biased, the equivalent circuit is that of Fig. 3-36. The external load resistance and reverse diode resistance may be appreciable compared to each other; if not, the larger one may be neglected. The time constant of the circuit of Fig. 3-36 should be large enough so that the capacitor voltage remains relatively constant for one period at the lowest frequency. Hence compromises are necessary.

3.9. In the slicer circuit of Fig. 3-37, the diodes have 200 Ω forward resistance and 500,000 Ω reverse resistance. The input voltage is a sine wave as indicated and the generator internal resistance is 600 Ω. Find the output voltage v_o and the maximum value of the current in the 10,000 Ω resistor.

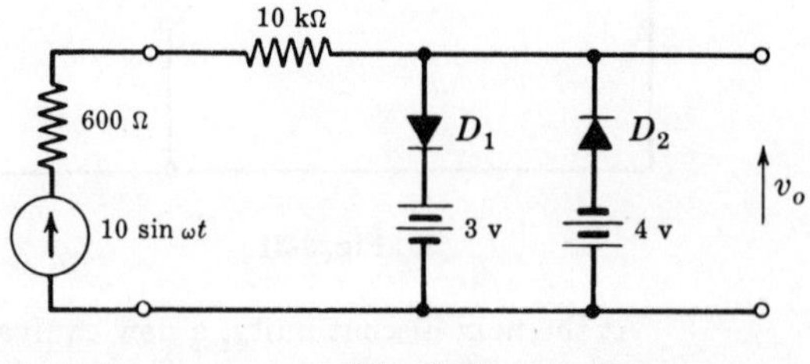

Fig. 3-37

By inspection of Fig. 3-37, the diode discontinuities occur when 10 sin ωt is approximately 3 and −4 volts. The circuit will be considered for each of the three regions.

(*a*) *Both diodes are reverse biased.* This region corresponds approximately to the time when $-4 \leq 10 \sin \omega t \leq 3$. Since the reverse resistance is not infinite, there is a non-zero current when the diodes are reverse biased. The equivalent circuit for this region of operation is shown in Fig. 3-38 below.

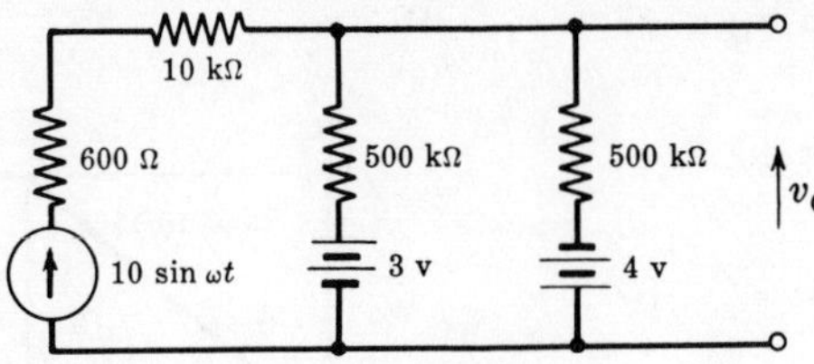

Fig. 3-38

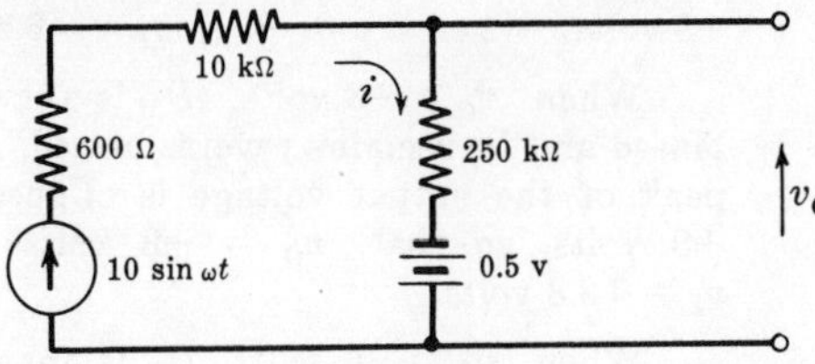

Fig. 3-39

Thévenin's theorem may be applied to the two diode branches, giving the circuit of Fig. 3-39. The loop equation may be written to find $i = (0.5 + 10 \sin \omega t)/260{,}000$ amps. Then

$$v_O = 250{,}000i - 0.5 = -.02 + 9.59 \sin \omega t \text{ volts}$$

(b) *Diode D_1 is forward biased.* This region is approximately the time when $10 \sin \omega t > 3$. The equivalent circuit is shown in Fig. 3-40. Thévenin's theorem may again be applied to the diode branches; this shows that the branch containing diode D_2 is negligible when slide rule accuracy is involved. The loop current $i = (-3 + 10 \sin \omega t)/10{,}800$ amps, and

$$v_O = 200i + 3 = 2.94 + 0.185 \sin \omega t \text{ volts}$$

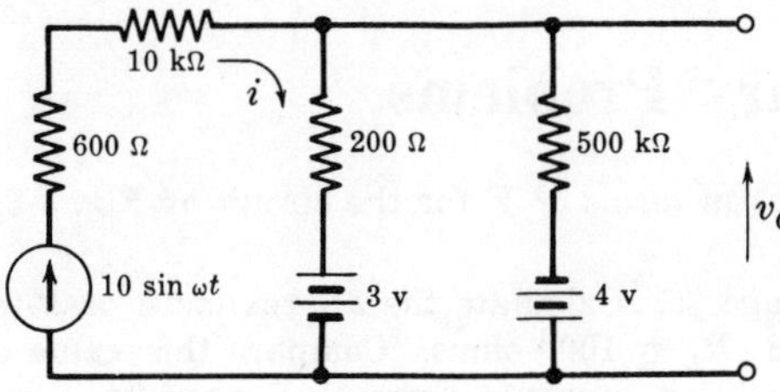

Fig. 3-40

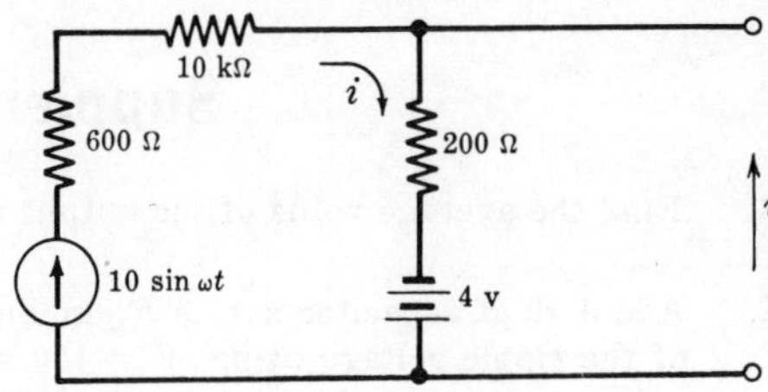

Fig. 3-41

(c) *Diode D_2 is forward biased.* This region is approximately the time when $10 \sin \omega t < -4$ v. Since the diode branch containing D_1 may be neglected, the equivalent circuit is that of Fig. 3-41. The loop current $i = (4 + 10 \sin \omega t)/10{,}800$ amps, and

$$v_O = 200i - 4 = -3.93 + 0.0185 \sin \omega t \text{ volts}$$

The output voltage v_O has been specified for the three regions of operation. If it is necessary to designate the points of discontinuity more exactly, the circuit of part (*a*) may be solved to find the exact times when the diode branch voltages are +3 and −4.

The maximum value of the current in the 10,000 ohm resistor is found by considering the maximum current for each of the three regions of operation. By inspection, this will be in region *b* when $\sin \omega t = +1$. Then $i_{(\max)} = (-3 + 10)/10{,}800 = 0.648$ ma.

3.10. In the slicer circuit of Fig. 3-17, page 39, let $V_1 = +8$ volts, $V_2 = +6$ volts (polarities as shown), and $R = 10$ kΩ. A terminating load resistance of 100 kΩ is connected to the terminals *A* and *B*. Sketch the output voltage v_O as a function of the input voltage v_I, giving the overall transfer characteristic of the circuit when terminated by a 100 kΩ resistor.

By inspection, both diodes are reverse biased when $-6 < v_O < +8$. Assuming ideal diodes, the currents are zero in the diode branches. Denoting the loop current by i, the loop equation for the circuit is

$$-v_I + 10{,}000i + 100{,}000i = 0 \quad \text{or} \quad i = v_I/110{,}000$$

The output voltage for this region of operation is

$$v_O = 100{,}000i = (10/11)v_I$$

Hence the overall transfer characteristic curve has slope 10/11 for the region between $-6 < v_O < +8$ volts. When $v_O = -6$ v, $v_I = -6.6$ v; when $v_O = +8$ v, $v_I = +8.8$ v.

When $v_O = +8$ volts, D_1 is forward biased and D_2 remains reverse biased. The peak of the output voltage is clipped at +8 volts, so that $v_O = +8$ volts for $v_I \geqq +8.8$ volts.

When $v_O = -6$ volts, D_2 is forward biased and D_1 is reverse biased. The base of the output voltage is clipped at −6 volts, so that $v_O = -6$ volts for $v_I \leqq -6.6$ volts.

The overall transfer characteristic has discontinuities at $v_O = -6$ v and +8 v (or $v_I = -6.6$ v and +8.8 v). The overall transfer characteristic curve is sketched in Fig. 3-42.

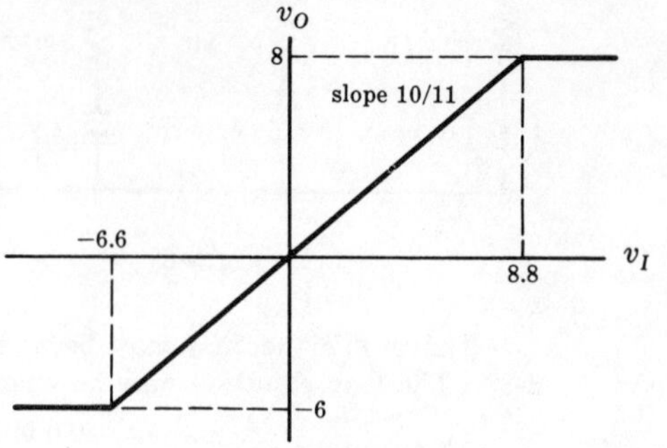

Fig. 3-42

Supplementary Problems

3.11. Find the average value of the output voltage v_O in terms of V for the circuit of Fig. 3-8, page 35.

3.12. Add a 40 μf capacitor across R_L in Fig. 3-8, page 35. Calculate the approximate peak-to-peak value of the ripple voltage using $V = 100$ volts and $R_L = 1000$ ohms. Compare this value of the ripple voltage with that obtained for the half-wave rectifier circuit in Example 2, page 34.

3.13. Given the circuit of Fig. 3-43. Sketch the instantaneous output voltage v_O for one complete cycle.

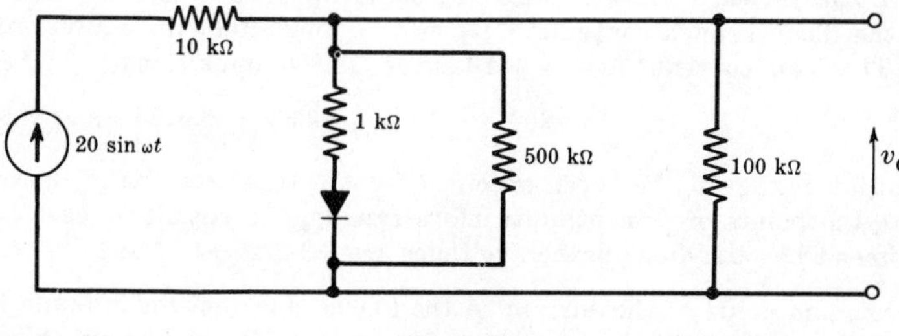

Fig. 3-43

3.14. Refer to the circuit of Fig. 3-12, page 36. Indicate the diode connections which will make A negative with respect to B and show the polarity of the capacitor voltages.

3.15. Using the voltage doubler circuit of Fig. 3-12, page 36, as a basis, show how additional diodes and capacitors can be added to connect this circuit into a voltage tripler.

3.16. Given the circuit of Fig. 3-14, page 37, with the sine wave generator replaced by a generator having a repetitive output with one period defined by

$$\begin{aligned} v(t) &= 1000t & 0 \leqq t < .01 \\ &= 10 - 1000(t - .01) & .01 < t < .03 \\ &= -10 + 1000(t - .03) & .03 < t < .04 \end{aligned}$$

Let $R = 1000\ \Omega$ and $V_1 = +2$ volts. Assuming D is an ideal diode, graph one cycle of v_O indicating maximum and minimum values. Also sketch one cycle of the current in the diode.

3.17. In the circuit of Fig. 3-16, page 38, let $V = 100$ volts, $V_1 = -20$ volts, and assume D is an ideal diode. (*a*) Choose a value of R so that the current in the diode does not exceed a peak value of 20 ma. (*b*) Calculate the average power dissipated in the resistor R.

3.18. In the circuit of Fig. 3-14, page 37, let $V = 40$ volts, $R = 10$ kΩ, $V_1 = +20$ volts. The diode forward resistance $r_{PF} = 100\ \Omega$ and the reverse resistance $r_{PR} = 400$ kΩ. The output terminals are connected to a circuit which contains a resistance of 100 kΩ connected across the terminals A and B. Graph one cycle of the output voltage v_O indicating the minimum and maximum values and the points of discontinuity of v_O.

3.19. Given the circuit of Fig. 3-20, page 40, with $V_1 = -2$ volts and $V = 10$ volts. Find the steady-state capacitor voltage v_C and sketch the output voltage v_O for several cycles assuming the switch is closed at $\omega t = \pi/2$ radians.

3.20. For the circuit of Fig. 3-44, graph the steady-state output voltage v_O.

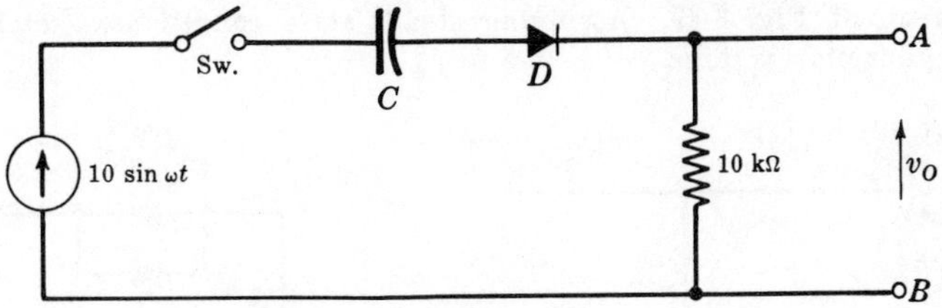

Fig. 3-44

3.21. Refer to the circuit of Fig. 3-18, page 39. Let $V = 20$ volts, $V_1 = 0$, D be an ideal diode, and $C = 0.01\ \mu$f. The output terminals A and B are connected to a circuit which has 100 kΩ resistance. If the frequency of the sine wave is 100 cps, find the maximum and minimum values of the capacitor voltage.

3.22. Given the circuits of Fig. 3-45(*a*) and (*b*). Graph the transfer characteristics of each circuit (v_O vs. v_I), assuming ideal diodes.

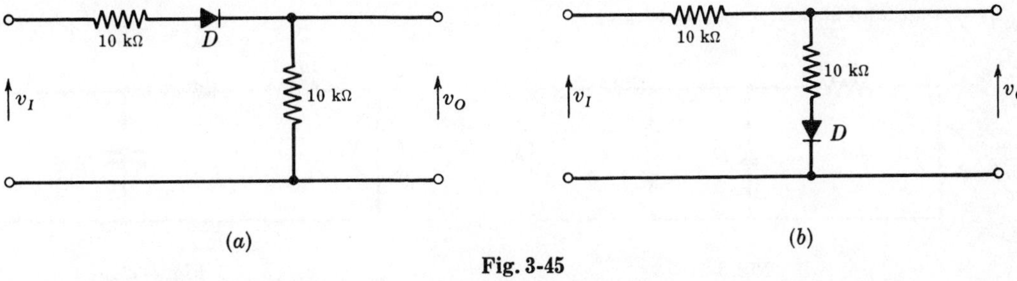

Fig. 3-45

3.23. Given the circuits of Fig. 3-46(*a*) and (*b*), sketch the steady-state output voltage v_O for at least one cycle of v_I. In both circuits $v_I = 10 \sin 200\,\pi t$ volts. Find the steady-state voltage across each capacitor if the diodes are ideal.

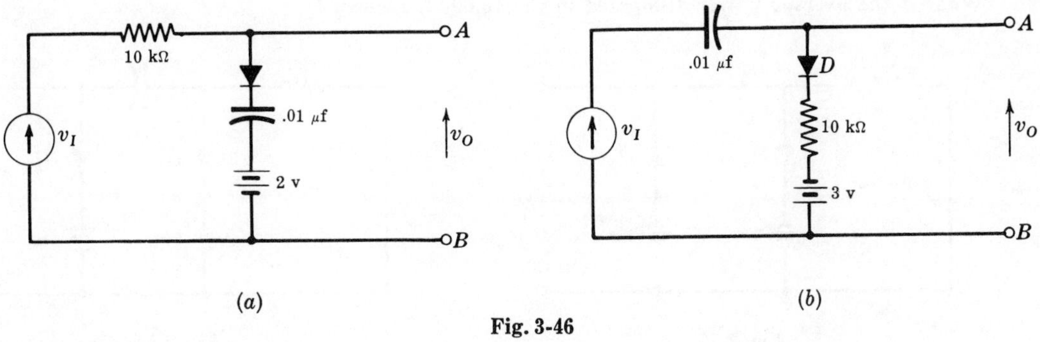

Fig. 3-46

3.24. Given the circuit of Fig. 3-47, graph the steady-state output voltage v_O for at least one cycle of the input voltage.

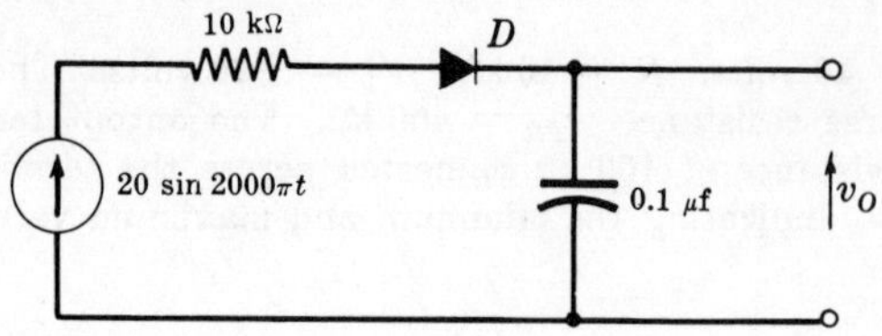

Fig. 3-47

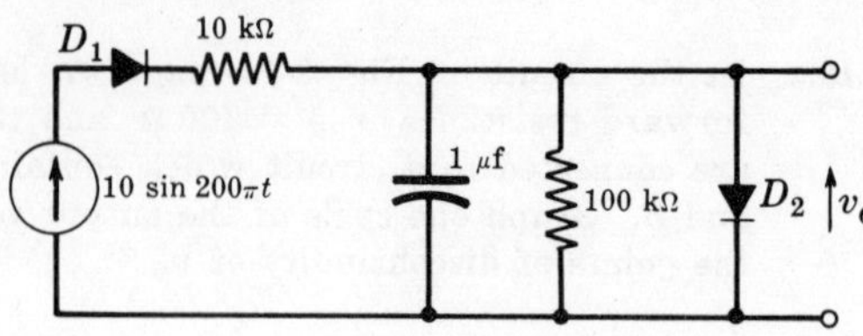

Fig. 3-48

3.25. For the circuit of Fig. 3-48, sketch the output voltage for one cycle of the input voltage assuming steady-state conditions and ideal diodes.

3.26. Given the circuit of Fig. 3-49. Assuming steady-state conditions, sketch the output voltage and calculate the capacitor voltage.

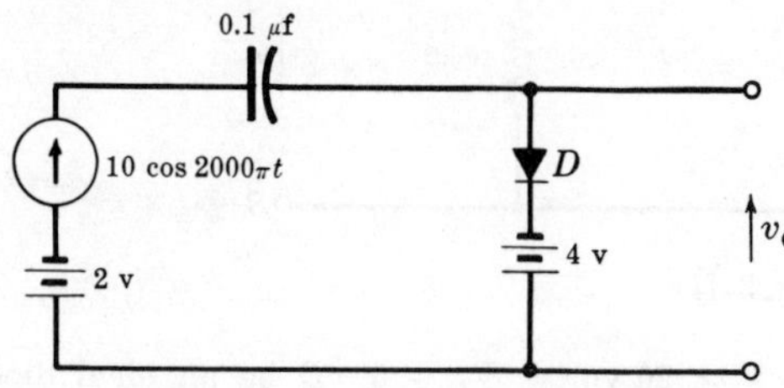

Fig. 3-49

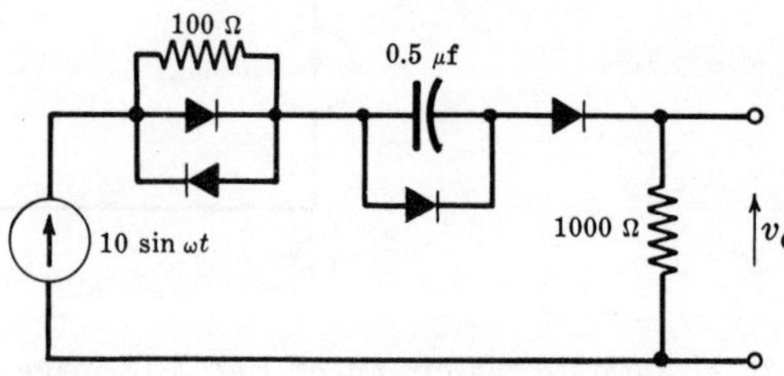

Fig. 3-50

3.27. For the circuit of Fig. 3-50, assume ideal diodes and graph the steady-state output voltage for at least one cycle. Can any circuit elements be eliminated without affecting the performance?

3.28. Given the circuit of Fig. 3-51. Assuming an ideal diode, calculate the average value of the output voltage and sketch one cycle of v_O.

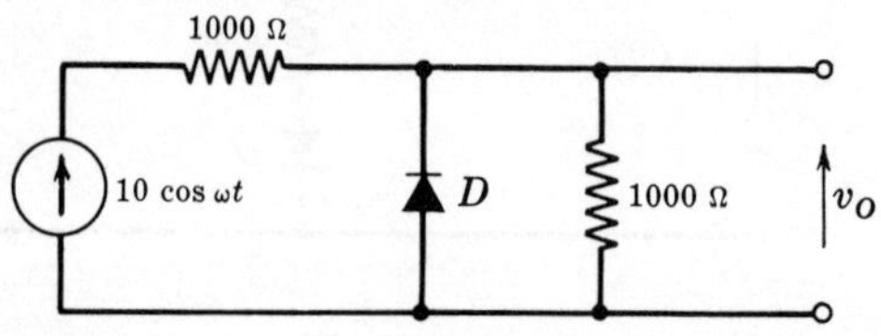

Fig. 3-51

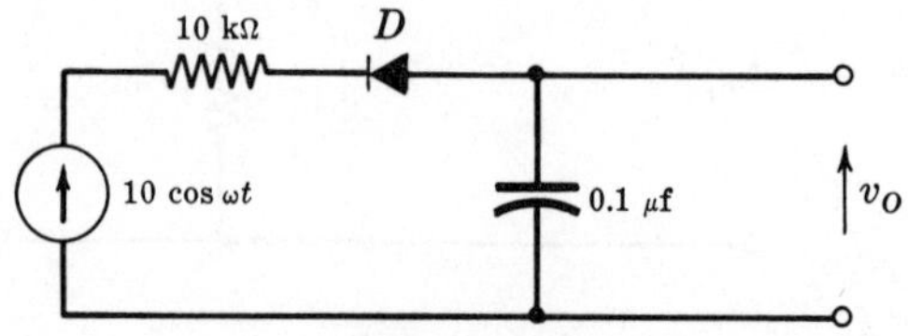

Fig. 3-52

3.29. Given the circuit of Fig. 3-52. Assuming an ideal diode, sketch the steady-state output voltage v_O.

3.30. Given the circuit of Fig. 3-53. Assuming ideal diodes, graph the steady-state output voltage v_O. What is the average power dissipated in the 10,000 Ω resistor?

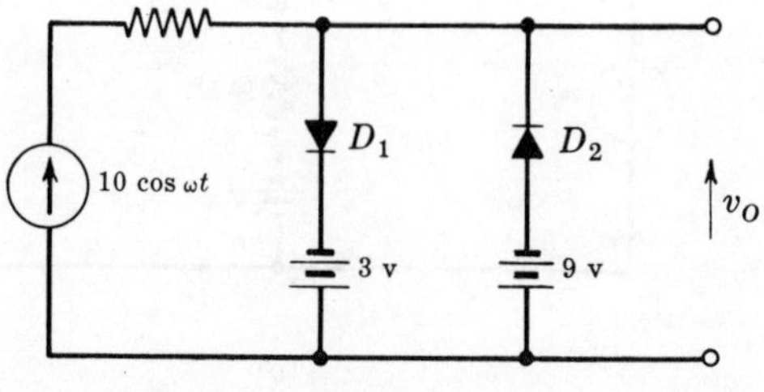

Fig. 3-53

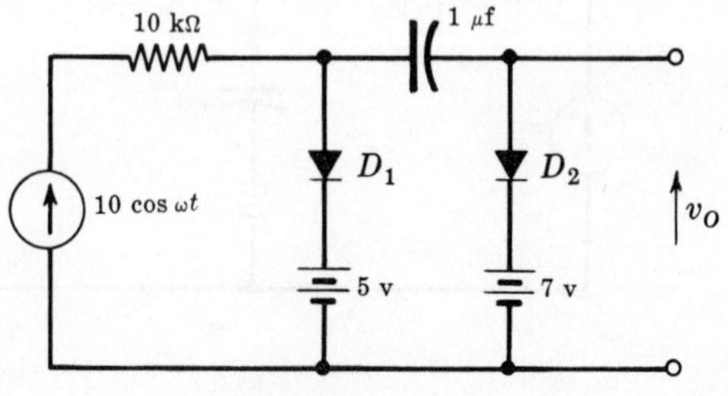

Fig. 3-54

3.31. For the circuit of Fig. 3-54 above, assume ideal diodes and sketch the steady-state output voltage v_O.

3.32. Given the circuit of Fig. 3-55. Assuming an ideal diode and that the *R-C* time constants are long compared to one period of the generator, graph one cycle of v_O.

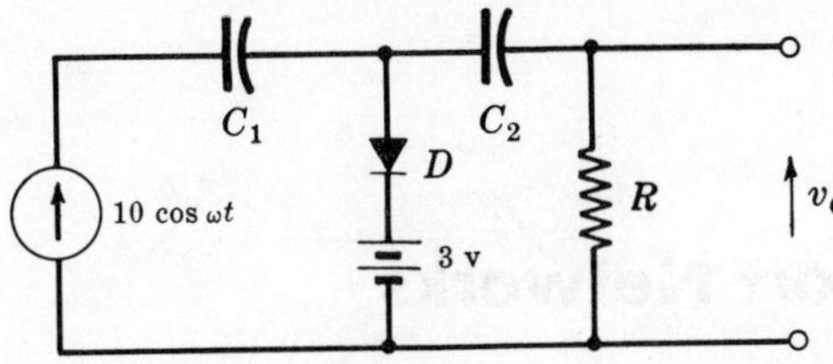

Fig. 3-55

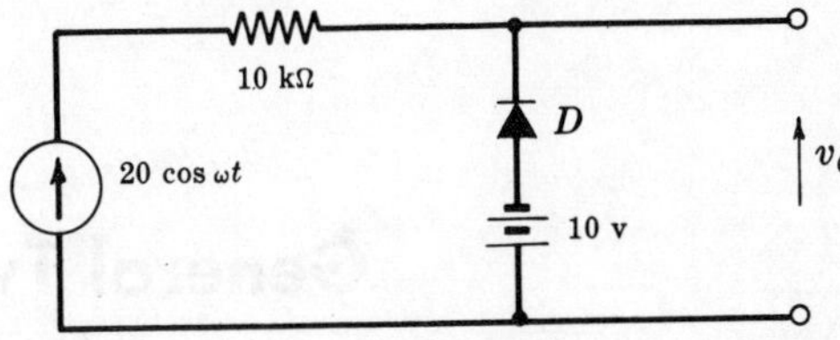

Fig. 3-56

3.33. For the circuit of Fig. 3-56, assume an ideal diode and sketch one cycle of v_O.

3.34. For the circuit of Fig. 3-57, assume ideal diodes and graph the transfer function v_O vs. v_I.

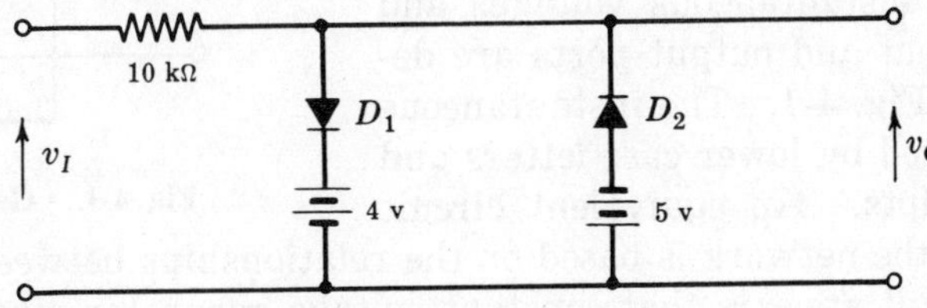

Fig. 3-57

3.35. Fig. 3-58 shows a portion of a circuit diagram. The output voltage v_O is observed on an oscilloscope (dc amplifiers) and is sketched in Fig. 3-59. Complete the circuit diagram by adding the elements needed to give the output v_O as shown in Fig. 3-59. (When the vertical amplifier of an oscilloscope is direct coupled, the beam deflection is proportional to the instantaneous input voltage. When the oscilloscope vertical amplifier includes capacitor coupling, the beam deflection is proportional to the instantaneous value of the varying component only. In other words, the average value of the applied voltage is not displayed.)

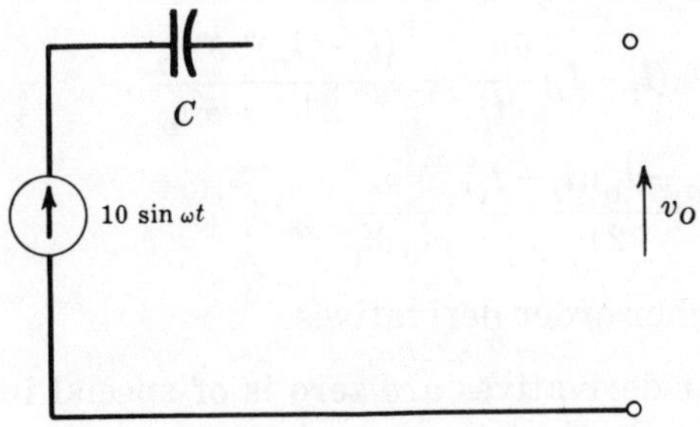

Fig. 3-58

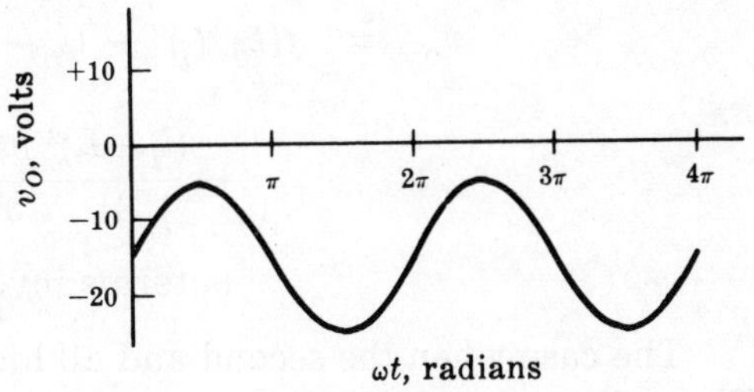

Fig. 3-59

Chapter 4

General Two Port Networks

4.1 TWO PORT r PARAMETERS

Consider any physically realizable two port network consisting of time-invariant, passive or active circuit elements. Following the established conventions and the notation introduced in Chapter 2, the instantaneous voltages and currents of the input and output ports are denoted as shown in Fig. 4-1. The instantaneous values are designated by lower case letters and upper case subscripts. An equivalent circuit representation for the network is based on the relationships between these voltages and currents. Historically it appears that constant voltage generator circuits were employed most frequently, so this representation will be developed first. The equivalent circuit is developed by considering each port separately.

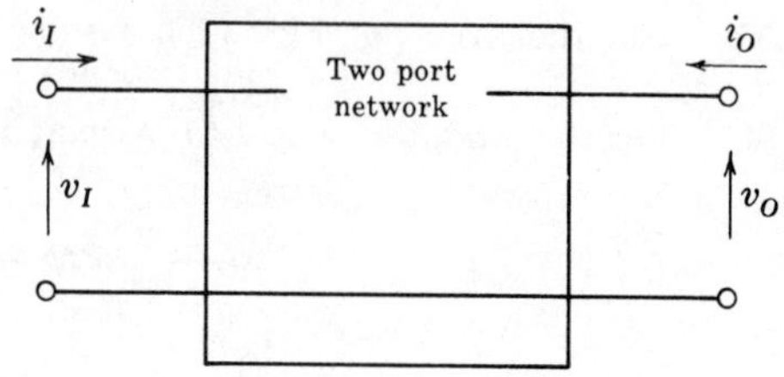

Fig. 4-1. General Two Port Network

Consider the output port first and write the functional relationship as

$$v_O = f(i_O, i_I)$$

Since i_I and v_I are not independent of each other, only one must be included in the functional relationship. Here i_I is used because of the particular representation that is desired. The function $v_O = f(i_O, i_I)$ is geometrically a surface and the ordered pair (I_O, I_I) is a point on the surface. Expanding $v_O = f(i_O, i_I)$ about the point (I_O, I_I) by Taylor's series,

$$v_O = f(I_O, I_I) + (i_O - I_O)\frac{\partial v_O}{\partial i_O} + (i_I - I_I)\frac{\partial v_O}{\partial i_I} + \frac{(i_O - I_O)^2}{2!}\frac{\partial^2 v_O}{\partial^2 i_O}$$

$$+ \frac{(i_I - I_I)^2}{2!}\frac{\partial^2 v_O}{\partial^2 i_I} + \frac{2(i_O - I_O)(i_I - I_I)}{2!}\frac{\partial^2 v_O}{\partial i_O\,\partial i_I}$$

$$+ \text{ terms involving higher order derivatives}$$

The case when the second and all higher order derivatives are zero is of special interest. Then all terms of the above equation are zero but the first three, and thus

$$v_O = f(I_O, I_I) + (i_O - I_O)\frac{\partial v_O}{\partial i_O} + (i_I - I_I)\frac{\partial v_O}{\partial i_I} \qquad (4.1)$$

The partial derivatives $\partial v_O/\partial i_O$ and $\partial v_O/\partial i_I$ are constant when the higher order derivatives are zero, which implies that a region about the point (I_O, I_I) is a plane. In practical situations this usually is not exactly true but is an adequate approximation. This is a generalization of the concept of "small signal" analysis which was discussed in Chapter 2, where "small signal", "linear approximation", or whatever name is used, meant restricting operation to the straight line portion of the curve. In the present situation involving functions of two variables, this means restricting operation to the plane portion of the surface.

By definition $\partial v_O/\partial i_O$ is the differential output resistance. The partial derivative $\partial v_O/\partial v_I$ involves a concept called the forward transfer resistance which has the dimensions of resistance and which expresses the output voltage as a function of the input current. There is also a reverse transfer function, so the two must be distinguished. To avoid writing the partial derivatives repeatedly, the following notation is used in this book:

$$\frac{\partial v_O}{\partial i_O} = r_o \qquad \text{and} \qquad \frac{\partial v_O}{\partial i_I} = r_f \tag{4.2}$$

The lower case r with a lower case subscript indicates a differential resistance; the o denotes output and the f means forward transfer. In some notation r_o is written as r_{22} and r_f as r_{21}. Letters are used as subscripts in this book because in some cases it is necessary to specify the reference terminals, so the number notation would then contain three subscripts.

Using this notation, equation (*4.1*) is written

$$v_O = f(I_O, I_I) + (i_O - I_O)r_o + (i_I - I_I)r_f$$

The first term on the right of the equal sign is the average value V_O (when linear operation is assumed),

$$v_O = V_O + (i_O - I_O)r_o + (i_I - I_I)r_f \tag{4.3}$$

When operation is restricted to the linear operating range, superposition may be applied to separate the instantaneous values into the average values and the instantaneous values of varying components. Writing $v_O = V_O + v_o$, $i_O = I_O + i_o$ and $i_I = I_I + i_i$, and substituting into (*4.3*),

$$v_O = V_O + v_o = V_O + (I_O + i_o - I_O)r_o + (I_I + i_i - I_I)r_f \tag{4.4}$$

The terms of (*4.4*) which involve instantaneous values of varying components are

$$v_o = i_o r_o + i_i r_f \qquad \text{or} \qquad -v_o + r_o i_o + r_f i_i = 0 \tag{4.5}$$

In a circuit representation which is equivalent to (*4.5*), v_o is the voltage rise across the output terminals going in the direction of the current i_o. The term $r_o i_o$ is represented as a resistance r_o in series with the non-reference output terminal (terminal into which i_o is directed). The term $r_f i_i$ is a voltage drop which is independent of either the voltage drop v_o or the current i_o in the output port; thus $r_f i_i$ is shown as a constant voltage generator. In calling this a constant voltage generator, it should be clear that this means the generator voltage is independent of i_o; it does not mean that the generator voltage is independent of time. In fact, since only varying components are considered, the generator voltage is zero if it is time independent. This is consistent with the voltage and current generator models introduced in elementary circuit courses. Hence the circuit of Fig. 4-2 is equivalent to equation (*4.5*) when only the terminal characteristics of the output port are considered. It is understood that this applies only for varying components and linear operation.

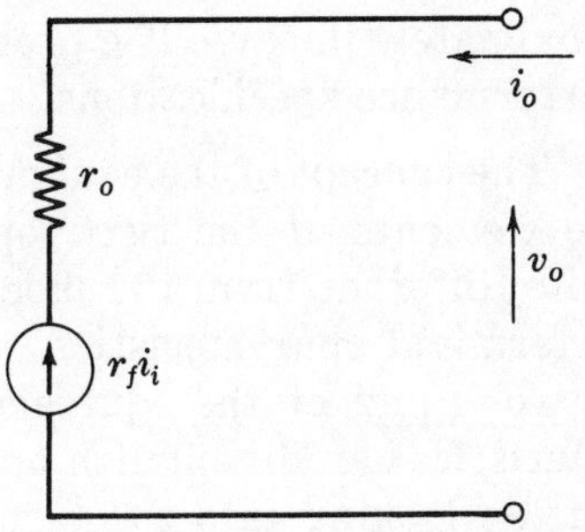

Fig. 4-2. Output Port Representation

Consider the input port, keeping in mind that the model being developed contains constant voltage generators and resistive parameters. Write the functional relationship as

$$v_I = f(i_O, i_I)$$

Following the procedure outlined for the output port and expanding v_I about the point (I_O, I_I) and writing only the first three terms of the Taylor's series expansion,

$$v_I = f(I_O, I_I) + (i_O - I_O)\frac{\partial v_I}{\partial i_O} + (i_I - I_I)\frac{\partial v_I}{\partial i_I} \quad (4.6)$$

Using the adopted notation scheme,

$$\partial v_I/\partial i_I = r_i \quad \text{and} \quad \partial v_I/\partial i_O = r_r \quad (4.7)$$

The resistance r_i is the differential input resistance and r_r is the differential reverse transfer resistance. In general, the subscript i denotes input and r designates reverse transfer function. If numbered subscripts are used, $r_{11} = r_i$ and $r_{12} = r_r$. Substituting (4.7) into (4.6),

$$v_I = V_I + (i_O - I_O)r_r + (i_I - I_I)r_i \quad (4.8)$$

Apply superposition and put $v_I = V_I + v_i$, $i_O = I_O + i_o$, and $i_I = I_I + i_i$ into (4.8). The terms which have varying components are

$$v_i = r_r i_o + r_i i_i \quad \text{or} \quad -v_i + r_i i_i + r_r i_o = 0 \quad (4.9)$$

Following the procedure used in developing the model for the output port, the circuit of Fig. 4-3 is drawn to represent the input port.

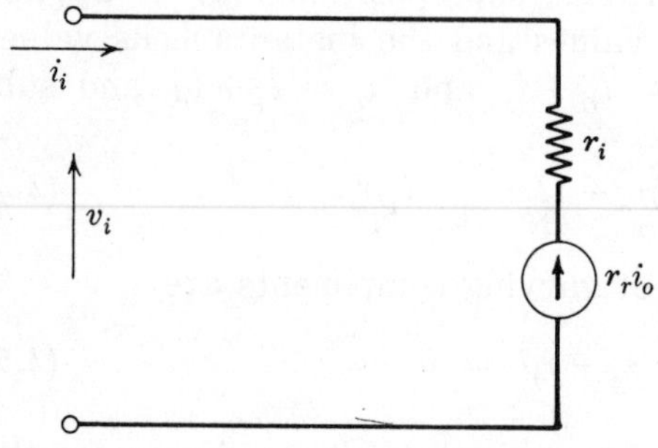

Fig. 4-3. Input Port Representation

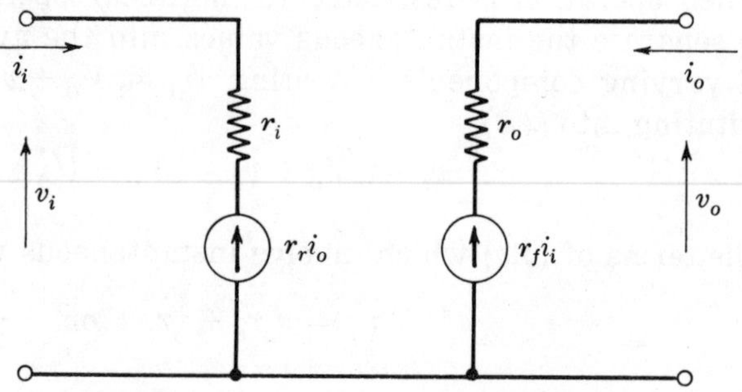

Fig. 4-4. Two Port Representation

The reference terminals of the input and output ports are common as shown in Fig. 4-1, hence the circuits of Fig. 4-2 and Fig. 4-3 may be combined as in Fig. 4-4. The circuit of Fig. 4-4 is the equivalent circuit for the two port network of Fig. 4-1. Of course, this applies only for varying components and for a range of operation which is linear or approximately linear. The problem of linearity is a practical one which must take into account performance specifications.

The concept of the equivalent circuit may seem confusing when first encountered because the elements of the two port network represented by the equivalent circuit are usually quite different from the models used in the equivalent circuit. The equivalence is limited to terminal characteristics. It may be said that given only the terminals of the physical network and of the equivalent circuit, no experiment involving electrical terminal characteristics can distinguish one from the other. In other words, the equivalence is from the outside looking into the terminals only, with the possibility of considerable difference on the inside.

The parameters of the equivalent circuit of Fig. 4-4 are sometimes called the r parameters because all four are resistances. This is a special case of the z parameters in which they are resistive. In this book one of the interests is in representing electronic control devices of the resistive class, so the parameters will be called the r parameters when employed to represent an electronic control device.

The r parameters are sometimes designated the open circuit parameters. Note that if the output port of the equivalent circuit of Fig. 4-4 is an open circuit, the output port terminal voltage is equal to the generator voltage; hence $v_o = r_f i_i$ when $i_o = 0$. Thus r_f can be found as the ratio v_o/i_i for $i_o = 0$. Considering the input port when $i_o = 0$, then $v_i = r_i i_i$; thus $r_i = v_i/i_i$ when $i_o = 0$. Similarly the input port may be made an open circuit, so $i_i = 0$; then $r_r = v_i/i_o$ and $r_o = v_o/i_o$. It should be understood that only the varying components are considered in the equivalent circuit.

Another concept in equivalent circuits which may be new to the reader is the *controlled generator*.

4.2 CONTROLLED GENERATORS

In most elementary electrical circuits courses, voltage and current generator models are either time independent (constant in the time sense) or specified as given functions of time. These are *independent generators.* A generator is independent if it is independent of the circuit variables (voltages and currents). In this sense a generator is *dependent* if its instantaneous magnitude is not independent of the circuit variables. These definitions need to be extended in order to classify all the generators encountered in equivalent electronic circuits.

In the equivalent circuit of Fig. 4-4, the instantaneous magnitude of each generator is specified in terms of a parameter (a differential transfer resistance) times the current in the opposite port. These are called *controlled generators* because the instantaneous magnitude is controlled by a circuit variable (a current in this case). The word controlled is used in this book even though some authors use controlled source and dependent source synonymously. It is necessary to distinguish between the controlled generators which are independent of the generator terminal voltage and current and those which are not. Generators will be classified as independent or dependent to distinguish these two types of controlled generators. This does not lead to any inconsistency with the definition of independent generators as given in elementary circuits books.

The controlled generators in the equivalent circuit of Fig. 4-4 are, in general, dependent generators. This may be seen by considering the generator in the output port of the circuit of Fig. 4-4. Unless the current i_i is independent of the magnitude of the generator in the input port (the magnitude is a function of i_o), i_i is not independent of the circuit variables i_o and v_o. If an ideal constant current generator is connected external to the input port so that i_i is independent of v_i, then the generator of the output port of Fig. 4-4 is a controlled generator which is independent.

The controlled generator of the output port of Fig. 4-4 is also independent if the parameter r_r is zero; then i_i is independent of v_o and i_o. This will be the case in most vacuum tube equivalent circuits.

In general, a controlled generator may be either a voltage or current generator. A voltage generator may be either voltage or current controlled. If a voltage generator is voltage controlled, the transfer function is dimensionless – a partial derivative of a voltage with respect to a voltage. If a voltage generator is current controlled, the transfer function is a differential transfer resistance as was the case in the circuit of Fig. 4-4.

A current generator also may be either voltage or current controlled. If it is voltage controlled, the transfer function is a differential transfer conductance. If it is current controlled, the transfer function is dimensionless – a partial derivative of a current with respect to a current.

In developing a two port equivalent circuit representation, the input port and the output port may each be designated either by a constant voltage or a constant current generator. Each generator may be either current or voltage controlled. In a practical situation, the most convenient representation is usually chosen. This may depend on the particular problem and the particular devices included. For example, a different representation may be more convenient for a transistor operating at low frequencies than for one operating in a much higher range of frequencies. Examples will follow in later sections where other equivalent two port representations will be developed.

Another problem which must be considered when controlled generators are involved is that, in general, Thévenin's theorem and Norton's theorem, as given in elementary courses, may not be applied. The statements of these theorems assume linear circuit elements and independent generators. For example, in Fig. 4-4 Thévenin's theorem may not be applied to the terminals of the output port unless the generators are treated as dependent generators.

This discussion does not contradict what was said in Section 4.1 about developing the equivalent circuit considering varying components on the linear operating range. It leads to the concept of output impedance which is not always explicitly defined. In this book, output impedance is considered in the context of a representation for the entire circuit looking into the output port. An equivalent one port representation is developed for the entire circuit looking into the output port; it includes the circuit external to the input port of the network.

4.3 OUTPUT IMPEDANCE

To introduce the concept of output impedance, consider the equivalent circuit of Fig. 4-4 with a generator connected to the input port as in Fig. 4-5. This representation involves only varying components; hence it is not the most general case. Consider a one port equivalent circuit model for the circuit looking into the output port at the terminals A and B. The loop equations

$$-v_{\text{gen}} + r_{\text{gen}} i_i + r_i i_i + r_r i_o = 0$$

$$-v_o + r_o i_o + r_f i_i = 0$$

contain two unknowns i_i and i_o. Since the present interest is in the output port, the equations are solved for

$$i_o = \frac{\begin{vmatrix} (r_{\text{gen}} + r_i) & v_{\text{gen}} \\ r_f & v_o \end{vmatrix}}{\begin{vmatrix} (r_{\text{gen}} + r_i) & r_r \\ r_f & r_o \end{vmatrix}} = \frac{(r_{\text{gen}} + r_i)v_o - r_f v_{\text{gen}}}{r_o(r_{\text{gen}} + r_i) - r_f r_r} \tag{4.10}$$

which exhibits the output current i_o as a function of v_o and v_{gen}; hence an equivalent constant current generator representation follows directly from this equation. Rewrite (*4.10*) as

$$-i_o + \frac{r_{\text{gen}} + r_i}{r_o(r_{\text{gen}} + r_i) - r_f r_r} v_o - \frac{r_f}{r_o(r_{\text{gen}} + r_i) - r_f r_r} v_{\text{gen}} = 0 \tag{4.11}$$

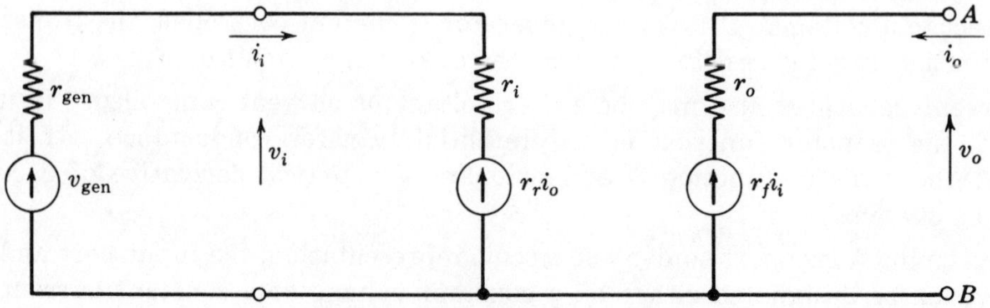

Fig. 4-5. Equivalent Circuit

The equivalent circuit for (*4.11*) is developed by representing each term as in Fig. 4-6. The first term i_o is the external current into node A. In the second term the coefficient of v_o is a linear conductance (r parameters assumed constant). The third term is independent of v_o and i_o, hence is shown as a voltage controlled, current generator (this is an independent generator). The coefficient of v_{gen} is a transfer conductance. The linear conductance g_{ab} is the output admittance and the reciprocal is the output impedance. In general, when constant voltage generator representations are used, loop equations are written, so the unknowns are currents and the driving functions are voltages. Solutions to the equations then include currents as functions of the driving functions which are voltages.

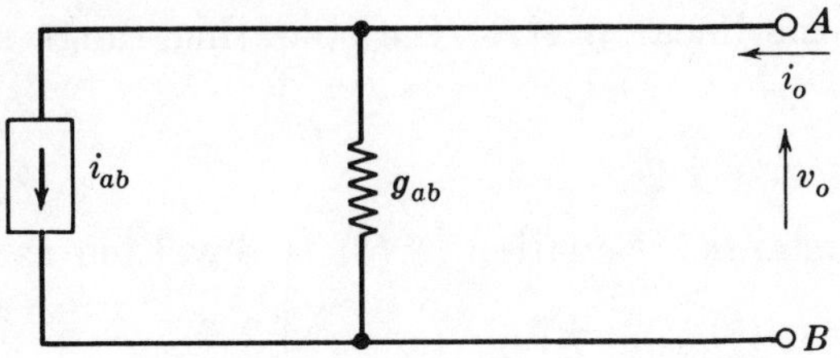

$$i_{ab} = \frac{-r_f}{r_o(r_{gen}+r_i) - r_f r_r} v_{gen}$$

$$g_{ab} = \frac{r_{gen}+r_i}{r_o(r_{gen}+r_i) - r_f r_r}$$

Fig. 4-6. One Port Representation

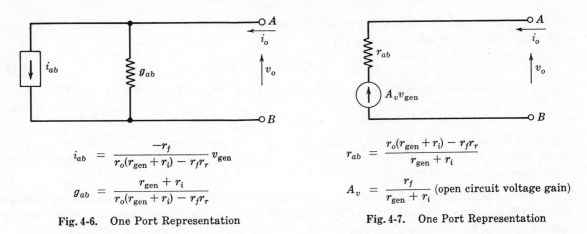

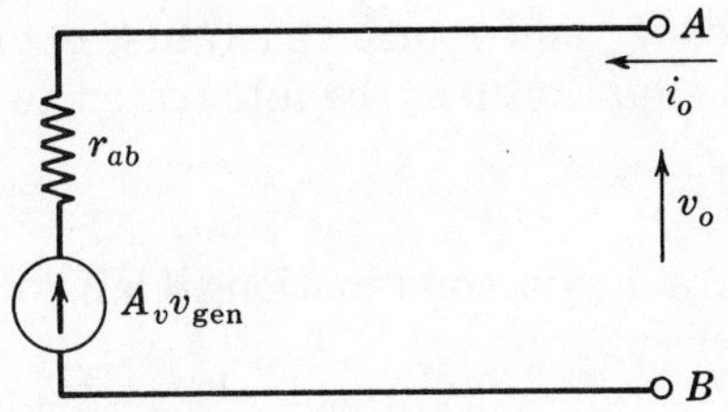

$$r_{ab} = \frac{r_o(r_{gen}+r_i) - r_f r_r}{r_{gen}+r_i}$$

$$A_v = \frac{r_f}{r_{gen}+r_i} \quad \text{(open circuit voltage gain)}$$

Fig. 4-7. One Port Representation

To develop an equivalent constant voltage generator circuit with a voltage controlled, voltage generator for the circuit of Fig. 4-5, rewrite (*4.11*) as

$$-v_o + \frac{r_o(r_{gen}+r_i) - r_f r_r}{(r_{gen}+r_i)} i_o + \frac{r_f}{r_{gen}+r_i} v_{gen} = 0 \qquad (4.12)$$

The equivalent circuit is developed by considering (*4.12*) term by term, giving the circuit of Fig. 4-7. The coefficient of i_o is the output impedance r_{ab} which is the reciprocal of g_{ab}. The coefficient of v_{gen} is the open loop voltage gain ($i_o = 0$).

Thévenin's theorem may be applied to the circuit of Fig. 4-6 to show that it is equivalent to that of Fig. 4-7. The generator in Fig. 4-6 may be considered an independent generator. The coefficient of v_{gen} is independent of the circuit variables v_o, i_o, v_i and i_i when linear operation is involved. The generator voltage v_{gen} is also independent of the circuit variables. The equivalence may also be confirmed by applying Norton's theorem to the circuit of Fig. 4-7 to show it is equivalent to that of Fig. 4-6.

The coefficient of v_{gen} in Fig. 4-7 is the open circuit voltage gain and is designated by A_v, where the A denotes amplification and the subscript v indicates voltage amplification. Care must be exercised to distinguish the parameters used in a representation for an electronic control device from those involving an amplifier circuit. For example, in Fig. 4-6 the coefficient of v_{gen} is a transfer conductance which differs from the transfer functions employed in the two port representations.

4.4 VOLTAGE CONTROLLED, VOLTAGE GENERATOR TWO PORT REPRESENTATION

A two port model for the network of Fig. 4-1 can be developed using voltage generators which are voltage controlled, instead of current controlled as in Fig. 4-4. In this case the functional relationship for the output port is

$$v_o = f(i_o, v_I)$$

When only the linear range of operation is considered, a simpler approach than Taylor's series may be used. This method is more convenient where there are more than two independent variables. The total differential of v_O is

$$dv_O = \frac{\partial v_O}{\partial i_O} di_O + \frac{\partial v_O}{\partial v_I} dv_I$$

By definition $\partial v_O/\partial i_O$ is the differential output resistance, denoted r_o. The forward transfer function $\partial v_O/\partial v_I$ is dimensionless and is a ratio of voltages, so it is symbolized μ_f. Thus

$$dv_O = r_o di_O + \mu_f dv_I$$

When r_o and μ_f are constants, operation is on the linear part of the operating range and this equation may be integrated to give

$$v_O = r_o i_O + \mu_f v_I + k \tag{4.13}$$

where k is a combination of all integration constants. Equation (*4.13*) is rewritten as

$$(1/k)v_O - (r_o/k)i_O - (\mu_f/k)v_I = +1$$

which is the equation of a plane in the standard intercept form. In the general case the plane portion of the surface may be limited to a small region around the average values, which are determined from $v_O = f(i_O, v_I)$.

Since operation is restricted to the linear operating range, superposition may be applied so that the instantaneous values may be written in terms of average values and instantaneous values of varying components. Thus

$$v_O = V_O + v_o = r_o(I_O + i_o) + \mu_f(V_I + v_i) + k$$

The terms involving average values only are

$$V_O = r_o I_O + \mu_f V_I + k$$

The terms containing varying components only are

$$v_o = r_o i_o + \mu_f v_i \quad \text{or} \quad -v_o + r_o i_o + \mu_f v_i = 0 \tag{4.14}$$

The functional relationship for the input port is

$$v_I = f(i_I, v_O)$$

and the total differential of v_I is

$$dv_I = \frac{\partial v_I}{\partial i_I} di_I + \frac{\partial v_I}{\partial v_O} dv_O$$

Let $\partial v_I/\partial i_I = r_i$ and $\partial v_I/\partial v_O = \mu_r$. Following the procedure used in arriving at equation (*4.14*) for the output port, the input port equation for varying components only is

$$-v_i + r_i i_i + \mu_r v_o = 0 \tag{4.15}$$

Following the procedure employed to develop the circuit of Fig. 4-4, equations (*4.14*) and (*4.15*) may be used to obtain the equivalent two port representation displayed in Fig. 4-8.

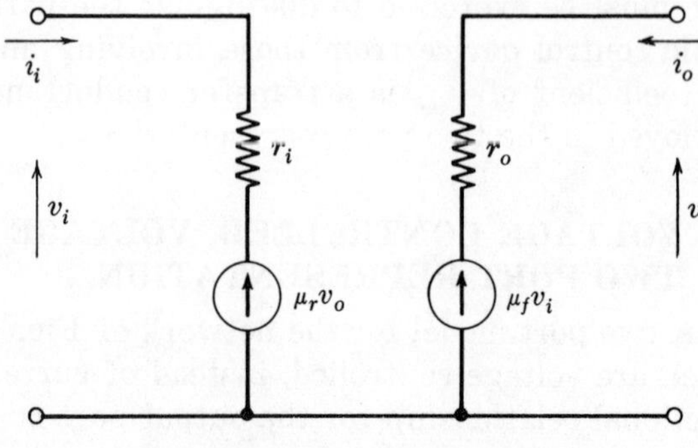

Fig. 4-8. Two Port Representation

4.5 TWO PORT g PARAMETERS

An equivalent circuit representation using voltage controlled, current generators for the general two port network of Fig. 4-1 will be considered next. In this case the functional relationship for the output port is

$$i_O = f(v_O, v_I)$$

The parameters are readily found by taking the total differential of i_O:

$$di_O = \frac{\partial i_O}{\partial v_O} dv_O + \frac{\partial i_O}{\partial v_I} dv_I$$

The parameters are all conductances and so are called the g parameters; they may be considered as a special case of the y parameters. Let $\partial i_O/\partial v_O = g_o$ and $\partial i_O/\partial v_I = g_f$; then

$$di_O = g_o\, dv_O + g_f\, dv_I$$

Again, consider the case where operation is restricted to the linear range so that g_o and g_f are constant; then

$$i_O = g_o v_O + g_f v_I + k$$

where k is a constant. Applying superposition and considering only the varying components,

$$-i_o + g_o v_o + g_f v_i = 0 \qquad (4.16)$$

Examine (*4.16*) term by term. The equivalent circuit for the output port is a constant current generator circuit having a conductance g_o in parallel with a current generator $g_f v_i$ as shown in Fig. 4-9. Since $g_f v_i$ has a plus sign, the current is away from the non-reference output terminal when v_i is positive; hence the arrow in the current generator is directed as shown in Fig. 4-9.

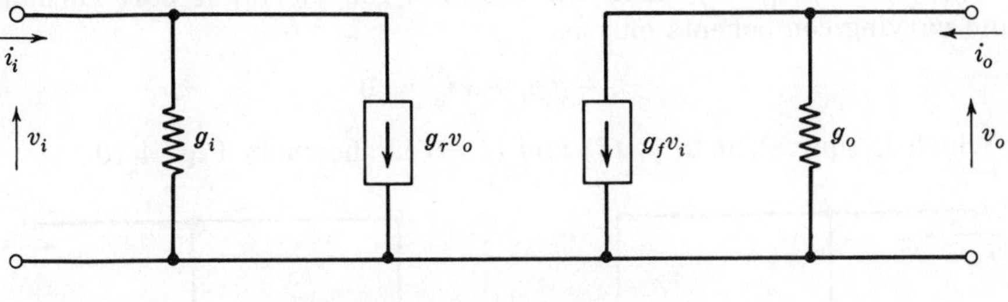

Fig. 4-9. Two Port Representation

The functional relationship for the input port is

$$i_I = f(v_I, v_O)$$

and the total differential of i_I is

$$di_I = \frac{\partial i_I}{\partial v_I} dv_I + \frac{\partial i_I}{\partial v_O} dv_O$$

Let $\partial i_I/\partial v_I = g_i$ and $\partial i_I/\partial v_O = g_r$; then

$$di_I = g_i\, dv_I + g_r\, dv_O$$

When g_i and g_r are constant,

$$i_I = g_i v_I + g_r v_O + k$$

Applying superposition and writing the equation for varying components only

$$-i_i + g_i v_i + g_r v_o = 0 \qquad (4.17)$$

When this equation is examined term by term, the equivalent circuit for the input port is shown in Fig. 4-9. The circuit of Fig. 4-9 is equivalent to the two port network of Fig. 4-1 so long as operation is restricted to the linear range and only varying components are considered.

The g parameters are sometimes called the short-circuit parameters. Note that if the output port of Fig. 4-9 is shorted, $i_o = g_f v_i$. Thus the forward transfer conductance may be found by taking the ratio of the short circuit current i_o to the input voltage v_i. Similarly, g_r may be obtained by taking the ratio of the short circuit input current i_i to v_o. The output conductance g_o can be found by considering the input shorted so that $v_i = 0$; then $g_o = i_o/v_o$. If the output port is shorted so that $v_o = 0$, then $g_i = i_i/v_i$.

4.6 CURRENT CONTROLLED, CURRENT GENERATOR TWO PORT REPRESENTATION

An equivalent circuit representation using current generators which are current controlled may be desired. In this case the functional relationship for the output port is

$$i_O = f(v_O, i_I) \quad \text{and hence} \quad di_O = \frac{\partial i_O}{\partial v_O} dv_O + \frac{\partial i_O}{\partial i_I} di_I$$

The differential output conductance is again denoted g_o. The forward transfer parameter $\partial i_O/\partial i_I$ is designated α_f. The equation for the output port considering linear operation and varying components only is

$$-i_o + g_o v_o + \alpha_f i_i = 0 \tag{4.18}$$

For the input port,

$$i_I = f(v_I, i_O) \quad \text{and} \quad di_I = \frac{\partial i_I}{\partial v_I} dv_I + \frac{\partial i_I}{\partial i_O} di_O$$

Let $\partial i_I/\partial v_I = g_i$ and $\partial i_I/\partial i_O = \alpha_r$; then the equation for the input port considering linear operation and varying components only is

$$-i_i + g_i v_i + \alpha_r i_o = 0 \tag{4.19}$$

The circuit which is equivalent to (*4.18*) and (*4.19*) is shown in Fig. 4-10.

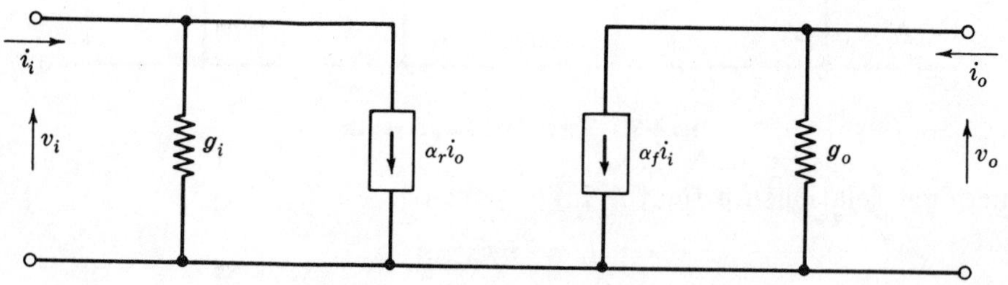

Fig. 4-10. Two Port Representation

4.7 TWO PORT h PARAMETERS

An equivalent circuit is usually chosen to represent a specific electronic control device or a specific two port network because it is the most convenient to use. A special hybrid equivalent circuit is used so frequently (especially for transistors) that it is called the hybrid circuit and the parameters are the h parameters. This hybrid circuit has a current controlled, current generator for the output port and a voltage controlled, voltage generator for the input port. The functional relationship for the output port is

$$i_O = f(v_O, i_I) \quad \text{and hence} \quad di_O = \frac{\partial i_O}{\partial v_O} dv_O + \frac{\partial i_O}{\partial i_I} di_I$$

Since the parameters are denoted as h parameters, let $\partial i_O/\partial v_O = h_o$ and $\partial i_O/\partial i_I = h_f$. The equation for the output port assuming linear operation and varying components only is

$$-i_o + h_o v_o + h_f i_i = 0 \qquad (4.20)$$

For the input port,

$$v_I = f(i_I, v_O) \qquad \text{and} \qquad dv_I = \frac{\partial v_I}{\partial i_I} di_I + \frac{\partial v_I}{\partial v_O} dv_O$$

Let $\partial v_I/\partial i_I = h_i$ and $\partial v_I/\partial v_O = h_r$; then the equation for the input port considering only varying components and assuming linear operation is

$$-v_i + h_i i_i + h_r v_o = 0 \qquad (4.21)$$

Equations (4.20) and (4.21) for the input and output ports may be represented by the equivalent circuit of Fig. 4-11. This may be shown by considering each equation term by term.

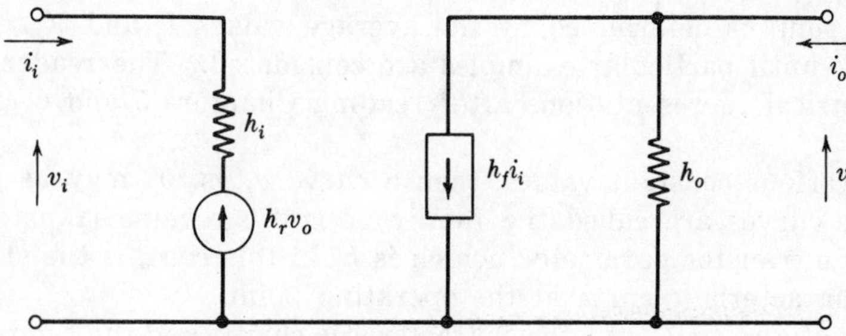

Fig. 4-11. Hybrid Equivalent Circuit

Other hybrid equivalent circuits are possible, but this particular circuit is employed to represent transistors. For example, another hybrid circuit might include a current controlled, voltage generator for the input port and a voltage controlled, current generator for the output port. Another example would be to use a current controlled, voltage generator for the output port. As new electronic control devices are developed, it may be that one of the examples mentioned or one of the other possible combinations of generators may be the most convenient.

4.8 GEOMETRICAL REPRESENTATIONS

In the preceding sections of this chapter the functional relationships for the input and the output ports were written in a particular form depending on the equivalent circuit desired. Each port of a two port network involves two independent variables; therefore a geometrical representation involves two surfaces. In general, the functional relationships are nonlinear; hence the surfaces representing the functional relationships are not planes. However, in most cases the pertinent functions are continuous and have continuous derivatives. If this is not so, special techniques must be employed.

In vacuum tube, transistor, and other electronic control device applications, the surface is approximated in two dimensions by displaying the curves formed by the intersection of the surface and a series of planes. For example, consider

$$v_I = f(i_I, v_O)$$

The surface may be visualized by using the coordinate system of Fig. 4-12. Consider the planes $v_O = k_l$ where k_l represents a sequence of equally spaced constant values. For each value of v_O there is a curve of v_I vs. i_I which is the intersection of the $v_O = k_l$ plane and the surface $v_I = f(i_I, v_O)$. This family of curves is called the static input characteristic curves. If a sufficient number and distribution of values of v_O are taken, the family of curves adequately represents the surface. The input parameter of the two port representation, which is h_i in this case, is the slope of the static input characteristic curve at the operating point. This follows from the definition of h_i.

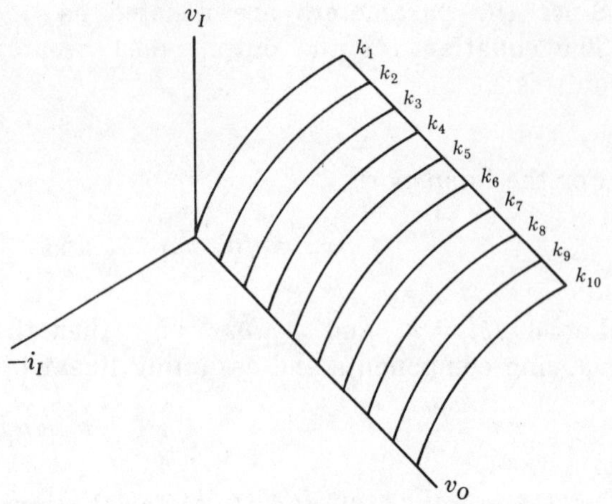

Fig. 4-12. Static Characteristics

The operating point is determined by the average values I_I and V_O. This is usually difficult to visualize until particular examples are considered. The reader may better appreciate the geometrical representations after reading Chapters 5 and 6 and then reading this section again.

If i_I takes on various constant values, then a curve v_I vs. v_O may be plotted for each value of i_I. These curves are called the family of static reverse transfer characteristic curves. The reverse transfer parameter, which is h_r in this case, is the slope of the static reverse transfer characteristic curve at the operating point.

The family of static reverse transfer characteristic curves and the family of static input characteristic curves involve the same information, but presented in a different way. As a matter of convenience in writing, it is customary to refer to the families of curves as reverse transfer characteristics and input characteristics. It is understood that this means the family of curves, unless otherwise specified. The reverse transfer characteristics may be plotted from the input characteristics, and vice versa. In a particular application the curves are plotted so that the parameters for the equivalent input circuit representation are the slopes of the input characteristic curve and the reverse transfer characteristic curve at the operating point.

The functional relationship for the output port may be considered in the same way as has been done for the input port. For example, in the equation $i_O = f(v_O, i_I)$ the dependent variable is i_O and the independent variables are v_O and i_I. The family of curves formed by the intersections of the planes $i_I = k_l$ and the surface $i_O = f(v_O, i_I)$ is the family of static output characteristic curves. The slope of the static output characteristic curve at the operating point gives the output parameter of the equivalent circuit, which is a conductance h_o in this example. The family of curves formed by the intersections of the planes $v_O = k_l$ and the surface $i_O = f(v_O, i_I)$ is the family of static forward transfer characteristic curves. The forward transfer function h_f is the slope of the static forward transfer characteristic curve at the operating point.

In practice, the equivalent circuit parameters of an electronic control device may be determined by using measuring instruments such as bridges rather than by graphical analysis. However, geometrical representations and graphical determination of the parameters best illustrate and emphasize the definitions of the parameters. A parameter is determined graphically by drawing a line tangent to the proper curve at the operating point and then finding the slope of the tangent line, taking large increments to improve the accuracy of the graphical measurement. This approach emphasizes the fact that the parameters are partial derivatives.

4.9 EQUIVALENT "tee" AND "pi" CIRCUITS

In addition to the general two port equivalent circuits which have parameters that are slopes of characteristic curves, there are equivalent circuits derived from these. The equivalent "tee" circuit shown in Fig. 4-13 is sometimes used to represent a transistor. The resistances r_1, r_2 and r_3 are usually denoted with lower case letters as subscripts; the letters indicate the terminals of the transistor to which they are connected. This circuit is derived from the r parameter equivalent circuit of Fig. 4-4. See Problem 4-6.

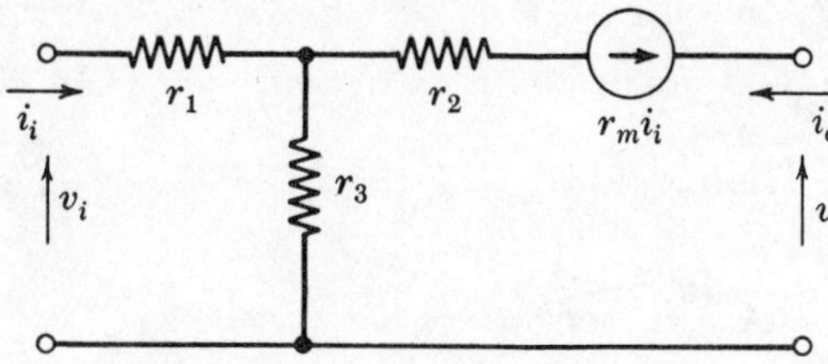

Fig. 4-13. Equivalent "tee" Circuit

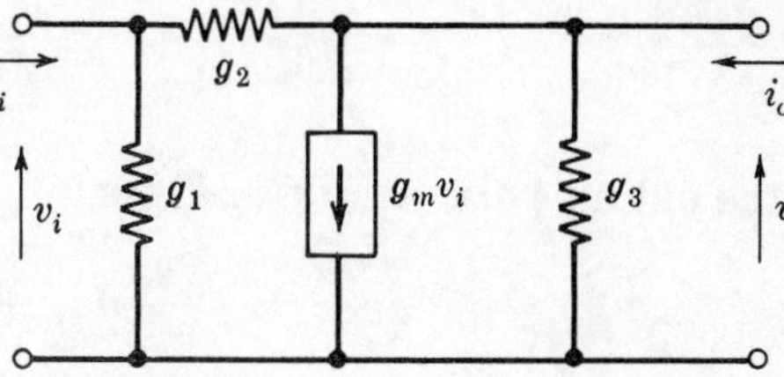

Fig. 4-14. Equivalent "pi" Circuit

The equivalent "pi" circuit of Fig. 4-14 is also employed to represent transistors. This circuit is derived from the g parameter equivalent circuit of Fig. 4-9. See Problem 4-7. The circuit of Fig. 4-14 has a current generator which is voltage controlled. However, it may be more convenient to use a current controlled generator in some applications. In other cases a voltage generator may be more convenient. There are so many possible equivalent circuits that only those which are to be used are emphasized.

4.10 TRANSMISSION PARAMETERS

A specific two port model which is useful when networks are cascaded (connected in a chain arrangement) is one which involves the *transmission parameters* (also called chain parameters or *ABCD* parameters). This representation has not been employed to represent electronic control devices but is useful in displaying amplifier circuits. Established conventions are a problem because of inconsistencies among the various fields in which they are used. The transmission parameters were first employed in power transmission applications where the current conventions are as in Fig. 4-15. However, to be consistent with the conventions of this book, the current references of Fig. 4-1, page 50, will be used. Transmission parameters involving i_O will be defined as the negative of the partial derivatives because of the reference direction of the current i_O. The notation used in power transmission applications and in many network applications involves numbered subscripts for the voltages and currents rather than letters. Letters are used in this book because of the need to distinguish instantaneous values, average values, and instantaneous values of varying components.

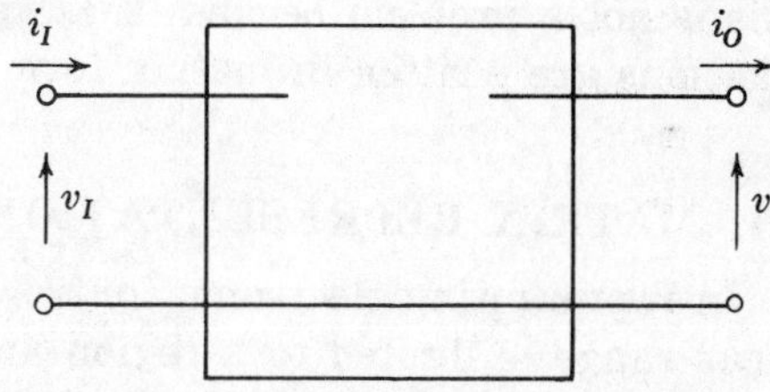

Fig. 4-15. General Two Port Network

The transmission parameters are defined as follows:

$$A = \frac{\partial v_I}{\partial v_O}, \quad B = -\frac{\partial v_I}{\partial i_O}, \quad C = \frac{\partial i_I}{\partial v_O}, \quad D = -\frac{\partial i_I}{\partial i_O}$$

Each parameter is a transfer function. In this representation the voltage and current at one port are related to the voltage and current at the other port, so there are two equations for each port.

For the input port, $v_I = f(v_O, i_O)$ and $i_I = f(v_O, i_O)$. Then

$$dv_I = \frac{\partial v_I}{\partial v_O} dv_O + \frac{\partial v_I}{\partial i_O} di_O = A\, dv_O - B\, di_O$$

$$di_I = \frac{\partial i_I}{\partial v_O} dv_O + \frac{\partial i_I}{\partial i_O} di_O = C\, dv_O - D\, di_O$$

When A, B, C and D are constant and only varying components are considered, these equations become

$$v_i = Av_o - Bi_o \tag{4.22}$$

$$i_i = Cv_o - Di_o \tag{4.23}$$

For the output port, $v_O = f(v_I, i_I)$ and $i_O = f(v_I, i_I)$. Then

$$dv_O = \frac{\partial v_O}{\partial v_I} dv_I + \frac{\partial v_O}{\partial i_I} di_I = \frac{1}{A} dv_I + \frac{1}{C} di_I$$

$$di_O = \frac{\partial i_O}{\partial v_I} dv_I + \frac{\partial i_O}{\partial i_I} di_I = -\frac{1}{B} dv_I - \frac{1}{D} di_I$$

Considering only varying components and assuming linear operation,

$$v_o = v_i/A + i_i/C = \mathcal{A}v_i + \mathcal{C}i_i$$

$$i_o = -v_i/B - i_i/D = -\mathcal{B}v_i - \mathcal{D}i_i$$

These equations contain parameters which are the inverse of the transmission parameters; hence they are called the *inverse transmission parameters* and denoted with script letters (sometimes they are indicated by primed quantities).

An equivalent circuit does not follow directly from the two equations for each port. This is not a problem because these parameters are frequently used for networks and the equations are written in matrix form.

4.11 MATRIX REPRESENTATIONS

In representing electronic control devices, geometrical models are emphasized since the linear range is limited to a region around an operating point. When networks consisting only of linear elements are studied, it is often convenient to consider a matrix representation. The matrix representation is also useful in equivalent circuit problems involving varying components of the voltages and currents.

Consider the equivalent circuit of Fig. 4-4, page 52, and the equations for the input and output ports:

$$r_i i_i + r_r i_o = v_i$$

$$r_f i_i + r_o i_o = v_o$$

In matrix form, these equations are written as

$$[r][i] = [v] \tag{4.24}$$

where $[r]$ is the matrix of the coefficients of the currents, $[i]$ is the column matrix of the unknown currents (dependent variables) and $[v]$ is the column matrix of the port voltages which are the driving functions (independent variables). When using a matrix representation, numbered subscripts may be more convenient for the parameters because the subscripts indicate the row and column of the specific element of the matrix.

Each side of (*4.24*) is multiplied by the inverse of the matrix $[r]$:

$$[r]^{-1}[r][i] = [r]^{-1}[v] \quad \text{or} \quad [i] = [r]^{-1}[v]$$

Following the rules for finding the inverse of a matrix,

$$[r]^{-1} = \begin{bmatrix} r_o/\Delta & -r_r/\Delta \\ -r_f/\Delta & r_i/\Delta \end{bmatrix}$$

where $\Delta = r_i r_o - r_f r_r$.

Consider the equivalent circuit of Fig. 4-9 and the circuit equations. Writing the equations in matrix form,

$$[i] = [g][v]$$

The matrix $[g]$ is the same as the inverse matrix $[r]^{-1}$. Hence corresponding elements are equal, which provides a transformation of parameters.

$$g_i = r_o/\Delta, \quad g_r = -r_r/\Delta, \quad g_f = -r_f/\Delta, \quad g_o = r_i/\Delta$$

It is much more convenient in some transformation problems to use matrices than to work with the equivalent circuits. Transformations can be made by writing the equations for the equivalent circuits and setting equivalent voltage and current equations equal to each other. See Problems 4.6 and 4.7.

Solved Problems

4.1. Given the two port network of Fig. 4-16. Represent this circuit using the r parameter equivalent circuit. Evaluate r_i, r_r, r_o and r_f.

The nodal equations for the two non-reference terminals are

$$+0.15v_i - 0.05v_o = i_i$$
$$-0.05v_i + 0.083v_o = i_o$$

from which

$$v_i = 8.3i_i + 5.0i_o$$

When $i_o = 0$, $r_i = v_i/i_i = 8.3\ \Omega$; when $i_i = 0$, $r_r = v_i/i_o = 5.0\ \Omega$.

Solving the nodal equations for v_o,

$$v_o = 15.0i_o + 5.0i_i$$

When $i_i = 0$, $r_o = v_o/i_o = 15.0\ \Omega$; when $i_o = 0$, $r_f = v_o/i_i = 5.0\ \Omega$.

When the network contains only passive elements, then $r_f = r_r$ as in this problem. The equivalent circuit is shown in Fig. 4-17.

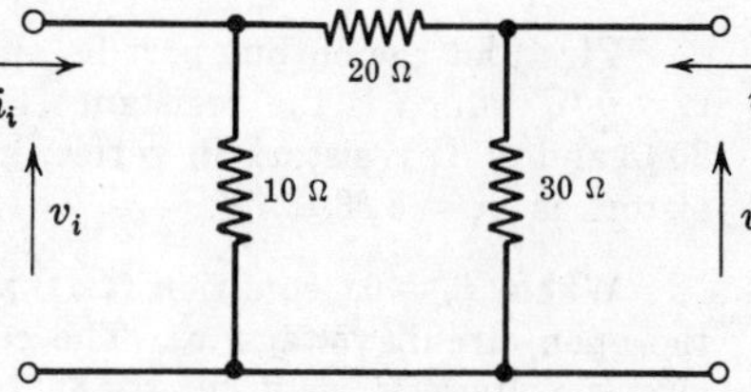

Fig. 4-16

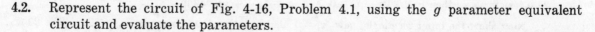

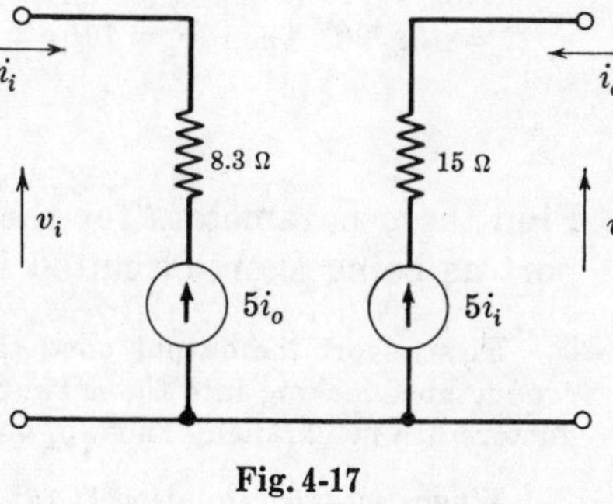

Fig. 4-17

4.2. Represent the circuit of Fig. 4-16, Problem 4.1, using the g parameter equivalent circuit and evaluate the parameters.

In general when a g parameter representation is desired, loop equations are written; then the driving functions (independent variables) are the voltages and the dependent variables are currents. Assuming a loop current i_1 in the middle loop with a clockwise direction, the loop equations are

$$10i_i - 10i_1 = v_i$$
$$-10i_i + 60i_1 + 30i_o = 0$$
$$+30i_1 + 30i_o = v_o$$

from which $i_i = 0.15v_i - 0.05v_o, \quad i_o = 0.083v_o - 0.05v_i$

When $v_i = 0$, $g_r = i_i/v_o = -0.05$ mho, $g_o = i_o/v_o = 0.083$ mho; when $v_o = 0$, $g_i = i_i/v_i = 0.15$ mho, $g_f = i_o/v_i = -0.05$ mho.

Note that $g_f = g_r$, since the network elements are all passive. In this example there are only two nodal equations, so the solutions for i_i and i_o are the same as the nodal equations of Problem 4.1. The equivalent circuit is displayed in Fig. 4-18.

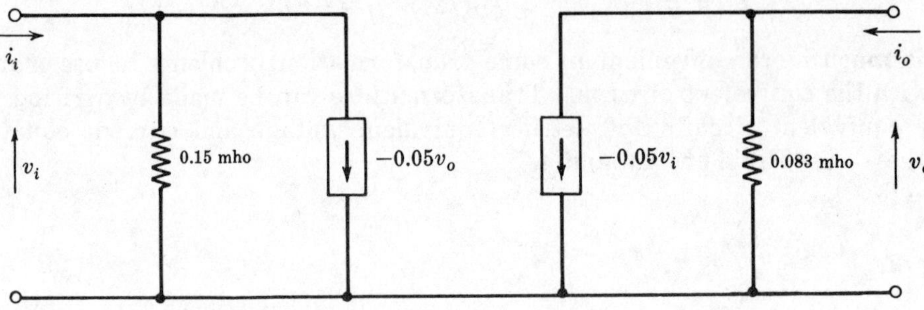

Fig. 4-18

4.3. Find the r parameters for the circuit of Fig. 4-16, Problem 4.1, by considering each port as open-circuited in turn and applying equations (*4.5*) and (*4.9*).

First, let the output port be open-circuited. Then $i_o = 0$ and equation (*4.9*), page 52, becomes $r_i = v_i/i_i$ which is the resistance looking into the input port (output open-circuited). Combine the 20 Ω and 30 Ω resistors in series, giving 50 Ω. The parallel combination of the 50 Ω and 10 Ω resistors is $r_i = 8.33$ Ω.

When $i_o = 0$, equation (*4.5*), page 51, is $r_f = v_o/i_i$. Assume an input current i_i and then find the open circuit voltage v_o. The current in the 30 Ω resistor is $i_{30} = v_i/50 = 8.33i_i/50$. The open-circuit voltage is $v_o = 30i_{30} = 30(8.33i_i/50)$, and $r_f = v_o/i_i = 5$ Ω.

Let the input port be open-circuited. Then $i_i = 0$ and equation (*4.5*) becomes $r_o = v_o/i_o$; this is the resistance looking into the output port. $r_o = 30(30)/(30+30) = 15.0$ Ω.

When $i_i = 0$, equation (*4.9*) becomes $r_r = v_i/i_o$. The current in the 10 Ω resistor is $i_{10} = v_o/30 = 15i_o/30$. Then $v_i = 10i_{10} = 5.0i_o$ and $r_r = v_i/i_o = 5.0$ Ω.

4.4. Find the g parameters for the circuit of Fig. 4-16, Problem 4.1, by considering each port as being short-circuited in turn and using equations (*4.16*) and (*4.17*), page 57.

First, short the output port; then $v_o = 0$ and equation (*4.17*) becomes $g_i = i_i/v_i$. This is the conductance looking into the network with the output port shorted, so that the 0.1 and 0.05 mho conductors are in parallel. Thus $g_i = 0.1 + 0.05 = 0.15$ mho.

When $v_o = 0$, equation (*4.16*) yields $g_f = i_o/v_i$. The current in the 30 Ω resistor is zero since v_o is zero, so i_o is in the 20 Ω resistor directed toward the input node. The voltage drop across the 20 Ω resistor is v_i; then $-(1/20)v_i - i_o = 0$ and $g_f = i_o/v_i = -0.05$ mho.

Next short the input terminals; now $v_i = 0$ and equation (*4.16*) becomes $g_o = i_o/v_o$. The 20 Ω and 30 Ω resistors are in parallel, so $g_o = 0.033 + 0.05 = 0.083$ mho.

When $v_i = 0$, equation (*4.17*) is $g_r = i_i/v_o$. The current in the 20 Ω resistor is i_i and the voltage drop is v_o (directions are opposite); then $-i_i - (1/20)v_o = 0$ and $g_r = i_i/v_o = -0.05$ mho.

4.5. Represent the network of Fig. 4-16, Problem 4.1, by writing the transmission parameters in matrix form.

From equations (*4.22*) and (*4.23*), page 62: when $i_o = 0$, $A = v_i/v_o$ and $C = i_i/v_o$; when $v_o = 0$, $B = -v_i/i_o$ and $D = -i_i/i_o$.

Let the output port be an open circuit; then $i_o = 0$. The nodal equation for the non-reference terminal of the output port is

$$+0.033v_o + 0.05(v_o - v_i) \;=\; 0 \qquad \text{from which} \qquad A \;=\; v_i/v_o \;=\; 1.66$$

Write a loop equation and let i_1 be the current in the 20 Ω resistor directed clockwise:

$$-v_i + 20i_1 + v_o \;=\; 0$$

Substituting $i_1 = v_o/30$ and $v_i = 8.3i_i$ (from Problem 4.1), we find $C = i_i/v_o = 0.20$. Shorting the terminals of the output port, $v_o = 0$ and $B = -v_i/i_o = 20$.

The nodal equation for the non-reference terminal of the input port is

$$-i_i + 0.1v_i + 0.05v_i \;=\; 0$$

Setting $v_i = -20i_o$, we find $D = -i_i/i_o = 3$.

Entering the numerical values into equations (*4.22*) and (*4.23*),

$$v_i \;=\; 1.66v_o - 20i_o$$
$$i_i \;=\; 0.2v_o - 3i_o$$

which in matrix form are written

$$\begin{bmatrix} v_i \\ i_i \end{bmatrix} \;=\; \begin{bmatrix} 1.66 & -20 \\ 0.2 & -3 \end{bmatrix} \begin{bmatrix} v_o \\ i_o \end{bmatrix}$$

4.6. Derive the equations for expressing the parameters of the equivalent "tee" circuit of Fig. 4-13, page 61, in terms of the r parameters of the equivalent circuit given in Fig. 4-4, page 52.

The loop equations for the circuit of Fig. 4-4 are

$$r_i i_i + r_r i_o \;=\; v_i$$
$$r_f i_i + r_o i_o \;=\; v_o$$

The loop equations for the circuit of Fig. 4-13 are

$$(r_1 + r_3)i_i + \qquad r_3 i_o \;=\; v_i$$
$$(r_m + r_3)i_i + (r_2 + r_3)i_o \;=\; v_o$$

Equating corresponding values of v_i and v_o,

$$(r_1 + r_3)i_i + \qquad r_3 i_o \;=\; r_i i_i + r_r i_o$$
$$(r_m + r_3)i_i + (r_2 + r_3)i_o \;=\; r_f i_i + r_o i_o$$

Rewrite these equations as

$$(r_1 + r_3 - r_i)i_i + \qquad (r_3 - r_r)i_o \;=\; 0$$
$$(r_m + r_3 - r_f)i_i + (r_2 + r_3 - r_o)i_o \;=\; 0$$

Since each coefficient must be zero (if the currents were zero, it would be trivial),

$$r_3 \;=\; r_r, \qquad r_1 \;=\; r_i - r_3 \;=\; r_i - r_r, \qquad r_2 \;=\; r_o - r_3 \;=\; r_o - r_r, \qquad r_m \;=\; r_f - r_3 \;=\; r_f - r_r$$

4.7. Derive the equations for expressing the parameters of the equivalent "pi" circuit of Fig. 4-14, page 61, in terms of the g parameters of the equivalent circuit of Fig. 4-9, page 57.

The nodal equations for the circuit of Fig. 4-9 are

$$g_i v_i + g_r v_o = i_i$$
$$g_f v_i + g_o v_o = i_o$$

The nodal equations for the circuit of Fig. 4-14 are

$$(g_1 + g_2)v_i - g_2 v_o = i_i$$
$$(g_m - g_2)v_i + (g_2 + g_3)v_o = i_o$$

Equating the two expressions for i_i and for i_o,

$$(g_i - g_1 - g_2)v_i + (g_r + g_2)v_o = 0$$
$$(g_f - g_m + g_2)v_i + (g_o - g_2 - g_3)v_o = 0$$

Since each coefficient must equal zero,

$$g_2 = -g_r, \quad g_1 = g_i - g_2 = g_i + g_r, \quad g_3 = g_o - g_2 = g_o + g_r, \quad g_m = g_f + g_2 = g_f - g_r$$

4.8. Given the two port network of Fig. 4-19. (*a*) Represent the network using the g parameter equivalent circuit. (*b*) Write the equations relating the voltages and currents using the g parameters in matrix form. Find the inverse of the matrix $[g]$ and draw the equivalent circuit using the r parameters.

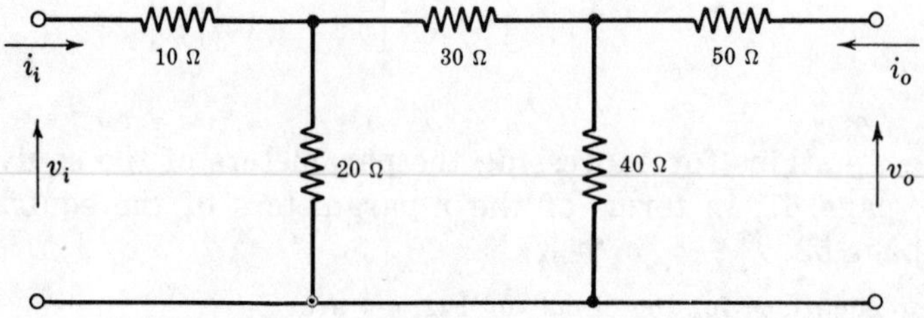

Fig. 4-19

(*a*) Choose a current i_1 directed clockwise in the loop that includes the 30 Ω resistor and write the loop equations.

$$30i_i - 20i_1 = v_i$$
$$-20i_i + 90i_1 + 40i_o = 0$$
$$+ 40i_1 + 90i_o = v_o$$

from which $$i_i = \frac{6500v_i - 800v_o}{159{,}000} \quad \text{and} \quad i_o = \frac{2300v_o - 800v_i}{159{,}000}$$

When $v_o = 0$: $g_i = i_i/v_i = 6500/159{,}000 = 0.0409$ mho, $g_f = i_o/v_i = -0.00503$ mho.

When $v_i = 0$: $g_r = i_i/v_o = -800/159{,}000 = -0.00503$ mho, $g_o = i_o/v_o = 0.0145$ mho.

(*b*) Entering the numerical values of the g parameters into the equations and writing in matrix form,

$$\begin{bmatrix} 0.0409 & -0.00503 \\ -0.00503 & 0.0145 \end{bmatrix} \begin{bmatrix} v_i \\ v_o \end{bmatrix} = \begin{bmatrix} i_i \\ i_o \end{bmatrix}$$

$$[r] = [g]^{-1} = \begin{bmatrix} 25.6 & 8.86 \\ 8.86 & 72.0 \end{bmatrix}$$

Thus $r_i = 25.6\ \Omega$, $r_r = 8.86\ \Omega$, $r_f = 8.86\ \Omega$, and $r_o = 72.0\ \Omega$.

4.9. Assume that the circuit of Fig. 4-11, page 59, has a terminating conductance G_L connected to the terminals of the output port. Derive an equation for the input resistance R_i when R_i is defined as v_i/i_i.

Writing a loop equation for the input port and a nodal equation for the output port,

$$h_i i_i + h_r v_o = v_i$$

$$h_f i_i + (h_o + G_L)v_o = 0$$

from which $$i_i = \frac{(h_o + G_L)v_i}{h_i(h_o + G_L) - h_f h_r} \quad \text{and} \quad R_i = v_i/i_i = h_i - \frac{h_f h_r}{h_o + G_L}$$

Since the output port is terminated with a passive element, the representation for the input port is a passive element which is a resistance in this case.

Supplementary Problems

4.10. Represent the circuit of Fig. 4-19, page 66, using the h parameters as given in Fig. 4-11, page 59. Specify the numerical values of the h parameters.

4.11. Represent the circuit of Fig. 4-16, page 63, using the equivalent circuit of Fig. 4-10, page 58, and calculate the numerical values of the parameters.

4.12. Derive the equations that express the h parameters of Fig. 4-11, page 59, in terms of the r parameters of Fig. 4-4, page 52.

4.13. The circuit of Fig. 4-11, page 59, has a generator connected to the input port as shown in Fig. 4-20. Draw the equivalent circuit looking into the output port (one port representation) using a constant voltage generator circuit (Thévenin equivalent) and then by using a constant current generator circuit (Norton equivalent). Assume that the magnitude of the generated voltage v_{gen} remains constant. Find the equations for the circuit parameters in terms of the parameters of the circuit of Fig. 4-20.

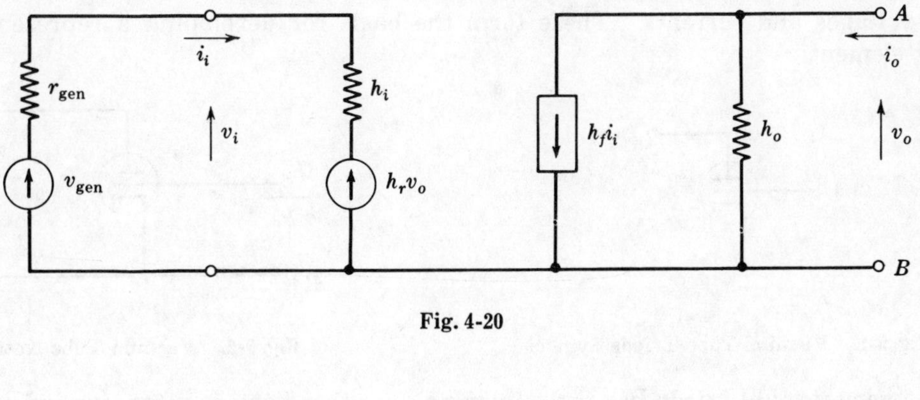

Fig. 4-20

Chapter 5

Two Port Electronic Control Devices, I

5.1 RESISTIVE CLASS OF ELECTRONIC CONTROL DEVICES

In the analysis and synthesis of circuits containing electronic control devices, the first consideration is a representation for each of the control devices. In this book, only electronic control devices of the resistive class are considered, i.e. the class of control devices involving resistive parameters. The present distinction is only between the types of devices in which the parameters are resistive as compared to those types in which they are reactive. The reactive types are those in which the parameters are inductive or capacitive.

The resistive class of control devices may be divided into subclasses. The subclass considered in this chapter is that whose input port is represented as an open circuit. Vacuum tube triodes, tetrodes and pentodes operating with the grid negative are among the more common examples. The circuit representations are developed as generally as possible, using specific examples as illustrations. The first example to be studied is the vacuum tube triode.

5.2 VACUUM TUBE TRIODE CHARACTERISTICS

The vacuum tube triode is considered a three element electronic control device. In addition to the plate and cathode elements, there is a third called the grid. The heater is assumed to be connected to a power supply, hence is not included in a discussion of the triode as an electronic control device. The symbol for the vacuum tube triode is given in Fig. 5-1. A distinction is sometimes made between the directly heated and the indirectly heated cathode as shown in Fig. 2-3, page 12, but this is not important in the present discussion. The electrical characteristics of the triode involve relationships between terminal voltages and currents. These form the basis for developing a representation as a circuit element.

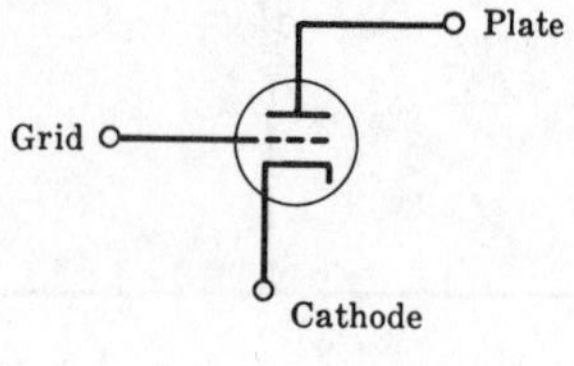

Fig. 5-1. Vacuum Tube Triode Symbol

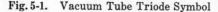

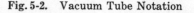

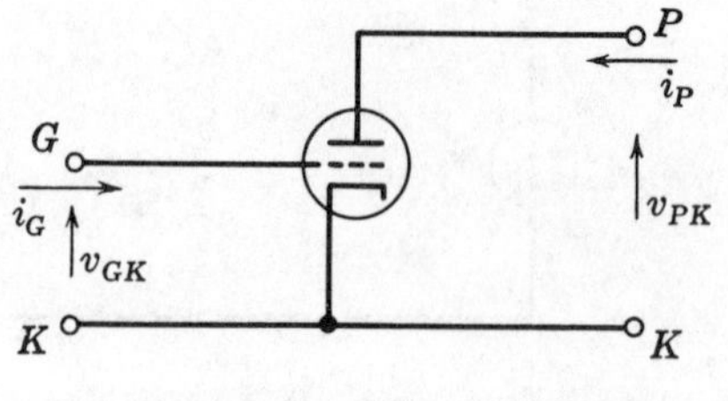

Fig. 5-2. Vacuum Tube Notation

The vacuum tube triode has three terminals, so there are at most two independent terminal voltages and two independent terminal currents (assuming no charge sources or sinks). Any one of the three elements may be employed as a common or reference element. However, in the case of the vacuum tube the cathode is usually chosen as the reference element. To be consistent with the conventions previously adopted in this book, the voltage drops from the non-reference elements to the reference element are considered along with the currents into the non-reference elements as indicated in Fig. 5-2. The notation for the voltage drops and currents is an extension of that used for diodes as listed in Appendix A.

Consider the output port of the triode, which is the plate and cathode terminal pair. Following the procedures outlined in Chapter 4, write the functional relationship for the voltages and currents so that the desired equivalent circuit is obtained. Considering instantaneous values, write

$$i_P = f(v_P, v_G) \tag{5.1}$$

This assumes the cathode temperature is in the normal operating range and that either the cathode temperature is constant or that operation is restricted to the space charge limited region of operation so that the plate current is independent of the cathode temperature. In general the relationship given in (*5.1*) is not linear, hence must be presented graphically. Because there are two independent variables, this relationship is geometrically a surface. This cannot be shown in two dimensions, so the surface must be approximated by families of curves as discussed in Chapter 4. For a vacuum tube triode, the *family of static plate characteristic curves* (often abbreviated "plate characteristics") is usually given. These curves are the intersections of the surface $i_P = f(v_P, v_G)$ and the planes $v_G =$ constants. The average plate characteristic curves for a 6C4 triode are graphed in Fig. 5-3. Since individual tubes are not identical but have characteristics which vary somewhat from tube to tube, averages of a number of tubes are plotted. Hence the characteristics of any one tube may differ somewhat from the averages, but this is not a serious problem unless a critical application is being considered. In the case of linear vacuum tube triode amplifiers, the interest is usually in the region of operation where the grid is negative with respect to the cathode, so the grid current is usually negligible.

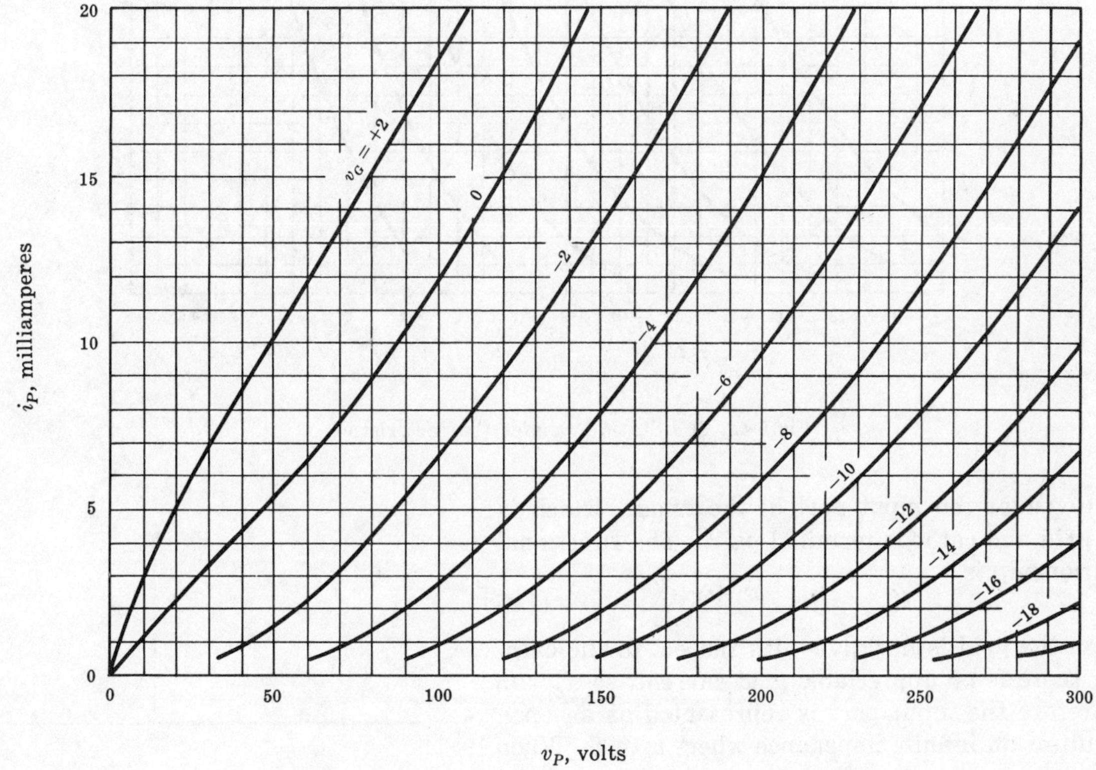

Fig. 5-3. 6C4 Static Plate Characteristics

The functional relationship of equation (*5.1*) can also be approximated by a family of curves resulting from the intersection of the surface $i_P = f(v_P, v_G)$ and the planes v_P = constants. This family of curves is called the *family of static forward transfer characteristic curves*. This is sometimes written as just the transfer characteristics; but this is done only when there is no ambiguity, since there is still another family of static forward transfer curves. The static forward transfer characteristic curves include the same information as the static plate characteristic curves. For example, the static forward transfer curves can be plotted from the static plate curves. Vertical lines are drawn on the plate characteristic curves for a chosen set of values of v_P. The points where the plate curves (v_G = a constant) intersect the particular vertical line (v_P = a constant) are used to plot one curve i_P vs. v_G. The family of static transfer characteristic curves for a 6C4 triode is shown in Fig. 5-4.

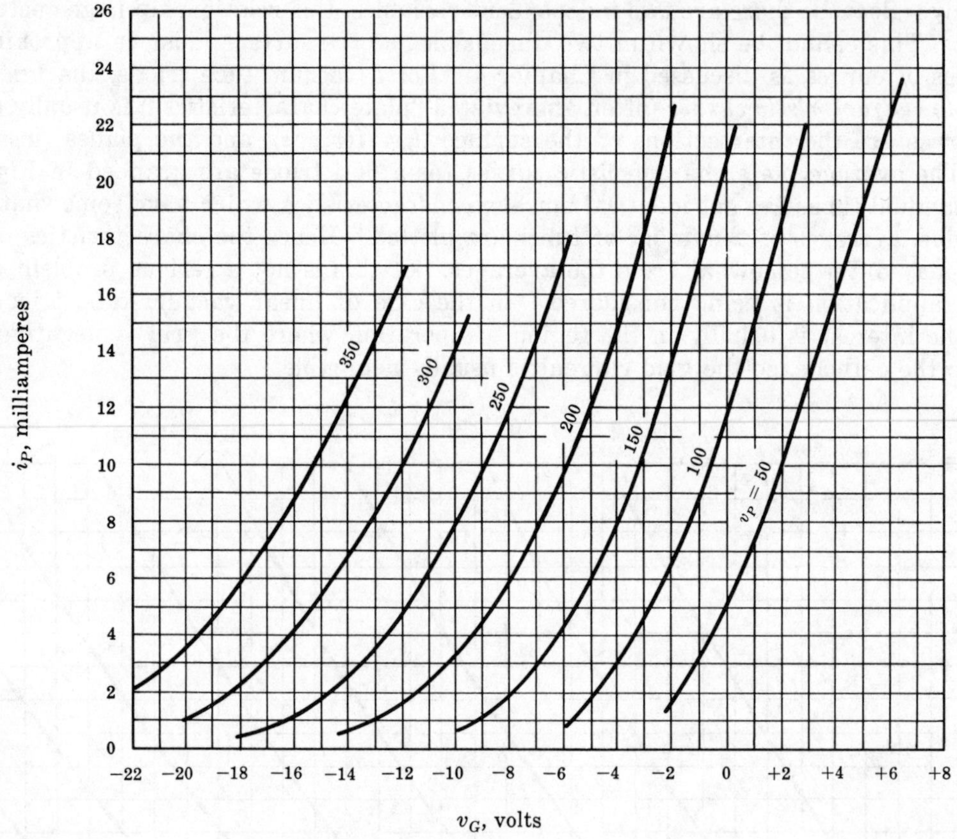

Fig. 5-4. **6C4 Static Transfer Characteristics**

Consider the input port of the triode, which is the grid and cathode terminal pair. The functional relationship is

$$i_G = f(v_G, v_P) \tag{5.2}$$

When the grid is negative with respect to the cathode, there is no appreciable grid current, so $i_G = 0$. Therefore the input port is represented as an open circuit or an infinite impedance when $v_G < 0$. When $v_G < 0$, the family of static grid characteristic curves all lie on the negative v_G axis as shown in Fig. 5-5.

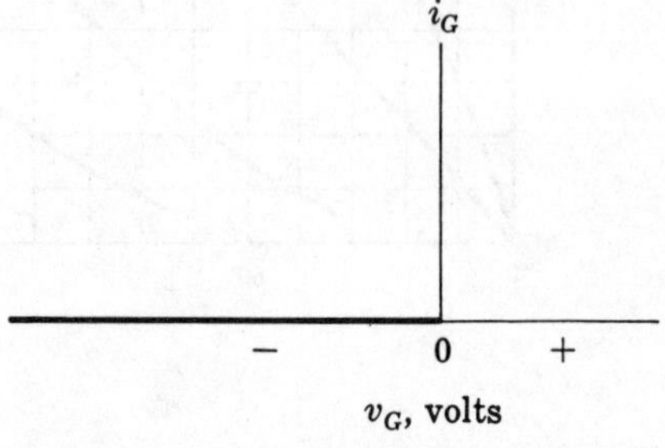

Fig. 5-5. **Static Grid Characteristics**

5.3 TRIODE EQUIVALENT CIRCUITS

The representation for the output port is developed by following the procedures outlined in Chapter 4. One approach involves expanding the function of equation (*5.1*) about a point (V_P, V_G) using Taylor's series.

$$i_P \;=\; f(V_P, V_G) \;+\; (v_P - V_P)\frac{\partial i_P}{\partial v_P} \;+\; (v_G - V_G)\frac{\partial i_P}{\partial v_G} \;+\; \cdots \qquad (5.3)$$

When operation is restricted to the linear range, the second and all higher derivatives are zero, so the only non-zero terms are those written in equation (*5.3*). By definition, $\partial i_P/\partial v_P = g_p$, the differential plate conductance, and $\partial i_P/\partial v_G = g_m$, the differential forward transfer conductance or mutual conductance. When the second and higher derivatives are zero, g_m and g_p are constants. Substituting g_m, g_p and $I_P = f(V_P, V_G)$ into (*5.3*),

$$i_P \;=\; I_P + (v_P - V_P)g_p + (v_G - V_G)g_m \qquad (5.4)$$

Since operation is restricted to the linear range, superposition may be applied, so the instantaneous values may be written as the average values plus the instantaneous values of the varying components. Substituting $v_P = V_P + v_p$, $v_G = V_G + v_g$, and $i_P = I_P + i_p$ into equation (*5.4*),

$$i_P \;=\; I_P + i_p \;=\; I_P + (V_P + v_p - V_P)g_p + (V_G + v_g - V_G)g_m \qquad (5.5)$$

Canceling the terms involving average values,

$$i_p \;=\; g_p v_p + g_m v_g \qquad \text{or} \qquad -i_p + g_p v_p + g_m v_g \;=\; 0 \qquad (5.6)$$

All terms of (*5.6*) contain varying components, so this is the equation for the output port of the triode when only varying components are considered within the linear operating range.

Examine each term of equation (*5.6*) (v_p and i_p references as in Fig. 5-2). The term $g_p v_p$ is a current in a linear conductor in the direction of the voltage drop v_p. The term $g_m v_g$ is independent of v_p and i_p, so it is shown as a voltage controlled, current generator. When v_g is positive this term has a + sign, so the arrow in the current generator is away from the plate node. Thus the equivalent circuit for the output port may be drawn as in Fig. 5-6.

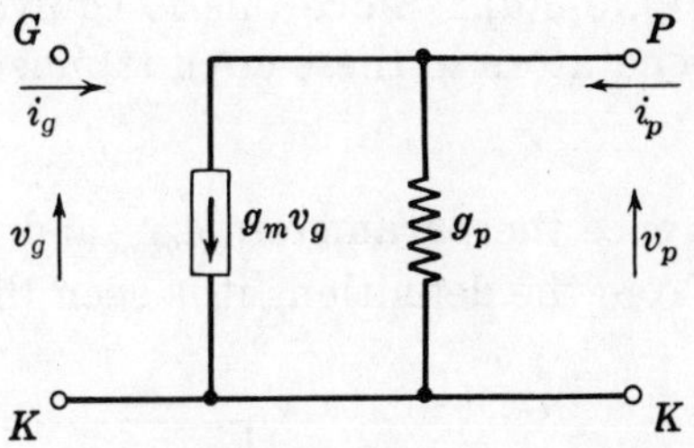

Fig. 5-6. Triode Equivalent Circuit

The input port is an open circuit (assuming $v_G < 0$), and its representation is added to complete Fig. 5-6. Fig. 5-6 exhibits an equivalent circuit for a vacuum tube triode when only varying components are considered and operation is limited to the linear range. This is sometimes called the "ac equivalent circuit". In practical situations operation is restricted to a range which is approximately linear, so this is sometimes said to involve "small signal analysis" or "linear approximation". This particular equivalent circuit contains a constant current generator circuit with a voltage controlled generator. Since $i_G = 0$, the triode plate circuit cannot be represented using current controlled generators.

A model may be developed which contains a constant voltage generator circuit in which the generator is voltage controlled. In this case, write the functional relationship as

$$v_P \;=\; f(i_P, v_G) \qquad (5.7)$$

and expand the function about an average value using Taylor's series:

$$v_P \;=\; f(I_P, V_G) \;+\; (i_P - I_P)\frac{\partial v_P}{\partial i_P} \;+\; (v_G - V_G)\frac{\partial v_P}{\partial v_G} \;+\; \cdots \qquad (5.8)$$

By definition, $\partial v_P/\partial i_P = r_p$, the differential plate resistance, and $\partial v_P/\partial v_G = -\mu$, a dimensionless quantity. It is somewhat confusing to have μ defined as the negative of a partial derivative. As will be seen later, $\partial v_P/\partial v_G$ is negative; so years ago μ was defined as the negative of the partial derivative, making μ a positive quantity. When r_p and μ are constants (linear range), superposition may be applied as before, and after substitution the final equation containing only varying components is

$$-v_p + r_p i_p - \mu v_g = 0 \tag{5.9}$$

Examine each term of this equation. v_p is the voltage rise from cathode to plate. The term $r_p i_p$ is the voltage drop across a linear resistor r_p in the direction of the current i_p. The term $-\mu v_g$ is independent of v_p and i_p, so it is shown as a voltage generator. A convention has been established showing the arrow directed opposite to the voltage drop. Thus the arrow is upward and the generator magnitude is $-\mu v_g$ as shown in Fig. 5-7. The input port is still an open circuit, so the grid terminal is shown as an open circuit. This equivalent circuit is, of course, only for varying components and linear operation. The equivalent circuit representation which is the most convenient should be used in a particular problem.

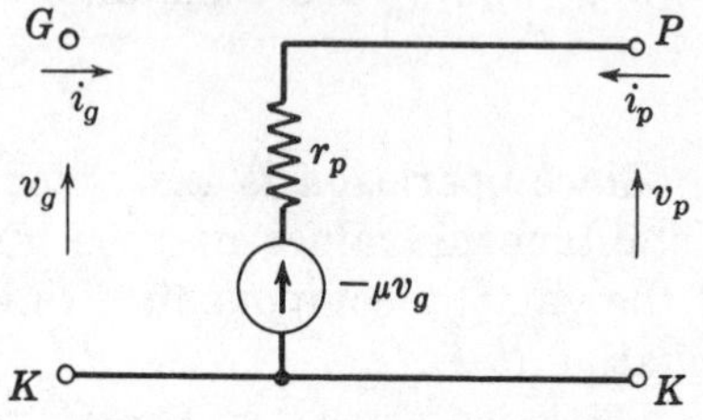

Fig. 5-7. Triode Equivalent Circuit

Since both the equivalent circuits of Fig. 5-6 and Fig. 5-7 represent the same thing, they should be equivalent to each other in some way. In the circuits of Fig. 5-6 and Fig. 5-7, v_g is independent of i_p and v_p because the reverse transfer function is zero, hence the generators are independent. This will not always be the case for vacuum tube equivalent circuits. Apply Thévenin's theorem to the circuit of Fig. 5-6 and obtain the circuit of Fig. 5-8(*a*). Since this is equivalent to the circuit of Fig. 5-7, the coefficients of the two generators in these circuits must be the same:

$$\mu = g_m/g_p = g_m r_p \tag{5.10}$$

Hence the parameters μ, r_p and g_m are not independent but are related by equation (*5.10*). From the definitions it is seen that $r_p = 1/g_p$.

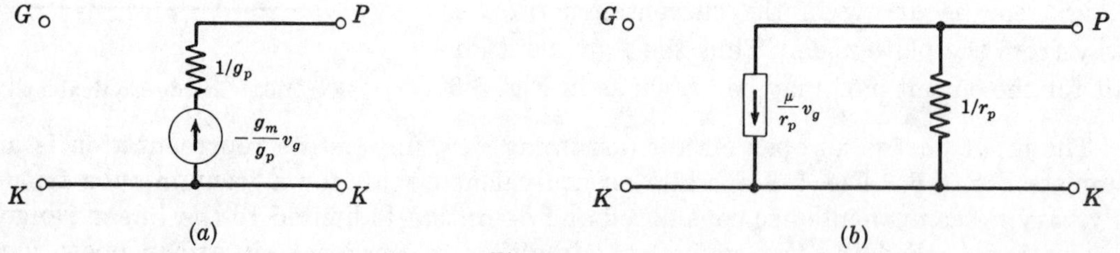

Fig. 5-8

Norton's theorem may be applied to the circuit of Fig. 5-7, giving the circuit of Fig. 5-8(*b*). Similarly, from Fig. 5-6 and Fig. 5-8(*b*), equation (*5.10*) follows: $g_m = \mu/r_p$ or $\mu = g_m r_p$.

5.4 OPERATING POINT AND TUBE PARAMETERS

The equivalent circuits which have been developed for vacuum tube triodes are for varying components only and for operation on the linear operating range. The parameters must be determined for a given operating point. The operating point or quiescent point involves the average values of the triode terminal voltages and currents. In general, the relationships between the voltages and currents are nonlinear, so the operating point is found graphically. Hence the particular circuit of interest must be examined. The circuit equations may be written in terms of the average values of the voltages and currents, and then the operating point found by plotting these equations on the static characteristics of the triode. When the operating point has been determined, the parameters may be found by graphically determining the slopes of the appropriate curves. Note that each parameter is defined as a partial derivative.

Consider the simple triode amplifier circuit of Fig. 5-9. When control devices of the resistive class are employed in amplifier circuits, they are usually placed in series with a linear resistor R_L and a constant voltage source V_{PP}. This circuit is similar to some of the diode circuits of Chapter 2; but the triode has three terminals, so there are transfer characteristics also to be considered. The characteristics of the vacuum tube triode, in this example a 6J5, are given graphically by the static plate characteristic curves of Fig. 5-10. The plate loop equation in terms of instantaneous values is

$$-V_{PP} + R_L i_P + v_P = 0 \tag{5.11}$$

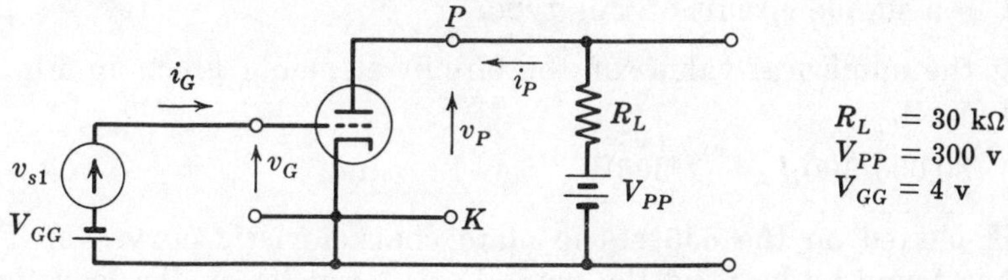

Fig. 5-9. Simple Triode Amplifier

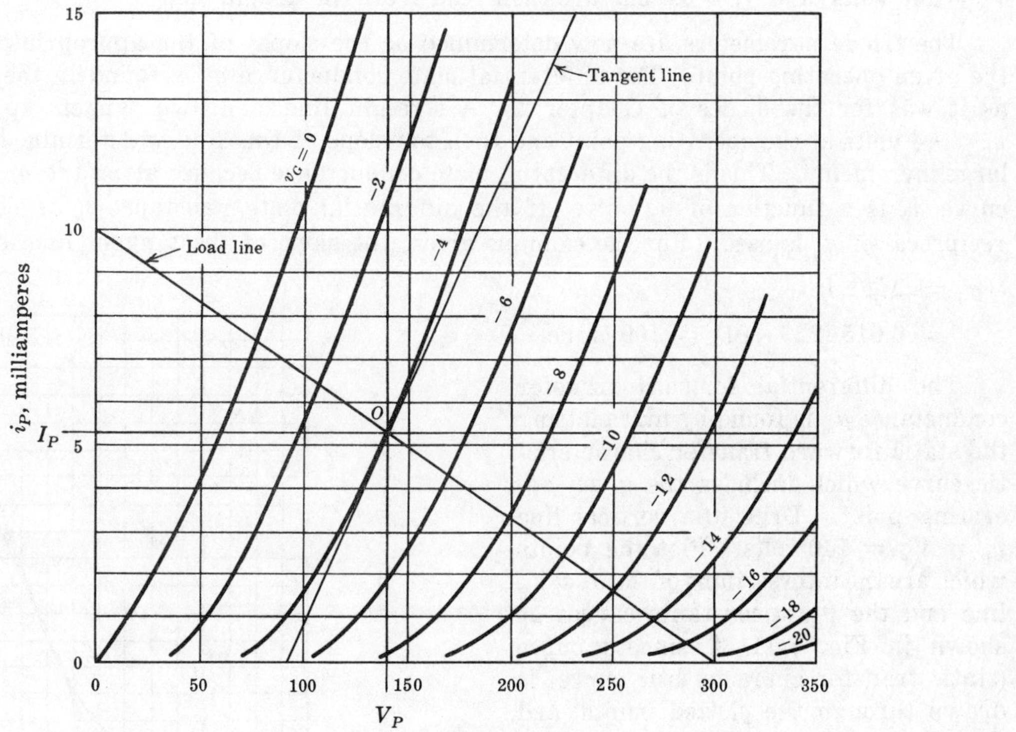

Fig. 5-10. 6J5 Plate Characteristics

Rewrite (5.11) as

$$(R_L/V_{PP})i_P + (1/V_{PP})v_P = +1 \tag{5.12}$$

which is the equation of a line in the standard intercept form and is the load line. Equation (*5.12*) is identical to equation (*2.11*), page 18, for the diode circuit of Fig. 2-13. The operating point is a particular point on the load line, so (*5.12*) is usually written in terms of the average values of i_P and v_P rather than the instantaneous values. The load line is plotted on the family of static plate characteristic curves, giving a solution to $i_P = f(v_P, v_G)$ and equation (*5.12*).

The grid loop equation in terms of average values (direction of i_G if it were not zero) is

$$+V_{GG} + 0 + V_G = 0 \qquad (5.13)$$

The signal source v_{si} will usually have an average value of zero and so is written as zero in (*5.13*). This is done to emphasize the fact that the average value must be included if it is not zero. By (*5.13*),

$$V_G = -V_{GG} \qquad (5.14)$$

The average value of the grid to cathode voltage drop is usually negative and is often called the "grid bias." The value of the grid supply voltage V_{GG} is chosen to give the desired operating point in a simple circuit of this type.

Substituting the numerical values of the circuit elements given in Fig. 5-9 into equations (*5.12*) and (*5.14*),

$$(30{,}000/300)I_P + (1/300)V_P = +1 \qquad \text{and} \qquad V_G = -4 \text{ volts}$$

The load line is plotted on the 6J5 static plate characteristic curves of Fig. 5-10. The operating point is found by locating the point $V_G = -4$ volts on the load line. This is the intersection of the plate characteristic curve $v_G = -4$ volts and the load line. The values $V_P = 140$ volts and $I_P = 5.3$ ma are then read from the graph.

The triode parameters are now determined as the slopes of the appropriate curves at the given operating point. The differential plate conductance g_p is found in the same way as it was for the diodes of Chapter 2. A straight line is drawn tangent to the curve $v_G = -4$ volts at the operating point and then the slope of this line is determined by taking large increments. This is the differential plate conductance because along the $v_G = -4$ volts curve, i_P is a function of v_P only. If the differential plate resistance r_p is desired, the reciprocal of g_p is used. For the example given, the slope of the tangent line is

$$g_p = \Delta i_P/\Delta v_P$$
$$= 0.015/(229 - 91) = 109\ \mu\text{mho}$$

The differential forward transfer conductance g_m is found by first plotting the static forward transfer characteristic curve which includes the given operating point. Draw the vertical line $v_P = V_P = 140$ volts. Plot the points which are the intersections of the $v_P = V_P$ line and the v_G = constants curves as shown in Fig. 5-11. A smooth curve (static transfer characteristic curve) is drawn through the plotted points, and the operating point is then located on this curve. The slope of the static transfer characteristic curve at the operating point is g_m. Since v_P is a constant along this curve, i_P is a function of v_G only

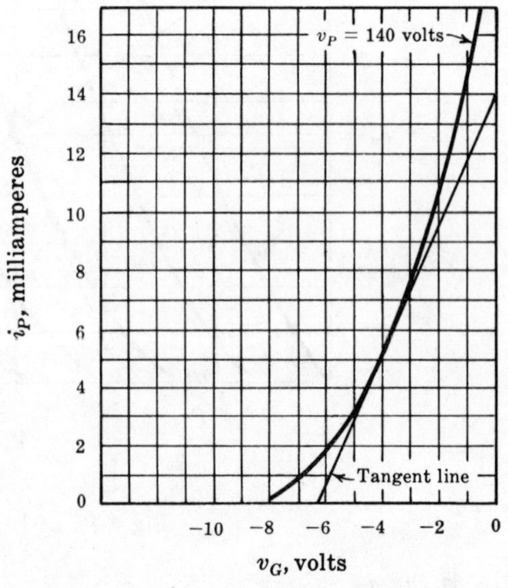

Fig. 5-11. 6J5 Static Transfer Characteristic Curve

along this curve. A line is drawn tangent to the transfer characteristic curve at the operating point and the slope of the tangent line is determined by taking large increments. For this example,

$$g_m = \Delta i_P/\Delta v_G = 0.0142/[0-(-6.5)] = 2190\ \mu\text{mhos}$$

If the equivalent constant voltage generator circuit is to be used, the parameter μ can be calculated knowing g_p and g_m. However, since μ involves a forward transfer characteristic, it may be determined graphically by plotting the appropriate transfer characteristic curve and then finding the slope at the operating point. Since by definition $\mu = -\partial v_P/\partial v_G$, the curve for $i_P = I_P = 5.3$ ma is plotted. This curve is plotted in the second quadrant because $v_G < 0$, as shown in Fig. 5-12. The entire family of these curves is often called the *family of constant current curves.* This terminology is used for vacuum tube characteristic curves only and was probably adopted to avoid ambiguity, since there are two families of static forward transfer characteristic curves. The slope of the transfer characteristic curve at the operating point is found graphically as before. From Fig. 5-12, v_P decreases as v_G increases; hence $\partial v_P/\partial v_G$ is negative. However, by definition μ is positive. For this example,

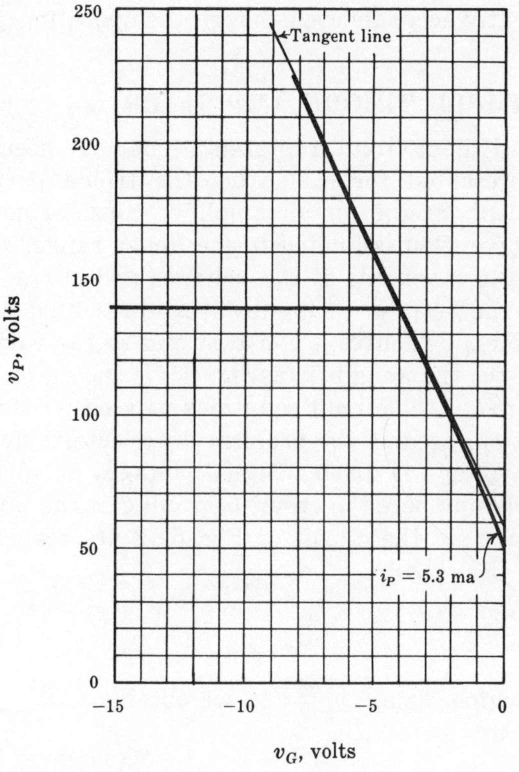

Fig. 5-12. 6J5 Constant Current Curve

$$\mu = -\Delta v_P/\Delta v_G = -(260-59)/(-10.0-0) = 20.1$$

The value of μ calculated using g_m and g_p is

$$\mu = g_m/g_p = (2190\times 10^{-6})/(109\times 10^{-6}) = 20.1$$

The two values of μ are usually somewhat different because of the poor accuracy when small graphs are used.

In this example the parameters were determined by plotting the appropriate curve and then finding the slope of the curve at the operating point. This emphasizes that the parameters are defined as partial derivatives. Since the characteristics are usually an average of a number of tubes and are not those of a particular tube, the parameters are estimates. Also to be considered is the problem of accuracy of the graphical determinations. For these reasons the parameters are sometimes estimated by taking incremental values of the appropriate voltages and currents. For the example given, g_p could have been estimated by choosing two points on the $v_G = -4$ volts static plate characteristic curve and finding the slope of the line through these points. If the points are close together, the accuracy is limited by the ability to read the graph to a sufficient number of significant figures. If the points are chosen farther apart, they may not be on the linear portion of the curve. However, in many practical problems a sufficient degree of accuracy may be achieved. If this procedure is followed it is not necessary to plot the transfer characteristic curves, since the incremental values can be taken directly from the plate characteristics by choosing points along the curves v_P = a constant, v_G = a constant, or i_P = a constant.

The parameters of individual electronic control devices may be determined for an established operating point by some direct measurement procedure employing electronic instruments. In many design problems the values of the parameters given in handbooks are sufficiently accurate even though the operating point may not be the same.

5.5 SIMPLE TRIODE AMPLIFIER

Equivalent circuit representations have been developed for the triode and methods have been presented for evaluating the triode parameters from characteristic curves. The equivalent circuit for an amplifier considering only varying components is now drawn. Since operation is limited to the linear range, superposition may be applied and then only varying components of the voltages and currents are considered in the equivalent circuit. Hence the dc power supplies are represented as short-circuits, since the terminal voltage of an ideal dc source is constant and so the varying component is zero. The ac equivalent circuit for the amplifier of Fig. 5-9 is shown in Fig. 5-13. After the equivalent circuit has been drawn, the circuit equations may be written and solved. For the example under consideration, assume the problem is to determine the output signal voltage when the input signal voltage is given. Signal voltages at various points in amplifiers will be rather arbitrarily numbered in order, beginning at the input terminal. The grid loop and plate nodal equations for the circuit of Fig. 5-13 are respectively

$$-v_{s1} + v_g = 0 \quad \text{or} \quad v_g = v_{s1} \tag{5.15}$$

$$g_m v_g + g_p v_p + G_L v_p = 0 \tag{5.16}$$

from which, using $v_p = v_{s2}$, we obtain

$$v_{s2} = \frac{-g_m}{g_p + G_L} v_{s1}$$

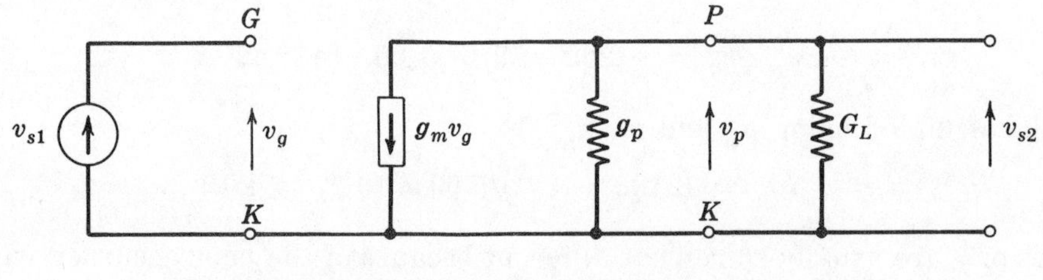

Fig. 5-13. Amplifier Equivalent Circuit

By definition, the voltage gain is

$$A_v = v_{s2}/v_{s1} = -g_m/(g_p + G_L) \tag{5.17}$$

For example, assume that $v_{s1} = 2 \sin \omega t$; then

$$v_{s2} = \{-2190 \times 10^{-6}/[(109 + 33.3) \times 10^{-6}]\} v_{s1}$$

$$= -15.4\, v_{s1} = -30.8 \sin \omega t = 30.8 \sin(\omega t + \pi) \text{ volts}$$

In general, the voltage gain is complex and is often expressed in polar form. The voltage gain is denoted A_v where the A indicates amplification and the subscript v specifies voltage amplification. In this example the voltage gain has magnitude 15.4 and a negative sign. The negative sign indicates a shift in phase of 180° or π radians between the input and output signal voltages.

A graphical solution to the simple amplifier circuit of Fig. 5-9 may provide additional insight into amplifiers containing resistive electronic control devices. A load line for the circuit of Fig. 5-9 is plotted on the 6J5 static plate characteristic curves of Fig. 5-14. The instantaneous grid to cathode voltage drop v_G may be considered point by point and the instantaneous values of i_P and v_P may be found on the graph. If linear operation is assumed, then the varying components of the voltages and currents are all sinusoidal when a sinusoidal signal voltage is applied to the input. The instantaneous value of v_G may be sketched. First draw a line perpendicular to the load line at the operating point. Knowing v_{s1}, the maximum and minimum values of v_G may be located on the load line. For the present example,

$$v_G \;=\; V_G + v_g \;=\; -4 + 2\sin\omega t \text{ volts}$$

When $\sin\omega t = 1$, $v_{G(\max)} = -2$ volts; when $\sin\omega t = -1$, $v_{G(\min)} = -6$ volts. Lines may be drawn perpendicular to the load line indicating $v_{G(\max)}$ and $v_{G(\min)}$ and then the sine wave may be sketched.

$$\text{When } v_G = -2\text{:} \quad i_P = i_{P(\max)} = 6.4 \text{ ma}, \; v_P = v_{P(\min)} = 107 \text{ volts}$$

$$\text{When } v_G = -6\text{:} \quad i_P = i_{P(\min)} = 4.3 \text{ ma}, \; v_P = v_{P(\max)} = 170 \text{ volts}$$

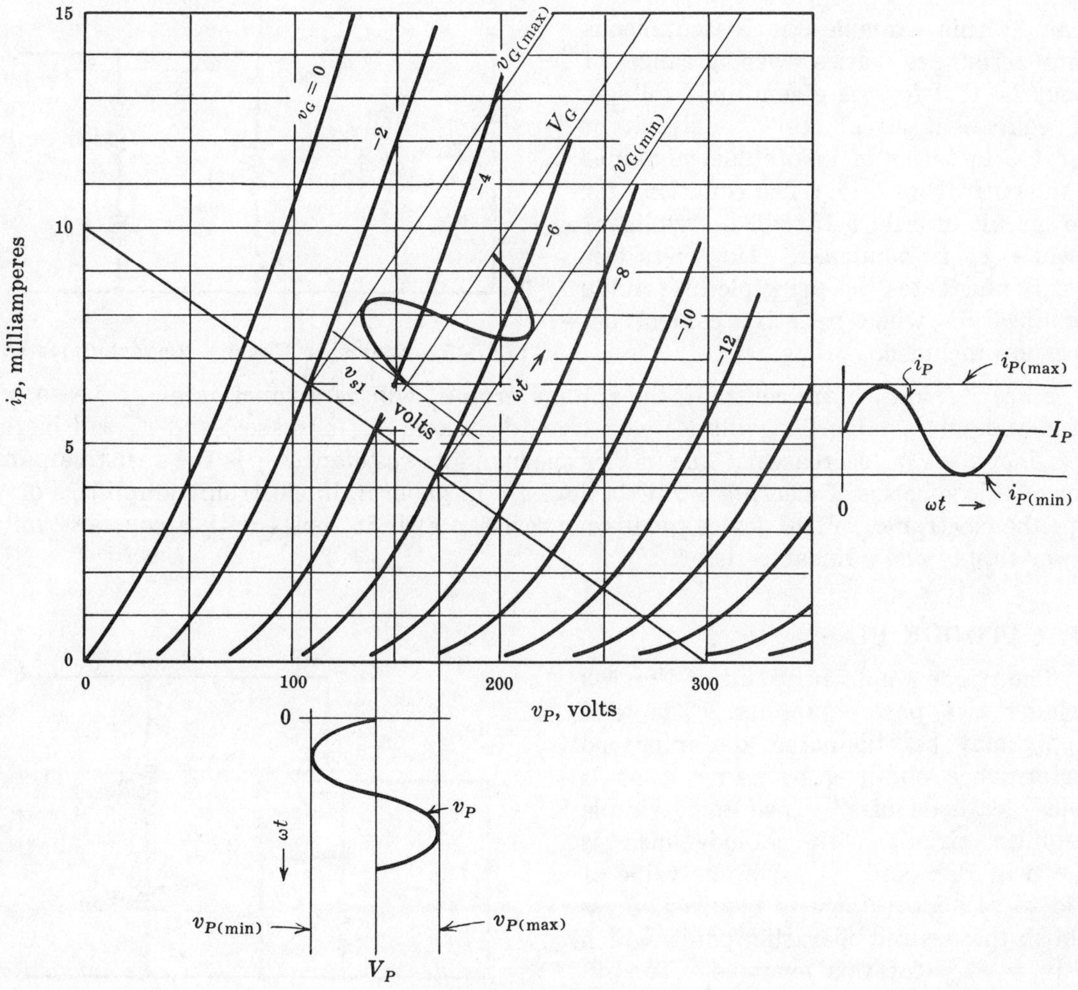

Fig. 5-14. Graphical Solution

The instantaneous values of i_P and v_P may be sketched as shown in Fig. 5-14. As may be seen on this graph, i_p is in phase with v_g but v_p is shifted in phase by π radians. The magnitude of the varying components may be calculated assuming linear operation and sinusoidal variations by taking the difference between the maximum and minimum values, giving the peak-to-peak values of the varying components. For this example,

$$v_p = \tfrac{1}{2}(v_{P(\min)} - v_{P(\max)}) \sin \omega t = \tfrac{1}{2}(107 - 170) \sin \omega t = -31.5 \sin \omega t \text{ volts}$$

$$i_p = \tfrac{1}{2}(i_{P(\max)} - i_{P(\min)}) \sin \omega t = \tfrac{1}{2}(6.4 - 4.3)10^{-3} \sin \omega t = 1.05 \times 10^{-3} \sin \omega t \text{ amps}$$

and the voltage gain $A_v = v_{s2}/v_{s1} = v_p/v_g = -15.7$. The value of A_v calculated from the equivalent circuit was -15.4.

It is of interest to consider the instantaneous resistance looking between the plate and the cathode terminals of the triode. For the triode, $v_{P(\max)}$ occurs when i_P is minimum and $v_{P(\min)}$ occurs when i_P is maximum. This is different from the diode circuits of Chapter 2, but the concept of instantaneous resistance is the same. When $v_G = -2$, the instantaneous plate resistance is

$$r_{P(\min)} = v_{P(\min)}/i_{P(\max)} = 107/0.0064 = 16{,}750\ \Omega$$

When $v_G = -6$, the instantaneous plate resistance is

$$r_{P(\max)} = v_{P(\max)}/i_{P(\min)} = 170/0.0043 = 39{,}500\ \Omega$$

Thus in this example the instantaneous plate resistance varies over a range of about 2.4 to 1 for the given input voltage. An equivalent circuit for the circuit of Fig. 5-9 in terms of instantaneous values is shown in Fig. 5-15. The equations for the circuit of Fig. 5-15 will be nonlinear because r_P is nonlinear. However, this circuit illustrates the principle of voltage amplification when resistive control devices are included.

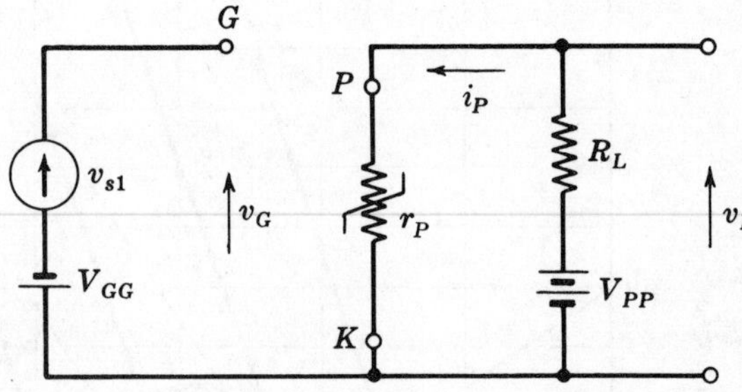

Fig. 5-15. Equivalent Circuit – Instantaneous Values

Since V_{PP} and R_L are constant, the voltage drop v_P will be a function of r_P. Also, r_P is a function of v_G. Hence v_P will decrease as r_P decreases (v_G increases) and v_P will increase as r_P increases (v_G decreases). The instantaneous plate resistance r_P is the resistive parameter of this subclass of electronic control device. In general, in electronic amplifiers of this type the electronic control device (nonlinear resistance) is in series with a constant voltage power supply and a linear resistor.

5.6 CATHODE BIAS

The triode amplifier circuit of Fig. 5-9 includes two power supplies. The grid supply may be eliminated and improved performance obtained by using what is called "cathode bias". The same simple amplifier circuit with cathode bias is shown in Fig. 5-16. An average value of grid to cathode voltage is required to establish the desired operating point, which is $V_G = -4$ v for this example. The instantaneous plate current $i_P = I_P + i_p$ is out of the cathode, so when a resistor is

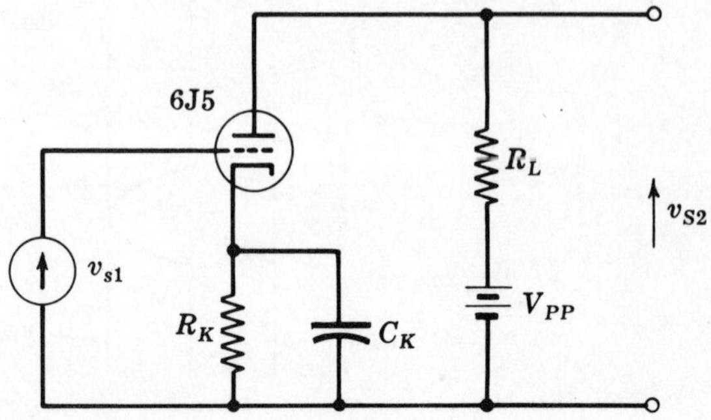

Fig. 5-16. Cathode Bias Amplifier

placed in series from cathode to ground (reference), the voltage drop across the cathode resistor R_K is

$$v_{\text{KGnd}} = R_K i_P = R_K I_P + R_K i_p \qquad (5.18)$$

The average value of v_{KGnd} is $R_K I_P$ which has the desired polarity, since the cathode is positive with respect to the ground and the average value of the grid to ground voltage is zero. There is also the varying voltage from cathode to ground $R_K i_p$ when only a resistor is used, hence a rather large capacitor C_K is added from the cathode to ground as shown in Fig. 5-16. Considering only varying components, the varying voltage is

$$v_{\text{kgnd}} = Z_K i_p \qquad (5.19)$$

where Z_K is the impedance of the parallel combination of the capacitor and resistor and is chosen so that the varying voltage is so small that it may be neglected. This will be given more consideration later.

The problem of choosing circuit parameters for a given amplifier is not a simple one, since all of the parameters must be specified and the criteria for optimizing the design must be considered. For this reason the present discussion is mostly concerned with the analysis of given circuits. To add cathode bias to the circuit of Fig. 5-9, the value of R_K may be determined approximately by substituting the values of V_G and I_P into equation (*5.18*).

$$V_{\text{KGnd}} = R_K I_P = 4 = R_K(0.0053), \qquad R_K = 4/0.0053 = 750\ \Omega$$

In practice the nearest manufactured value is used, which in this case is 680 Ω. However, since this is an academic problem, a value of 700 Ω will be taken so that the operating point remains approximately the same. The value of R_L is also not an available value but was chosen to make the i_P intercept an even 10 ma. Writing the plate loop equation for the circuit of Fig. 5-16 in terms of average values,

$$-V_{PP} + R_L I_P + V_P + R_K I_P = 0 \qquad \text{or} \qquad \frac{R_L + R_K}{V_{PP}} I_P + \frac{1}{V_{PP}} V_P = +1 \qquad (5.20)$$

Because of the cathode resistance R_K, the i_P intercept of the load line is slightly different from that given by equation (*5.12*). If R_L is large compared to R_K, then $R_L + R_K$ is approximately equal to R_L. In this example the difference between 30,000 Ω and 30,700 Ω is not significant. However, this will not always obtain. The small difference in the slope of the load line accounts for the difference between the instantaneous plate to ground voltage and the instantaneous plate to cathode voltage.

The grid loop equation in terms of average values is

$$0 + V_G + R_K I_P = 0 \qquad (5.21)$$

where the first zero indicates that the average value of the external generator is zero. The operating point is the point which satisfies equations (*5.20*) and (*5.21*) and the relationship for the tube $i_P = f(v_P, v_G)$. The load line is a solution for $i_P = f(v_P, v_G)$ and equation (*5.20*). Equation (*5.21*) may also be plotted on the plate characteristic curves as shown in Fig. 5-17. Values of V_G are chosen so that they lie on $v_G =$ a constant curves and then the corresponding values of I_P are calculated using equation (*5.21*). For example, let $V_G = -2$; then $I_P = 2/700 = 2.86$ ma. A smooth curve is drawn through all the plotted points as shown in Fig. 5-17 below; this curve is called the *grid curve*. The operating point is the intersection of the plate load line and the grid curve as shown in Fig. 5-17. For this circuit,

$$V_P = 135 \text{ volts}, \quad I_P = 5.3 \text{ ma}, \quad \text{and} \quad V_G = -3.75 \text{ volts}$$

V_G is read from the graph by interpolating linearly between the two adjacent $v_G =$ a constant curves. The tube parameters may be found as before. One parameter is determined by plotting the appropriate forward transfer characteristic curve and finding the slope at the operating point. Since there is no $v_G =$ a constant curve through the operating point, the slope of the $v_G =$ a constant curve nearest the operating point may be used because the slope should be constant over the linear operating range. An alternative would be to sketch the $v_G = V_G$ curve by linear interpolation and then find the slope of this curve at the operating point.

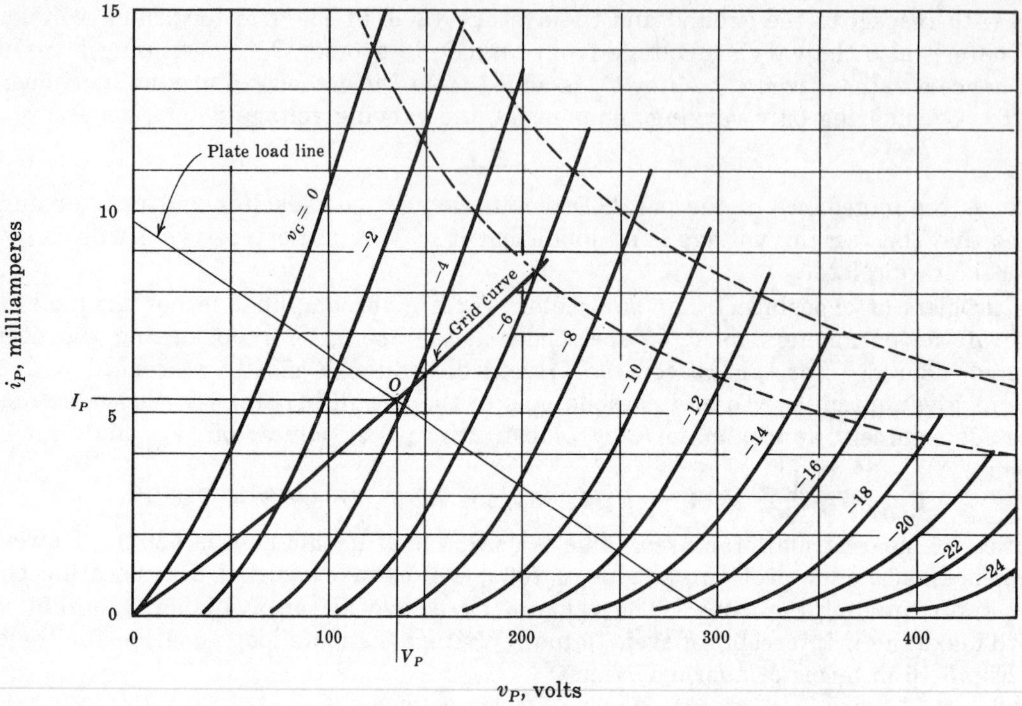

Fig. 5-17. 6J5 Static Plate Characteristics

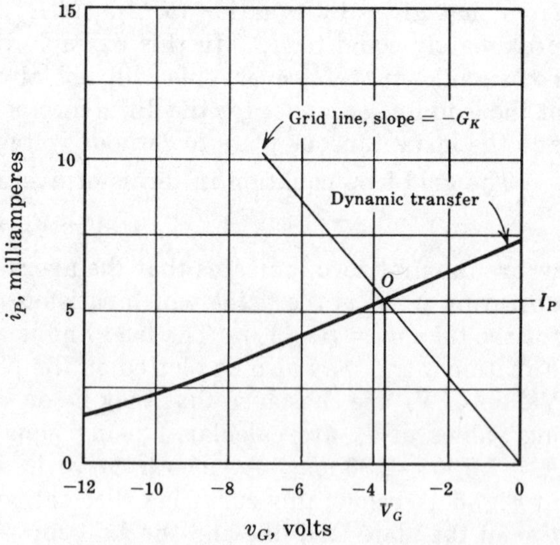

Fig. 5-18. Dynamic Transfer Characteristic Curve

In Fig. 5-18, equation (*5.21*) is plotted on a graph of i_P vs. v_G; it is a straight line through the origin with slope $-1/R_K$. The plate load line may be transferred from the static plate characteristic curves to the graph of i_P vs. v_G by plotting enough points from the load line to define the curve. The intersections of the plate load line and the static plate characteristics are plotted point by point. For example, when $v_G = 0$, $i_P = 7.3$ ma. The plate load line is not a straight line on the i_P vs. v_G graph. This curve is used frequently enough so that it has been called the dynamic transfer characteristic curve. The dynamic transfer curve takes into account the particular tube in a particular circuit. It differs from the static transfer characteristic curve in that v_P is not constant but varies depending on the current i_P. The static transfer characteristic curve for $v_P = V_P$ will intersect the dynamic transfer characteristic curve at the operating point.

Another advantage of cathode bias is that it tends to stabilize the operating point. If something should occur which would tend to increase the plate current, the "bias" voltage V_G would decrease (become more negative), causing the plate current to decrease. This may be seen by noting the points along the load line which are the intersections with the static plate characteristic curves.

The amplifier circuit (with cathode bias) of Fig. 5-16 is very similar to the amplifier circuit of Fig. 5-9. Considering only varying components in the frequency range where Z_K is very small, the equivalent circuit is drawn as in Fig. 5-13. The same equivalent circuit is used to represent the triode and the operating point is nearly the same as that for the circuit of Fig. 5-9, so the triode parameters are approximately the same. Hence for all practical purposes the voltage gain of the amplifier with cathode bias including a cathode capacitor, is the same when the frequency of the input signal is high.

5.7 VOLTAGE GENERATOR EQUIVALENT CIRCUIT

The voltage gain for the amplifier circuit of Fig. 5-13 was found to be

$$A_v = v_{s2}/v_{s1} = -g_m/(g_p + G_L) \qquad (5.17)$$

In some cases it is more convenient to represent the amplifier by using a constant voltage equivalent circuit as shown in Fig. 5-19. The grid loop equation is still

$$-v_{s1} + v_g = 0 \quad \text{or} \quad v_g = v_{s1} \qquad (5.22)$$

and the plate loop equation is

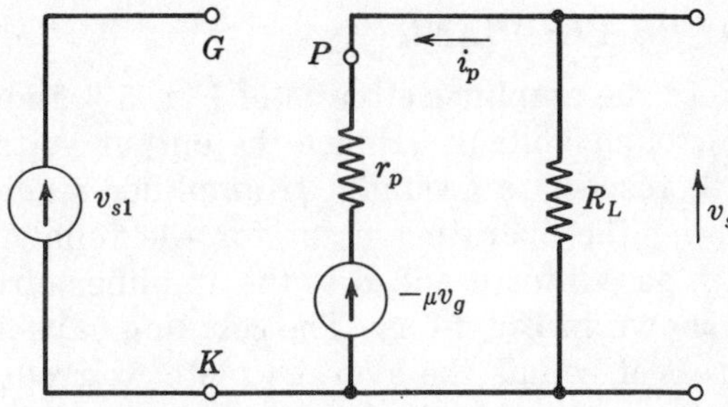

Fig. 5-19. Amplifier Equivalent Circuit

$$R_L i_p + r_p i_p - \mu v_g = 0 \qquad (5.23)$$

From the two equations, $i_p = \mu v_{s1}/(R_L + r_p)$. Then $v_{s2} = -R_L i_p$ and

$$A_v = -\mu R_L/(R_L + r_p) \qquad (5.24)$$

To show that equations (5.24) and (5.17) are the same, divide the numerator and denominator of (5.24) by $R_L r_p$ and obtain (5.17).

5.8 UNBYPASSED CATHODE

Consider the amplifier circuit of Fig. 5-16 without the cathode bypass capacitor C_K. The equivalent circuit is shown in Fig. 5-20 for this case. The equivalent circuit for the triode tube is drawn first, and then the remainder of the circuit is sketched considering only varying components. Here the constant voltage equivalent circuit is employed to represent the tube.

To find the voltage gain $A_v = v_{s2}/v_{s1}$, write the grid loop and plate loop equations

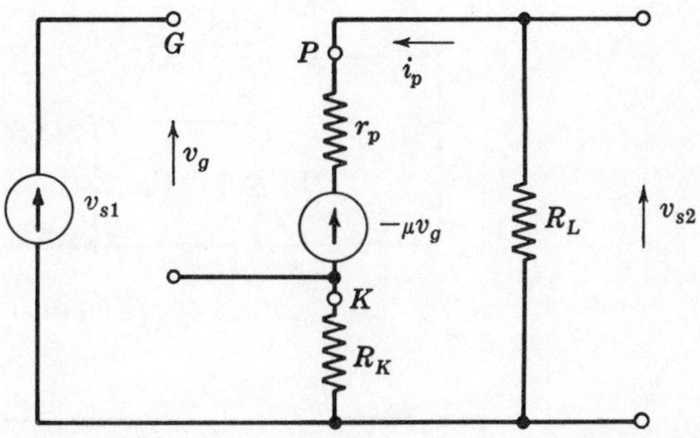

Fig. 5-20. Unbypassed Cathode Amplifier Circuit

$$-v_{s1} + v_g + R_K i_p = 0 \qquad (5.25)$$

$$R_L i_p + r_p i_p - \mu v_g + R_K i_p = 0 \qquad (5.26)$$

and solve for $i_p = \mu v_{s1}/[R_L + r_p + (1+\mu)R_K]$. Then $v_{s2} = -R_L i_p$ and

$$A_v = -\mu R_L/[R_L + r_p + (1+\mu)R_K] \qquad (5.27)$$

Note that when the cathode is unbypassed (capacitor omitted) the denominator of the voltage gain equation has the additional term $(1+\mu)R_K$, hence the gain is lower.

When a capacitor is in the cathode circuit, (*5.25*) would be written as

$$-v_{s1} + v_g + Z_K i_p = 0 \quad \text{or} \quad v_g = v_{s1} - Z_K i_p \tag{5.28}$$

The magnitude of the impedance Z_K is the highest at the lowest frequency, hence C_K should be chosen so that $Z_K i_p$ is very small compared to v_{sl} for the lowest frequency. This will be examined in more detail when overall frequency response is considered.

Using the values for the present example ($\mu = 20.1$, $r_p = 9{,}180\ \Omega$, $R_L = 30{,}000\ \Omega$, $R_K = 700\ \Omega$), we find $A_v = -11.2$. The voltage gain without the cathode capacitor has magnitude 11.2 as compared to 15.4 with the capacitor.

5.9 *RC* COUPLING

In the amplifier circuits of Fig. 5-9 and 5-16, the output signal voltage is the tube plate to ground voltage. Hence the output voltage includes an average value which is positive with respect to ground. If amplifier stages are cascaded, this creates problems in establishing the operating point for the following amplifier tube. A simple capacitor-resistor high pass filter is added to the amplifier circuit to eliminate the average value at the output as shown in Fig. 5-21. The coupling capacitor C_C charges so that the average voltage drop across it equals the average plate to ground voltage. The resistor across the output terminals is denoted R_G because this is frequently the grid resistance of the next amplifier stage. For the present, consider only signal frequencies which are high enough so that the impedance of the coupling capacitor is negligibly small. Then the equivalent circuit has the added resistor R_G as shown in Fig. 5-22. The grid loop equation is the same as equation (*5.22*). The plate nodal equation is now

$$+g_m v_g + g_p v_{s2} + G_L v_{s2} + G_G v_{s2} = 0 \tag{5.29}$$

Substitute $v_g = v_{s1}$ and solve for the voltage gain

$$A_v = \frac{-g_m}{g_p + G_L + G_G} \tag{5.30}$$

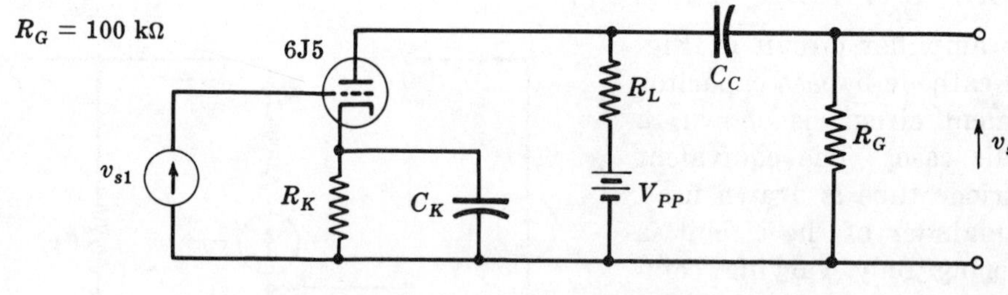

Fig. 5-21. Amplifier with Coupling Network

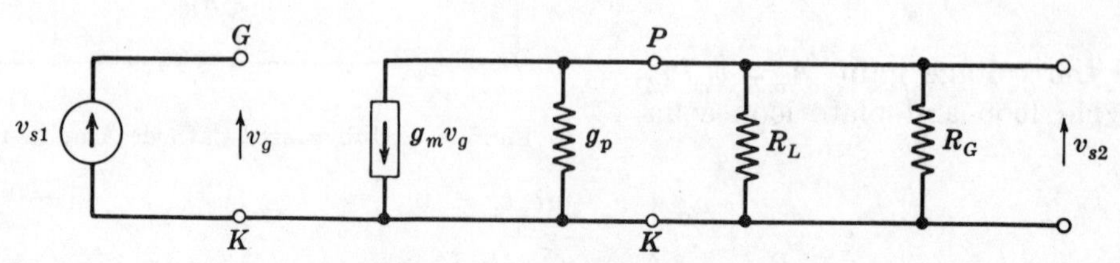

Fig. 5-22. Equivalent Circuit

From Fig. 5-21 and 5-22 it is seen that the effective resistance external to the tube plate is no longer R_L, but is the parallel combination of R_L and R_G. It is necessary to consider the "ac" load line as distinguished from the "dc" load line. The dc load line involves the resistance external to the tube in the plate loop considering only average values. This has R_L and R_K in series as shown before. However, for varying components (ac signals), the resistance external to the tube is the parallel combination of R_L and R_G (with the cathode bypass capacitor in the circuit). The ac load line is drawn through the operating point with slope $-(G_L + G_G)$. This is shown in Fig. 5-23 for the circuit of Fig. 5-21, using the same values for the circuit elements as before but adding $R_G = 100{,}000\ \Omega$. An easy way to draw the ac load line is to first draw a line with slope $-(G_L + G_G)$, and then draw the line through the operating point which is parallel to this line.

A dynamic transfer characteristic curve may now be drawn to include the effect of R_G in the circuit. To distinguish the two dynamic transfer characteristic curves, the one involving points along the dc load line is said to be based on the dc load line, while the one involving points along the ac load line is said to be based on the ac load line.

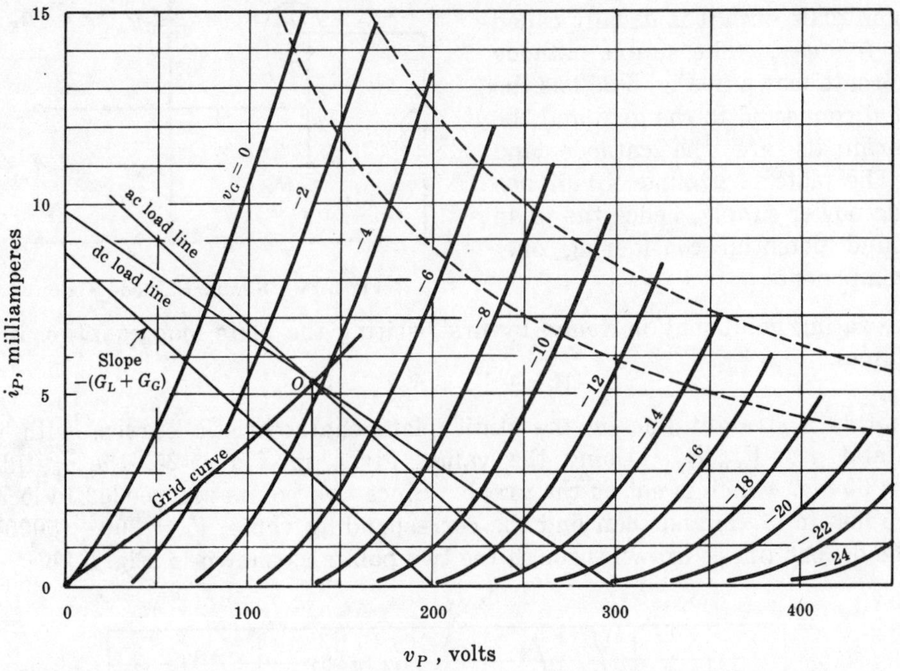

Fig. 5-23. 6J5 Static Plate Characteristics

The ac load line is used when varying signals are considered. The transfer characteristic curves for this example are shown in Fig. 5-24. The static transfer and the two dynamic transfer curves all pass through the operating point.

The linear operating range may be determined from the dynamic transfer characteristic curve based on the ac load line. The linear operating range is that range of v_G and i_P for which the dynamic transfer characteristic curve based on the ac load line is a straight line or approximately a straight line. For the present example, this curve is approximately a straight line for the range $-8 < v_G < 0$.

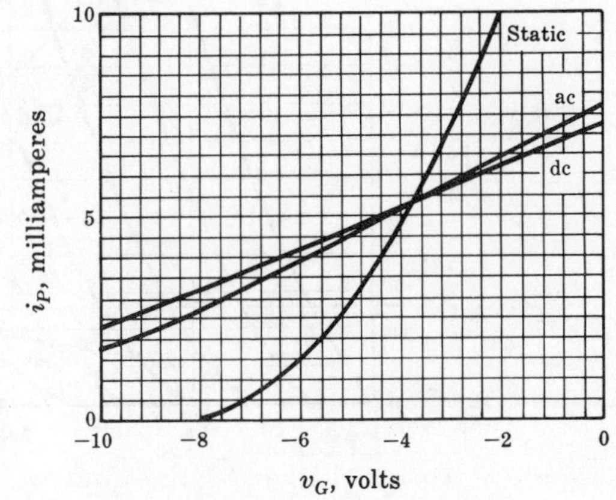

Fig. 5-24. Transfer Characteristic Curves

Since the amplifier circuit of Fig. 5-21 includes two capacitors, it is in general frequency dependent. Shunt and stray circuit capacitances have not been shown. In this chapter it is assumed that the *mid-band* range of frequencies is involved. This is the range of frequencies where the circuit parameters are all resistive, i.e. coupling and bypass capacitors have negligibly small impedances and shunt capacitances have negligibly large impedances. This will be discussed further in Chapter 7.

5.10 CATHODE FOLLOWER AMPLIFIER

The reference terminal for the vacuum tube is the cathode terminal, so the equivalent circuit has been developed with the cathode as the reference element. The simple amplifier circuits considered in previous sections have the cathode as the common terminal, with the input signal applied to the grid and the output signal taken from the plate. Circuits having the plate or grid as common elements are also frequently encountered. The common plate circuit is usually called a *cathode follower*. The simple cathode follower circuit shown in Fig. 5-25 has the input signal connected to the grid and the output terminals are the cathode and ground. The plate to ground circuit has only the dc power supply, hence the plate is at ground potential considering only varying components.

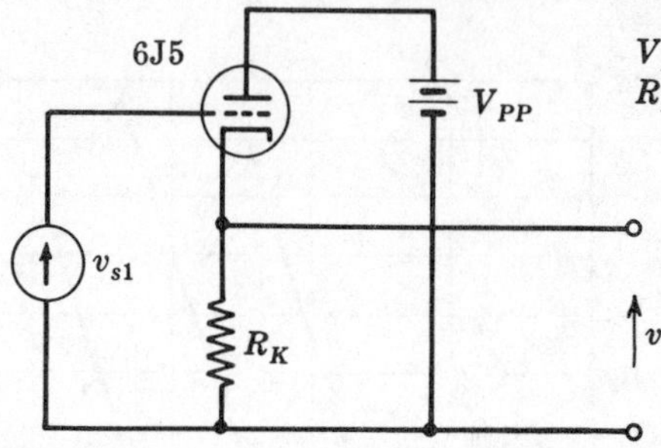

Fig. 5-25. Simple Cathode Follower Circuit

The operating point may be found by first writing the plate loop equation in terms of average values,

$$-V_{PP} + V_P + R_K I_P = 0 \tag{5.31}$$

which plots as a straight line on the static plate characteristic curves, with intercepts $v_P = V_{PP}$ and $i_P = V_{PP}/R_K$. Using the values given in Fig. 5-25, the i_P intercept is $300/5000 = 60$ ma, which is not on the graph. Since two points are needed to locate a line, let $I_P = 15$ ma in (*5.31*) and then find the corresponding value $V_P = 300 - (5000)(0.015) = 225$ v. The dc load line is drawn through the two points as shown in Fig. 5-26.

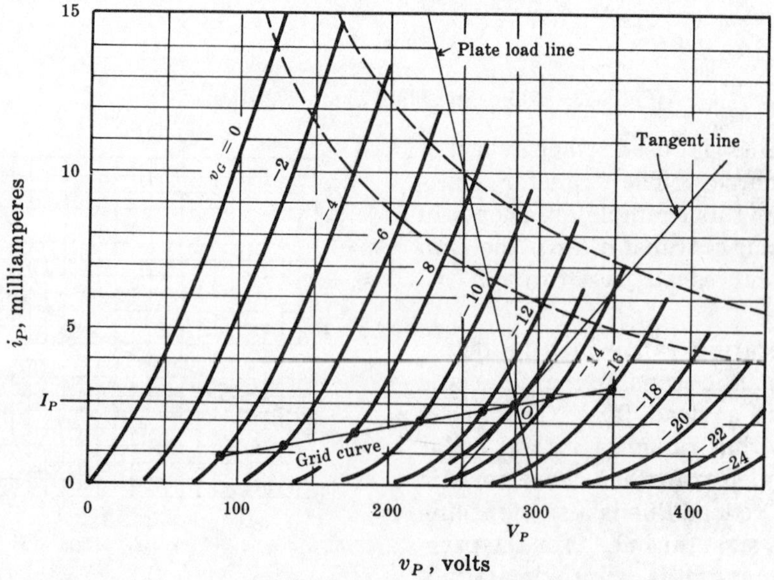

Fig. 5-26. 6J5 Static Plate Characteristics

The grid loop equation in terms of average values is

$$0 + V_G + R_K I_P = 0 \tag{5.32}$$

A sufficient number of points may be plotted and the curve representing the grid equation may be drawn on the plate characteristics as shown in Fig. 5-26. Values of v_G are chosen so that they fall on the $v_G =$ a constant curves, and then the corresponding values of i_P are calculated. The operating point is the intersection of the two curves. For this example, $V_P = 287$ v, $I_P = 2.6$ ma, and $V_G = -13$ v.

The parameters for the tube may be found by using the procedures previously outlined. Here the slopes of the two adjacent $v_G =$ a constant curves are not the same, so a $v_G = -13$ v curve is sketched by interpolating between the $v_G = -12$ v and $v_G = -14$ v curves as shown in Fig. 5-26. The slope of this curve is found at the operating point and is

$$g_p = 0.01/(413 - 240) = 57.8 \ \mu\text{mho}$$

The static transfer characteristic curve for this operating point is plotted in Fig. 5-27. The slope of the line tangent to the static transfer characteristic curve of Fig. 5-27 at the operating point is

$$g_m = 0.0064/[-10 - (-15.2)] = 1230 \ \mu\text{mho}$$

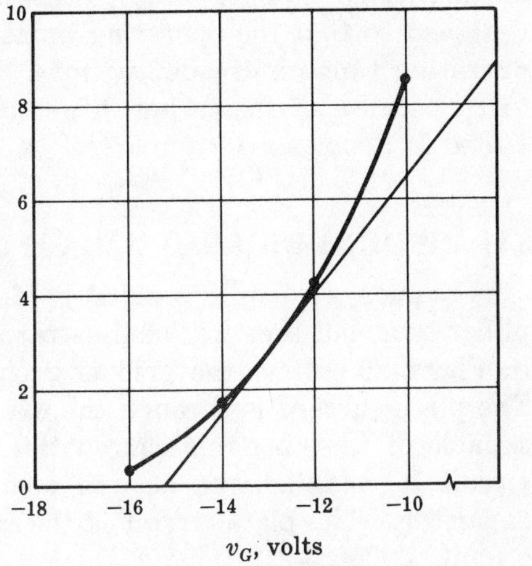

Fig. 5-27. Static Transfer Characteristic Curve

The equivalent circuit for the simple cathode follower circuit of Fig. 5-25 is shown in Fig. 5-28(*a*) with the tube indicated by a constant current generator representation. This circuit is redrawn in Fig. 5-28(*b*). The voltage gain for this circuit is found by writing the grid loop and cathode nodal equations

$$-v_{s1} + v_g + v_{s2} = 0 \tag{5.33}$$

$$-g_m v_g + g_p v_{s2} + G_K v_{s2} = 0 \tag{5.34}$$

and solving to obtain

$$A_v = v_{s2}/v_{s1} = g_m/(g_p + G_K + g_m) \tag{5.35}$$

For this example, $A_v = \dfrac{1230 \times 10^{-6}}{(58 + 200 + 1230) \times 10^{-6}} = 0.828.$

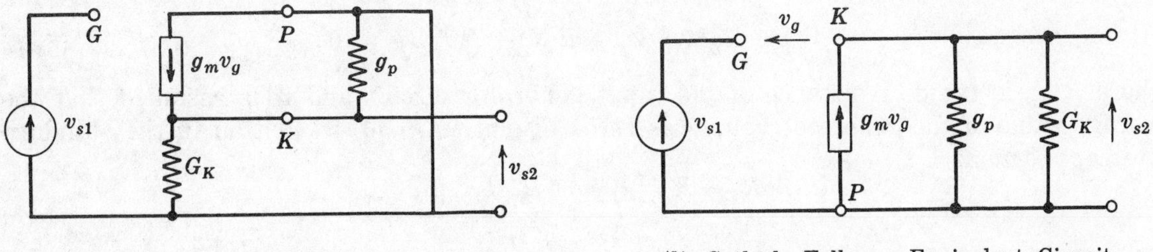

(*a*) Cathode Follower Equivalent Circuit (*b*) Cathode Follower Equivalent Circuit

Fig. 5-28

The voltage gain is positive, since there is no phase shift between the input and output signal voltages in a cathode follower. The output signal voltage is taken from the cathode terminal, so it is equal to $R_K i_p$. The varying component of the plate current is in phase with the grid voltage, so the output signal voltage is in phase with the input signal voltage. Since all quantities in equation (*5.35*) are positive, the voltage gain is always less than 1.

If g_m is very large compared to $g_p + G_k$, the voltage gain will approach 1. It is, however, possible to have a power gain (output signal power greater than input signal power) greater than 1. Even though the voltage gain is less than 1, there are many applications where cathode followers are useful because of the high input impedance and low output impedance. This will be discussed in Chapter 7.

The cathode follower circuit of Fig. 5-25 has the disadvantage that the triode is operated at a low average plate current. This is not a desirable operating point in terms of linear operating range or of the values of the tube parameters. Practical cathode followers are usually designed so that the operating point involves a higher average plate current. The linear operating range is frequently not a problem, since the actual value of v_g is small compared to v_{s1} because of the rather large voltage drop across the cathode resistor. This is illustrated in Problems 5.1 and 5.2.

5.11   GROUNDED GRID AMPLIFIER

Common grid or *grounded grid triode amplifier* circuits are often of the tuned amplifier type, but they will be illustrated by considering an untuned amplifier circuit as shown in Fig. 5-29. Here the grid is grounded and the input signal is applied to the cathode. The plate current is through the external generator, so the external input circuit must be capable of carrying this current (dc current path). A cathode resistor is included in the circuit to establish the desired operating point, and is shown bypassed with a cathode capacitor. The plate circuit is the same as that shown for the common cathode amplifier circuit of Fig. 5-21.

$R_L = 30\ \text{k}\Omega$
$R_G = 100\ \text{k}\Omega$
$R_K + R_{\text{GEN}} = 700$
$V_{PP} = 300\ \text{v}$

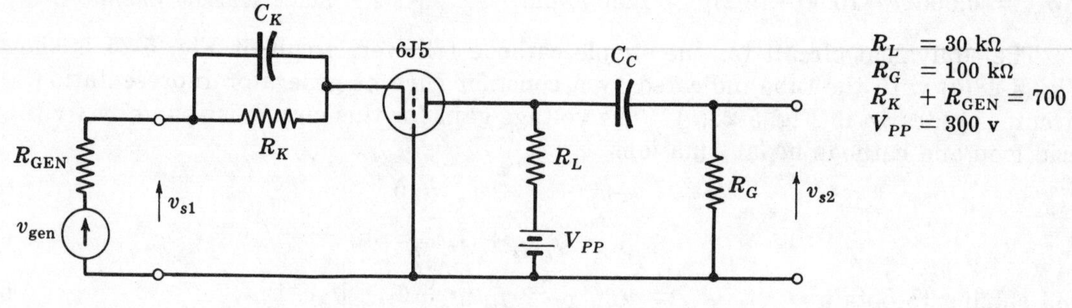

Fig. 5-29.   Common Grid Amplifier

The operating point for the triode is found by writing the plate-cathode loop equation in terms of average values,

$$-V_{PP} + R_L I_P + V_P + R_K I_P + R_{\text{GEN}} I_P = 0 \qquad (5.36)$$

where R_{GEN} is the dc resistance of the input generator circuit and it is assumed that the average value of the generator voltage is zero. Equation (5.36) is written in the standard intercept form as

$$\frac{R_L + R_K + R_{\text{GEN}}}{V_{PP}} I_P + \frac{1}{V_{PP}} V_P = +1 \qquad (5.37)$$

Entering the numerical values listed in Fig. 5-29,

$$\frac{30,700}{300} I_P + \frac{1}{300} V_P = +1 \qquad (5.38)$$

The grid-cathode loop equation in terms of average values is

$$V_G + R_K I_P + R_{\text{GEN}} I_P = 0 \qquad (5.39)$$

Substituting the numerical values,

$$V_G + 700I_P = 0 \tag{5.40}$$

Equations (*5.38*) and (*5.40*) were plotted on the 6J5 static plate characteristic curves of Fig. 5-17. The operating point for the triode in this circuit is the same as in the common cathode amplifier circuit, when the same values of circuit parameters are used and R_K is chosen so that $R_{\text{GEN}} + R_K$ has the same value as the cathode resistor in the common cathode amplifier. The parameters for the 6J5 are then the same as those previously determined for this operating point ($\mu = 20.1$ and $r_p = 9180\ \Omega$).

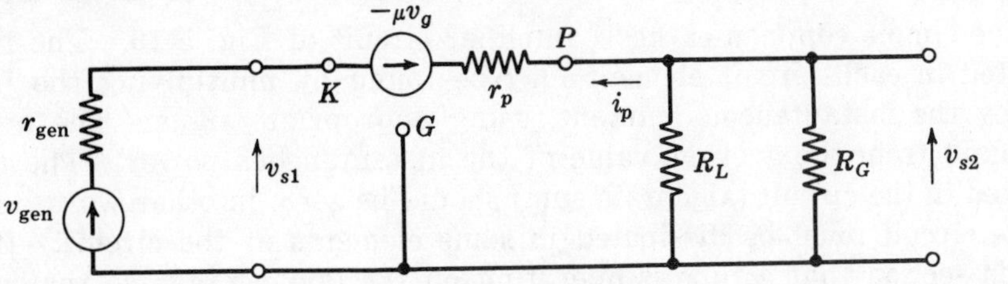

Fig. 5-30. Equivalent Circuit of the Common Grid Amplifier

The equivalent circuit is given in Fig. 5-30 assuming the signal frequency is high enough so that the impedance of each of the capacitors is negligibly small. The generator resistance is shown with lower case letters and lower case subscripts in Fig. 5-30 to emphasize that this is the effective "ac" resistance. The external generator resistance must be included in the representation because there is a non-zero varying current in the generator. In the common cathode amplifier and common plate amplifier, the generator current is zero when the grid is negative. Then it is not necessary to include the generator resistance, since the voltage drop is zero when the current is zero.

The equation for the signal voltage gain is to be derived. Let $R_L^* = R_L R_G/(R_L + R_G)$. Then the plate-cathode and grid-cathode loop equations are respectively

$$R_L^* i_p + r_p i_p - \mu v_g + r_{\text{gen}} i_p + v_{\text{gen}} = 0 \tag{5.41}$$

$$+v_g + r_{\text{gen}} i_p + v_{\text{gen}} = 0 \tag{5.42}$$

from which

$$i_p = \frac{-(\mu+1)v_{\text{gen}}}{R_L^* + r_p + (\mu+1)r_{\text{gen}}} \tag{5.43}$$

and the output signal voltage $v_{s2} = -R_L^* i_p$. The open circuit voltage gain is

$$A_v = \frac{v_{s2}}{v_{\text{gen}}} = \frac{(\mu+1)R_L^*}{R_L^* + r_p + (\mu+1)r_{\text{gen}}} \tag{5.44}$$

Using the numerical values given in Fig. 5-29 and assuming $r_{\text{gen}} = 600\ \Omega$, the voltage gain of this particular amplifier is $A_v = +10.8$.

The gain expression has a plus sign, indicating there is no phase shift between the input and output signal voltages. The phase relationship between v_g and v_p is unchanged, but in the common grid amplifier the external generator is connected in the grid-cathode circuit with opposite polarity. This is seen by comparing the circuits of Fig. 5-29 and Fig. 5-21.

In this example the voltage gain was specified as the ratio of the output signal voltage to the generator signal voltage v_{gen}. In applications where an external signal source is connected to the input terminals, the input terminal voltage is usually measured. The voltage gain is then specified in terms of the terminal voltage rather than the generator voltage. It is important that the problem be correctly specified when output impedances are calculated in later sections.

5.12 POWER DISSIPATION

Representations have been developed for the vacuum tube triode operating on the linear range considering only varying components. Procedures have been outlined for determining the operating points and the parameters of the vacuum tube at a given operating point. Another consideration is the power dissipated in each circuit element of the amplifier circuit. In a resistive circuit element, the product of the instantaneous voltage drop and instantaneous current in the direction of the voltage drop is the instantaneous power dissipated and is positive. This convention will be followed so that if the power in an element is negative, it means this element is supplying power to the circuit.

Consider the simple common cathode amplifier circuit of Fig. 5-16. The instantaneous power dissipated in each circuit element may be found by multiplying the instantaneous voltage drop by the instantaneous current, using appropriate signs. The average power can then be found from the average value of the instantaneous power. The total average power dissipated in the circuit (algebraic sum) should be zero; in other words, all the power supplied to the circuit must be dissipated in some elements of the circuit. It is assumed throughout this section that a tube is operating on the linear range so that superposition may be applied. The instantaneous power dissipated in the plate power supply is

$$p_{PS} = V_{PP}(-i_P) = -V_{PP}i_P = -V_{PP}I_P - V_{PP}i_p \tag{5.45}$$

If the varying component of i_P is sinusoidal, the average value of i_p is zero and the average power dissipated in the plate power supply is

$$P_{PS} = -V_{PP}I_P \tag{5.46}$$

The instantaneous power dissipated in the plate load resistor R_L is

$$p_{R_L} = R_L i_P^2 = R_L(I_P + i_p)^2 = R_L I_P^2 + 2R_L I_P i_p + R_L i_p^2 \tag{5.47}$$

Assuming i_p is sinusoidal and may be written as $i_p = I_o \sin \omega t$, the average power in R_L for an integer number of cycles is

$$P_{R_L} = \frac{1}{2\pi/\omega}\int_0^{2\pi/\omega} (R_L I_P^2 + 2R_L I_P I_o \sin \omega t + R_L I_o^2 \sin^2 \omega t)\, dt \tag{5.48}$$

Substituting $\sin^2 \omega t = \frac{1}{2} - \frac{1}{2}\cos 2\omega t$ and integrating, the average power is

$$P_{R_L} = R_L I_P^2 + R_L(I_o/\sqrt{2})^2 \tag{5.49}$$

The first term is proportional to the square of the average value of the plate current and the second term is proportional to the square of the rms value of the varying component of the plate current.

It was shown earlier that the varying component of the plate current i_p is in phase with the grid signal voltage v_g and that v_p is π radians out of phase with v_g and i_p. Assuming $i_p = I_o \sin \omega t$, write $v_p = V_o \sin(\omega t + \pi) = -V_o \sin \omega t$. The instantaneous plate power dissipated in the tube is

$$p_P = v_P i_P = (V_P - V_o \sin \omega t)(I_P + I_o \sin \omega t) \tag{5.50}$$

The average plate power is

$$P_P = V_P I_P - V_o I_o/2 = V_P I_P - (V_o/\sqrt{2})(I_o/\sqrt{2}) \tag{5.51}$$

which involves the product of the average voltage and average current minus the product of the rms value of the signal voltage and rms value of the signal current. This is significant because the actual power dissipated in the tube is decreased as the signal power output is increased. In low power tubes, the plate dissipation is usually specified as $V_P I_P$ which is the maximum dissipation occurring when there is no signal. Since $V_P I_P$ = a constant for a given tube, this may be plotted as a hyperbola on the static plate characteristic curves. The operating point should lie below and to the left of this hyperbola.

When the signal frequency is high so that the varying component of the cathode to ground voltage is zero, the varying component of the current in R_K is zero, and thus the instantaneous power dissipated in R_K is equal to the average power:

$$P_{R_K} = R_K i_P^2 = R_K I_P^2 \quad (5.52)$$

The instantaneous power dissipated in all elements is positive except for the power supply. Hence all the power is supplied by the plate power supply. However, considering only varying components (signals), the vacuum tube supplies an average signal power as shown by equation (*5.51*). This is consistent with the representation developed for the triode which is a generator circuit when only varying components are considered. It emphasizes the fact that the vacuum tube triode acts as a generator considering only varying components, but all the power is supplied from the plate power supply. However, because of the control action of the triode, the triode appears as a power source when only varying components are considered.

If superposition can be applied, the average power dissipated in each element may be calculated from the equivalent circuit considering average values only and the one considering varying components only. The equivalent circuit for the circuit of Fig. 5-16 considering average values only (dc), is given in Fig. 5-31. The instantaneous power in each element of Fig. 5-31 is the same as the average power, since only average values are considered. The average power dissipated in each element of Fig. 5-31 may be calculated,

$$P_{PS} = -V_{PP} I_P, \quad P_{R_L} = R_L I_P^2, \quad P_P = V_P I_P, \quad P_{R_K} = R_K I_P^2$$

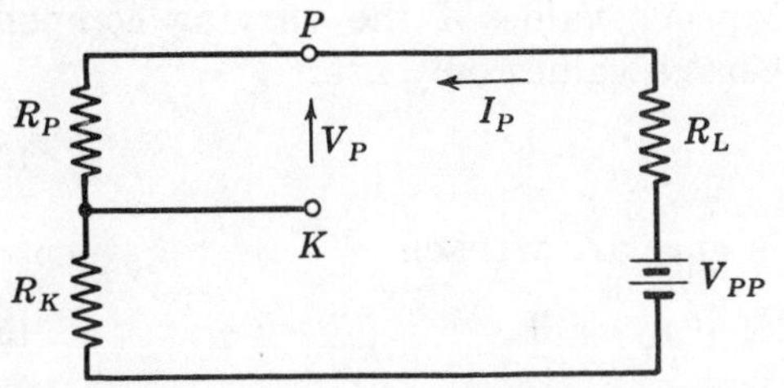

Fig. 5-31. dc Equivalent Circuit

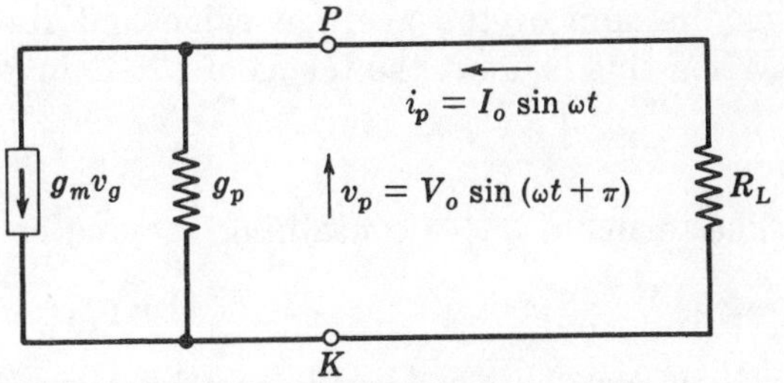

Fig. 5-32. ac Equivalent Circuit

The equivalent circuit for the circuit of Fig. 5-16 considering only varying components (ac) is displayed in Fig. 5-32. The average power dissipated in each element of the circuit of Fig. 5-32 may be calculated,

$$P_{R_L} = R_L (I_o/\sqrt{2})^2 \quad \text{and} \quad P_P = -(V_o/\sqrt{2})(I_o/\sqrt{2}) \quad (5.53)$$

The average power for each element determined for each of the equivalent circuits may be added, giving the total average power previously calculated. In general, the instantaneous power cannot be found in this manner.

5.13 VACUUM TUBE TETRODES

A vacuum tube tetrode is a four element device which has two grids as shown in Fig. 5-33 below. In vacuum tubes which have more than one grid, the grids are numbered consecutively beginning with the grid nearest the cathode. This is grid No. 1 or the control grid. Grid No. 2 is often called the screen grid. In most circuits with a common cathode, the input signal is applied to the control grid and the output signal is taken from the plate terminal. The control grid is usually negative with respect to the cathode, so the input is represented as an infinite resistance as in the triode. The screen grid is usually positive with respect to the cathode, so the screen grid current is not zero.

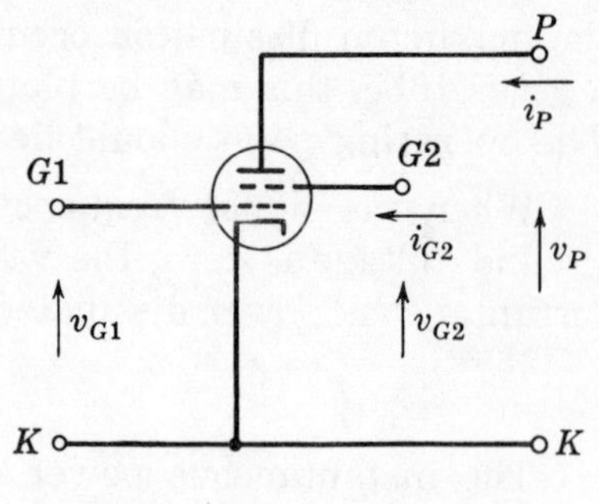

Fig. 5-33. Vacuum Tube Tetrode

To develop a representation for the tetrode looking into the output port, which is the plate and cathode terminal pair, the functional relationship is written

$$i_P = f(v_{G1}, v_{G2}, v_P) \tag{5.54}$$

which assumes the plate current is independent of cathode temperature. Since there are three independent variables, take the total differential of i_P:

$$di_P = \frac{\partial i_P}{\partial v_{G1}} dv_{G1} + \frac{\partial i_P}{\partial v_{G2}} dv_{G2} + \frac{\partial i_P}{\partial v_P} dv_P \tag{5.55}$$

Let $g_{m1} = \partial i_P / \partial v_{G1}$, $g_{m2} = \partial i_P / \partial v_{G2}$, and $g_p = \partial i_P / \partial v_P$ and write (5.55) as

$$di_P = g_{m1}\, dv_{G1} + g_{m2}\, dv_{G2} + g_p\, dv_P \tag{5.56}$$

On the linear operating range, g_{m1}, g_{m2} and g_p are constant; hence

$$i_P = g_{m1} v_{G1} + g_{m2} v_{G2} + g_p v_P + k \tag{5.57}$$

where k is a combination of all integration constants.

Since operation is linear, superposition is applied and an instantaneous value written as the sum of the average value and the instantaneous value of the varying component. When this is done, the terms of (5.57) involving average values only are

$$I_P = g_{m1} V_{G1} + g_{m2} V_{G2} + g_p V_P + k \tag{5.58}$$

The terms of (5.57) containing varying components only are written

$$-i_p + g_{m1} v_{g1} + g_{m2} v_{g2} + g_p v_p = 0 \tag{5.59}$$

which may be considered term by term, giving the equivalent circuit of Fig. 5-34. This equivalent circuit has two controlled generators but otherwise is the same as that developed for the triode. In most applications the voltage from the screen grid to the cathode is constant, hence the varying component v_{g2} is zero. Then the second generator magnitude is zero even though g_{m2} is not zero; thus the equivalent circuit reduces to that developed for the triode. The gain formulas derived for triode circuits can be applied to tetrode circuits when the screen grid to cathode voltage is constant.

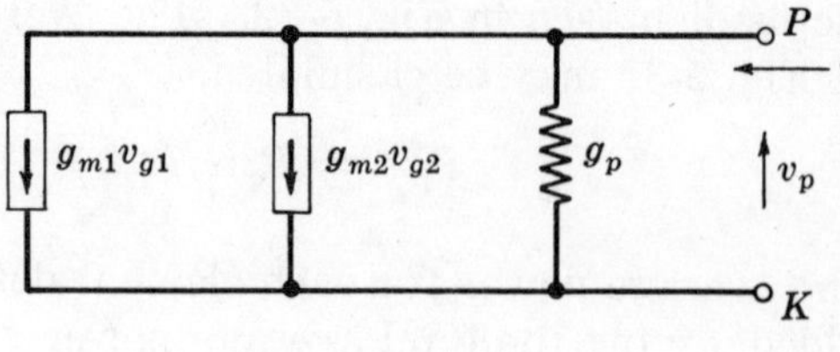

Fig. 5-34. Equivalent Circuit for a Tetrode

In the vacuum tube tetrode, the plate current is a function of three variables as given in equation (5.54). A geometrical representation for this functional relationship involves four dimensions. This problem is usually solved by choosing a particular value for the screen grid voltage v_{G2} and then assuming v_{G2} constant. For this one value $v_{G2} = V_{G2}$, the plate current is a function of only two variables:

$$i_P = f(v_{G1}, v_P) \quad \text{for } v_{G2} = V_{G2}, \text{ a constant} \tag{5.60}$$

This functional relationship may be approximated by a family of static plate characteristic curves which are plotted in the same manner as was done for the triode. The family of tetrode plate curves applies only for the given value of screen grid voltage v_{G2}. In some cases, more than one family of static plate characteristic curves is plotted; each family is graphed for a different value of v_{G2}.

When v_{G2} is a constant, the operating point for the tetrode may be found using the procedures developed for the triode. The plate and grid loop equations are written in terms of average values and then plotted. There is one added problem in a circuit with cathode bias. The cathode current includes the screen grid current i_{G2} which must be estimated from given characteristic curves or from data supplied. The parameters of the tetrode may be determined for a given operating point by following the procedures outlined for the triode. There is usually not enough information given for determining g_{m2}, and it is not needed because it may usually be assumed that v_{G2} is a constant. See Problem 5.6.

In general, vacuum tube tetrodes are not employed as frequently as other types of vacuum tubes. For many applications the vacuum tube pentodes have all the desirable features of tetrodes without several of the disadvantages. For this reason more consideration will be given to vacuum tube pentodes. One of the disadvantages of triodes is that they have relatively high plate to grid capacitance. This was reduced in tetrodes by adding the screen grid, but tetrodes have considerable "secondary emission" from the plate element, especially for plate voltages below the screen grid voltage. The secondary emission problem was solved by adding another grid near to the plate, making a total of five elements, so these tubes are called pentodes.

5.14 VACUUM TUBE PENTODES

A vacuum tube pentode is a five element control device which has three grids as shown in Fig. 5-35. Grid 1 is the control grid, grid 2 is the screen grid and grid 3 is the suppressor grid. The suppressor grid is usually connected to the cathode or to the common ground; thus the suppressor grid current is negligibly small. In most applications the control grid is negative, hence the control grid current is also negligible. The screen grid is positive, so the screen grid current is appreciable in most applications.

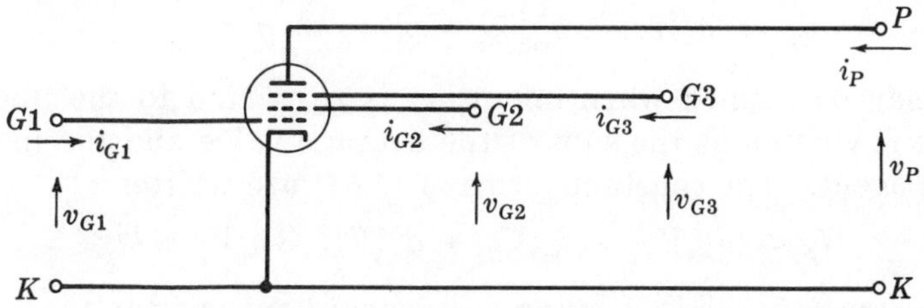

Fig. 5-35. Vacuum Tube Pentode

The output port of the pentode is usually the plate and cathode terminal pair. The functional relationship for the output port is

$$i_P = f(v_{G1}, v_{G2}, v_{G3}, v_P) \tag{5.61}$$

To represent this geometrically requires five dimensions. For the pentode, the family of static plate characteristic curves is usually plotted for $v_{G3} = 0$ and $v_{G2} = V_{G2}$ (a constant) as shown in Fig. 5-36 below. A circuit model for the output port can be developed in a manner similar to that outlined for tetrodes. The total differential of i_P is

$$di_P = \frac{\partial i_P}{\partial v_{G1}} dv_{G1} + \frac{\partial i_P}{\partial v_{G2}} dv_{G2} + \frac{\partial i_P}{\partial v_{G3}} dv_{G3} + \frac{\partial i_P}{\partial v_P} dv_P \tag{5.62}$$

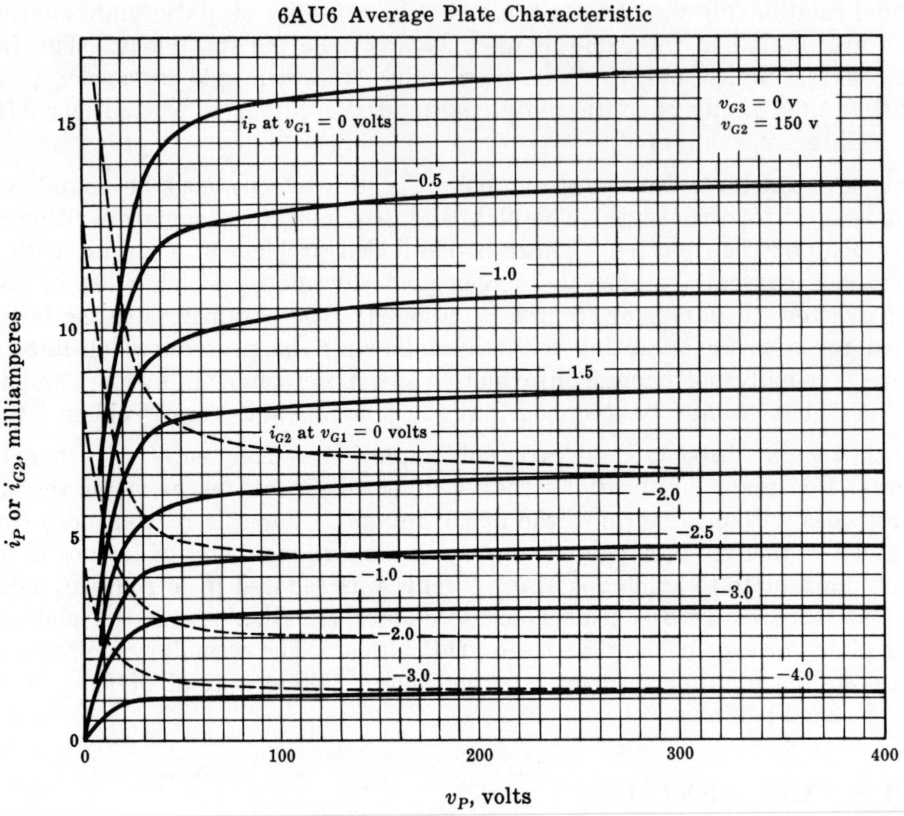

Fig. 5-36. Pentode Plate Characteristics

Substituting $g_{m1} = \partial i_P/\partial v_{G1}$, $g_{m2} = \partial i_P/\partial v_{G2}$, $g_{m3} = \partial i_P/\partial v_{G3}$ and $g_p = \partial i_P/\partial v_P$,

$$di_P = g_{m1}\,dv_{G1} + g_{m2}\,dv_{G2} + g_{m3}\,dv_{G3} + g_p\,dv_P \qquad (5.63)$$

On the linear operating range, g_{m1}, g_{m2}, g_{m3} and g_p are constant; then

$$i_P = g_{m1}v_{G1} + g_{m2}v_{G2} + g_{m3}v_{G3} + g_pv_P + k \qquad (5.64)$$

Superposition may be applied when operation is restricted to the linear range. Each instantaneous value is written as the sum of the average value and the instantaneous value of the varying component. The constant terms of (*5.64*) are written

$$I_P = g_{m1}V_{G1} + g_{m2}V_{G2} + g_{m3}V_{G3} + g_pV_P + k \qquad (5.65)$$

The terms of (*5.64*) containing only varying components are written

$$-i_p + g_{m1}v_{g1} + g_{m2}v_{g2} + g_{m3}v_{g3} + g_pv_p = 0 \qquad (5.66)$$

An equivalent circuit representation for (*5.66*) is constructed by considering the equation term by term. There are three terms in (*5.66*) which involve controlled generators as shown in Fig. 5-37. This equivalent circuit has one more controlled generator than the tetrode equivalent circuit.

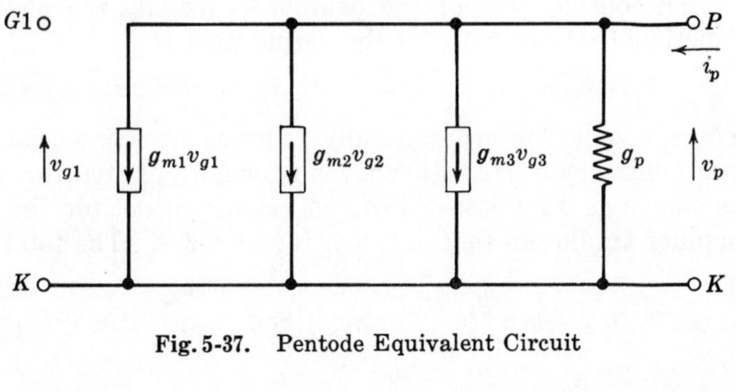

Fig. 5-37. Pentode Equivalent Circuit

The input port of the pentode is usually the grid and cathode terminal pair. The control grid G_1 is usually negative with respect to the cathode, so the control grid current is negligible. Hence the input port is represented as an infinite resistance or an open circuit as in Fig. 5-37.

In most applications involving pentodes, the suppressor grid voltage is either zero or a constant and the screen grid voltage is a constant. Hence the varying components v_{g2} and v_{g3} are zero, so that the magnitude of each of the two controlled generators is zero even though g_{m2} and g_{m3} are not zero. In these cases the equivalent circuit for pentodes reduces to the equivalent circuit with one controlled generator, as for triodes. Then the gain formula derived for triode circuits may be applied to pentode circuits; however, the range of values for pentode parameters is different from the range for triode parameters.

A simple pentode amplifier circuit is shown in Fig. 5-38. The operating point for the pentode is found by using procedures similar to those followed for triode amplifier circuits.

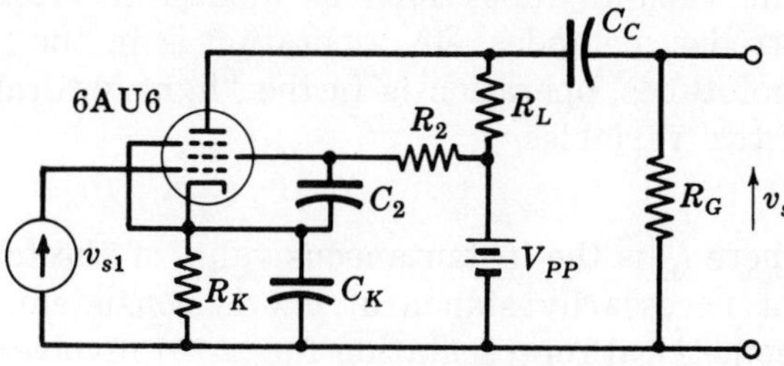

Fig. 5-38. Pentode Amplifier

The screen grid voltage v_{G2} must be the value used in plotting the family of static plate characteristic curves; otherwise, the family of plate characteristic curves is just an approximation. This is not as serious a problem as it might appear, because the curves used are usually averages for a number of tubes. It is only in special applications that curves are plotted for given tubes and then employed for calculations.

The plate loop equation based on average values is

$$-V_{PP} + R_L I_P + V_P + R_K(I_P + I_{G2}) = 0 \tag{5.67}$$

In pentodes the cathode current is the sum of the screen grid current and plate current. Unless $I_{G2} < I_P$, the screen grid current must be determined or estimated. In most problems the desired operating point is known approximately, so the value of R_K is chosen to give an operating point near the desired value. The screen grid current can be estimated from i_{G2} curves which are sometimes plotted on the family of static plate characteristic curves. If i_{G2} is assumed constant, then (5.67) may be written

$$(R_L + R_K)I_P + V_P = V_{PP} - R_K I_{G2} \tag{5.68}$$

which is the equation of the dc load line. The term $R_K I_{G2}$ is usually small compared to V_{PP} and may then be neglected. Assuming the average value of the external generator v_{s1} is zero, the control grid loop equation based on average values is

$$+V_{G1} + R_K(I_P + I_{G2}) = 0 \tag{5.69}$$

Unless $I_{G2} < I_P$, equation (5.69) must include the screen grid current I_{G2}.

The screen grid voltage V_{G2} is usually lower than the plate supply voltage V_{PP}. Since the screen grid current is not zero, the screen grid voltage may be obtained from the plate supply by a series resistance as shown in Fig. 5-38. The screen grid current will be a function of the control grid voltage v_{G1}, hence the screen grid voltage will not have a negligibly small varying component unless a capacitor C_2 is added. The combination of the resistor R_2 and the capacitor C_2 form a low pass filter (see Chapter 7). It may be noted qualitatively that the voltage across the capacitor terminals will tend to remain constant. It is necessary to choose the value of C_2 large enough so that the varying component of the screen grid voltage is negligible at the lowest signal frequency.

5.15 PHOTOTUBES

Phototubes are another example of two port electronic control devices of the resistive class in which the input port is an infinite resistance. The symbols for the vacuum and gas phototubes are shown in Fig. 5-39. Phototubes have only two elements, a plate and a cathode. Phototubes resemble diodes physically, but their electrical characteristics resemble those of triodes. In nearly all applications involving thermionic vacuum tubes such as diodes, triodes, tetrodes, pentodes, etc., operation is in the "space-charge limited region". In the case of phototubes, operation is in the "light saturation region". The plate current is a function of two variables,

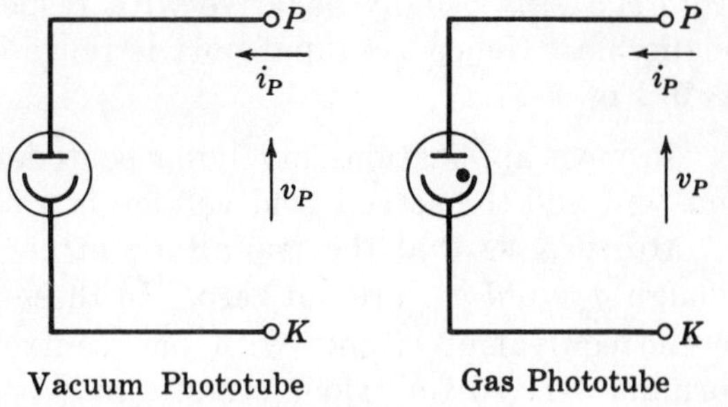

Fig. 5-39

$$i_P = f(v_P, l_K) \tag{5.70}$$

where l_K is the instantaneous value of the total light flux on the cathode. This notation is not necessarily standard, but is consistent with the notation adopted in this book. A geometrical representation for (*5.70*) involves a surface, so families of characteristic curves are plotted to represent the surface. A family of static plate characteristic curves for a vacuum phototube is shown in Fig. 5-40. The unit of light flux is the lumen. It may also be noted that the plate current is given in microamperes because phototubes involve much lower values of current than conventional thermionic vacuum tubes.

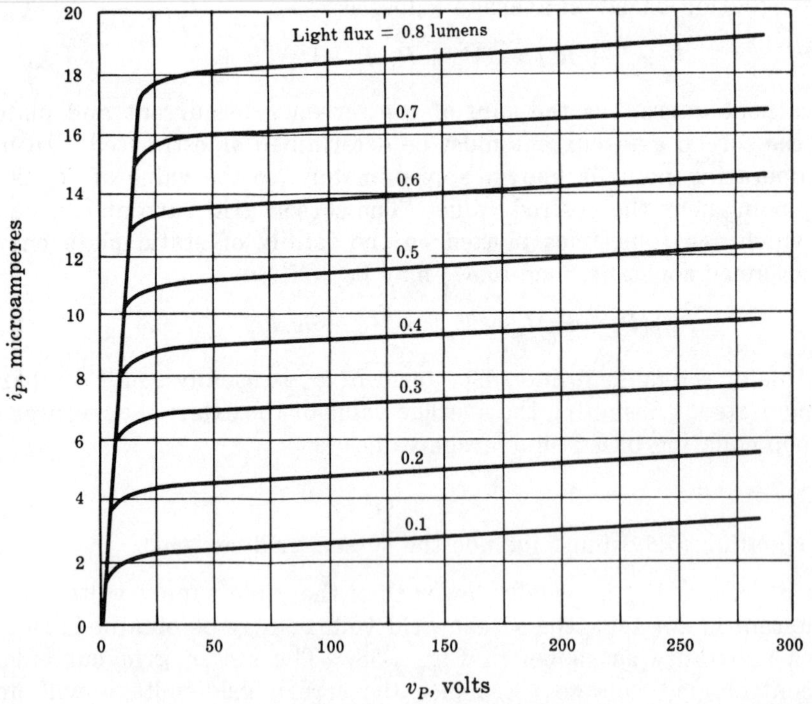

Fig. 5-40. Static Plate Characteristics of a Vacuum Phototube

A representation for a phototube may be developed in a manner similar to that used for the thermionic vacuum tube triode. Expanding equation (*5.70*) by Taylor's series, assuming linear operation so only the first three terms are shown,

$$i_P \;=\; f(V_P, L_K) + (v_P - V_P)\frac{\partial i_P}{\partial v_P} + (l_K - L_K)\frac{\partial i_P}{\partial l_K} \tag{5.71}$$

Substituting the average value plus the instantaneous value of the varying component for the instantaneous value, we obtain

$$i_P \;=\; I_P + i_p \;=\; f(V_P, L_K) + v_p\frac{\partial i_P}{\partial v_P} + l_k\frac{\partial i_P}{\partial l_K} \tag{5.72}$$

The average values are again determined graphically. The terms containing only varying components are written as

$$-i_p + \frac{\partial i_P}{\partial v_P}v_p + \frac{\partial i_P}{\partial l_K}l_k \;=\; 0 \tag{5.73}$$

The coefficient of v_p is $\partial i_P/\partial v_P$ which is by definition the differential plate conductance g_p. The coefficient of l_k is $\partial i_P/\partial l_K$ which is the derivative of the plate current with respect to the instantaneous light flux on the cathode. This is sometimes called the "luminous sensitivity", but there is no standard notation such as obtains for transfer resistance or transfer conductance. For this reason, no letter symbol will be used for this transfer characteristic. The equivalent circuit for a phototube considering only varying components is shown in Fig. 5-41.

The concept of port may be generalized so that the input port of the phototube is considered to be the cathode as a light receiver. The input signal is the instantaneous light flux incident on the cathode of the phototube. It is assumed that the instantaneous light flux on the cathode is independent of the plate voltage and plate current and that the light source is not "loaded" by the phototube. Hence the input port of the phototube is represented as an infinite resistance.

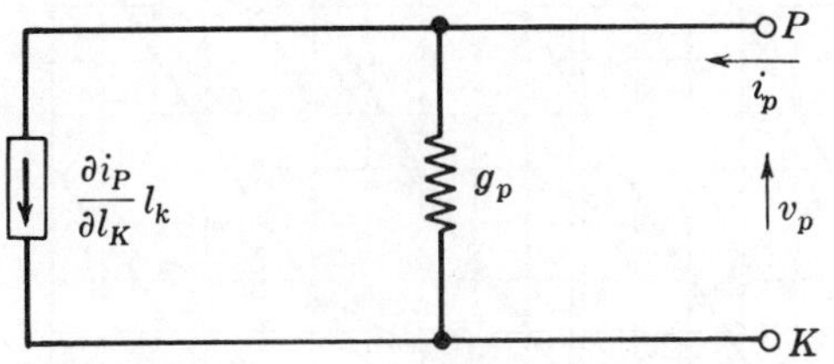

Fig. 5-41. Phototube Equivalent Circuit

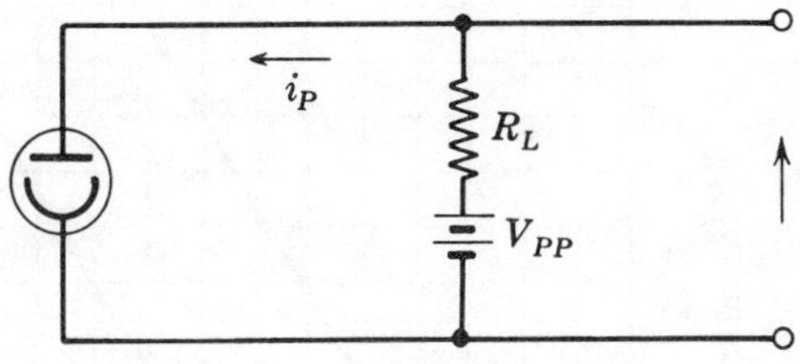

Fig. 5-42. Simple Phototube Circuit

A simple vacuum phototube circuit is given in Fig. 5-42. The plate loop equation in terms of instantaneous values is

$$-V_{PP} + R_L i_P + v_P \;=\; 0 \tag{5.74}$$

which is the equation of a straight line (load line) that may be plotted on the static plate characteristics for the phototube. See Problem 5.43.

The operating point is determined by the average value of the light flux L_K incident on the cathode. After the operating point is found, a static transfer characteristic curve may be drawn for the given operating point. The slope of the static transfer characteristic curve at the operating point is the transfer characteristic $\partial i_P/\partial l_K$. In applications where linear operation is involved, the equivalent circuit may be used.

Frequently, phototubes are employed where switching is involved, and thus operation is not linear. In such cases the linear equivalent circuit is usually not very useful.

Solved Problems

5.1. The cathode follower circuit of Fig. 5-25, page 84, has been changed to that of Fig, 5-43 which allows the tube to be biased for a more favorable operating point. Choose the resistor R_1 so that V_G is approximately -6 volts. (*a*) Find V_P and I_P for this value of R_1. (*b*) Determine g_m and g_p graphically.

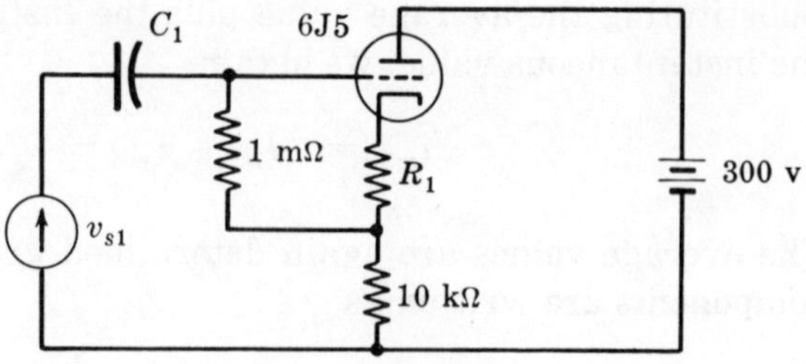

Fig. 5-43

Since $V_G = -6$ volts,. the voltage drop across R_1 must be $+6$ volts in the direction of i_P. The current in the 1 megohm grid resistor is negligible because the grid is negative. First, determine the dc plate load line by writing the plate loop equation in terms of average values:

$$-300 + V_P + 6 + 10{,}000 I_P = 0 \quad \text{or} \quad V_P + 10{,}000 I_P = 294 \qquad (5.75)$$

which may be plotted using the v_P intercept 294 and the point $i_P = 20$ ma. The corresponding value of v_P for $i_P = 20$ ma is found by substituting into (*5.75*):

$$v_P = 294 - (10{,}000)(0.02) = 94 \text{ volts}$$

The dc plate load line is now plotted as in Fig. 5-44. The intersection of the $v_G = -6$ volts plate characteristic curve and the dc load line is the operating point for these conditions. From the graph, $I_P = 9.6$ ma and $V_P = 198$ volts.

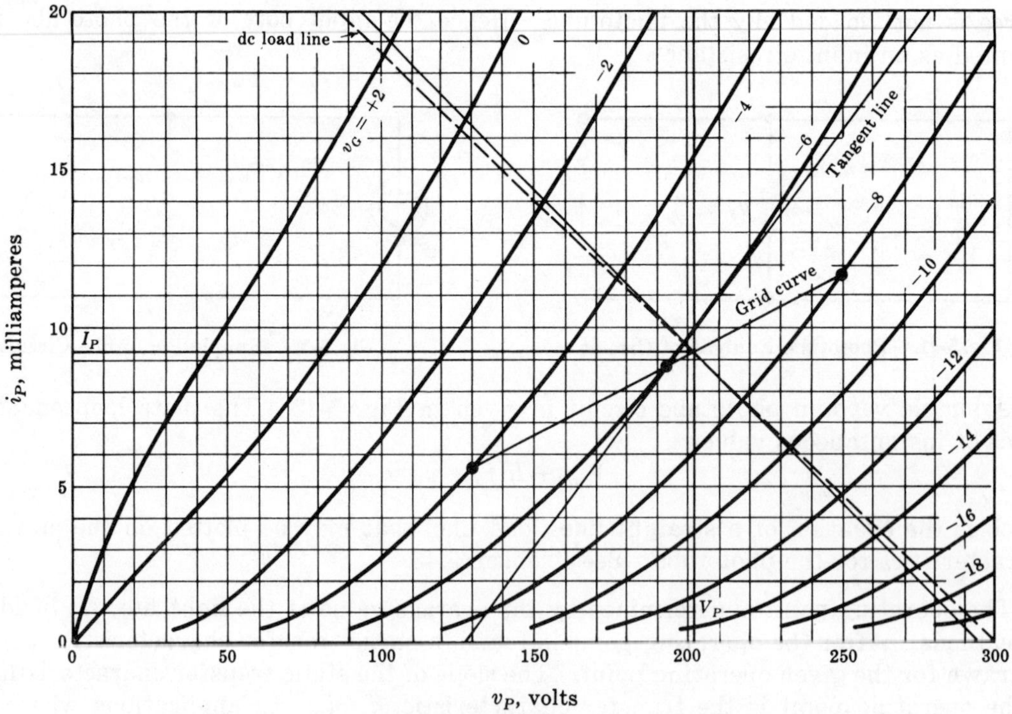

Fig. 5-44. 6C4 Static Plate Characteristics

R_1 may now be calculated knowing the current in R_1 and the voltage drop: $R_1 = 6/0.0096 = 625\ \Omega$. Since this is not a standard value, let $R_1 = 680\ \Omega$. The dc load line using this value of R_1 will not be appreciably different near the operating point, as may be shown by considering the plate loop equation using $R_1 = 680\ \Omega$.

$$V_P + 10{,}680 I_P = 300$$

Two points on the dc load line are $i_P = 0$, $v_P = 300$; and for $i_P = 20$ ma,

$$v_P = 300 - (10{,}680)(0.02) = 86.4 \text{ volts}$$

This dc load line is shown as a dashed line in Fig. 5-44. The grid loop equation based on average values is

$$0 + V_G + 680 I_P = 0$$

Plot the grid curve point by point on the static plate characteristic curves of Fig. 5-44. For $v_G = -6$ v, $I_P = 6/680 = 8.83$ ma; for $v_G = -8$ v, $I_P = 8/680 = 11.77$ ma; for $v_G = -4$ v, $I_P = 4/680 = 5.88$ ma. The curve through these three points is sufficiently close to a straight line, so a line may be drawn through these points, giving a portion of the grid curve. The operating point is the intersection of the grid curve and the dc load line shown as a dashed line in Fig. 5-44.

(a) The final values are: $V_G = -6.3$ volts, $I_P = 9.3$ ma, $V_P = 203$ volts.

(b) To find g_p, take the slope of the nearest v_G = a constant curve, which is $v_G = -6$ volts. The slope of the line tangent to the plate characteristic curve is $g_p = (0.02 - 0)/(279 - 127) = 131\ \mu$mho.

To find g_m, plot the static transfer characteristic curve and then find the slope at the operating point as shown in Fig. 5-45: $g_m = 0.015/[-3.4 - (-10.6)] = 2080\ \mu$mho.

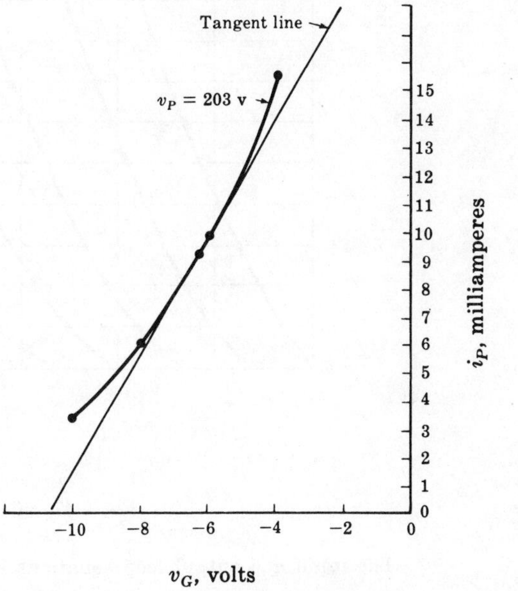

Fig. 5-45. Static Transfer Characteristic Curve

5.2. Given the cathode follower amplifier circuit of Fig. 5-46 and the 6J5 static plate characteristics of Fig. 5-47. Solve this circuit graphically for the open circuit transfer characteristics in terms of instantaneous values.

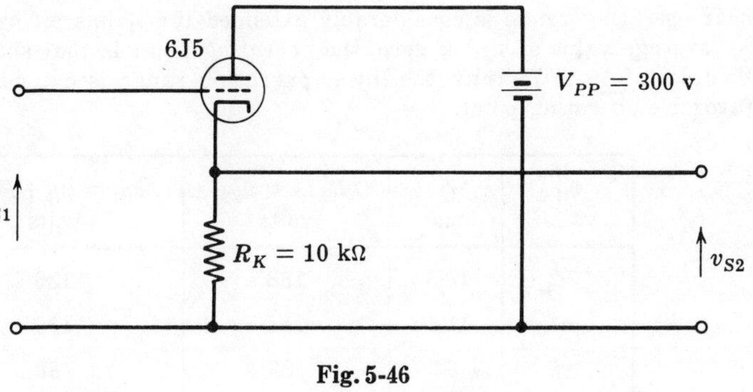

Fig. 5-46

The plate loop equation in terms of instantaneous voltages and currents is

$$-V_{PP} + v_P + R_K i_P = 0$$

Substituting numerical values,

$$(1/300)v_P + (10{,}000/300)i_P = +1$$

which is plotted on the static plate characteristics as in Fig. 5-47, using the v_P intercept and the $i_P = 15$ ma point. The corresponding value of v_P is $v_P = 300 - (10{,}000)(0.015) = 150$ volts. Any pair of instantaneous values (i_P, v_P) must lie on the load line graphed in Fig. 5-47 below.

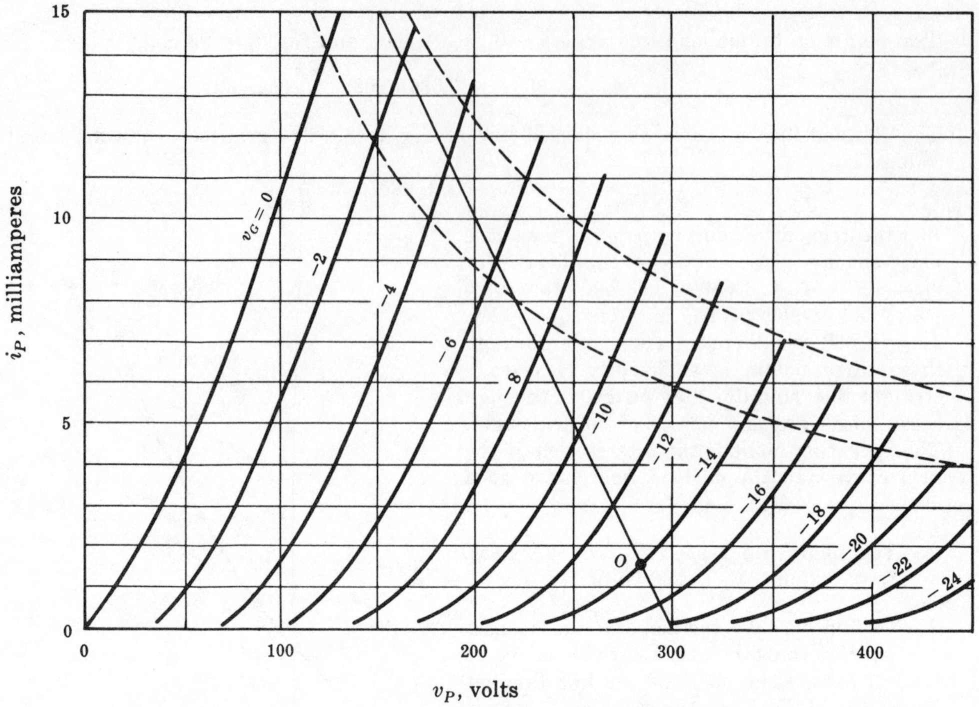

Fig. 5-47. 6J5 Static Plate Characteristics

The input and output loop equations in terms of instantaneous values are

$$-v_{S1} + v_G + R_K i_P = 0 \quad \text{or} \quad v_{S1} = v_G + R_K i_P \tag{5.76}$$

$$-v_{S2} + R_K i_P = 0 \quad \text{or} \quad v_{S2} = R_K i_P \tag{5.77}$$

A table is made using instantaneous values as shown below. The values of i_P for given values of v_G are taken from the intersections of the static plate characteristic curves and the load line. For each value of i_P the value of $R_K i_P$ can be computed, giving the instantaneous output voltage v_{S2}. Using equation (*5.76*), the corresponding value of the instantaneous input voltage v_{S1} is calculated. The curve showing v_{S2} as a function of v_{S1} is plotted as shown in Fig. 5-48. This curve indicates that the linear operating range is considerably extended if v_{S1} has an average value greater than zero. If V_{S1} (average value of v_{S1}) is zero, the operating point is that shown as O in Fig. 5-48. If V_{S1} is in the order of 50 to 75 volts, the linear operating range is extended and the tube is biased at a more favorable operating point.

v_G volts	i_P ma	$R_K i_P = v_{S2}$ volts	$v_{S1} = v_G + R_K i_P$ volts
−2	13.8	138	136
−4	11.4	114	110
−6	8.9	89	83
−8	6.7	67	59
−10	4.6	46	36
−12	2.9	29	17
−14	1.5	15	1
−16	0.5	5	−21

TABLE 5.1

The operating point for the given circuit was found by locating the point $V_{S1} = 0$ on the load line. The corresponding value of V_{S2} is 15 volts. Substitute into (5.77) to find $I_P = V_{S2}/R_K = 15/10{,}000 = 1.5$ ma. This point is located on the load line of Fig. 5-47, giving

$$V_G = -14 \text{ volts}$$

and $$V_P = 284 \text{ volts}$$

The small signal voltage gain is the slope of the transfer characteristic curve at the operating point. For this circuit,

$$A_v = \partial v_{S2}/\partial v_{S1}$$

$$= 100/[110-(-20)] = 0.77$$

The accuracy is limited because the operating point is not on the straight line portion of the transfer characteristic curve and the plotted points are relatively far apart.

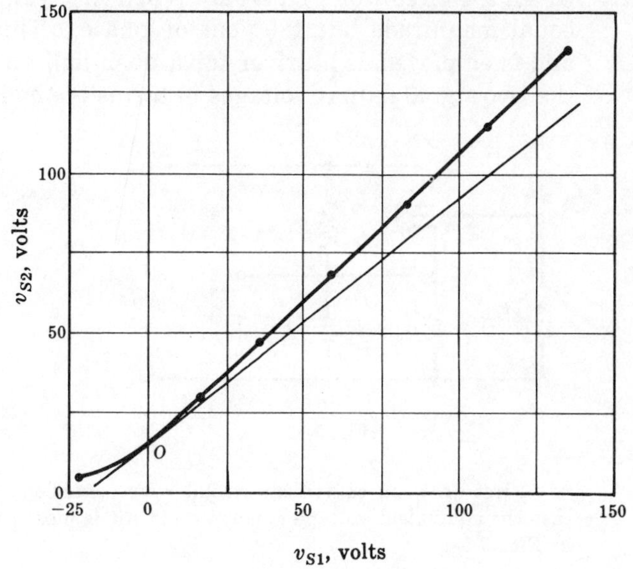

Fig. 5-48. Overall Cathode Follower Transfer Characteristics

5.3. Given the parallel combination of two tubes shown in Fig. 5-49. The tubes are operating on the linear portion of the operating range and their parameters are

$$T_1\text{:}\quad \mu_1 = 30, \quad r_{p1} = 12\ \text{k}\Omega$$

$$T_2\text{:}\quad \mu_2 = 60, \quad r_{p2} = 50\ \text{k}\Omega$$

Find the effective μ and r_p for the parallel combination.

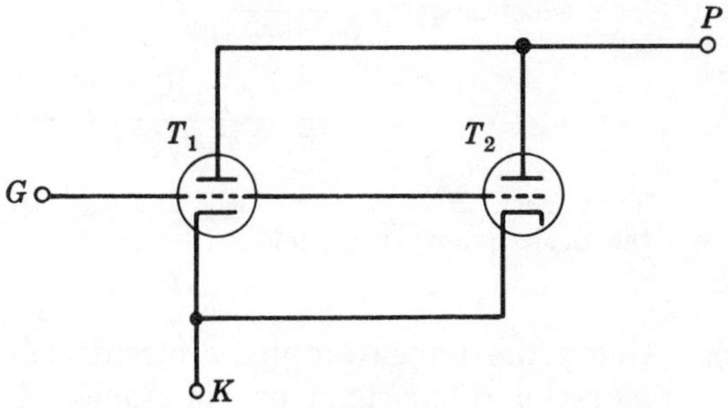

Fig. 5-49

Draw the ac equivalent circuit for the two tubes in parallel as shown in Fig. 5-50. The equation for the current loop "i" is

$$+60v_g + 50{,}000i + 12{,}000i - 30v_g = 0 \quad \text{or} \quad i = -484\times10^{-6}\,v_g$$

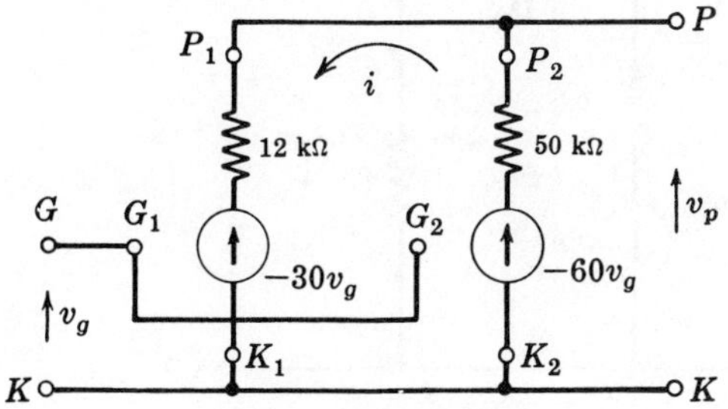

Fig. 5-50

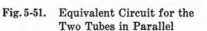

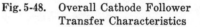

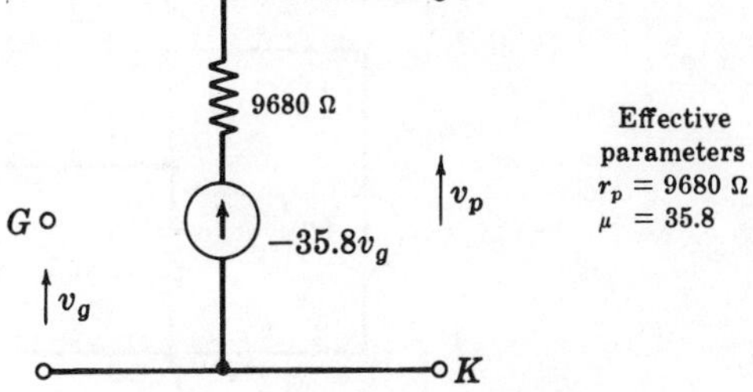

Fig. 5-51. Equivalent Circuit for the Two Tubes in Parallel

The two generators are independent because the generated voltage is independent of v_p and i_p. In the triode, the reverse transfer function is zero when the grid is negative. Thévenin's theorem may be used. The open-circuit voltage drop is $v_p = -50{,}000i - 60v_g = 24.2v_g - 60v_g = -35.8v_g$. The effective internal impedance (generator impedance zero) is the parallel combination:

$$R_{eff} = (12{,}000)(50{,}000)/62{,}000 = 9680\ \Omega$$

5.4. Given the circuit of Fig. 5-52. When $R_L = R_K$, the two output signal voltages are of equal magnitude but 180° out of phase. This circuit is a *split-load phase inverter* and is employed as a driver for a push-pull amplifier stage. Derive the equations for the two signal output voltages in terms of the input signal v_{s1}.

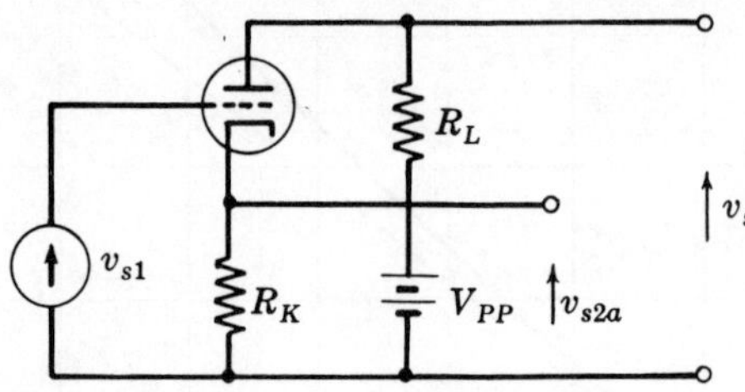

Fig. 5-52

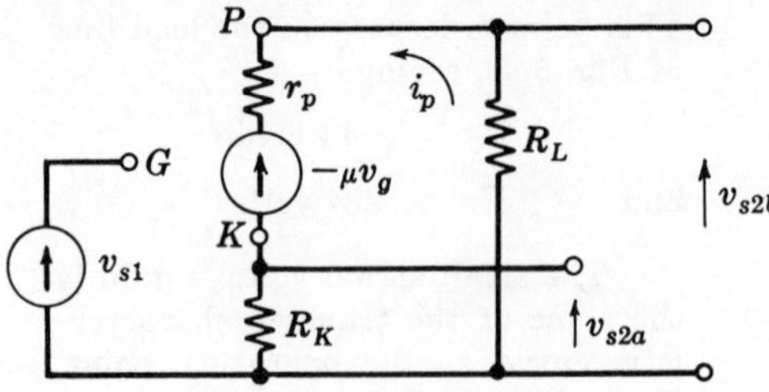

Fig. 5-53

First draw an equivalent circuit representation considering varying components only. In this case the equivalent voltage generator circuit is most convenient for representing the triode as shown in Fig. 5-53.

The grid loop and plate loop equations are

$$-v_{s1} + v_g + R_K i_p = 0 \qquad \text{and} \qquad R_L i_p + r_p i_p - \mu v_g + R_K i_p = 0$$

from which $i_p = \dfrac{+\mu v_{s1}}{R_L + r_p + R_K + \mu R_K}$. Note on the circuit diagram of Fig. 5-53 that

$$v_{s2a} = R_K i_p = \frac{\mu R_K}{R_L + r_p + (\mu+1)R_K} v_{s1}, \qquad v_{s2b} = -R_L i_p = \frac{-\mu R_L}{R_L + r_p + (\mu+1)R_K} v_{s1}$$

Thus when $R_L = R_K$, then v_{s2a} and v_{s2b} are the same in magnitude but opposite in phase, giving the desired output signals.

5.5. Given the triode amplifier circuit of Fig. 5-54. The power supply was poorly designed and has considerable ripple. Under normal operating conditions the instantaneous terminal voltage of the power supply is $v_{PP} = 250 + 2 \sin 377t$ volts. The tube parameters are $\mu = 30$ and $r_p = 20$ kΩ. If $v_{s1} = 1.0 \sin 2000\,\pi t$ volts, find the instantaneous value of the varying component of the output voltage v_{s2}.

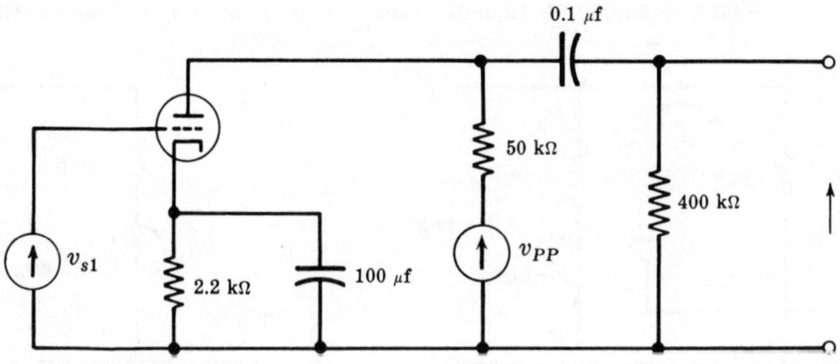

Fig. 5-54

Assume the tube is operating on the linear portion of the operating range so that superposition may be applied and an equivalent circuit used to represent the tube. First check to see if the cathode is effectively bypassed at 60 cps by computing the reactance of the capacitor at this frequency.

$$X_{C_K} = 1/\omega C_K = 1/[(377)(100 \times 10^{-6})] = 26.5\ \Omega$$

The cathode impedance may be assumed negligibly small.

Draw the equivalent circuit considering varying components, assuming the cathode is grounded as in Fig. 5-55. Apply superposition and first consider the 1000 cps signal frequency. The 0.1 μf coupling capacitor has reactance $X_{C_C} = 1/[(2000\pi)(0.1 \times 10^{-6})] = 1590\ \Omega$, which is negligible compared to the 400 kΩ and the effective resistance looking to the left. The circuit considering the 1000 cps signal is shown in Fig. 5-56.

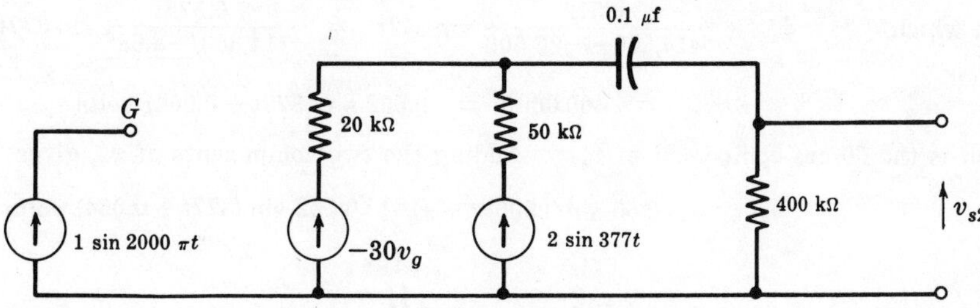

Fig. 5-55. Equivalent Circuit

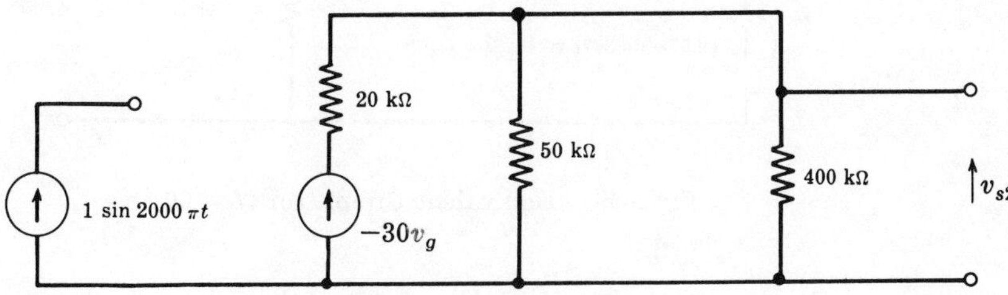

Fig. 5-56. Equivalent Circuit for $f = 1000$ cps

The 1000 cps component of the voltage v_{s2} is more conveniently found by using the constant current representation for the tube as shown in Fig. 5-57.

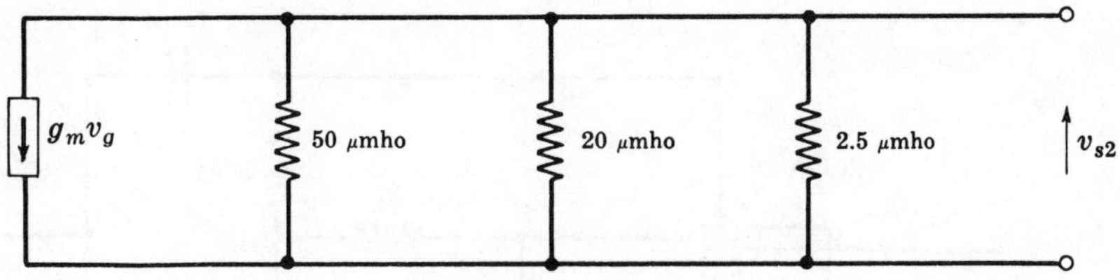

Fig. 5-57. Equivalent Circuit for $f = 1000$ cps

The plate node equation is, using $g_m = \mu/r_p = 30/20{,}000 = 1500\ \mu$mho,

$$1500 \times 10^{-6}\, v_g + 72.5 \times 10^{-6}\, v_{s2} = 0 \quad \text{or} \quad v_{s2} = -20.65 v_g = -20.65 \sin 2000\pi t \text{ volts}$$

This is the 1000 cps component of v_{s2}.

Consider the 60 cps ripple frequency. The equivalent circuit is shown in Fig. 5-58. The 60 cps component for v_{s2} may be found either by writing two loop equations and solving or by applying Thévenin's theorem to the circuit on the left of the terminals a and b. Use Thévenin's theorem and open the circuit at terminal a. Assume the current i of Fig. 5-58 and write the loop equation:

$$-2 \sin 377t + 50{,}000i + 20{,}000i = 0$$

or $\quad i = (2/70{,}000) \sin 377t$

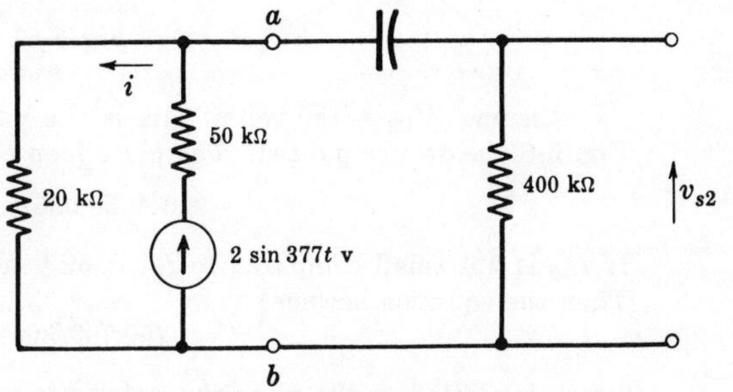

Fig. 5-58. Equivalent Circuit for $f = 60$ cps

The open circuit voltage $v_{ab} = 20{,}000i = 0.572 \sin 377t$ volts. The effective resistance $R_{ab} = (20{,}000)(50{,}000)/(70{,}000) = 14{,}300\ \Omega$.

The circuit may be redrawn as shown in Fig. 5-59. The loop equation is

$$-0.572 \sin 377t + 14{,}300i + \frac{-j}{377(0.1 \times 10^{-6})} i + 400{,}000i = 0$$

from which
$$i = \frac{0.572}{414{,}300 - j26{,}500} \sin 377t = \frac{0.572}{414{,}350\angle -3.68^\circ} \sin 377t$$

and
$$v_{s2} = 400{,}000i = 0.552 \sin(377t + 0.064) \text{ volts}$$

which is the 60 cps component of v_{s2}. Adding the two components of v_{s2} gives

$$v_{s2} = 20.65 \sin(2000\pi t + \pi) + 0.552 \sin(377t + 0.064) \text{ volts}$$

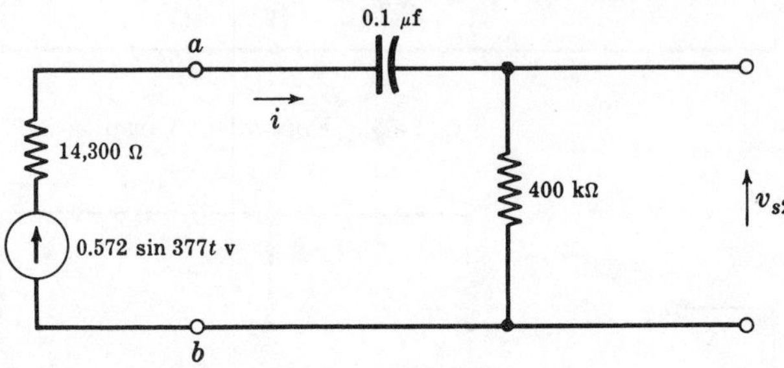

Fig. 5-59. Equivalent Circuit for $f = 60$ cps

5.6. Given the pentode amplifier circuit of Fig. 5-60 and the static plate characteristics for the 6AU6 in Fig. 5-61. (*a*) Determine the operating point for the circuit as given, assuming $V_{G2} = 150$ volts. (*b*) Find V_{G1}, V_P and I_P. (*c*) Choose R_2 so that $V_{G2} = 150$ volts. (*d*) Estimate g_m and g_p graphically. Calculate the approximate value of μ. (*e*) Compute the voltage gain $A_v = v_{s2}/v_{s1}$.

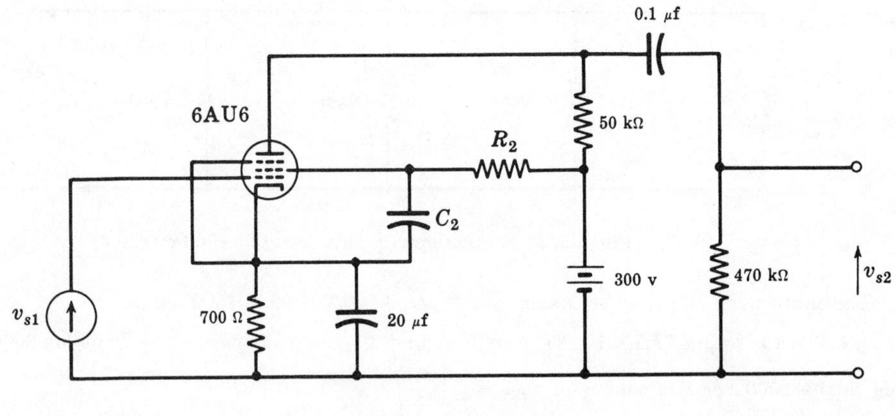

Fig. 5-60. Pentode Amplifier

Assume $V_{G2} = 150$ volts; this is the value for which the static plate characteristic curves of Fig. 5-61 below are plotted. The plate loop equation based on average values is

$$-300 + 50{,}000I_P + V_P + 700(I_P + I_{G2}) = 0$$

If I_{G2} is not small compared to I_P, it may not be neglected. As a first approximation, neglect I_{G2}. Then the equation becomes

$$(50{,}700/300)I_P + (1/300)V_P = +1$$

which is plotted on the plate characteristic curves of Fig. 5-61 using the intercepts $v_P = 300$ v and $i_P = 5.92$ ma.

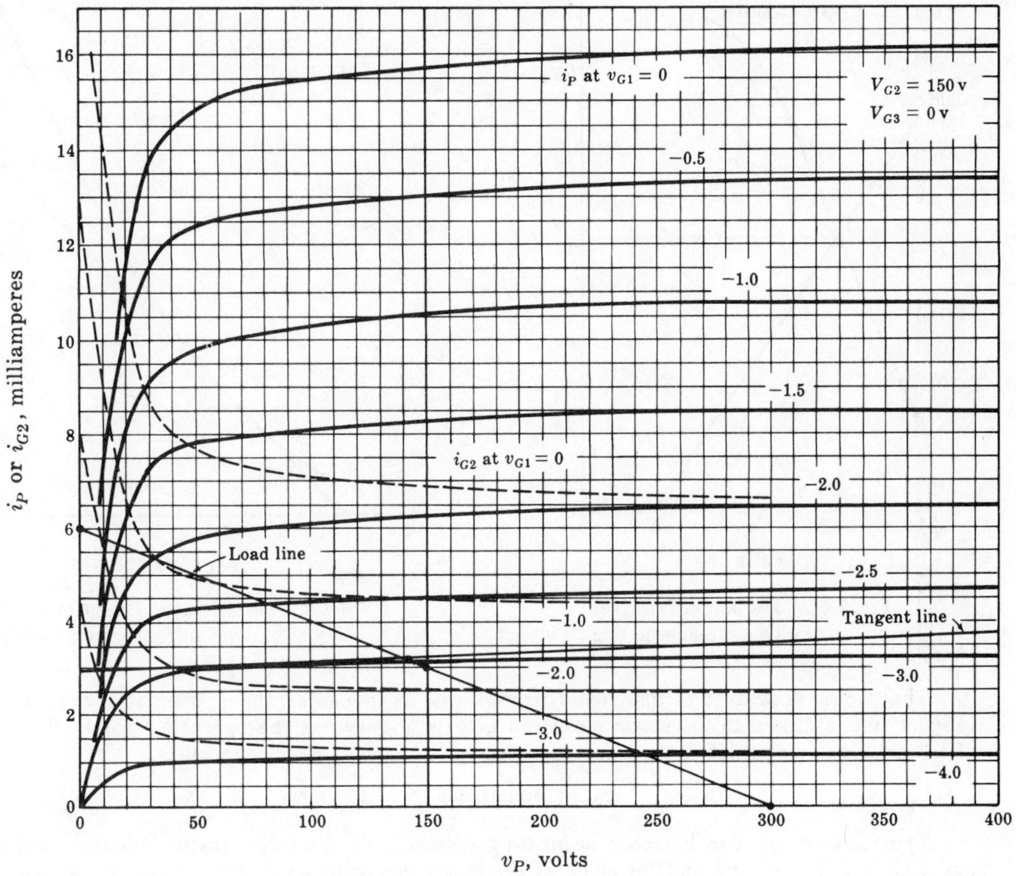

Fig. 5-61. 6AU6 Static Plate Characteristics

Next consider the control grid equation based on average values:

$$V_{G1} + 700(I_P + I_{G2}) = 0$$

In this equation I_{G2} cannot be neglected. Approximations must be used because of the number of variables involved. Note that the $v_{G1} = -3.0$ v curve intersects the load line near the center of the load line. Assume $v_{G1} = -3$ v; then the approximate value of I_{G2} may be found from the dashed curve for i_{G2} for $v_{G1} = -3$ v. For $v_P > 50$ v, this is approximately 1.2 ma. Assuming $I_{G2} =$ 1.2 ma (a constant), the grid equation becomes

$$V_{G1} + 700I_P + 700(0.0012) = 0 \quad \text{or} \quad V_{G1} + 700I_P + 0.84 = 0$$

Because of the shapes of the plate characteristic curves, it is difficult to plot the grid curve on this graph. The dynamic transfer characteristic curve based on the dc load line may be plotted as shown in Fig. 5-62 below. The grid curve as given in the grid equation is plotted on this graph so that the operating point may be found as shown. For the assumed value of I_{G2}, we have $V_{G1} =$ -3.05 volts, $I_P = 3.10$ ma, and $V_P = 150$ volts. Accuracy is rather limited because of the small slope of the plate curves. The value of I_{G2} for $v_P = 150$ and $V_{G1} = -3.05$ volts is approximately 1.2 ma as assumed. The value of R_2 is found by writing the loop equation for the screen grid circuit, considering average values:

$$-300 + R_2 I_{G2} + V_{G2} + 3.05 = 0$$

Substitute $V_{G2} = 150$ and $I_{G2} = 0.0012$ and obtain $R_2 = 123{,}000\ \Omega$. Since this is not a standard value, take $R_2 = 150{,}000\ \Omega$. In an amplifier of this type the screen grid voltage would probably be lower than 150 since the average plate voltage is 150, but the value of $V_{G2} = 150$ volts was chosen because the plate characteristic curves are plotted for this value of V_{G2}.

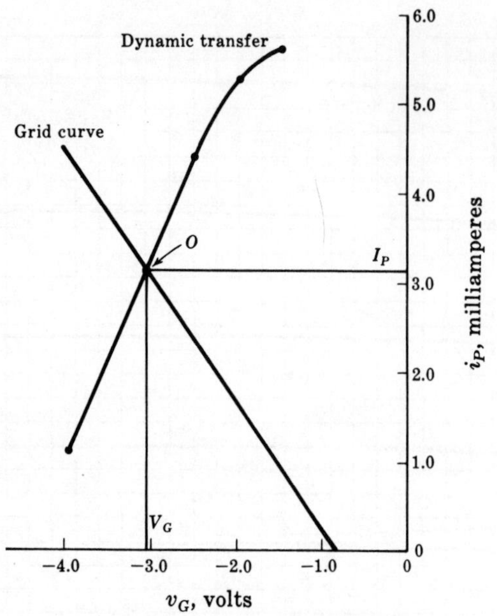

Fig. 5-62. Dynamic Transfer Characteristic Curve

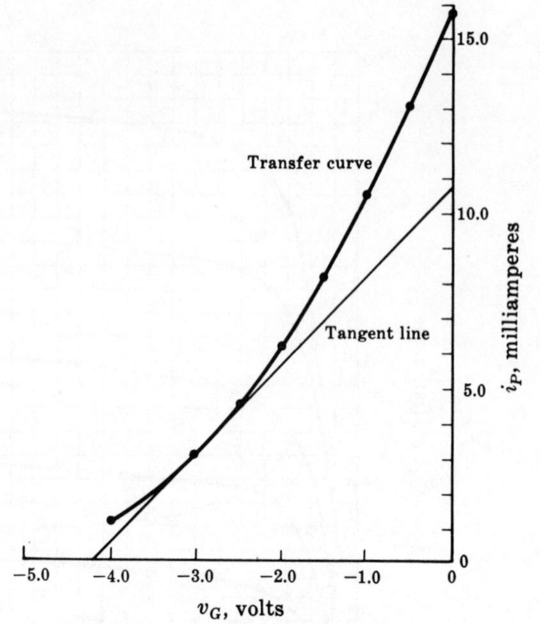

Fig. 5-63. Static Transfer Characteristic Curve

The pentode parameters are found graphically as slopes of the appropriate curves. First, a tangent line is drawn to the static plate characteristic curve at the operating point. The accuracy of the slope determination is very poor because the increment Δi_P is so small.

$$g_p = 0.0004/350 = 1.14 \ \mu\text{mho}$$

The value of g_m can be more accurately determined. Plot the static transfer characteristic curve as in Fig. 5-63 and find the slope at the operating point as shown.

$$g_m = 0.0107/4.2 = 2550 \ \mu\text{mho}$$

The calculated value of μ is $\mu = g_m/g_p = 2240$.

Draw the equivalent circuit as in Fig. 5-64. Using numerical values, the plate nodal equation is

$$2550 \times 10^{-6} v_g + (1.14 + 20 + 2.13) \times 10^{-6} v_{s2} = 0$$

Since $v_{s1} = v_g$, $A_v = v_{s2}/v_g = -109.5$.

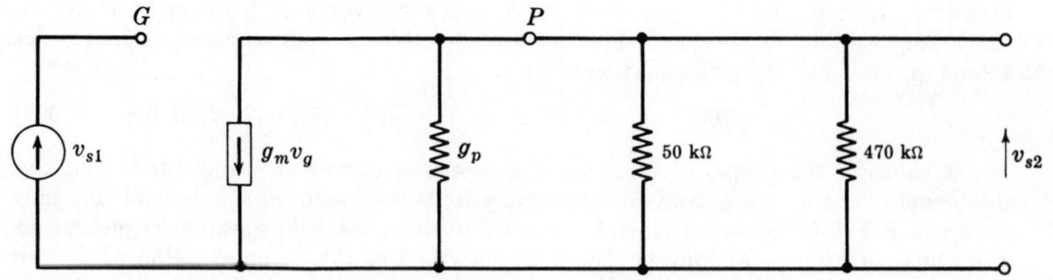

Fig. 5-64. Amplifier Equivalent Circuit

5.7. The cathode follower and the grounded grid amplifiers may be combined in such a way as to produce the *cathode coupled amplifier* shown in Fig. 5-65. Frequently, identical triodes are used and in many cases a double triode (two identical triodes in one envelope) is employed. Derive the equation for the mid-band voltage gain of the given circuit.

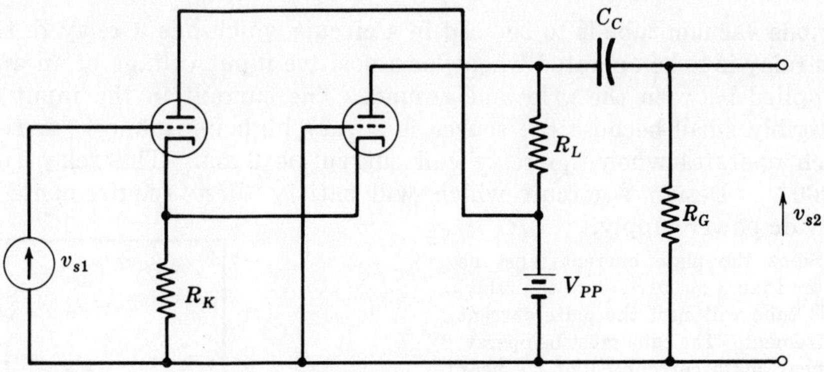

Fig. 5-65. Cathode Coupled Amplifier

First draw the ac equivalent circuit representing the tubes as shown in Fig. 5-66, with R_L and R_G combined as $R_L^* = R_L R_G/(R_L + R_G)$. Assume the parameters are identical for both tubes. The input loop and cathode nodal equations are respectively

$$-v_{s1} + v_{g1} + v_k = 0 \tag{5.78}$$

$$-g_m v_{g1} + g_p v_k + G_K v_k - i_{p2} = 0 \tag{5.79}$$

The loop equations for the grid G_2 and plate P_2 are respectively

$$v_{g2} + v_k = 0 \tag{5.80}$$

$$R_L^* i_{p2} + r_p i_{p2} - \mu v_{g2} + v_k = 0 \tag{5.81}$$

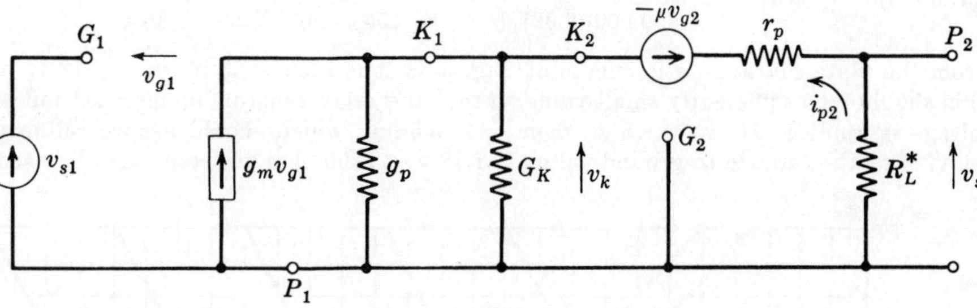

Fig. 5-66. Equivalent Circuit

Substitute (*5.78*) into (*5.79*) to eliminate the variable v_{g1}. Put (*5.80*) into (*5.81*) to eliminate the variable v_{g2}. Solve the two resulting equations for i_{p2} in terms of v_{s1} and obtain

$$i_{p2} = \frac{-g_m(\mu+1)v_{s1}}{(g_p + G_K)(r_p + R_L^*) + \mu + 1 + g_m(r_p + R_L^*)}$$

Since $v_{s2} = -R_L^* i_{p2}$,

$$A_v = v_{s2}/v_{s1} = \frac{g_m(\mu+1)R_L^*}{(g_p + G_K)(r_p + R_L^*) + \mu + 1 + g_m(r_p + R_L^*)}$$

or, in another form,

$$A_v = \frac{g_m(\mu+1)R_L^*}{2(\mu+1) + R_L^*(g_m + g_p) + G_K(r_p + R_L^*)}$$

The voltage gain is positive, indicating that there is no phase shift in the mid-band range of frequencies. The above expression may be simplified as follows: when $\mu \gg 1$, then $\mu + 1 \cong \mu$; when $g_m \gg g_p$, then $g_m + g_p \cong g_m$. The gain is then

$$A_v = \frac{g_m \mu R_L^*}{2\mu + g_m R_L^* + G_K(r_p + R_L^*)} \qquad \text{or} \qquad A_v = \frac{\mu R_L^*}{R_L^* + 2r_p + r_p(r_p + R_L^*)/\mu R_K}$$

5.8. A triode vacuum tube is to be used in a circuit which has a relay in the plate circuit. This relay is to be operated whenever a positive input voltage of an appropriate value is applied between the grid and ground. The current in the input circuit must be negligibly small because the source is a very high impedance. A relay is available which operates when the relay coil current is 3 ma. The relay coil resistance is 10,000 Ω. Design a circuit which will satisfy these requirements using a single 250 v dc power supply.

Since the plate current must be greater than 3 ma to operate the relay, a 6C4 tube will meet the plate current requirement. The tube must be operating near plate current cutoff (i_P near zero) when the grid to ground voltage is zero. Cathode bias requires a sufficient magnitude of current in the cathode resistor to produce the desired magnitude of V_G. To provide sufficient current in the cathode resistor, a second resistor R_1 may be added from the cathode terminal to the power supply as shown in Fig. 5-67.

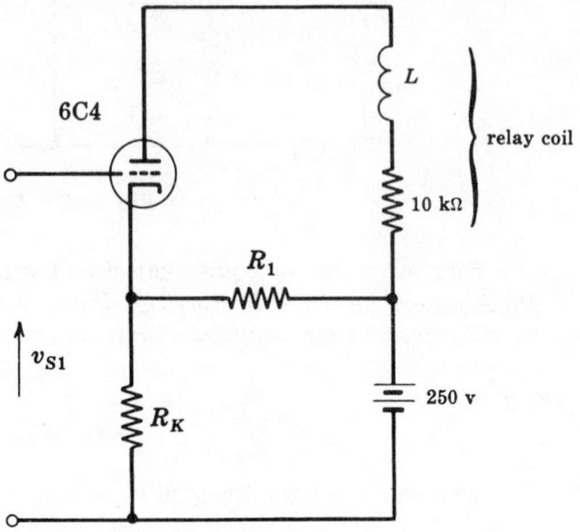

Fig. 5-67

As a starting point, plot the dc load line neglecting R_K. The equation of the load line is then

$$(10{,}000/250)i_P + (1/250)v_P = +1 \quad (5.82)$$

Since the 25 ma intercept is not on the graph, use the point $i_P = 20$ ma and find the corresponding voltage v_P:

$$(10{,}000)(0.02) + v_P = 250 \quad \text{or} \quad v_P = 50 \text{ v}$$

From the plate characteristic curves of Fig. 5-68 it is seen that if $v_G = -18$ v, then $i_P \cong 0.3$ ma. This should be a sufficiently small value so that the relay remains unoperated unless a positive input voltage is applied. If $v_G = -8$ v, then $i_P = 5.5$ ma which should assure reliable operation of the relay. For the cathode to ground voltage of 18 v, a 2200 ohm resistor could be used if the current is

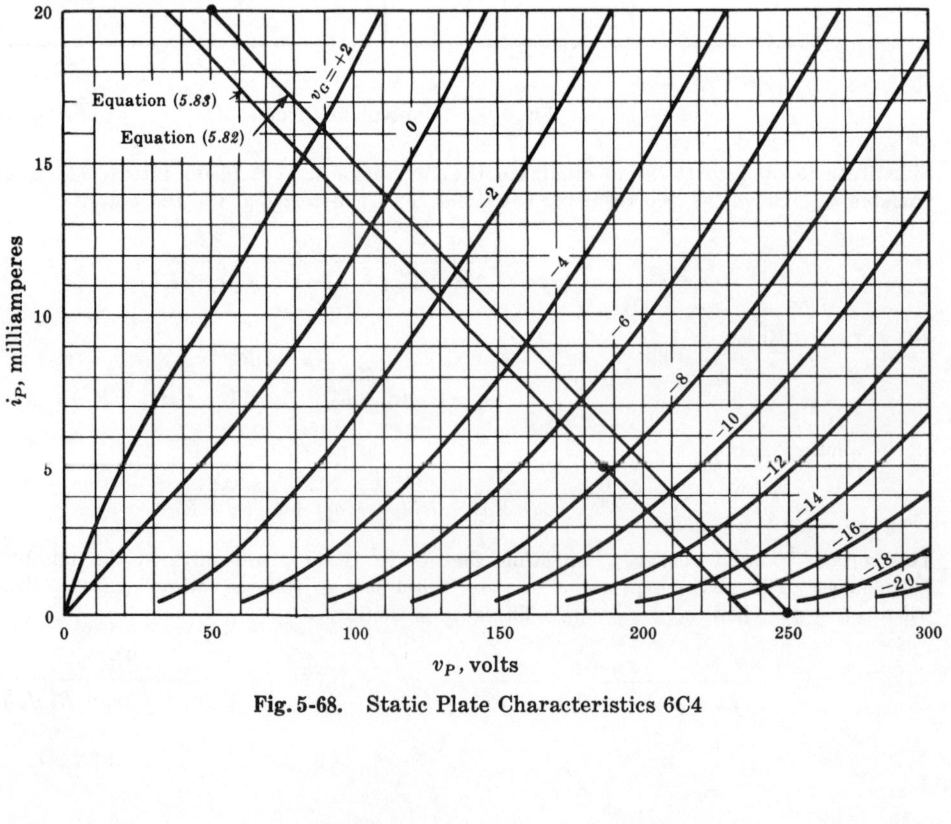

Fig. 5-68. Static Plate Characteristics 6C4

18/2200 = 8.2 ma. Since under these conditions the tube plate current is about 0.3 ma, the current in the resistor R_1 must be about 8 ma. The value of R_1 should then be $R_1 = (250 - 18)/0.008 = 29{,}000\ \Omega$; since this is not a standard value, use 33,000 Ω. The current in R_1 may be approximated by solving the loop equation assuming $i_P = 0.3$ ma:

$$-250 + 33{,}000i + 2200(i + 0.0003) = 0 \qquad \text{or} \qquad i = 7.1 \text{ ma}$$

The voltage from cathode to ground is $v_{KGnd} = 2200(7.1 + 0.3)10^{-3} = 16.3$ volts. Using this value, the plate loop equation is

$$-250 + 10{,}000i_P + v_P + 16.3 = 0 \qquad \text{or} \qquad (10{,}000/234)i_P + (1/234)v_P = +1 \qquad (5.83)$$

This line is also plotted on the plate curves of Fig. 5-68. The zero input plate current is approximately 0.5 ma. The grid to ground voltage required for reliable operation of the relay (assuming $i_P = 5$ ma) can be approximated. When the positive voltage is applied to the grid the plate current increases, so the load line shown is only an approximation. The $i_P = 5$ ma line intersects the load line at $v_G = -7.7$ volts. Using $v_G = -7.7$ volts and $i_P = 5$ ma, the input voltage may be approximated:

$$-v_{S1} + (-7.7) + 2200(0.005 + 0.0071) = 0 \qquad \text{or} \qquad v_{S1} = 18.9 \text{ v}$$

which should provide reliable operation of the relay.

This problem emphasizes that a tube cannot be biased to cutoff by using conventional cathode bias.

Supplementary Problems

5.9. Given the circuit of Fig. 5-69 and the static plate characteristic curves for a 6J5 tube in Fig. 5-70 below.

(a) Find the operating point. Determine V_G, V_P and I_P.

(b) Calculate the static plate conductance G_P.

(c) For the given operating point, find g_p and g_m. Graph any transfer characteristic curves needed.

(d) Find μ graphically, plotting any needed curves. Taking the values of g_p and g_m from part (c), calculate μ and compare with the value determined graphically. Explain any differences.

(e) Draw the equivalent circuit considering only average values. Include the values of the circuit elements.

(f) Construct the equivalent circuit considering only varying components ("ac" equivalent circuit), using a voltage generator circuit and then a current generator circuit for the triode. Include the values for the circuit elements.

(g) Compute the voltage gain A_v. Do not substitute into the formula but write the necessary equations and solve them.

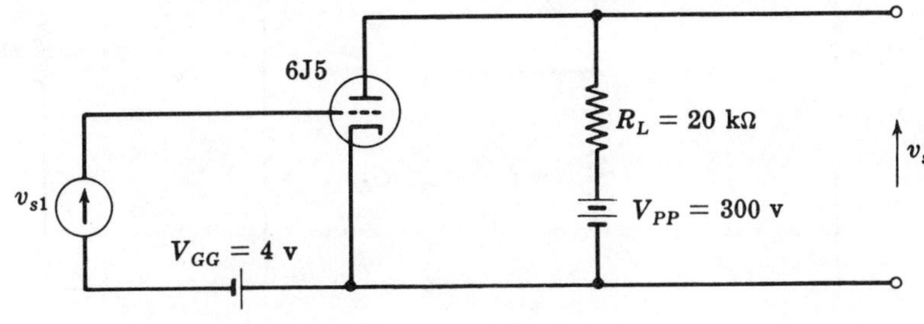

Fig. 5-69

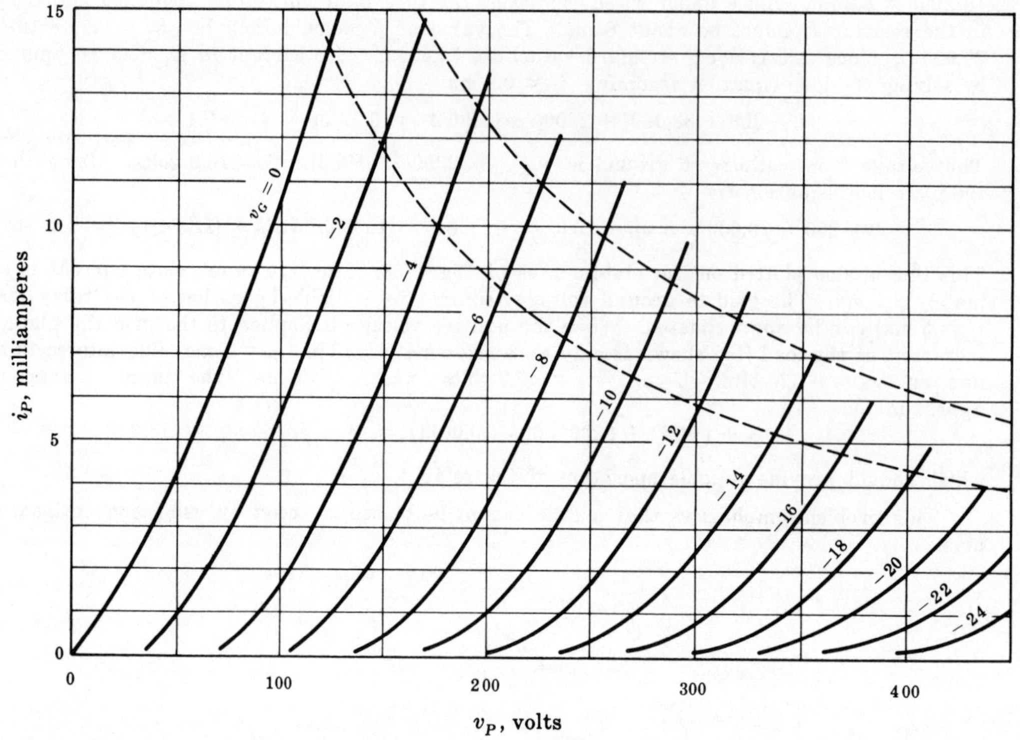

Fig. 5-70. 6J5 Plate Characteristics

5.10. Given the circuit of Fig. 5-71 and the static plate characteristic curves for a 6AV6 tube in Fig. 5-72 below.

(*a*) Determine V_G, V_P and I_P.

(*b*) Find μ and r_p for the given operating point, plotting any needed curves.

(*c*) Plot the dynamic transfer characteristic curve based on the dc load line and the one based on the ac load line, assuming C_K large.

(*d*) Assuming C_K is large so that the impedance from cathode to ground is negligible, construct the ac equivalent circuit and compute the voltage gain $A_v = v_{s2}/v_{s1}$.

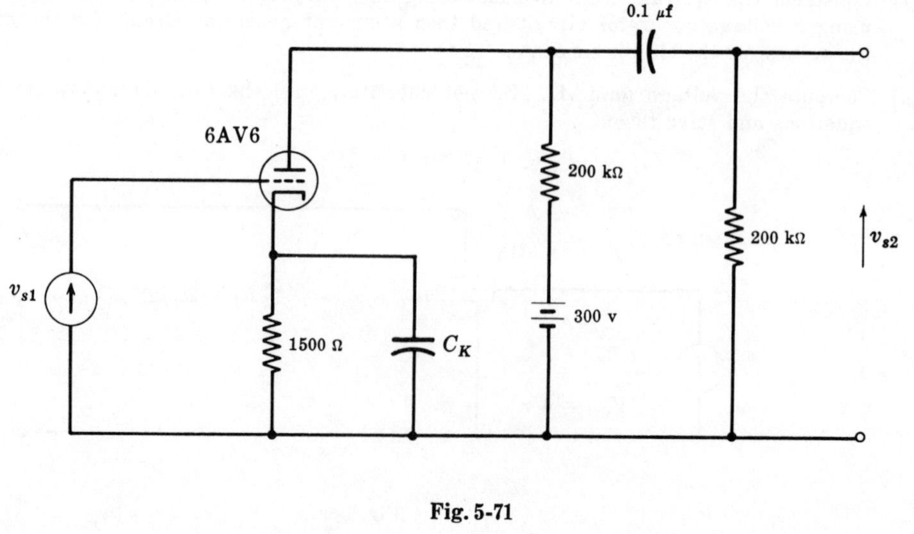

Fig. 5-71

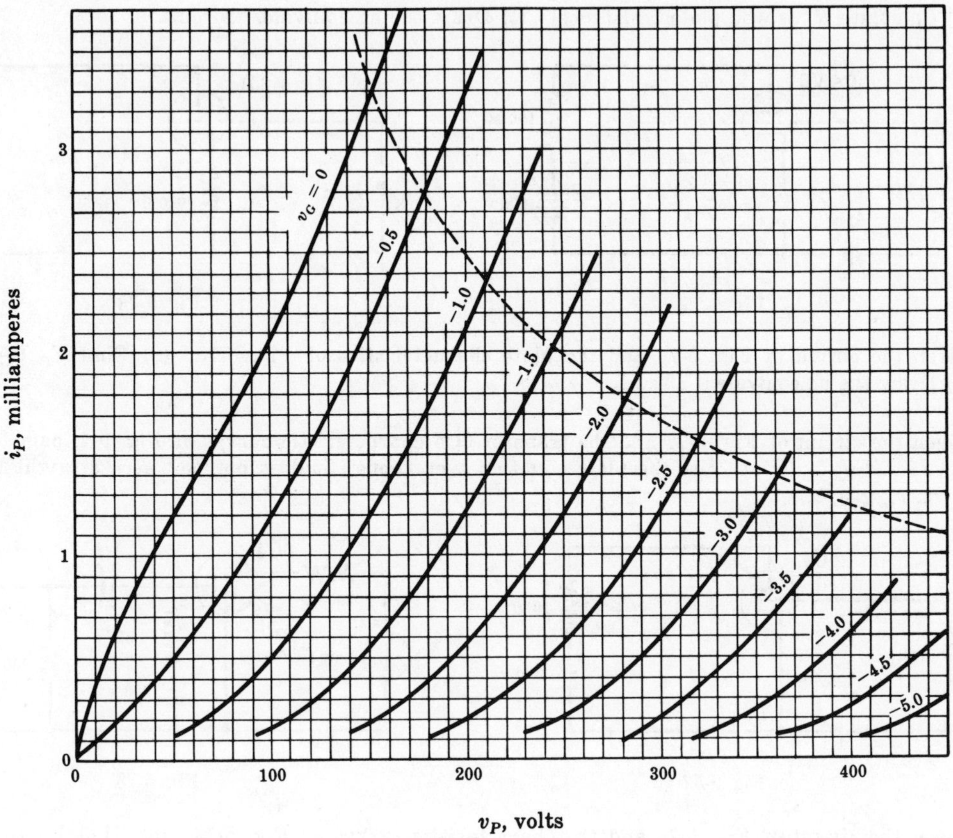

Fig. 5-72. 6AV6 Plate Characteristics

5.11. Given the circuit of Fig. 5-73 and using the characteristic curves of Fig. 5-74, find V_{PP} and V_{GG}. Take $V_P = 150$ volts and $V_G = -4$ volts.

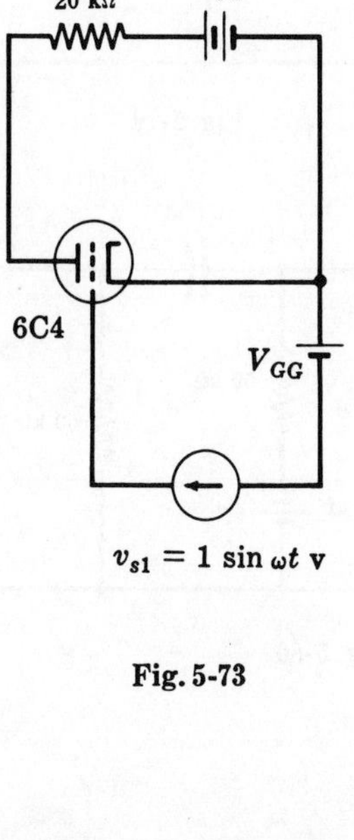

Fig. 5-73

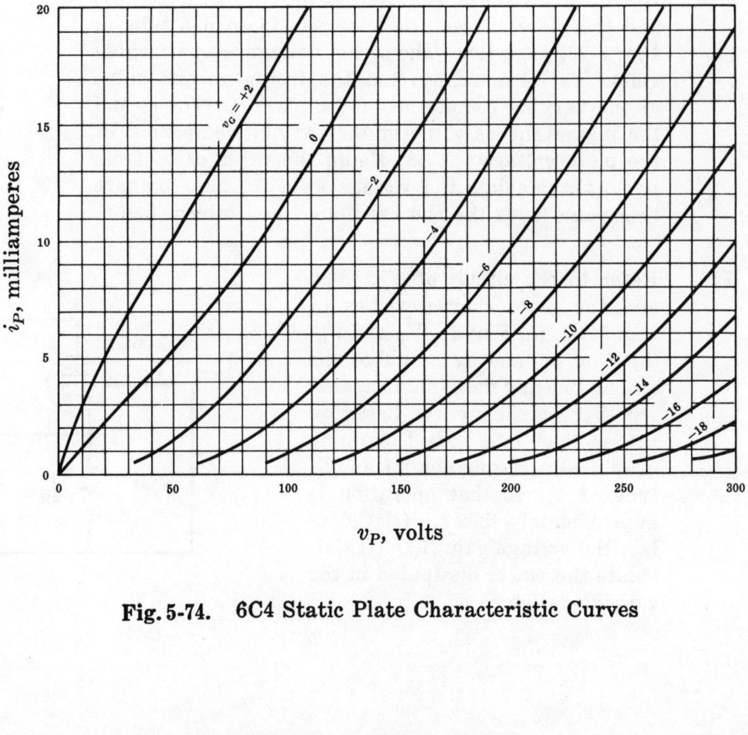

Fig. 5-74. 6C4 Static Plate Characteristic Curves

5.12. Given the circuit of Fig. 5-75 and the plate characteristics of Fig. 5-72. (*a*) Find V_P, I_P and V_G. (*b*) Calculate R_P (static plate resistance). (*c*) Draw the ac equivalent circuit.

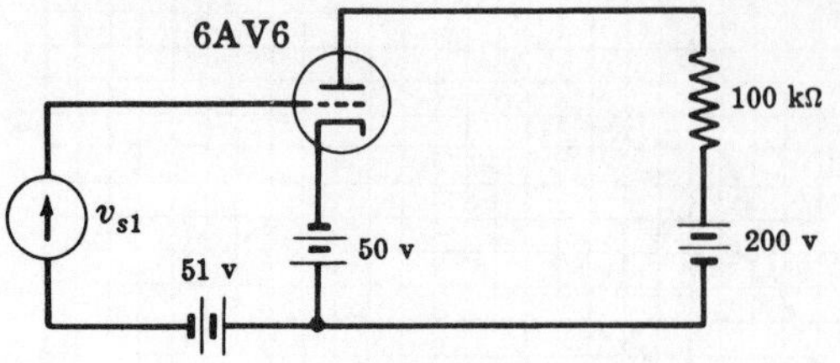

Fig. 5-75

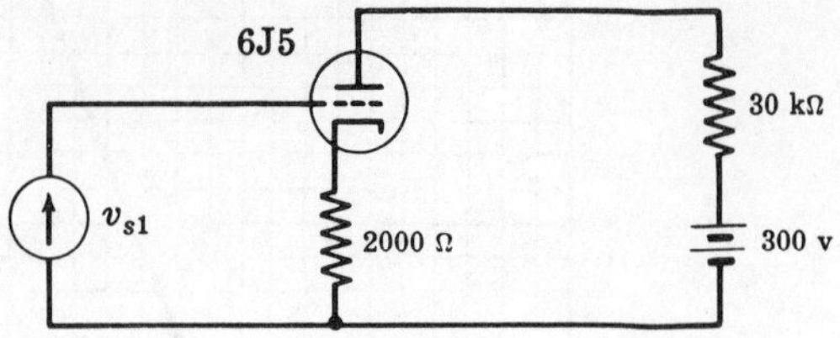

Fig. 5-76

5.13. Given the circuit of Fig. 5-76 and the plate characteristics of Fig. 5-70. (*a*) Find V_P, V_G and I_P. (*b*) Draw the ac equivalent circuit.

5.14. Given the circuit of Fig. 5-77 and the transfer 6C4 characteristic curves of Fig. 5-4, page 70. Find V_P, I_P and V_G. (*Hint.* Use the plate equation even though it does not plot as a straight line on a i_P vs. v_G graph.)

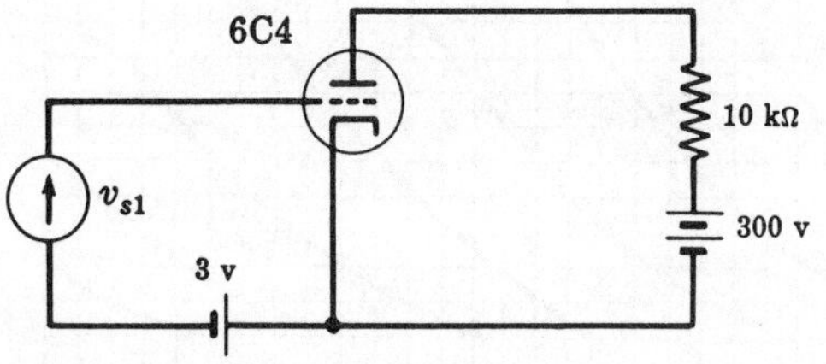

Fig. 5-77

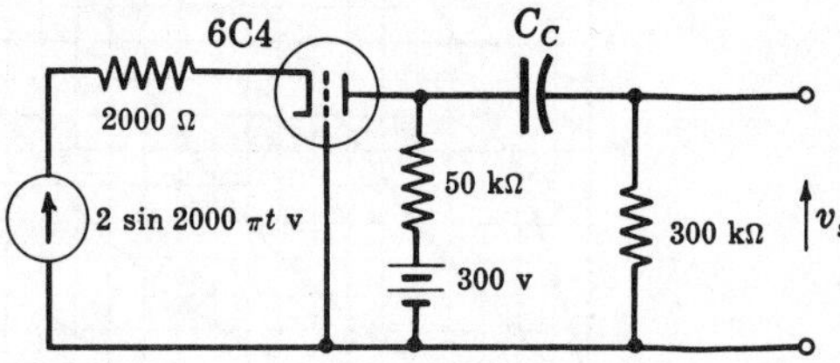

Fig. 5-78

5.15. Given the circuit of Fig. 5-78 and the characteristic curves of Fig. 5-74. (*a*) Find V_P, I_P and V_G. (*b*) Find μ and r_p for this operating point. (*c*) Draw the ac equivalent circuit assuming C_C is large so that the impedance is negligibly small. (*d*) Determine the signal output voltage v_{s2}.

5.16. Refer to the circuit of Fig. 5-79 and the plate characteristic curves of Fig. 5-74. (*a*) Find V_P, I_P and V_G. (*b*) Determine g_m and g_p for the given operating point.

5.17. Given the circuit of Fig. 5-69 (Problem 5.9, page 107) and the plate characteristic curves for a 6J5 tube in Fig. 5-70, page 108. The signal voltage $v_{s1} = 1 \sin \omega t$ volts. (*a*) Graphically find the instantaneous value of the varying component of the plate current i_p and the instantaneous value of the varying component of the plate voltage v_p. (*b*) Using the value of v_p from part (*a*), calculate the voltage gain A_v and compare this value with the one calculated in Problem 5.9(*g*).

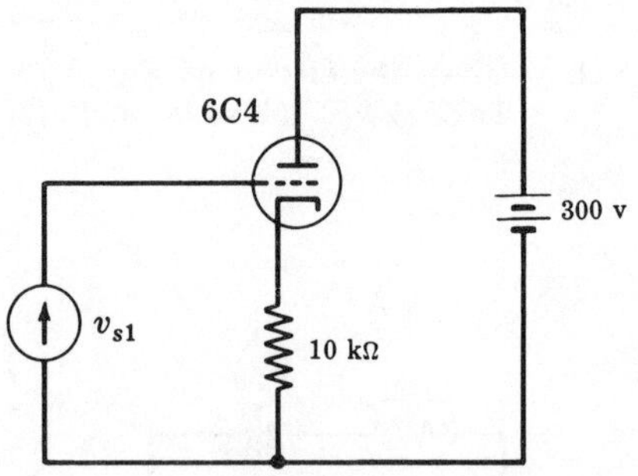

Fig. 5-79

5.18. Refer to the circuit of Fig. 5-80 and the plate characteristics of Fig. 5-74. (*a*) Find I_P, V_P and V_G. (*b*) Find g_m and g_p. (*c*) Plot the ac load line and the dynamic transfer characteristic curve based on the ac load line. (*d*) Determine graphically the maximum amplitude of v_{s1} so that operation is approximately linear. (*e*) Calculate the voltage gain A_v. (*f*) Calculate the power dissipated in the cathode resistor.

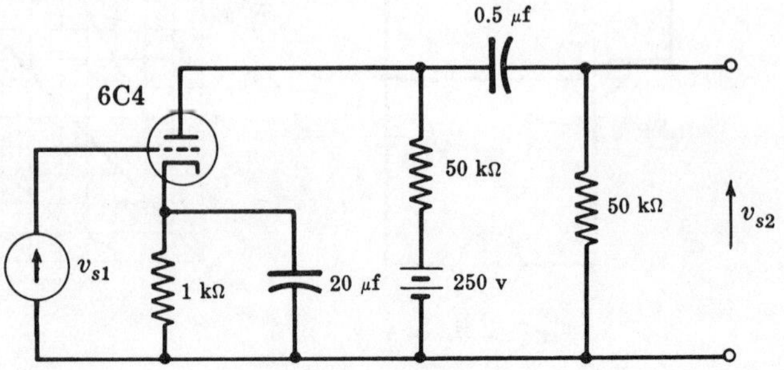

Fig. 5-80

5.19. Given the circuit of Fig. 5-81 and the tube parameters for the given operating point: $\mu = 20$, $r_p = 10$ kΩ. Take $v_{s1} = 0.1 \sin 2000\,\pi t$ volts. Work the following by writing the appropriate equations and solving rather than substituting into formulas. (*a*) Find v_{s2}. (*b*) Add a 30 μf electrolytic capacitor from cathode to ground. Show the polarity of the capacitor (voltage across it). What is v_{s2} when the cathode capacitor has been added? (*c*) Calculate the power dissipated in the 200 kΩ resistor for part (*a*) and part (*b*).

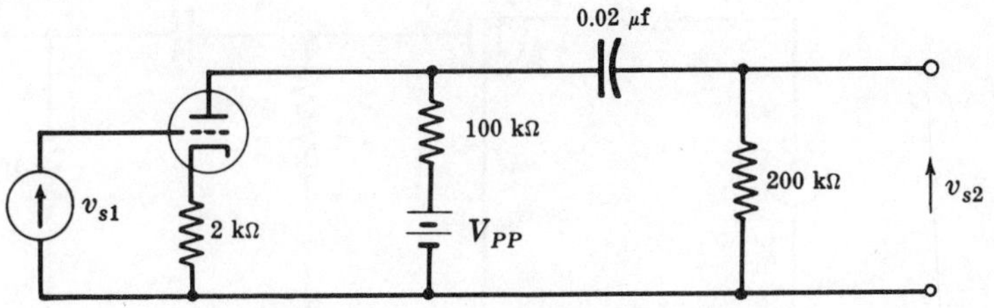

Fig. 5-81

5.20. Given the circuit of Fig. 5-82 and the plate characteristic curves of Fig. 5-74, page 109. Take $v_{s1} = 2 \sin \omega t$ volts. (*a*) Find V_P, I_P and V_G. (*b*) Account for all the power dissipated in the circuit and all the power supplied to the circuit. (*c*) Determine $r_{P(\max)}$ and $r_{P(\min)}$.

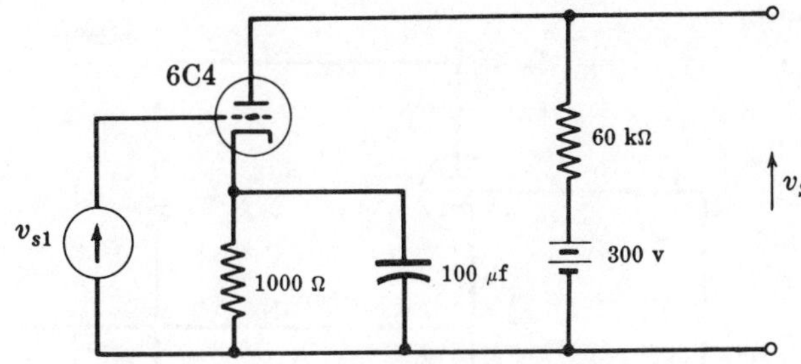

Fig. 5-82

5.21. Given the circuit of Fig. 5-83 and the plate characteristic curves of Fig. 5-74, page 109. (*a*) Determine the value of V_{S1} (the average value of v_{S1}) so that V_G is −4 volts. (*b*) Find V_P and I_P for the above value of V_G. (*c*) Choose V_2 so that the average value of v_{S2} (V_{S2}) is zero.

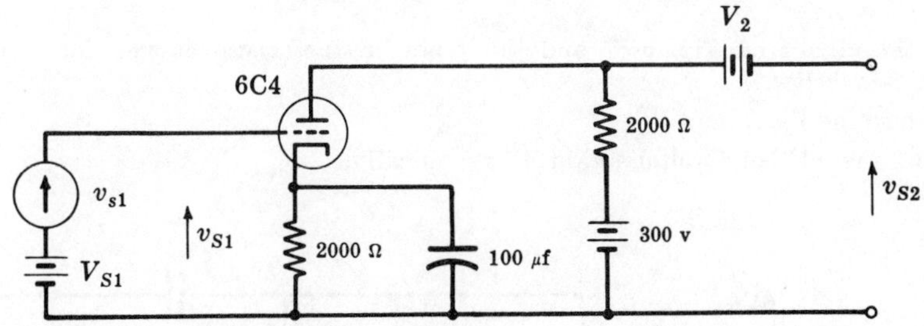

Fig. 5-83

5.22. Given the circuit of Fig. 5-84 and the 6C4 plate characteristics of Fig. 5-74, page 109. Take $v_{s1} = 0.5 \sin \omega t$ volts.

(*a*) Find V_P, I_P and V_G.

(*b*) Draw the equivalent circuit considering only the varying components.

(*c*) Determine μ and r_p for the tube.

(*d*) Find v_P maximum and v_P minimum for the circuit as given. (*Hint.* Find $v_{G(\min)}$ and $v_{G(\max)}$ first.)

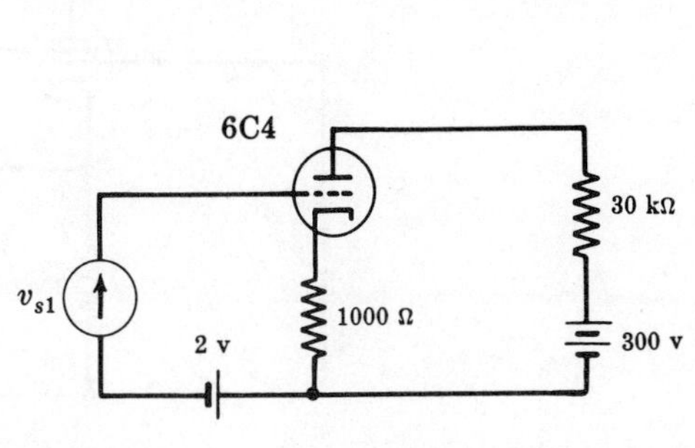

Fig. 5-84

5.23. Given the circuit of Fig. 5-85 and the parameters for the tube at the operating point: $\mu = 30$, $r_p = 8000\ \Omega$.

(a) Find the mid-band voltage gain $A_v = v_{s2}/v_{s1}$.

(b) Add a 30 μf capacitor across the 1000 Ω resistor and find A_v.

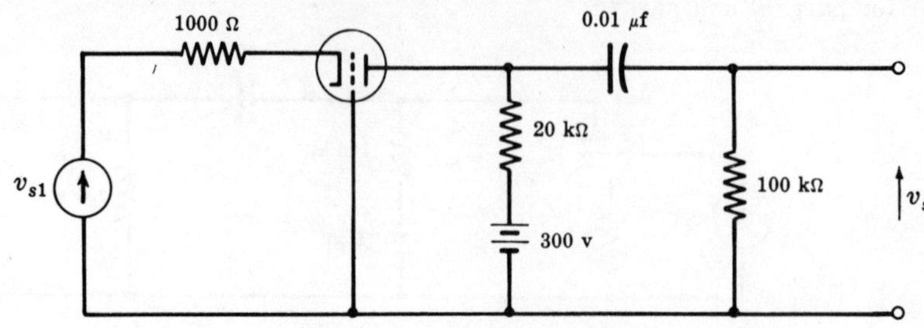

Fig. 5-85

5.24. Given the circuit of Fig. 5-86 and the tube parameters $\mu = 20$, $r_p = 10$ kΩ.

(a) What is the signal voltage gain of the circuit as given?

(b) What is the signal voltage gain if the 30 kΩ resistor is shorted out? Assume the tube parameters remain the same.

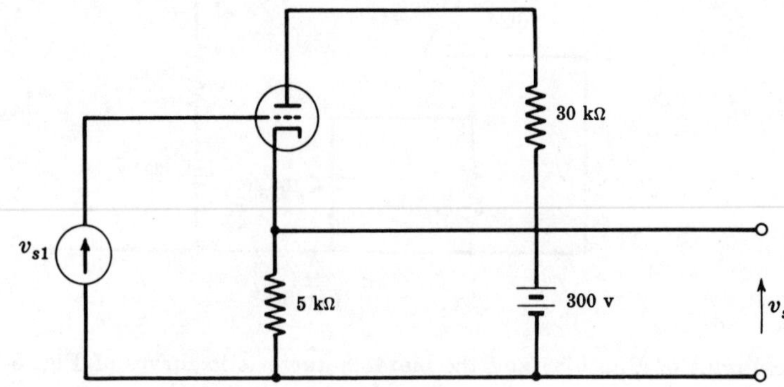

Fig. 5-86

5.25. Given the circuit of Fig. 5-87 and the plate characteristic curves for a 6V6 (triode connected) in Fig. 5-88 below.

(a) Determine V_P, I_P and V_G.

(b) Find the mid-band voltage gain A_v graphically.

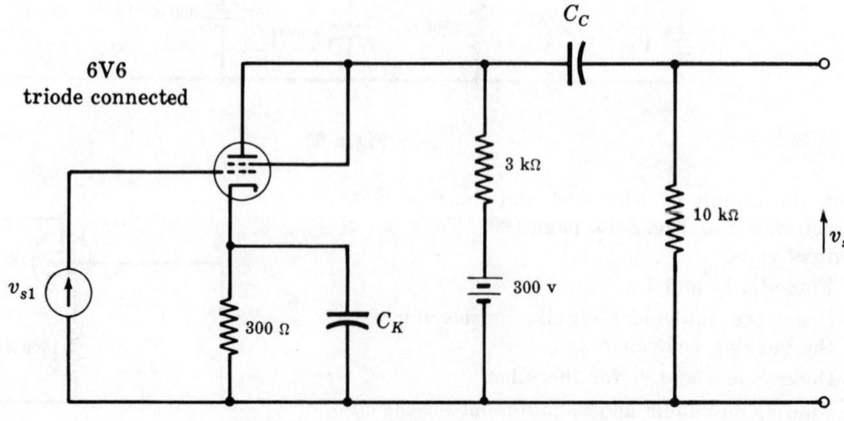

Fig. 5-87

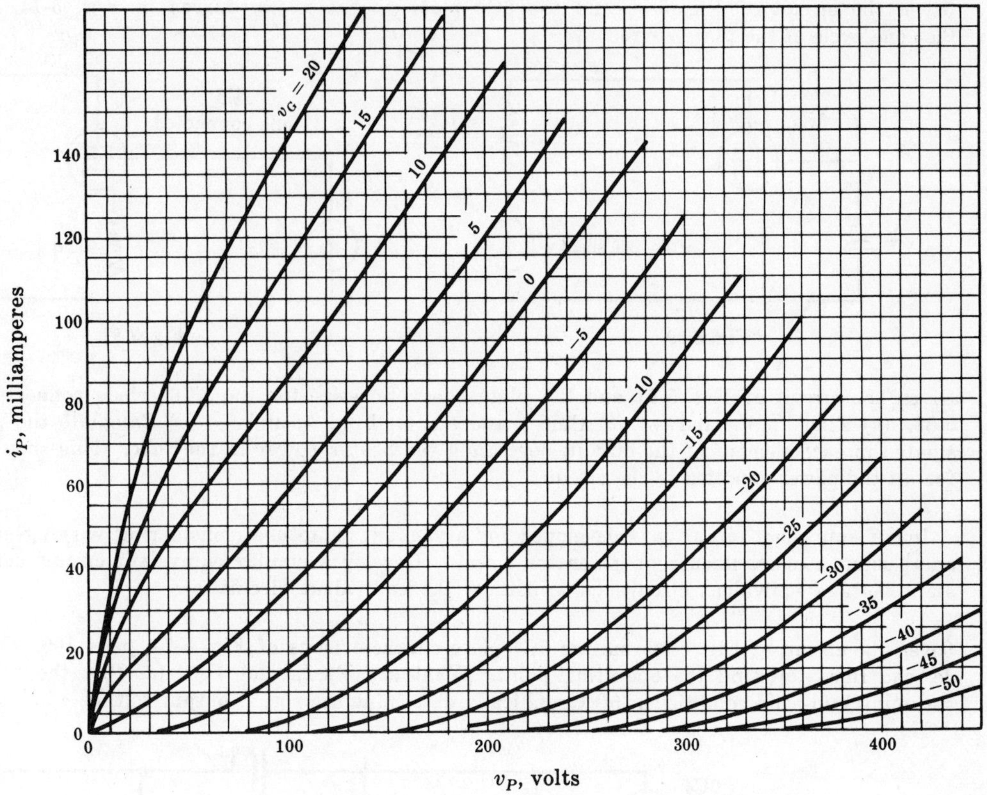

Fig. 5-88. 6V6 Triode Connected Static Plate Characteristics

5.26. Two triode tubes are connected in parallel. The individual tube parameters are: $\mu_1 = 20$ and $r_{p1} = 10$ kΩ; $\mu_2 = 30$ and $r_{p2} = 20$ kΩ. Show the equivalent circuit for the parallel combination, including values for the parameters.

5.27. Consider the cathode follower circuit of Fig. 5-25. Draw the equivalent circuit for this cathode follower circuit, representing the tube by a constant voltage generator circuit. Derive the equation for the voltage gain A_v in terms of μ, r_p and R_K. Show by substitution that your equation is the equivalent of equation (*5.35*), page 85.

5.28. Given the circuit of Fig. 5-89 and the 6C4 plate characteristics in Fig. 5-74, page 109. Assume D is an ideal diode. The input signal voltage $v_{S1} = 4 \sin 2000\,\pi t$. Plot one cycle of the instantaneous output voltage v_{S2}. (*Hint.* First determine the grid to ground voltage, keeping in mind that the grid current is zero when $v_G < 0$.)

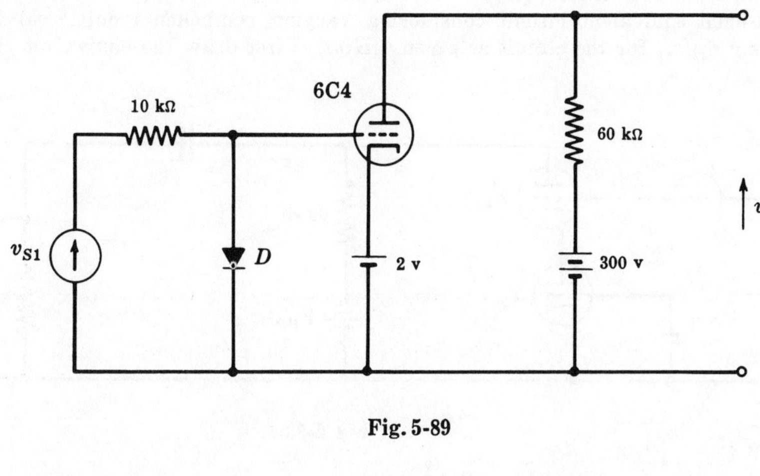

Fig. 5-89

5.29. Given the circuit of Fig. 5-90 and the 6C4 plate characteristic curves in Fig. 5-74, page 109. Plot one cycle of the plate current i_P.

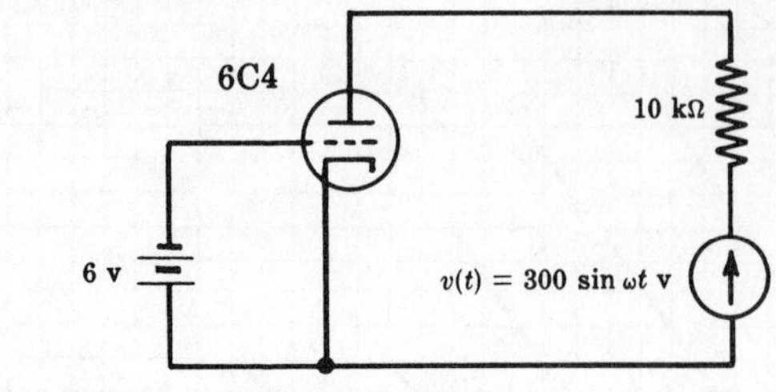

Fig. 5-90

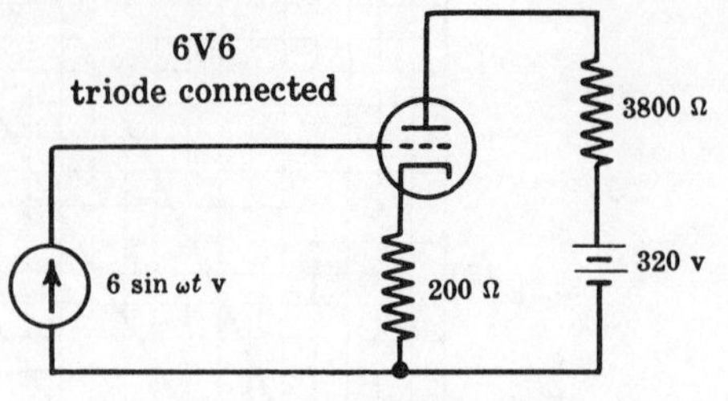

Fig. 5-91

5.30. Given the circuit of Fig. 5-91 and the plate characteristics for the 6V6 triode connected in Fig. 5-88. (*a*) Find V_P, V_G and I_P. (*b*) Find μ and r_p. (*c*) Find v_p and i_p. (*d*) Calculate the power dissipated in each element of the circuit, neglecting the heater power in the tube. This should account for all the power supplied to the circuit.

5.31. A linear equivalent circuit was developed for a vacuum phototube (considering varying components only) using a constant current generator circuit. Develop a similar equivalent circuit using a constant voltage equivalent circuit and construct the equivalent circuit.

5.32. Given the circuit of Fig. 5-92 and the 6C4 plate characteristics of Fig. 5-74, page 109. (*a*) Plot the dc load line. (*b*) Find the operating point. What are V_P, I_P and V_G? (*c*) Plot the ac load line assuming mid-band conditions. (*d*) Graphically determine the signal output voltage v_{s2}.

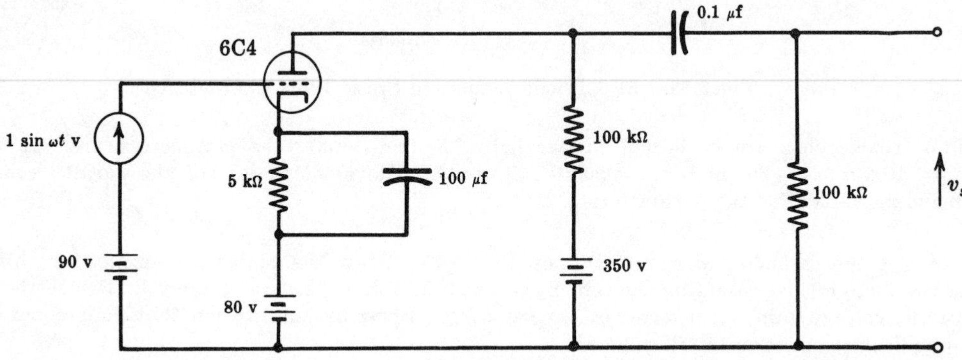

Fig. 5-92

5.33. Given the circuit of Fig. 5-93 (which is not necessarily practical). Assume the batteries in the circuit are such that at the established operating points for the tubes, the tube parameters are: $\mu = 10$ and $r_p = 2000\ \Omega$ for T_1; $g_m = 5000\ \mu$mho and $g_p = 1000\ \mu$mho for T_2. (*a*) Draw the simplest mid-band equivalent circuit considering varying components only. (*b*) Find the mid-band voltage gain v_{s2}/v_{s1} for the circuit as given. (*Hint.* First draw the equivalent circuit for T_2.)

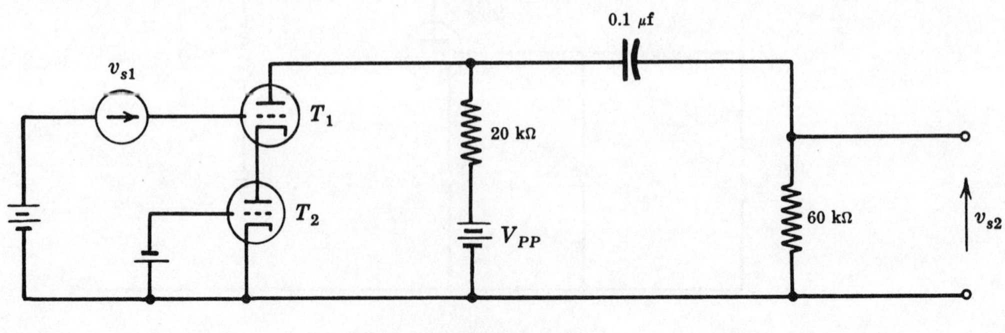

Fig. 5-93

5.34. Given the circuit of Fig. 5-94. The tube parameters are $\mu = 20$ and $r_p = 10{,}000\ \Omega$. Assume D is an ideal diode and sketch at least one cycle of the voltage v_{out}.

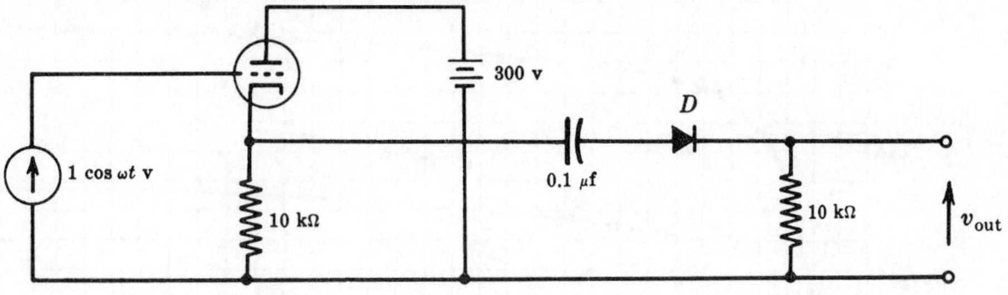

Fig. 5-94

5.35. Work Problem 5.3, page 99, by using the constant current generator representations for the two tubes. This illustrates that the most convenient representation should be used.

5.36. Given the circuit of Fig. 5-95 and the static plate characteristics for a 6AV6 tube in Fig. 5-72, page 109. Take $v_{s1} = 1 \sin \omega t$ volts. What are the maximum and minimum values of the instantaneous plate resistance for the given conditions?

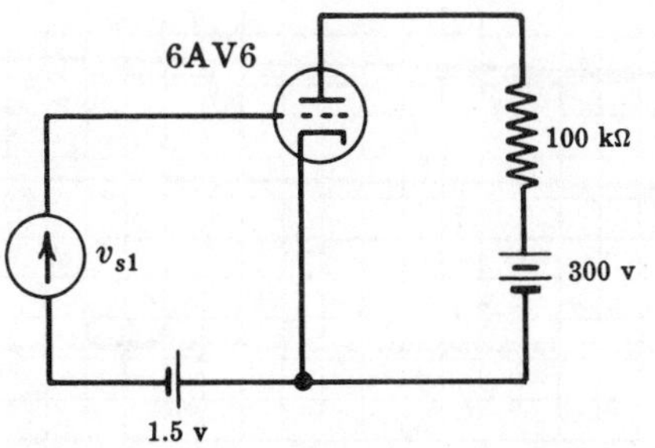

Fig. 5-95

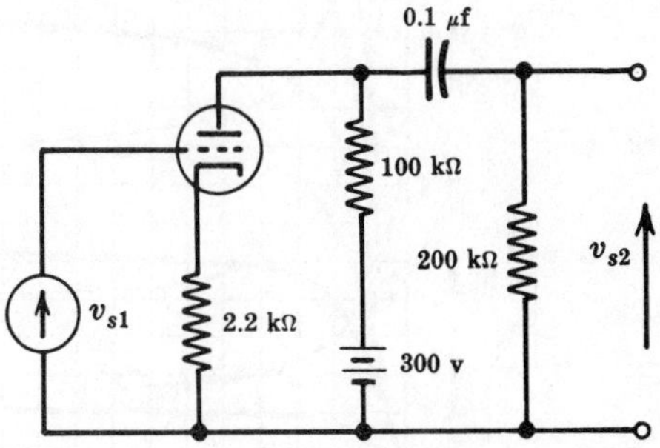

Fig. 5-96

5.37. Given the circuit of Fig. 5-96 and the tube parameters $g_m = 3100\ \mu$mho and $r_p = 10$ kΩ. (*a*) Determine the mid-band voltage gain $A_v = v_{s2}/v_{s1}$. (*b*) Add a 100 μf capacitor from cathode to ground and find the voltage gain. Compare with the answer to part (*a*).

5.38. Given the circuit of Fig. 5-97 and the plate characteristic curves for a 6AU6 tube in Fig. 5-98 below. (*a*) Estimate V_P, I_P, V_{G1}, V_{G2} and I_{G2} using the static plate characteristic curves. (*b*) Discuss some of the sources of error in part (*a*). (*c*) Determine g_m for the given operating point. Plot any needed curves. (*d*) Assuming r_p is infinite, calculate the mid-band voltage gain for the circuit as given.

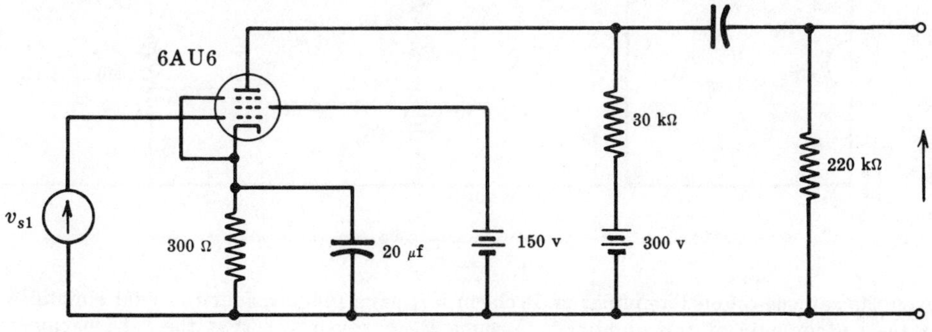

Fig. 5-97

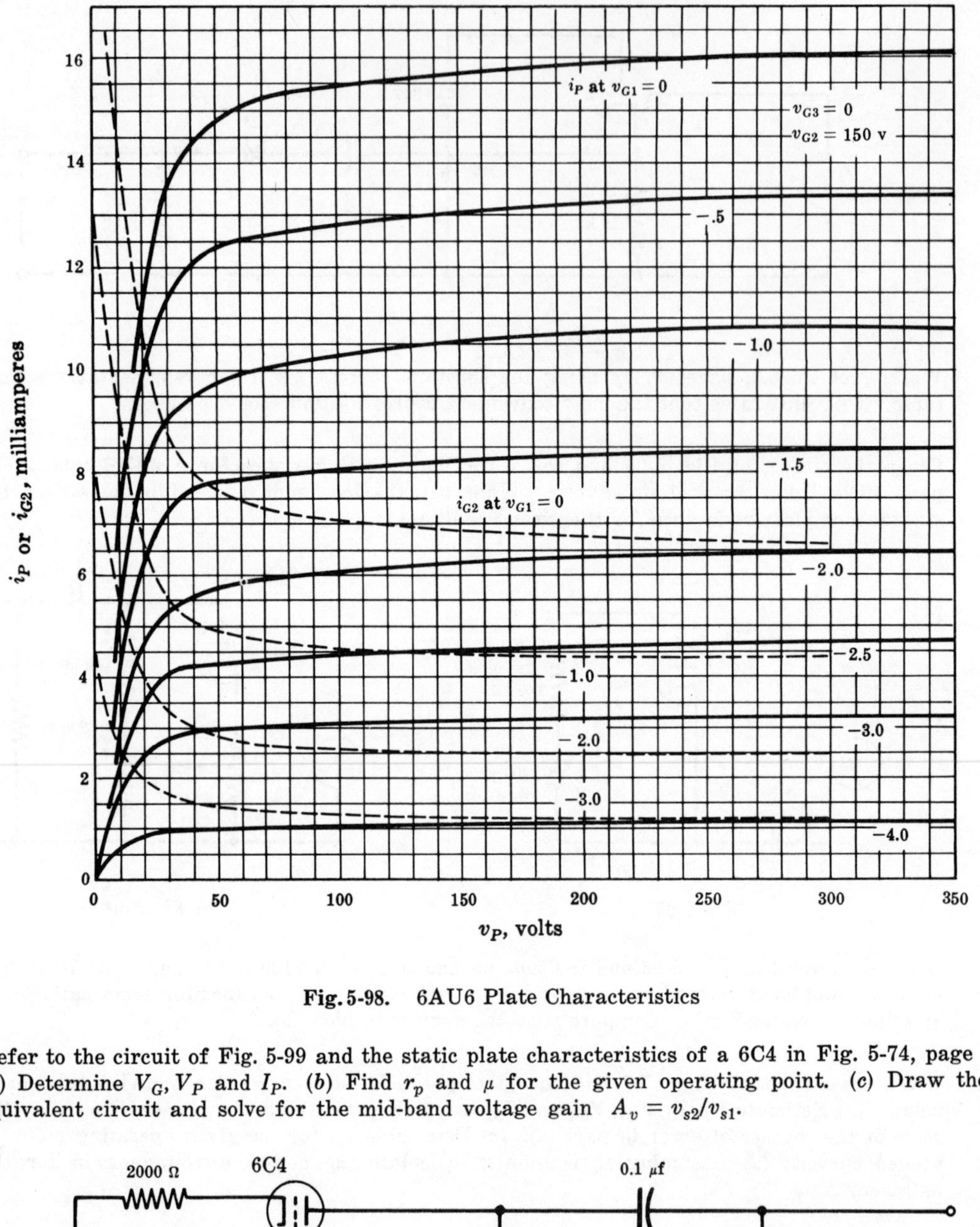

Fig. 5-98. 6AU6 Plate Characteristics

5.39. Refer to the circuit of Fig. 5-99 and the static plate characteristics of a 6C4 in Fig. 5-74, page 109. (*a*) Determine V_G, V_P and I_P. (*b*) Find r_p and μ for the given operating point. (*c*) Draw the ac equivalent circuit and solve for the mid-band voltage gain $A_v = v_{s2}/v_{s1}$.

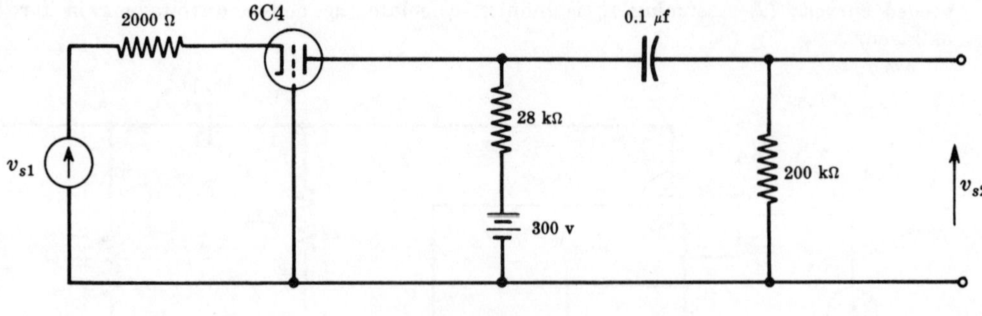

Fig. 5-99

5.40. Given the cathode-coupled amplifier of Problem 5.7, page 104, consider the final simplified expression for the voltage gain of this amplifier. Assume R_K is given and that the tube parameters are the same for both tubes and are constant. Find the value of R_L^* (combination of R_L and R_G) in terms of the remaining parameters which will maximize the voltage gain.

5.41. Given the cathode-coupled amplifier of Problem 5.7, page 104, and the following circuit parameters: $R_K = 2200\ \Omega$, $R_L = 100\ \text{k}\Omega$, $R_G = 470\ \text{k}\Omega$. The parameters for each tube are $\mu = 70$, $r_p = 45\ \text{k}\Omega$. Calculate the voltage gain of this amplifier by writing the circuit equations and solving, rather than substituting into the formula.

5.42. Given the amplifier circuit of Fig. 5-100 and the plate characteristic curves for a 6AV6 tube in Fig. 5-72, page 109. The "grid bias" (V_G) is to be approximately -1.0 volt. Choose the value of R_K which will give this operating point and specify the power rating of this resistor. It should be assumed that only 30% of the rated power can be safely dissipated in the resistor.

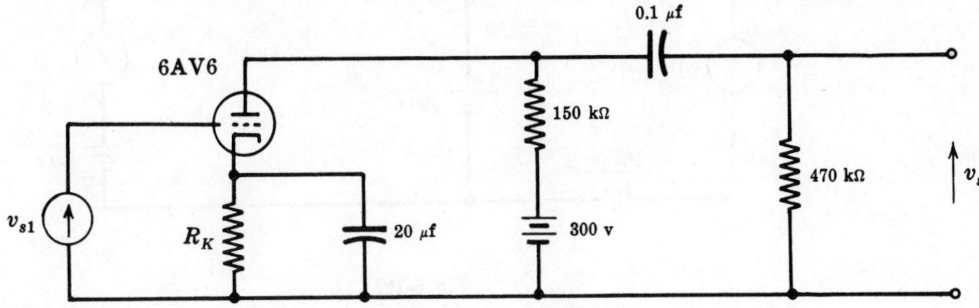

Fig. 5-100

5.43. Given the circuit of Fig. 5-101 and the plate characteristic curves for the phototube in Fig. 5-40, page 94, and the 6C4 plate characteristics in Fig. 5-74, page 109. Assume the phototube is in a projector which has the sound signal on film. Assume a test film is being used and that the instantaneous light flux on the cathode of the phototube is

$$l_K = 0.5 + 0.2 \sin 2000\pi t \text{ lumen}$$

(a) Find the instantaneous value of the output voltage v_{S2}.

(b) Determine the overall transfer function (gain) considering varying components only; this will be v_{s2}/l_k. (*Hint.* See Solved Problem 5.2, page 97.)

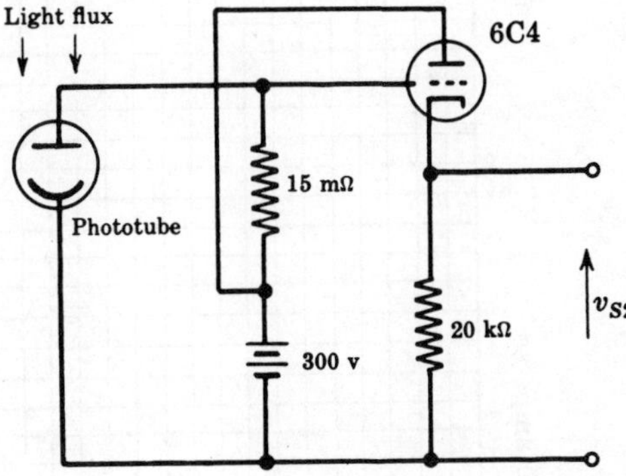

Fig. 5-101

5.44. Consider the final circuit of Problem 5.8, page 106. Assume the applied grid to ground voltage v_{S1} is +20 volts when the relay is to be operated. (a) Specify the power rating of each of the two resistors, considering the conditions in which the maximum power is dissipated. Assume only 50% of the rated power may be safely dissipated in the resistors. (b) Calculate the power dissipated in the relay coil when the 20 volt signal is applied.

5.45. Given the cathode follower circuit of Fig. 5-102 and the tube parameters $\mu = 70$ and $r_p = 44\ \text{k}\Omega$. (a) Draw the ac equivalent circuit. (b) Calculate the voltage gain $A_v = v_{s2}/v_{s1}$.

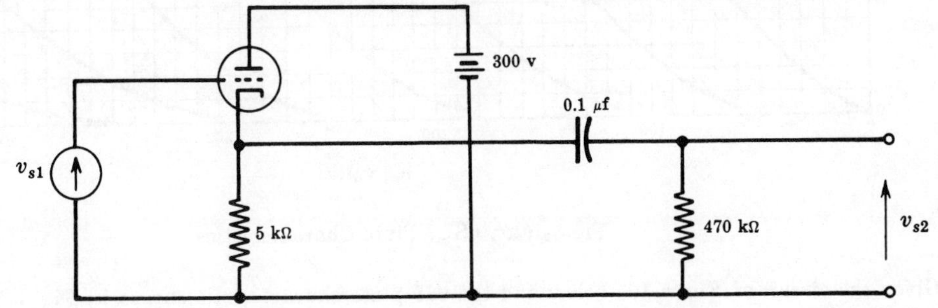

Fig. 5-102

5.46. Given the circuit of Fig. 5-103 and the static plate characteristics of one triode section of a 6SL7 tube in Fig. 5-104. The milliammeter in the plate circuit reads $I_P = 1.48$ ma. What is the value of V_{GG}?

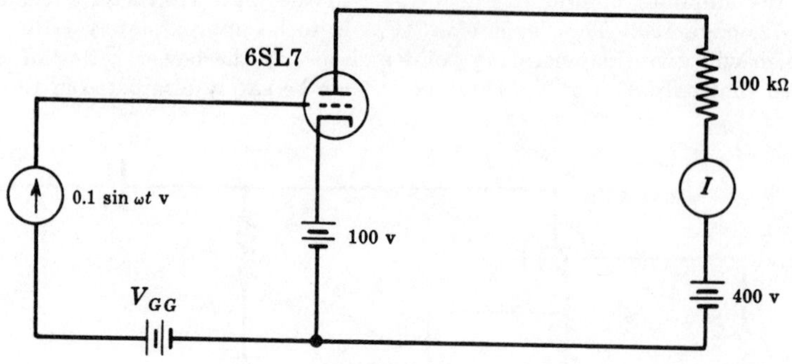

Fig. 5-103

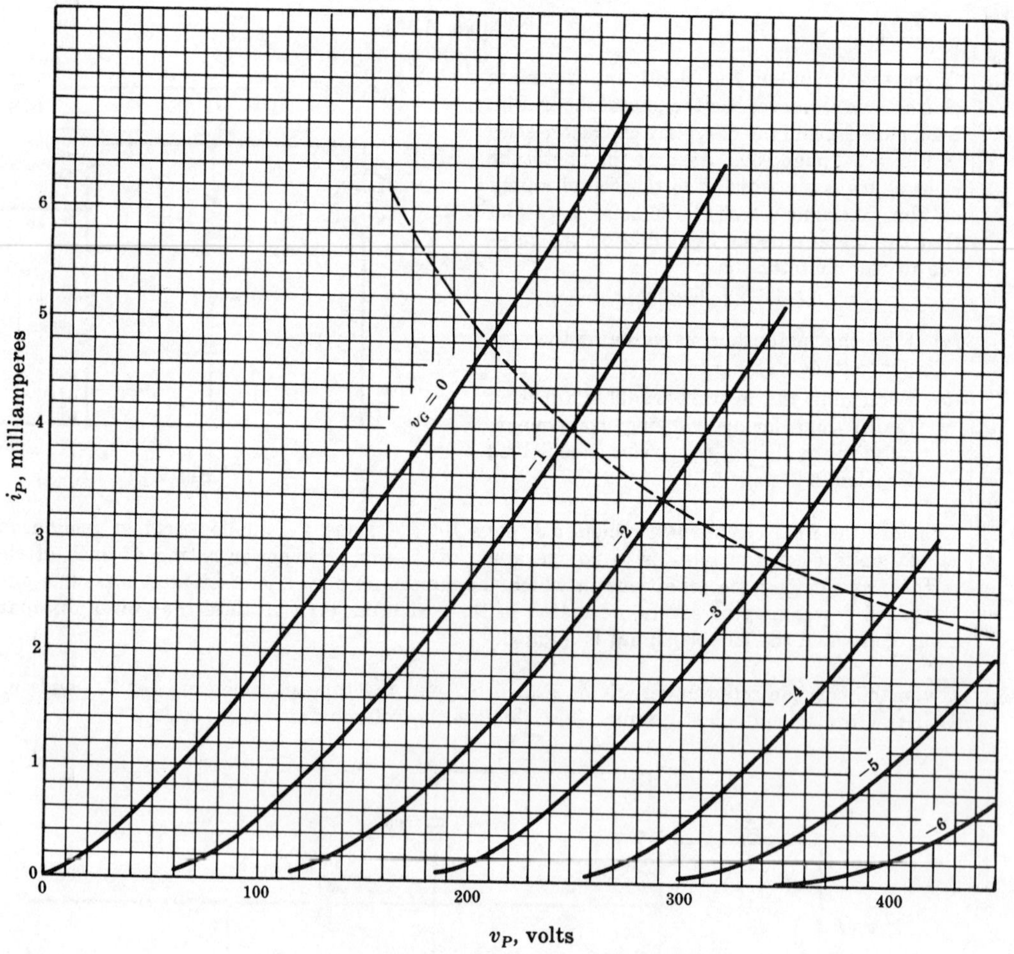

Fig. 5-104. 6SL7 Plate Characteristics

5.47. Given the circuit of Fig. 5-105 below and the 6J5 plate characteristic curves in Fig. 5-70, page 108.

(a) The voltage V_G is measured to be −3.0 volts. What is the value of R_K?

(b) For the above conditions, find V_P and I_P.

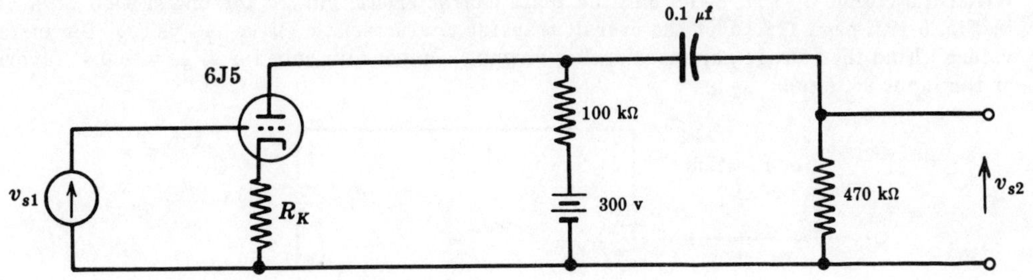

Fig. 5-105

5.48. Given the following equation for the plate voltage of a triode vacuum tube operating on the linear operating range: $v_P = r_p i_P - \mu v_G + k$, where k is a constant. The grid bias for plate current cutoff (plate current approaches zero) is approximately equal to $-v_P/\mu$. Show that this is a valid approximation and indicate the assumptions involved.

5.49. Given the circuit of Fig. 5-106. The two tubes are identical, with $\mu = 20$ and $r_p = 10$ kΩ. Find the overall voltage gain $A_v = v_{s3}/v_{s1}$ (considering varying components only).

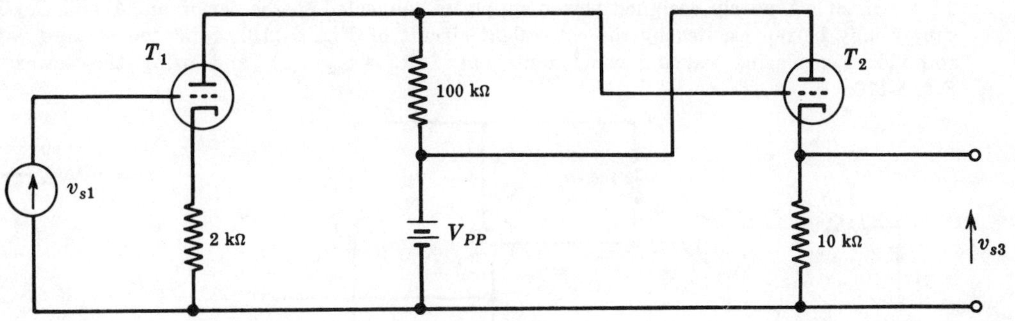

Fig. 5-106

5.50. Two triode tubes are to be connected in parallel. The parameters of the two tubes are

$$T_1\text{:}\ g_{m1} = 3000\ \mu\text{mho},\quad r_{p1} = 6700\ \Omega \qquad T_2\text{:}\ g_{m2} = 1900\ \mu\text{mho},\quad r_{p2} = 8500\ \Omega$$

Find the effective values of the parameters μ, r_p and g_m for the parallel combination.

5.51. Given the circuit of Fig. 5-107 and the parameters for the pentode tube: $g_m = 5100$ μmho, $r_p = 500$ kΩ. Assuming linear operation and mid-band frequencies, calculate (a) the voltage gain $A_v = v_{s2}/v_{s1}$ and (b) the value of μ for the pentode. (c) Assuming the tube parameters remain the same but that R_L is changed from 100 kΩ to 2 kΩ, find the voltage gain.

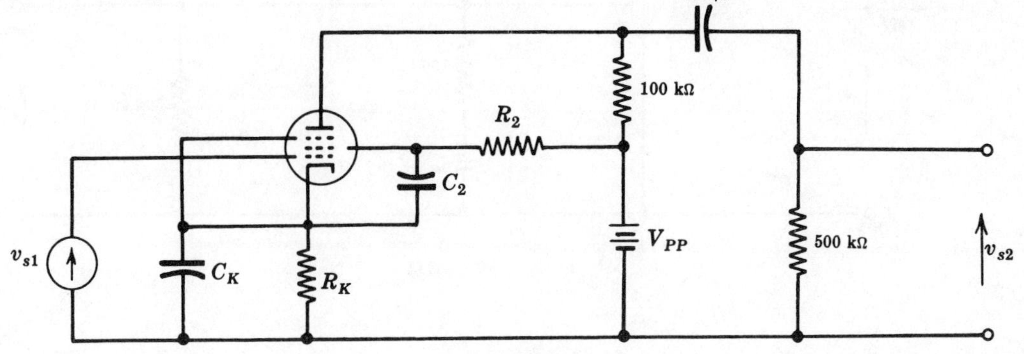

Fig. 5-107

5.52. Given the circuit of Fig. 5-108 and the plate characteristic curves for one section of a 6SL7 tube in Fig. 5-104, page 118. Plot the overall transfer characteristic curve v_{S2} vs. v_{S1} for instantaneous values. Find the voltage gain considering varying components only for $V_{S1} = 100$ v (average value of the input voltage).

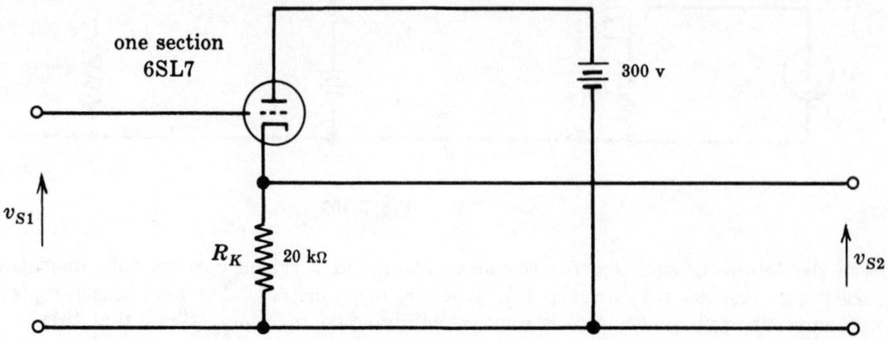

Fig. 5-108

5.53. Given the circuit of Fig. 5-109 and the tube parameters $\mu = 20$, $r_p = 10$ kΩ. The parameters are the same for both tubes. v_{gen} is a sinusoidal source and the generator resistance is included in the 2 kΩ resistor. A poorly designed power supply is connected to the terminals A and B. The power supply may be represented by the equivalent circuit of Fig. 5-110. Find the voltage gain for the amplifier, considering varying components only $(A_v = v_{s2}/v_{gen})$ and using the power supply of Fig. 5-110.

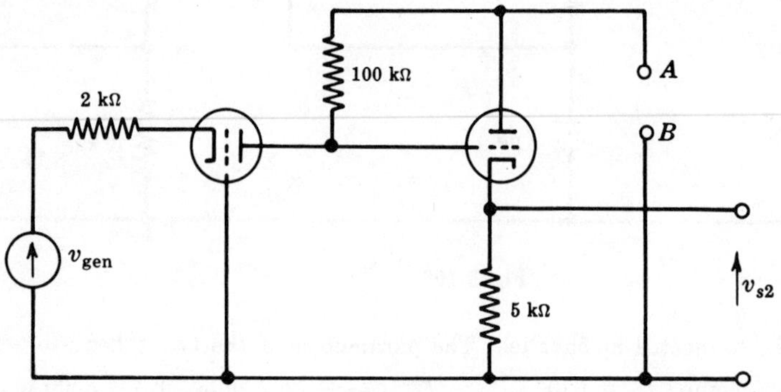

Fig. 5-109

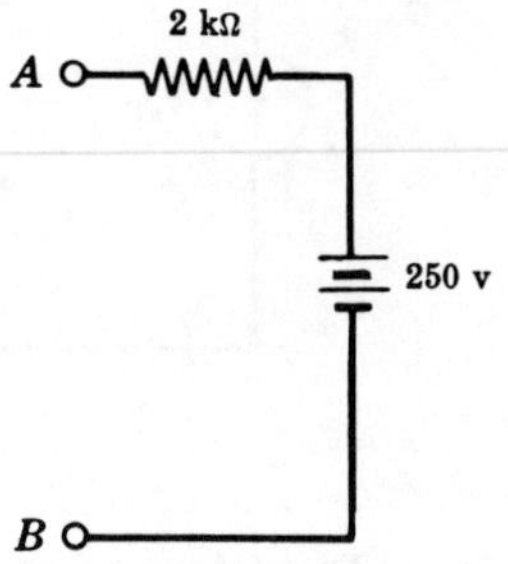

Fig. 5-110

5.54. Given the circuit of Fig. 5-111 and the tube parameters $\mu = 20$ and $r_p = 10$ kΩ. It is assumed that there is provision in the input circuit for the cathode current. Represent the amplifier for the mid-band range of frequencies using the h parameters. Specify the value of each h parameter.

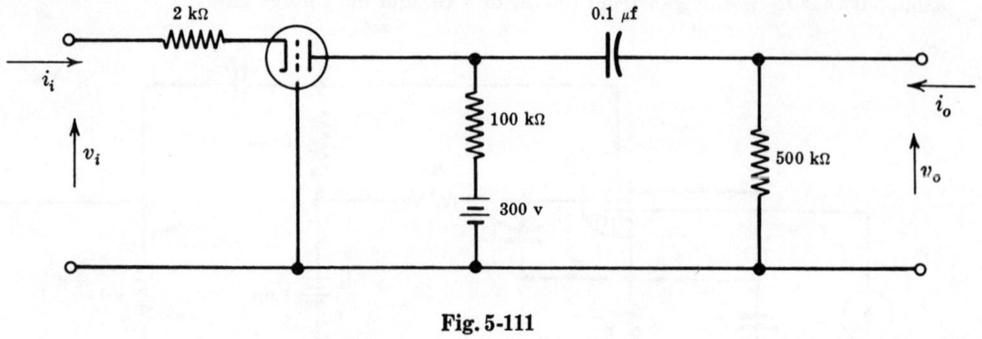

Fig. 5-111

Chapter 6

Two Port Electronic Control Devices, II

6.1 TRANSISTOR SYMBOLS

This chapter is a continuation of the discussion of two port electronic control devices of the resistive class but involves the subclass of electronic control devices for which the input port is not represented as an open circuit. The representations for the electronic control devices will involve two controlled generators as shown for a general two port network in Chapter 4. It must be emphasized that these models apply only to the cases where operation is on the linear range and where only varying components are considered. In general, when instantaneous voltages and currents are considered, a representation involves nonlinear models. One of the most common control devices of this subclass is the transistor. The physical principles of operation of transistors are quite different from those of vacuum tubes, but the major interest in this book is in circuit representations and applications. In this sense, the main difference is that the input port of the transistor is not represented as an infinite resistance.

For the present purposes, a transistor may be considered as a solid state electronic control device with two rectifying *P-N* junctions. A transistor has three terminals: the emitter, base, and collector. The symbols for transistors are shown in Fig. 6-1. The first transistors were of the "point contact" type. Transistor symbols probably were developed to represent this type of transistor physically. The terminals were identified according to the function of the element in the point contact type of transistor. The point contact transistors have been superseded by other types of transistors of which the junction type is probably the most common.

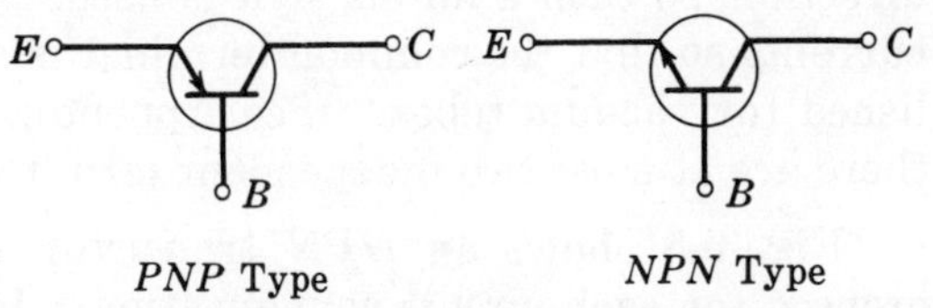

Fig. 6-1. Transistor Symbols

A junction transistor has two rectifying *P-N* junctions formed from extrinsic semiconductor materials of the *P* type and *N* type. The extrinsic semiconductor materials are identified as *N* type if the impurity atoms have five valence electrons, and are *P* type if the impurity atoms have three valence electrons. A *P-N* junction is the interface between a *P* type and *N* type of semiconductor material. The processes of forming junctions and the physical principles are beyond the scope of this book.

Junction transistors are designated as *PNP* or *NPN*, depending on the physical arrangement of the semiconductor materials as shown diagrammatically in Fig. 6-2. It is necessary

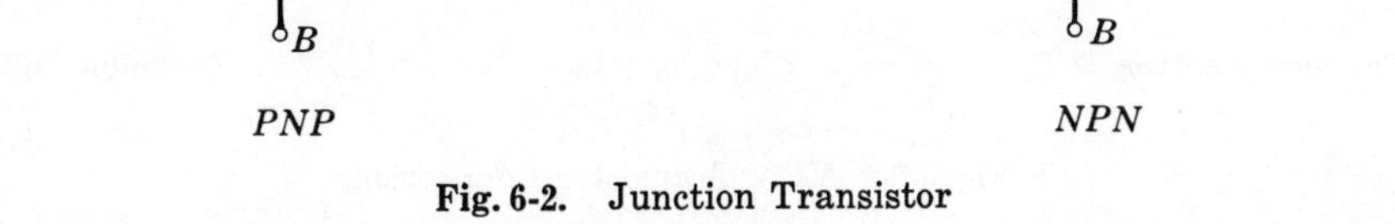

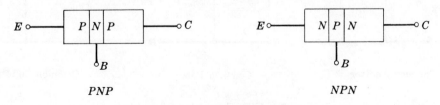

Fig. 6-2. Junction Transistor

to distinguish between the two types because of the problem of power supply polarity. In applications where linear operation is desired, the emitter-base junction is forward biased and the collector-base junction is reverse biased. A *P-N* junction is forward biased when the *P* region is positive with respect to the *N* region. A *P-N* junction has low resistance when it is forward biased and high resistance when reverse biased. When the *P* region is negative with respect to the *N* region, the junction is reverse biased. Hence the power supply polarity for a *PNP* transistor is opposite to that for an *NPN* transistor. Since it is necessary to distinguish between the two types, an arrow is placed on the emitter of the transistor symbol to indicate the actual current direction in the emitter.

In *PNP* transistors, the emitter region is *P* material and the base region is *N* material. When a junction is forward biased, the *P* type is positive, so the direction of current is into the emitter for *PNP* transistors. As shown in Fig. 6-1, the arrow on the emitter of the *PNP* symbol points inward. In *NPN* transistors, the emitter region is *N* material and the base is *P* material. When this junction is forward biased, the current is out of the emitter, so the arrow on the emitter of the *NPN* symbol points outward. The magnitude of the emitter current is the sum of the magnitudes of the base current and collector current (but for brief transient periods). Thus the direction of the base current and collector current can also be determined from the arrow on the emitter symbol.

6.2 TRANSISTOR EQUIVALENT CIRCUITS

In considering transistors as circuit elements, the same procedures will be followed as were used in developing equivalent circuits for other electronic control devices. In general, one terminal is chosen as reference or common terminal. For transistors, this may be any one of the three terminals; however, the emitter is most frequently the common terminal. The current reference directions are into the non-reference terminals. The voltage drops are positive going from the non-reference terminals to the reference terminal. This is true for both *PNP* and *NPN* types. The actual current direction may be opposite to the reference direction, so then a minus sign is used. Two subscripts are used in denoting voltages and currents so that the common terminal is specified. The notation follows the pattern established for vacuum tubes. (See Appendix B.) Since transistors have only three terminals, there are at most two independent terminal voltages and two independent terminal currents.

Fig. 6-3 shows an *NPN* transistor with the instantaneous voltage and current references for each circuit configuration. In *NPN* common emitter connections, the voltages v_{BE} and v_{CE} are positive as are the currents i_{BE} and i_{CE}. This is seen from the direction of the arrow on the emitter of the transistor symbol or can be determined from the bias polarity for each junction. In *NPN* common base connections, v_{CB} and i_{CB} are positive, but i_{EB} and v_{EB} are negative. In *NPN* common collector connections, i_{BC} and v_{BC} are positive, and v_{EC} and i_{EC} are negative.

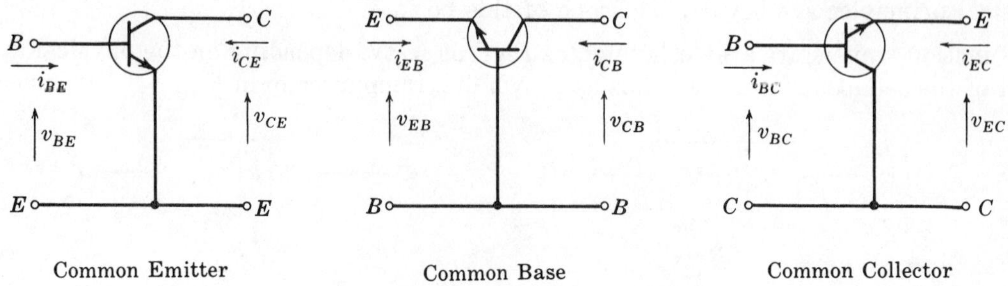

Fig. 6-3. *NPN* Transistor Connections

A *PNP* transistor with instantaneous voltage and current references is shown in Fig. 6-4. Because the references are the same, all the actual instantaneous voltages and currents have signs which are opposite to the corresponding ones for *NPN* transistors. For example, in the *PNP* common base connection, i_{CB} and v_{CB} are negative and i_{EB} and v_{EB} are positive.

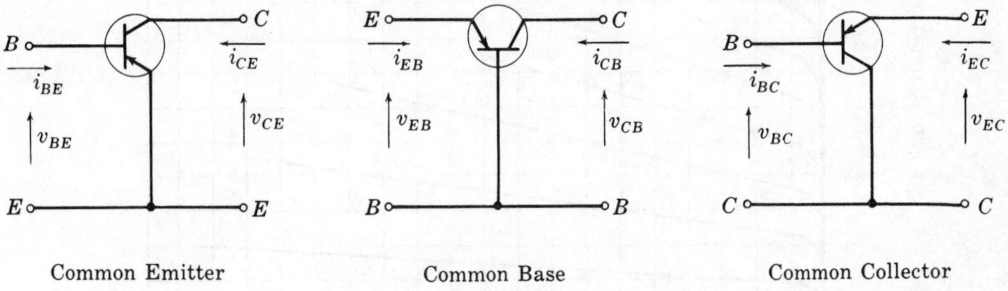

Fig. 6-4. *PNP* **Transistor Connections**

When considering only the terminal characteristics of transistors, it is not necessary to examine the internal conduction processes. In the external circuit, only electrons are moving, so an external current involves the flow of electrons which is opposite to the direction of the current. It is actually true that the only charges which physically move within a transistor are electrons, but the bound ions in the crystalline semiconductor materials must be accounted for.

The electrical terminal characteristics of transistors involve relationships between voltages and currents. Consider the output port of a transistor with the emitter common. The functional relationships between the voltages and currents are written in the way that is most convenient for developing the desired equivalent circuit. One of the more common equivalent circuits for transistors is the hybrid circuit discussed in Chapter 4 and shown in Fig. 4-11, page 59. With this in mind, the functional relationship is written as

$$i_{CE} = f(v_{CE}, i_{BE}) \tag{6.1}$$

Since v_{BE} and i_{BE} are not independent, only one must be included. Transistor characteristics are in general dependent on the internal temperature, but temperature dependence is neglected because this complicates the problem of developing a representation. It is therefore assumed that the temperature is constant or that the voltages and currents are independent of temperature. The relationship given in equation (*6.1*) is represented geometrically as a surface. For a specific transistor this surface may be approximated by a family of static collector characteristic curves as illustrated in Fig. 6-5 below where the family of collector curves is given for a *PNP* transistor.

Consider the input port of a transistor with the emitter common. Since the hybrid equivalent circuit is being developed, the functional relationship is

$$v_{BE} = f(i_{BE}, v_{CE}) \tag{6.2}$$

This is also represented geometrically as a surface which may be approximated by a family of static base characteristic curves as illustrated in Fig. 6-6 below. In transistors there is generally a non-zero signal current in the input port, so that the transistor representation includes two controlled generators.

The equivalent circuit for the output port is obtained by expanding equation (*6.1*) about an operating point by Taylor's series. When operation is within the linear range, the second and higher derivatives are zero; then

$$i_{CE} = f(V_{CE}, I_{BE}) + (v_{CE} - V_{CE})\frac{\partial i_{CE}}{\partial v_{CE}} + (i_{BE} - I_{BE})\frac{\partial i_{CE}}{\partial i_{BE}} \tag{6.3}$$

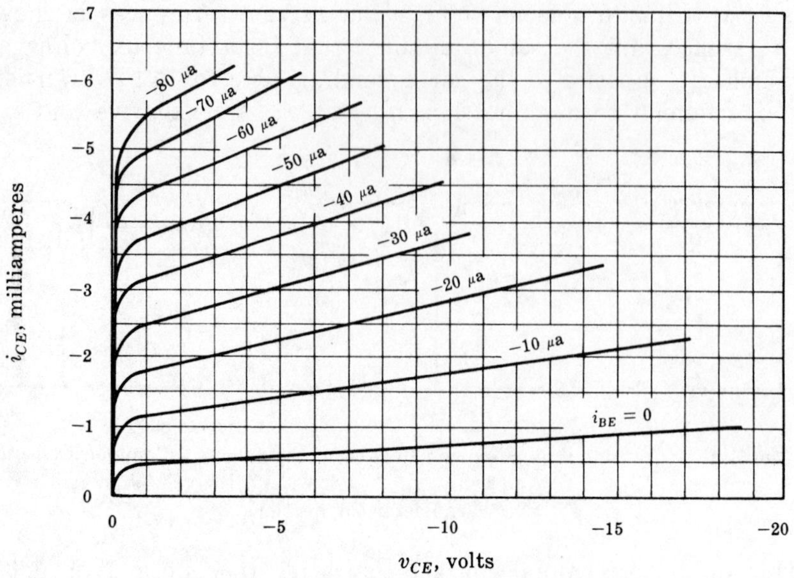

Fig. 6-5. Static Collector Characteristic Curves for a *PNP* Transistor – Common Emitter

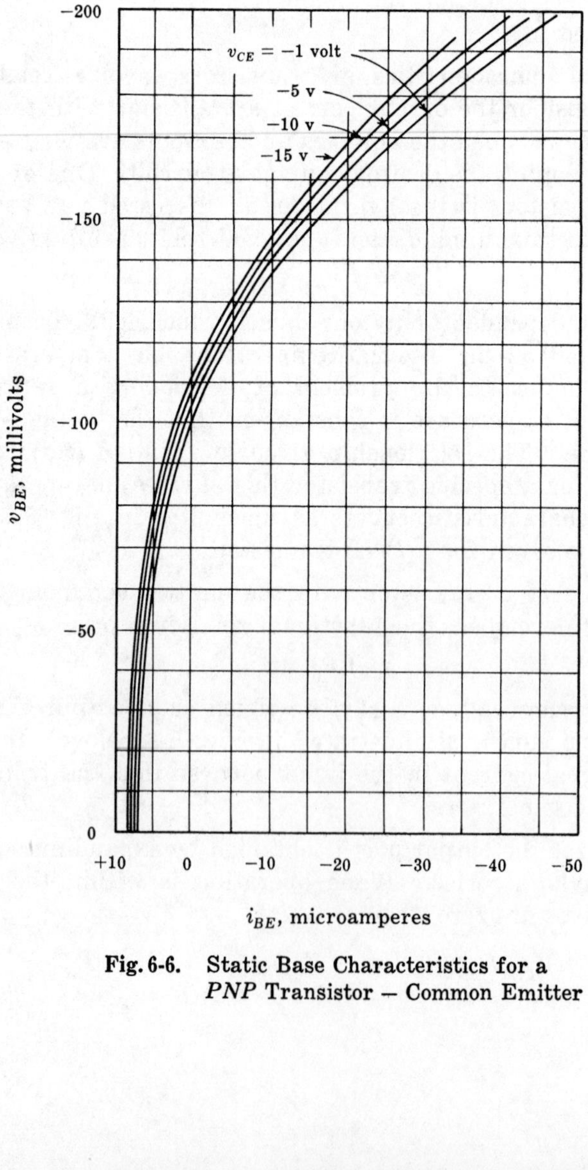

Fig. 6-6. Static Base Characteristics for a *PNP* Transistor – Common Emitter

For linear operation, the instantaneous values may be written as the sum of the average values and instantaneous values of the varying components, giving

$$i_{CE} = I_{CE} + i_{ce} = f(V_{CE}, I_{BE}) + (V_{CE} + v_{ce} - V_{CE})\frac{\partial i_{CE}}{\partial v_{CE}} + (I_{BE} + i_{be} - I_{BE})\frac{\partial i_{CE}}{\partial i_{BE}} \quad (6.4)$$

The operating point (which involves average values) is found graphically. The terms containing only varying components are written

$$-i_{ce} + (\partial i_{CE}/\partial v_{CE})v_{ce} + (\partial i_{CE}/\partial i_{BE})i_{be} = 0 \quad (6.5)$$

In the hybrid circuit for transistors, the parameters are identified as

$$\partial i_{CE}/\partial v_{CE} = h_{oe} = h_{22e} \quad \text{and} \quad \partial i_{CE}/\partial i_{BE} = h_{fe} = h_{21e} \quad (6.6)$$

The parameter h_{oe} is the differential collector conductance. Letter subscripts are used because it is convenient to include the letter (*e* in this case) specifying the common terminal. The letter *o* indicates output; hence a parameter with a subscript *o* will be either the differential output resistance or conductance, depending on the equivalent circuit examined. The parameter h_{fe} is the forward transfer parameter (*f* indicates forward), which in the hybrid circuit is dimensionless. Substituting into equation (*6.5*),

$$-i_{ce} + h_{oe}v_{ce} + h_{fe}i_{be} = 0 \quad (6.7)$$

which may be represented as a current controlled, current generator equivalent circuit.

The input port equivalent circuit is developed by expanding equation (*6.2*) about an operating point by Taylor's series. Assuming linear operation,

$$v_{BE} = f(I_{BE}, V_{CE}) + (i_{BE} - I_{BE})\frac{\partial v_{BE}}{\partial i_{BE}} + (v_{CE} - V_{CE})\frac{\partial v_{BE}}{\partial v_{CE}} \quad (6.8)$$

Since linear operation is assumed, the terms involving varying components only are written

$$-v_{be} + \frac{\partial v_{BE}}{\partial i_{BE}}i_{be} + \frac{\partial v_{BE}}{\partial v_{CE}}v_{ce} = 0 \quad (6.9)$$

The input port parameters are denoted

$$\partial v_{BE}/\partial i_{BE} = h_{11e} = h_{ie} \quad \text{and} \quad \partial v_{BE}/\partial v_{CE} = h_{12e} = h_{re}$$

Substituting into equation (*6.9*),

$$-v_{be} + h_{ie}i_{be} + h_{re}v_{ce} = 0 \quad (6.10)$$

The subscript *i* designates input parameter, which in this hybrid circuit (common emitter) is the differential base resistance. The subscript *r* indicates reverse transfer parameter, which is a dimensionless voltage ratio (partial derivative) in this hybrid circuit. Equation (*6.10*) may be represented as a voltage controlled, voltage generator equivalent circuit.

Equations (*6.7*) and (*6.10*) may each be considered term by term, giving the complete transistor equivalent circuit of Fig. 6-7. This equivalent circuit represents the transistor for varying components only.

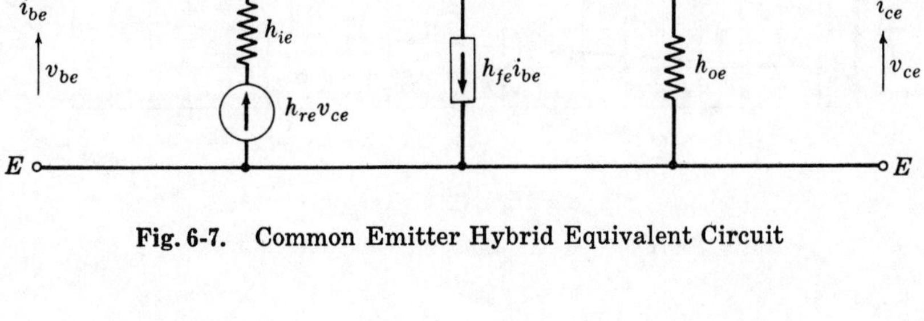

Fig. 6-7. Common Emitter Hybrid Equivalent Circuit

6.3 COMMON EMITTER AMPLIFIER

The operating point of a transistor is found graphically, because of the nonlinear relationships between voltages and currents when large ranges are involved. The complete circuit is considered and then the transistor input and output port equations are written based on average values (dc). As a simple example, consider the circuit of Fig. 6-8 which shows a *PNP* transistor in a common emitter circuit. When the emitter is common, both power supplies have the same polarity and hence the base power supply may be eliminated as will be shown later. Since the transistor is a *PNP* type, all the transistor voltages and currents are negative in the common emitter connection. However, the equations are written in terms of the reference polarities.

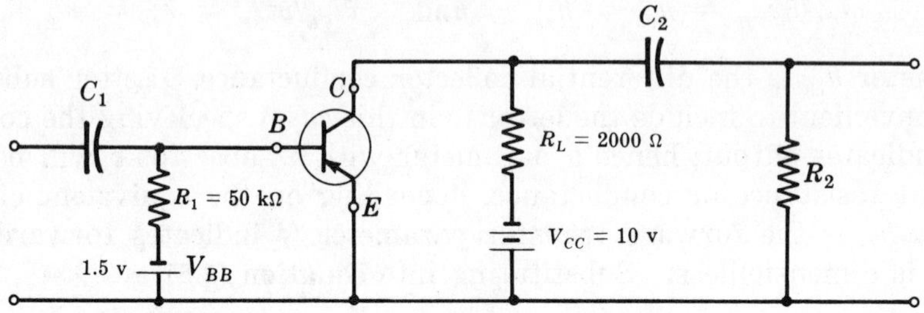

Fig. 6-8. Common Emitter Amplifier Circuit

The collector loop equation based on average (dc) values going in the *reference* direction for the collector current is

$$+V_{CC} + R_L I_{CE} + V_{CE} = 0 \quad \text{or} \quad -(R_L/V_{CC})I_{CE} - (1/V_{CC})V_{CE} = +1 \qquad (6.11)$$

which is the equation of a straight line in the intercept form with an i_{CE} intercept $-V_{CC}/R_L$ and a v_{CE} intercept $-V_{CC}$. To illustrate numerically, consider the values given in Fig. 6-8 and the transistor characteristics given in Figs. 6-9 and 6-10. The i_{CE} intercept is $-10/2000 = -5$ ma and the v_{CE} intercept is -10 v. The collector load line is drawn as in Fig. 6-9. At this time it is only known that the operating point is on this load line, and the exact point must be determined by examining the base equation.

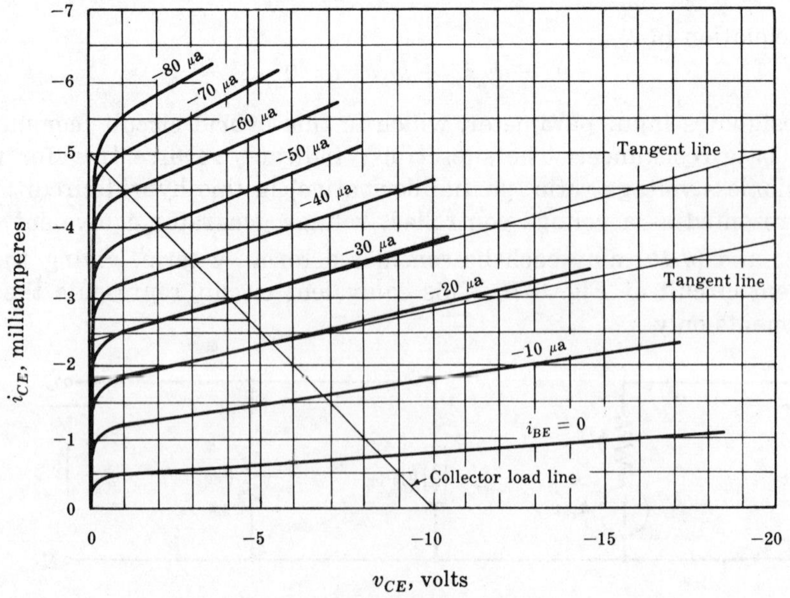

Fig. 6-9. Static Collector Characteristics

The base loop equation based on average values going in the *reference* direction for the base current is

$$+V_{BB} + R_1 I_{BE} + V_{BE} = 0 \quad \text{or} \quad -(R_1/V_{BB})I_{BE} - (1/V_{BB})V_{BE} = +1 \qquad (6.12)$$

which is in the intercept form with an i_{BE} intercept $-V_{BB}/R_1$ and a v_{BE} intercept $-V_{BB}$. Using the values given in Fig. 6-8, the i_{BE} intercept is $-1.5/50{,}000 = -30\ \mu A$ and the v_{BE} intercept is -1.5 volts. The base load line is plotted on the static base characteristic curves of Fig. 6-10. There is one problem: the -1.5 volts intercept on the v_{BE} axis is not on the graph. However, only two points are needed to locate a line; hence instead of the v_{BE} intercept, the point $v_{BE} = -200$ mv is used so that the two points are far apart, giving the best accuracy. The corresponding value of i_{BE} is found by substituting in equation (*6.12*):

$$1.5 + 50{,}000 i_{BE} + (-0.2) = 0$$

or

$$i_{BE} = -26\ \mu A$$

The base load line may now be plotted using the i_{BE} intercept and the point $i_{BE} = -26\ \mu A$ and $v_{BE} = -200$ mv as shown in Fig. 6-10. It is known that the operating point will lie on the base load line and also on the collector load line, hence the operating point is the point common to both lines. In general the procedure is to transfer one of the load lines (then not necessarily a straight line) to the other family of curves. Since the base characteristic curves are more clustered, the accuracy will be improved if the collector load line is transferred to the base characteristic curves on a point by point basis, giving the curve shown in Fig. 6-10. It is actually necessary only to draw that portion of the curve which intersects the base load line. In plotting the collector load line on the static base characteristic curves, it is necessary to linearly interpolate between the base curves for different values of collector voltage. For the numerical example, the operating point involves an average base current $I_{BE} = -27\ \mu A$ and an average base voltage $V_{BE} = -175$ mv. Better accuracy is obtained by plotting this point on the collector characteristic curves and then finding the collector voltage and current from the static collector characteristic curves as shown in Fig. 6-9. Again it is necessary to linearly interpolate between collector characteristic curves for different values of base current as indicated in Fig. 6-9. At the operating point, $V_{CE} = -4.6$ v and $I_{CE} = -2.7$ ma.

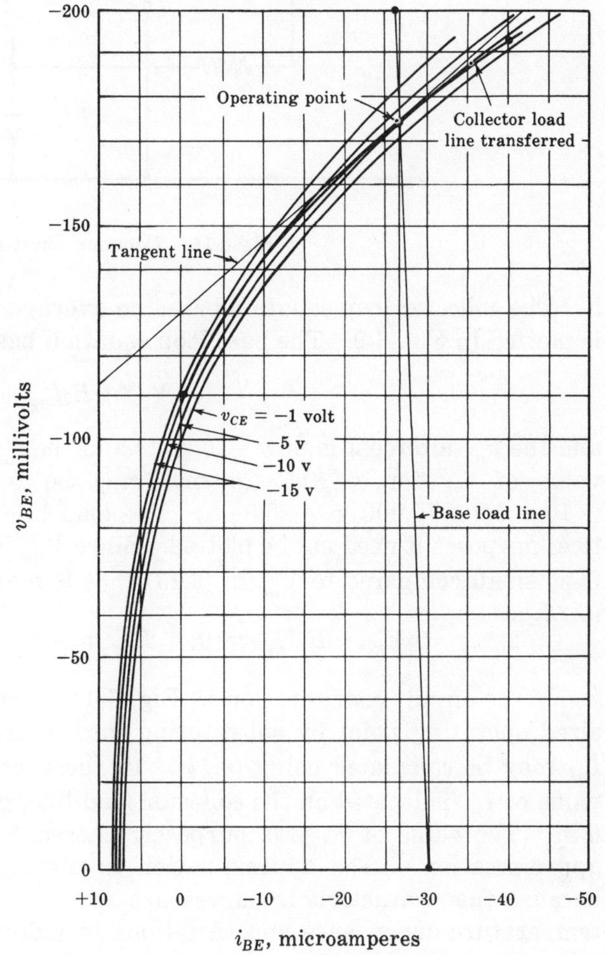

Fig. 6-10. Static Base Characteristics

When the common emitter connection is used, both the base and collector power supplies have the same polarity with respect to ground. The circuit of Fig. 6-8 was used to illustrate a more general approach; in practice there is only one power supply for a common emitter circuit as shown in Fig. 6-11. The value of R_1 in this circuit is larger because of the higher power supply voltage.

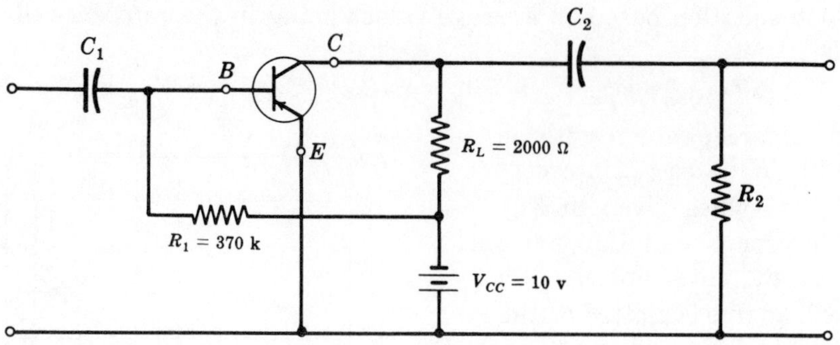

Fig. 6-11. Common Emitter Amplifier Circuit

The collector loop equation based on average values is the same as equation (*6.11*) which is plotted in Fig. 6-9. The base loop equation based on average values is now

$$+V_{CC} + R_1 I_{BE} + V_{BE} = 0 \qquad (6.13)$$

and the v_{BE} intercept is now -10 volts and the i_{BE} intercept is $-10/370{,}000 = -27\ \mu$A. The value of i_{BE} for $v_{BE} = -200$ mv may be found by substituting into (*6.13*): $I_{BE} = (-10 + 0.2)/370{,}000 = -26.5\ \mu$A. The load line is approximately vertical, so for all practical purposes it need not be plotted. Since V_{BE} is less than 0.2 volts and V_{CC} is 10 volts, V_{BE} is so small compared to V_{CC} in (*6.13*) that it may be neglected. Using this approximation, (*6.13*) is

$$V_{CC} + R_1 I_{BE} = 0, \quad I_{BE} = -V_{CC}/R_1 \quad \text{or} \quad R_1 = -V_{CC}/I_{BE} \qquad (6.14)$$

If the circuit configuration of Fig. 6-11 is used, then R_1 could be chosen to give the desired operating point by substituting into (*6.14*). If the values of R_1 and V_{CC} are given, I_{BE} may be calculated using (*6.14*). In the example, $I_{BE} = -10/370{,}000 = -27\ \mu$A. This value of I_{BE} is located on the collector load line, giving the operating point as shown in Fig. 6-9. (The value of R_1 was purposely chosen to give the same operating point using the approximation.) The approximation involved in (*6.14*) does not cause significant error because the characteristic curves are averages for a number of transistors and neglect temperature dependence and variations in individual transistors.

The circuit of Fig. 6-11 is simple and useful as an illustration but has serious limitations in practice. The base current is kept nearly constant, but this means only that the difference in magnitude between the emitter and collector currents is held constant. The collector current could increase beyond the maximum safe value; hence other circuits are used where "bias stability" is an important consideration. Because the base to emitter resistance is low, the series combination of the 370 kΩ resistor and 10 v battery approximates a constant current source. If the base current is held constant, the average collector current will vary from transistor to transistor, depending on its characteristics.

Having found the operating point for the transistor, the transistor parameters may be determined. The parameter h_{oe} is the differential collector conductance and is found by drawing a line tangent to the static collector characteristic curve at the operating point and then finding the slope of the tangent line using large increments. If the operating point does not fall on a collector characteristic curve, then either a curve is sketched through the operating point by linearly interpolating between the two adjacent curves or the slopes of the adjacent curves are found at the intersections with the load line. The slopes of the two adjacent collector characteristic curves are usually approximately the same as the slope of any inbetween curve unless the transistor is intended for low signal levels. In the example under consideration, the slopes of the collector curves for

$i_{BE} = -20\ \mu A$ and $i_{BE} = -30\ \mu A$ are found from the tangent lines drawn in Fig. 6-9. The slope of the line tangent to the $i_{BE} = -20\ \mu A$ curve is

$$\text{slope} = (-3.65 + 1.75)10^{-3}/(-20 - 0) = 95\ \mu\text{mho}$$

The slope of the line tangent to the $i_{BE} = -30\ \mu A$ curve is

$$\text{slope} = (-4.85 + 2.35)10^{-3}/(-20 - 0) = 125\ \mu\text{mho}$$

A value for h_{oe} for the given operating point can be found by assuming the slope changes linearly between the $i_{BE} = -20\ \mu A$ and $i_{BE} = -30\ \mu A$ curves, and then interpolating linearly between the slopes of the two curves. The change in slope is 30 μmho. Since $I_{BE} = -27\ \mu A$ taking 0.7(30 μmho) gives 21 μmho. Hence $h_{oe} = (95 + 21)\ \mu\text{mho} = 116\ \mu\text{mho}$.

An estimate of h_{oe} could have been based on the slopes of the two curves, since the characteristic curves are averages and the accuracy of graphical methods is limited. In practice the parameters of a transistor at a given operating point are often determined with electronic instruments.

The parameter h_{fe} is found by plotting the static forward transfer characteristic curve for the given operating point. A vertical line is drawn through the operating point on the collector characteristic curves as shown in Fig. 6-9. The static forward transfer characteristic curve for $V_{CE} = -4.7$ volts is plotted in Fig. 6-12. The operating point is located on this curve and then the slope of the forward transfer characteristic curve at the operating point is found graphically.

$$h_{fe} = [(-5.25 + 0.65)10^{-3}]/[(-60 - 0)10^{-6}] = 76.6$$

The parameter h_{ie} is the differential base resistance and is determined graphically by the slope of the static base characteristic curve at the operating point as shown in Fig. 6-10.

$$h_{ie} = (-0.20 + 0.115)/[(-41 - 10)10^{-6}] = 1665\ \Omega$$

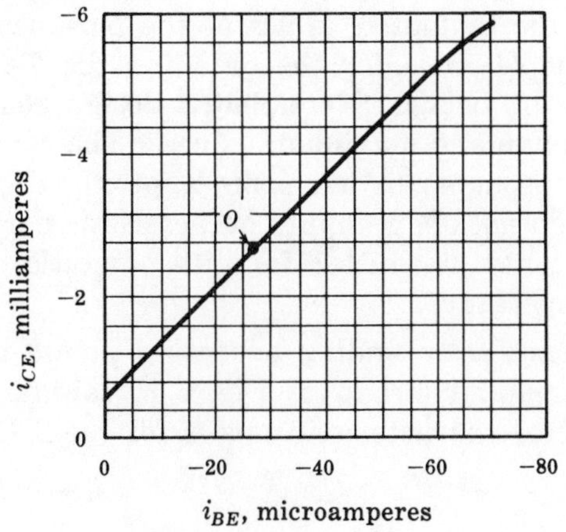

Fig. 6-12. Static Forward Transfer Characteristic Curve

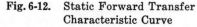

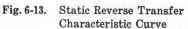

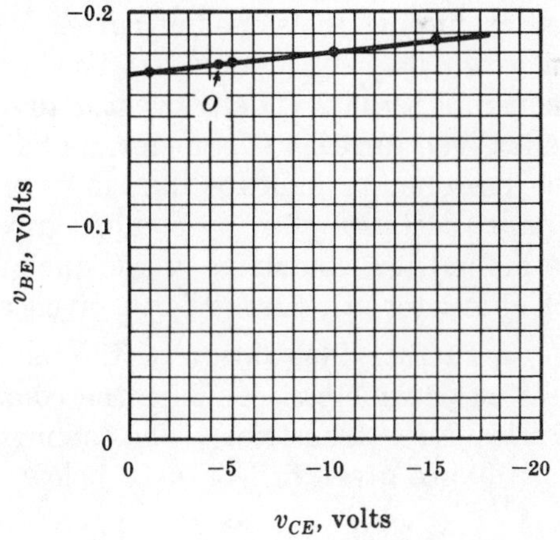

Fig. 6-13. Static Reverse Transfer Characteristic Curve

The parameter h_{re} is the reverse transfer parameter and is the slope of the reverse transfer characteristic curve at the operating point. The vertical line $i_{BE} = I_{BE} = -27\ \mu A$ is drawn on the static base characteristic curves and then the reverse transfer characteristic curve is plotted. The base characteristic curves are so close together that it is difficult to obtain reasonable accuracy. The static reverse transfer characteristic curve is plotted in Fig. 6-13. The slope of this curve at the operating point is

$$h_{re} = (-0.185 + 0.172)/(-15 + 1) = 928 \times 10^{-6}$$

The equivalent circuit (considering only varying components) for the amplifier of Fig. 6-11 may be drawn as shown in Fig. 6-14. In drawing the equivalent circuit, the transistor is replaced by its equivalent circuit. The remainder of the circuit is drawn after applying superposition and then considering the circuit for varying components only. For example, the one terminal of the resistor R_1 is connected to the power supply V_{CC} which has constant terminal voltage, so this terminal is at ground potential considering only varying components.

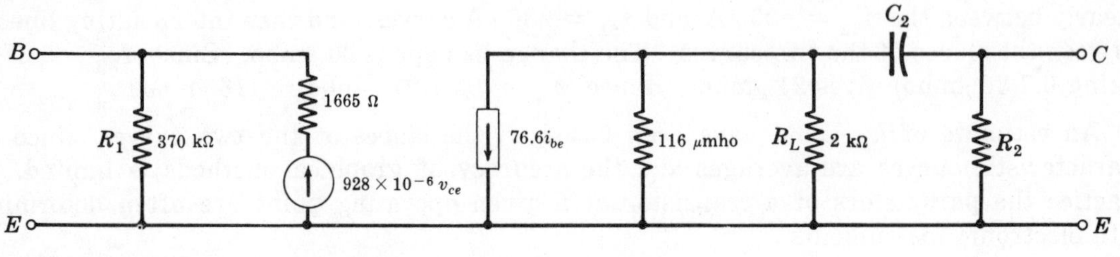

Fig. 6-14. Equivalent Transistor Circuit

The procedure for finding the operating point of the transistor in the circuit of Fig. 6-11 is the same if the transistor is the *NPN* type. The only difference is that the instantaneous and average voltages and currents have opposite signs for the *NPN* type. When only varying components are considered, the circuits are the same for *NPN* and *PNP* types of transistors.

6.4 COMMON BASE AMPLIFIER

One of the purposes of the present discussion is to emphasize the definitions of the transistor equivalent circuit parameters. When these are partial derivatives, the slopes of curves are involved. In practice the values of the parameters are not necessarily obtained by finding the slopes of curves. Frequently the tabulated values of the parameters for a given operating point are used. As an example, consider the data for the Texas Instruments Type 2N117 transistor given in Table 6.1 below. Note that a design center value is given for each parameter, and also the minimum and maximum values. The ranges of the parameters indicate that in practice it is reasonable to use some approximations. The data sheets supplied by manufacturers usually do not show input characteristic curves (base or emitter), but have the common base and common emitter families of static collector characteristic curves for a given ambient temperature.

As a further illustration of the methods for graphically finding operating points and transistor parameters, consider the common base amplifier circuit of Fig. 6-15 containing the 2N117 *NPN* transistor. The common base collector characteristic curves for the 2N117 transistor are given in Fig. 6-16 below.

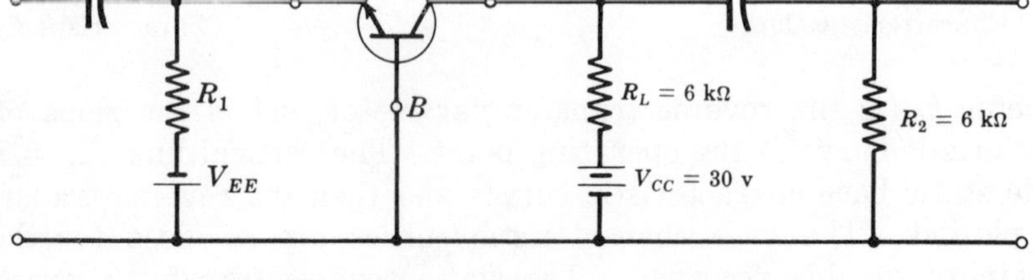

Fig. 6-15. Common Base Amplifier Circuit

absolute maximum ratings at 25°C ambient [except where advanced temperatures are indicated]

Collector Voltage Referred to Base	45 V
Emitter Voltage Referred to Base	1 V
Collector Current	25 mA
Emitter Current	−25 mA
Collector Dissipation	150 mW
at 100°C	100 mW
at 150°C	50 mW

junction temperature

Maximum Range	−65°C to +175°C

common base design characteristics at Tj = 25°C [except where advanced temperatures are indicated]

		test conditions		min.	design center	max.	unit
BV_{CBO}	Collector Breakdown Voltage	$I_C = 50\mu A$	$I_E = 0$	45	—	—	Volt
I_{CBO}	Collector Cutoff Current}	$V_{CB} = 30V$	$I_E = 0$	—	—	2	μA
	at 100° C}	$V_{CB} = 5V$	$I_E = 0$	—	—	10	μA
	at 150° C}	$V_{CB} = 5V$	$I_E = 0$	—	—	50	μA
h_{ib}	Input Impedance	$V_{CB} = 5V$	$I_E = -1mA$	30	42	80	Ohm
h_{ob}	Output Admittance	$V_{CB} = 5V$	$I_E = -1mA$	0.0	0.4	1.2	μmho
h_{rb}	Feedback Voltage Ratio	$V_{CB} = 5V$	$I_E = -1mA$	25	120	500	$X10^{-6}$
h_{fb}	Current Transfer Ratio	$V_{CB} = 5V$	$I_E = -1mA$	−0.9	−0.925	−0.953	—
PG_e	Power Gain*†	$V_{CE} = 20V$	$I_E = -2mA$	—	35	—	db
NF	Noise Figure*‡	$V_{CE} = 5V$	$I_E = -1mA$	—	20	—	db
$f_{\alpha b}$	Frequency Cutoff	$V_{CB} = 5V$	$I_E = -1mA$	—	4	—	mc
C_{ob}	Output Capacitance (1mc)	$V_{CB} = 5V$	$I_E = -1mA$	—	7	—	μμf
R_{cs}	Saturation Resistance*	$I_B = 2.2mA$	$I_C = 5mA$	—	100	200	Ohm

*Common Emitter †R_g = 1k; R_L = 20k ‡Conventional Noise—Compared to 1000 ohm resistor, 1000 cps and 1 cycle band width

TABLE 6.1

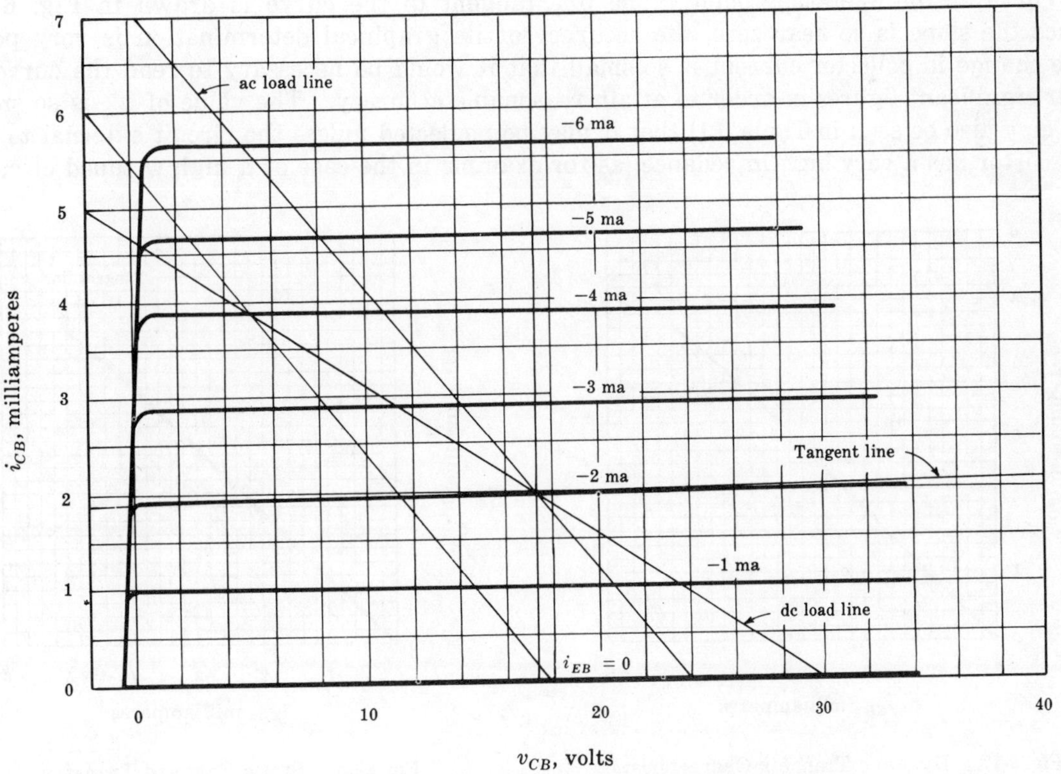

Fig. 6-16. 2N117 Collector Characteristics – Common Base

The differences between common emitter and common base collector characteristic curves may be seen by making comparisons between the two sets given. The magnitude of the emitter current is nearly the same as that of the collector current; hence the common base collector characteristic curves are almost parallel to the v_{CB} axis.

The operating point for the transistor of Fig. 6-15 is to be found. The collector loop equation based on average values is

$$-V_{CC} + R_L I_{CB} + V_{CB} = 0 \quad \text{or} \quad (R_L/V_{CC})I_{CB} + (1/V_{CC})V_{CB} = +1 \qquad (6.15)$$

Using the values given in Fig. 6-15, the i_{CB} intercept is $30/6000 = 5$ ma and the v_{CB} intercept is 30 v. The collector load line is plotted in Fig. 6-16. Usually the input circuit (emitter in this case) is designed to give approximately the desired operating point.

There are many considerations affecting the choice of operating point. In some cases maximum linear operating range is most important; in other cases maximum gain or efficiency may be more significant. In this example assume the maximum linear operating range is the most important consideration. The linear operating range is maximum when the operating point is in the middle of the linear range. However, the operating point must be known before the ac load line can be plotted. As an approximation, plot the dynamic transfer characteristic curve based on the dc load line as shown in Fig. 6-17. The center of the linear operating range of the dynamic transfer characteristic curve is approximately $i_{EB} = 2$ ma. Using this as the operating point, the ac load line is drawn through the operating point with a slope of $-1/3000$ as shown in Fig. 6-16. The dynamic transfer characteristic curve based on the ac load line may now be plotted. Since the static collector characteristic curves are almost parallel to the v_{CB} axis, the dynamic transfer characteristic curve based on the ac load line is for all practical purposes the same as that based on the dc load line.

The differential collector conductance h_{ob} is the slope of the static collector characteristic curve at the operating point. The line tangent to the curve is drawn in Fig. 6-16. Since the slope is so near zero, the accuracy of its graphical determination is very poor. The change in collector current is so small that it would be necessary to read the curve to four significant figures in order to attain reasonable accuracy. The value of h_{ob} is so small (as may also be seen in Table 6.1) that it may be neglected unless the circuit external to the transistor has a very high impedance (as for example in the case of a high Q tuned circuit).

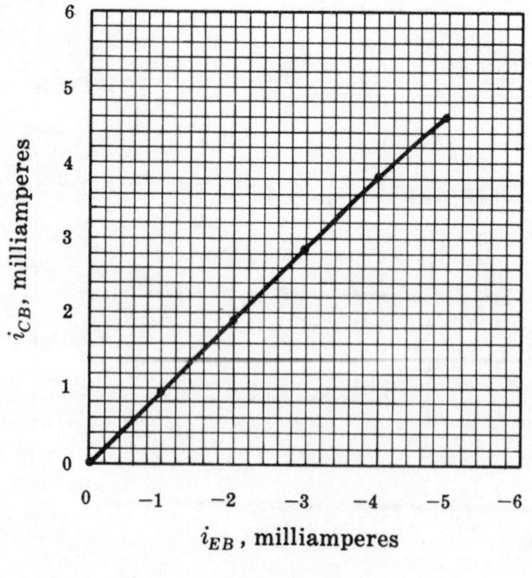

Fig. 6-17. Dynamic Transfer Characteristic Based on the dc Load Line

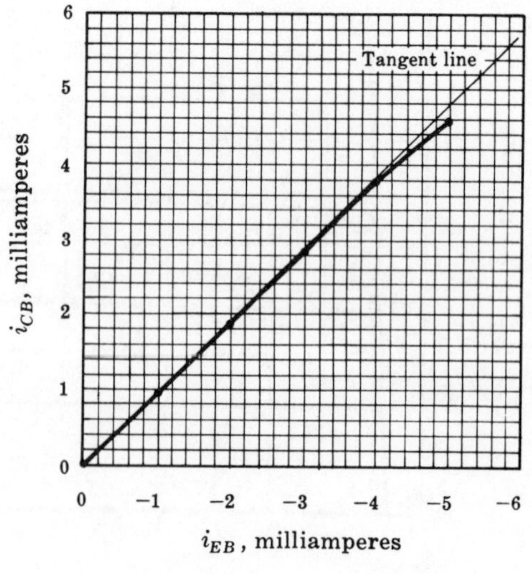

Fig. 6-18. Static Forward Transfer Characteristic Curve

The static forward transfer characteristic curve may be plotted as in Fig. 6-18 above. The slope of this curve at the operating point is the parameter h_{fb}. Since the magnitude of the collector current is always (except possibly for brief transient periods) slightly under the magnitude of the emitter current, the magnitude of h_{fb} will be less than 1. The sign of h_{fb} is negative because the emitter current is negative and the collector current is positive as shown in Fig. 6-18. In this example, $h_{fb} = (4.85-0)/(-5-0) = -0.97$.

The static emitter characteristics are not given, so the parameters for the input circuit cannot be determined graphically. The values of the parameters for another operating point are listed in Table 6.1. Note the range of values for the parameters h_{ib} and h_{rb}. The tabulated value is usually sufficiently accurate. Transistor data sheets may include graphs which may be used to calculate the value of transistor parameters at other operating points. Fig. 6-19 below includes a graph which gives the relative value of the parameters for any collector voltage relative to $v_{CB} = 5$ volts. Another graph gives the relative values of the parameters for emitter currents relative to $i_{EB} = -1$ ma. In addition, there is a graph which gives relative values of the parameters for junction temperatures relative to 25° C.

In the common base connection, the input parameter h_{ib} is a low resistance, about 42 ohms for the type 2N117 as given in Table 6.1. The input resistance to the transistor circuit will be determined in later sections, but this low value of h_{ib} suggests that the input resistance will be low. Since in the circuit of Fig. 6-15 the average value of the emitter current is −2.0 ma, a low value of R_1 is suggested so that V_{EE} is not unreasonably large. Let V_{EE} be 1.5 v. The exact value of V_{EB} cannot be determined without static emitter characteristic curves, which are not available. However, the emitter loop equation based on average values is

$$+V_{EE} + R_1 I_{EB} + V_{EB} = 0 \tag{6.16}$$

It is not necessarily true that V_{EB} will be small compared to $V_{EE} = 1.5$ volts. Assume V_{EB} is −0.2 v; then for $I_{EB} \cong -2$ ma, a value $R_1 = 650$ ohms is calculated by substituting into equation (*6.16*). Since 650 ohm resistors are not available, a nominal value of 680 would probably be used, giving an operating point which is in the order of $I_{EB} = -2$ ma, depending on the characteristics of the individual transistor and exact values of other circuit elements.

Thus in practice many approximations are necessary. The definitions of the transistor parameters involve partial derivatives, but in practice the parameters are usually not evaluated graphically. In design problems, final values of circuit elements and operating points are often determined empirically. There are no completely formal approaches to design; experience is most valuable.

6.5 STABILITY OF THE OPERATING POINT

Temperature effects were neglected in the development of the equivalent circuits for transistors considering only varying components. This is necessary if simple equivalent circuits are desired. To illustrate the problem, consider a *PNP* transistor with the emitter common. Assume that the functional relationship for the collector current includes the transistor junction temperature:

$$i_{CE} = f(i_{BE}, v_{CE}, t_J) \tag{6.17}$$

The total differential of i_{CE} is

$$di_{CE} = (\partial i_{CE}/\partial i_{BE})\, di_{BE} + (\partial i_{CE}/\partial v_{CE})\, dv_{CE} + (\partial i_{CE}/\partial t_J)\, dt_J \tag{6.18}$$

When operation is restricted to the linear range so that each partial derivative is a constant, then (*6.18*) may be integrated, giving

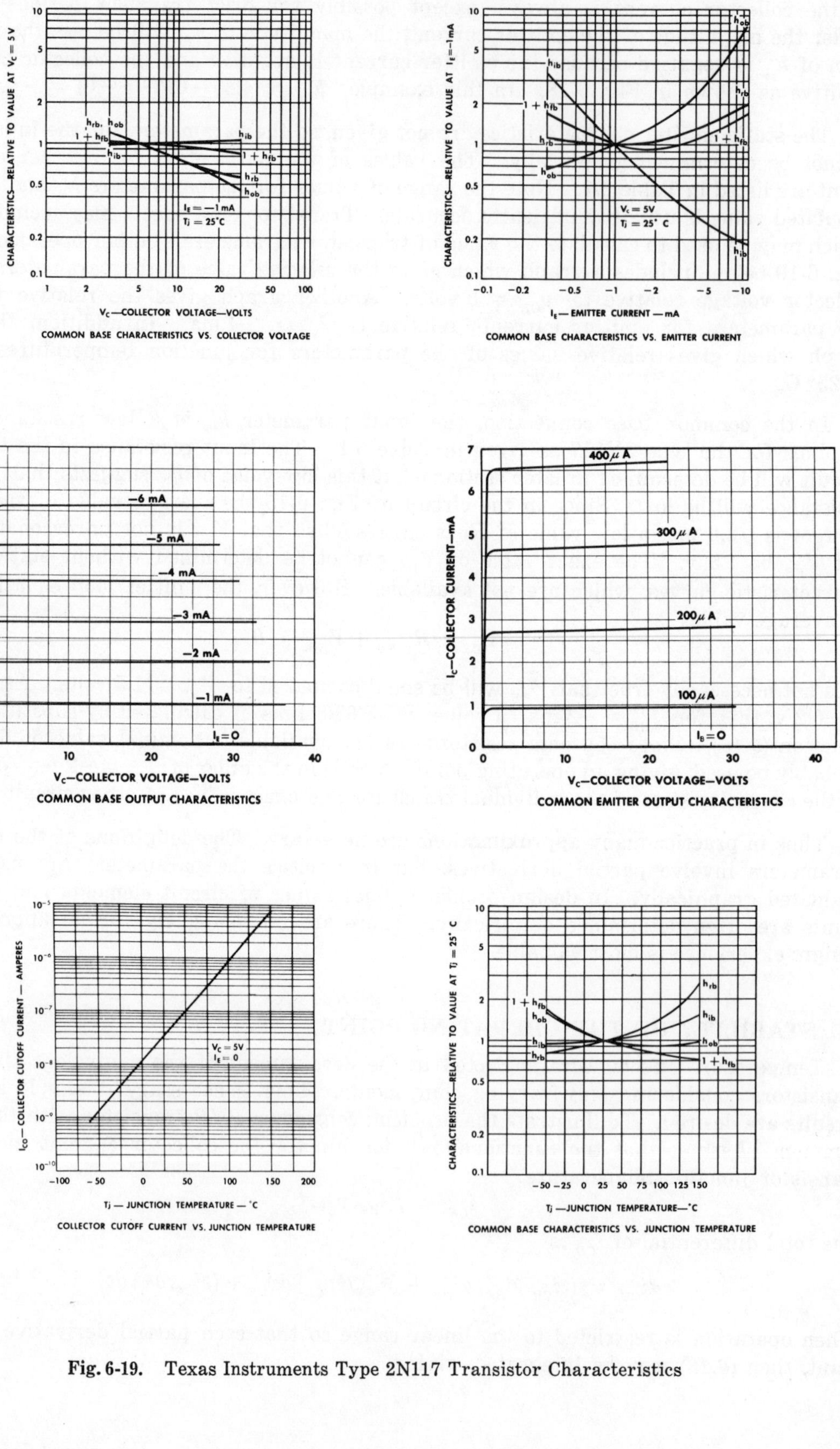

Fig. 6-19. Texas Instruments Type 2N117 Transistor Characteristics

$$i_{CE} = h_{fe} i_{BE} + h_{oe} v_{CE} + (\partial i_{CE}/\partial t_J) t_J + k_c \qquad (6.19)$$

where k_c is a combination of all constants of integration. Since linearity is assumed, superposition may be applied and the instantaneous values written as the sum of the average values and instantaneous values of the varying components.

The terms involving constants are

$$I_{CE} = h_{fe} I_{BE} + h_{oe} V_{CE} + (\partial i_{CE}/\partial t_J) T_J + k_c \qquad (6.20)$$

Following the adopted notation, t_J is the instantaneous temperature, T_J the average value, and t_j the instantaneous value of the varying component. It is probably meaningless to consider t_j, since the temperature does not change rapidly, because of the mass involved. Equation (6.20) does show that the average collector current changes with temperature unless $\partial i_{CE}/\partial t_J$ is zero. This relationship probably is not very useful, because in practice the temperature range of interest is large and involves a temperature relationship which is quite nonlinear. Since the temperature does not change rapidly, there is little to be gained by drawing the equivalent collector circuit for varying components which includes a generator accounting for the temperature variations about an average value. The effect of temperature variation is considered as involving a shift in the operating point.

Similarly the functional relationship is written for the base circuit:

$$v_{BE} = f(i_{BE}, v_{CE}, t_J) \qquad (6.21)$$

The total differential of v_{BE} may be written and then superposition applied, so the constant terms are

$$V_{BE} = h_{ie} I_{BE} + h_{re} V_{CE} + (\partial v_{BE}/\partial t_J) T_J + k_b \qquad (6.22)$$

where k_b is a combination of all constants of integration. This also may not be a very useful relationship if the temperature range is large. It does show, however, that the operating point will change with a change in temperature unless both $\partial v_{BE}/\partial t_J$ and $\partial i_{CE}/\partial t_J$ are zero.

The problem of temperature dependence is not as simple as this discussion might make it appear. There are two junctions in each transistor and it is not necessarily true that the temperature of both is the same. The junction temperature depends on the ambient temperature and the power dissipated in the region near the junction. The temperature at points interior to the transistor is not constant and depends on the provision made for the removal of thermal energy (heat sinks).

The power dissipation in a transistor circuit may be calculated in a manner similar to that employed in vacuum tube circuits (see Section 5.12). The instantaneous collector power dissipation is defined as

$$p_C = v_{CE} i_{CE}$$

Assuming a low power transistor so that the signal power output is very small, the average collector power dissipated is approximately

$$P_C = V_{CE} I_{CE} \qquad (6.23)$$

which may be plotted as a hyperbola on the family of collector characteristic curves. There is a hyperbolic curve for each pair of values of V_{CE} and I_{CE}.

As an example consider the simplified circuit of Fig. 6-20 and the static collector characteristic curves given in Fig. 6-21 below. The dc load line is plotted as shown in Fig. 6-21. Assume the base circuit is such that the operating point for the

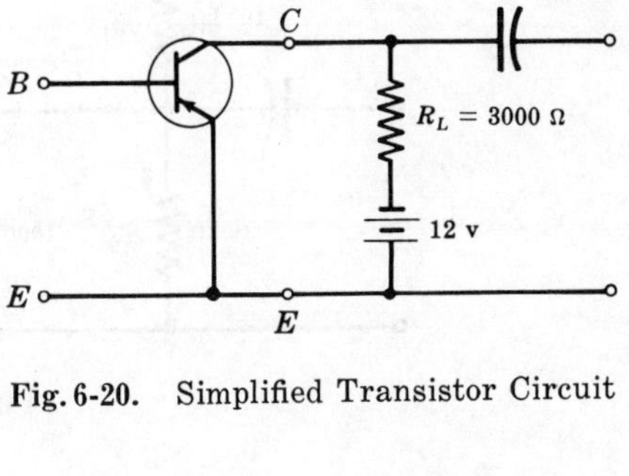

Fig. 6-20. Simplified Transistor Circuit

transistor is $V_{CE} = -6$ v and $I_{CE} = -2$ ma; for these values, $P_C = V_{CE}I_{CE} = (-6)(-0.002) = 0.012$ watts. This hyperbola is plotted on the collector curves of Fig. 6-21. Note that if the operating point moves downward along the load line, the collector dissipation decreases; if the operating point moves upward along the load line, the collector dissipation also decreases.

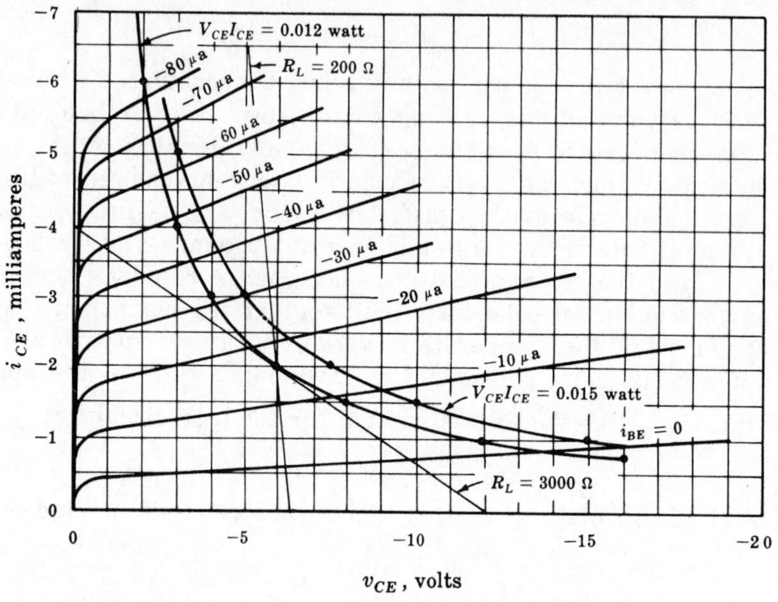

Fig. 6-21. Static Collector Characteristic Curves

Consider the same type of circuit but assume the resistance R_L (dc) is low, as might obtain if there is a transformer in the collector circuit or if R_L (itself) is very low. Assume the dc resistance R_L is 200 ohms and the operating point is the same as in Fig. 6-21. As the operating point now moves upward along the new load line ($R_L = 200\ \Omega$), the collector dissipation increases. This is seen more clearly if several more hyperbolas are plotted for different values of $V_{CE}I_{CE}$. Since the collector current increases with temperature, a condition known as "thermal runaway" may develop, causing the collector current to increase beyond safe values.

The transistor circuit of Fig. 6-22 was developed to provide for "bias stability". The resistance R_E in the emitter circuit serves a purpose similar to the cathode resistor in the vacuum tube circuit. As the magnitude of the emitter current increases, the base to emitter voltage changes in such a manner as to cause a decrease in the collector current. Since V_{BE} is usually small (order of millivolts), the average value of the base to ground voltage must be other than zero so that the emitter resistance can be large enough to provide the desired

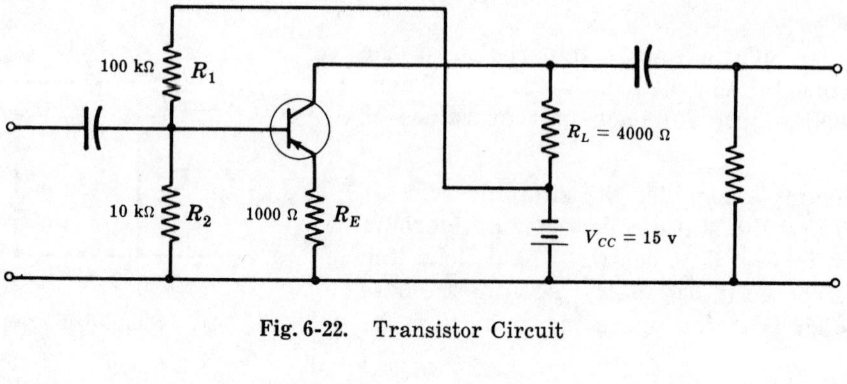

Fig. 6-22. Transistor Circuit

stability. If the emitter current is in the order of milliamperes and V_{BE} in the order of millivolts, R_E would be very small unless the base to ground average voltage is not zero.

The collector loop equation based on average values is

$$+V_{CC} + R_L I_{CE} + V_{CE} + R_E(I_{CE} + I_{BE}) = 0 \qquad (6.24)$$

Thévenin's theorem may be applied to the base circuit of Fig. 6-23(*a*) to give the equivalent circuit shown in Fig. 6-23(*b*).

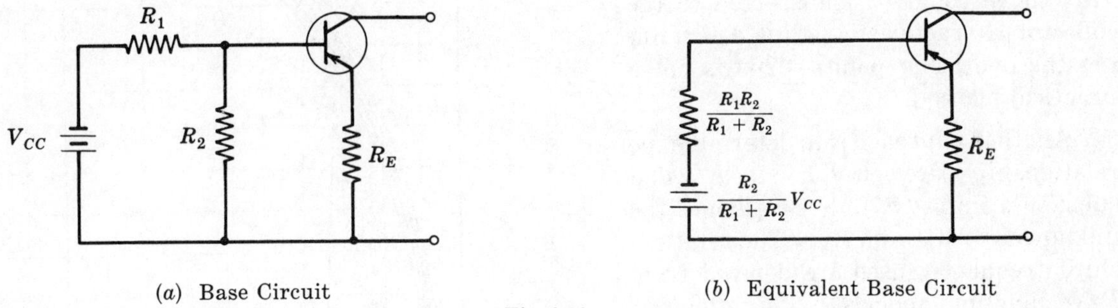

Fig. 6-23

The base loop equation for the base equivalent circuit of Fig. 6-23(*b*) based on average values is

$$+\frac{R_2}{R_1 + R_2} V_{CC} + \frac{R_1 R_2}{R_1 + R_2} I_{BE} + V_{BE} + R_E(I_{CE} + I_{BE}) = 0 \qquad (6.25)$$

To illustrate with a specific example, consider the circuit and values given in Fig. 6-22, the collector characteristics for the transistor given in Fig. 6-24, and the base characteristics given in Fig. 6-25. The values for the circuit parameters are substituted into equation (*6.24*) to determine the collector load line. Since i_{BE} is in the order of tens of microamperes and i_{CE} in the order of milliamperes, it is assumed that $I_{CE} + I_{BE} \cong I_{CE}$. The collector equation is

$$+15 + 4000 I_{CE} + V_{CE} + 1000 I_{CE} = 0 \quad \text{or} \quad -(5000/15) I_{CE} - (1/15) V_{CE} = +1 \qquad (6.26)$$

which is plotted in Fig. 6-24. The operating point will lie on the collector load line.

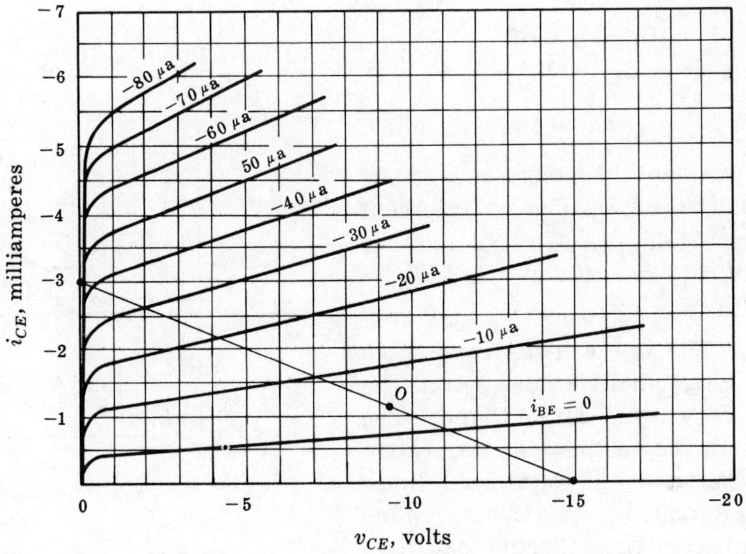

Fig. 6-24. Static Collector Characteristic Curves

The other equation to be satisfied is (*6.25*). There are three unknowns in (*6.25*): I_{BE}, V_{BE} and I_{CE}. Hence (*6.25*) involves a surface rather than a curve (straight line on the proper plot). The intersection of the surface defined by (*6.25*) and the surface representing the input characteristics, defines a curve. This curve could be transferred to the collector characteristics, thus determining the operating point. This is not a practical method.

Another approach is to determine the relationship between I_{CE} and I_{BE} and substitute into (*6.25*), thus reducing the unknowns to V_{BE} and I_{BE}. The relationships frequently used are derived from *P-N* junction models. The dynamic transfer characteristic curve based on the dc load line may be plotted, giving the relationship between i_{CE} and i_{BE} for this circuit as shown in Fig. 6-26. The portion of this curve up to $i_{BE} = -20\ \mu A$ can be approximated as a straight line as shown in Fig. 6-26. The equation of this line is

$$i_{CE} = -0.00078 + 69 i_{BE} \qquad (6.27)$$

which may be substituted into (*6.25*), giving

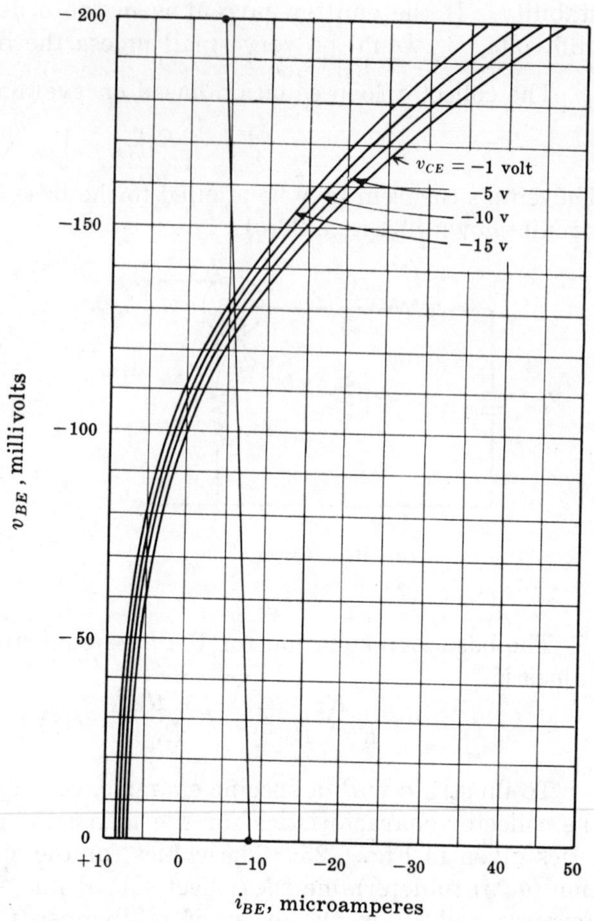

Fig. 6-25. Static Base Characteristic Curves

$$\frac{R_2}{R_1+R_2}V_{CC} + \frac{R_1R_2}{R_1+R_2}I_{BE} + V_{BE} + R_E(-0.00078 + 69I_{BE} + I_{BE}) = 0 \qquad (6.28)$$

Substituting numerical values,

$$\frac{10{,}000}{110{,}000}(15) + \frac{(10)(100)\times 10^6}{110{,}000}I_{BE} + V_{BE} + 1000(-0.00078 + 70I_{BE}) = 0$$

or

$$79{,}100 I_{BE} + V_{BE} = -0.588 \qquad (6.29)$$

which may be plotted on the static base characteristics given in Fig. 6-25. The i_{BE} intercept is $-0.588/79{,}100 = -7.43\ \mu A$. To find another point of the line, let $v_{BE} = -0.2$ volt; then $i_{BE} = -(0.588 + 0.2)/79{,}100 = -4.9\ \mu A$.

The dc collector load line can be transferred to the static base characteristic curves and the intersection of the two curves can be found. The operating point can be estimated accurately enough from the collector and base load lines, giving $I_{BE} = -5.3\ \mu A$ and $V_{BE} = -129$ mv. This value of I_{BE} is located on the collector load line, giving $V_{CE} = -9.3$ v and $I_{CE} = -1.2$ ma.

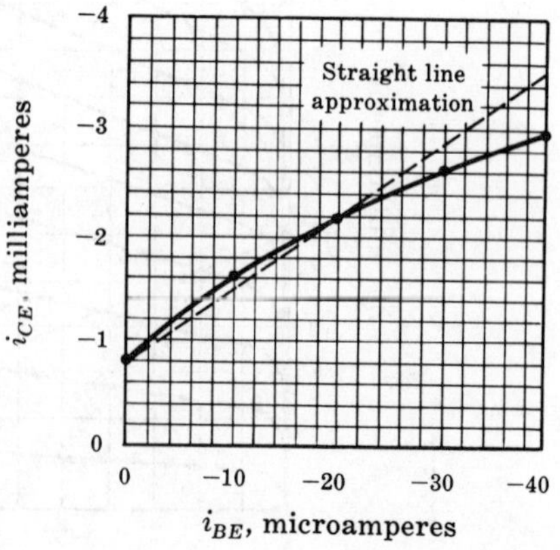

Fig. 6-26. Dynamic Transfer Characteristic Curve

A procedure for finding the operating point for the circuit of Fig. 6-22 has been outlined. To demonstrate how this circuit provides for stabilization of the operating point, more information is needed than is given in the families of static base and collector characteristic curves. These are plotted for a given temperature and do not show temperature dependence.

6.6 COMMON EMITTER AMPLIFIER TRANSFER FUNCTIONS

After the operating point is found for the transistor and the parameters are determined, the complete equivalent circuit considering varying components may be drawn. The circuit equations may then be written and solved. The transistor amplifier circuit of Fig. 6-27 is frequently used. Frequency response will be discussed in Chapter 7, so for the present it is assumed that operation is in the mid-band range of frequencies. This means that the capacitors C_1, C_2 and C_E are so large that their impedances are negligibly small in this range of frequencies. The equivalent circuit considering only varying components is then as shown in Fig. 6-28. The common terminal of resistors R_1 and R_L is at ground potential for varying components, since the terminal voltage of V_{CC} is constant. A little practice may be required to draw the equivalent circuits by inspection. This circuit is called the mid-band equivalent circuit. In this case the resistance R_3 will be assumed to include all the effective load resistance at the output terminals.

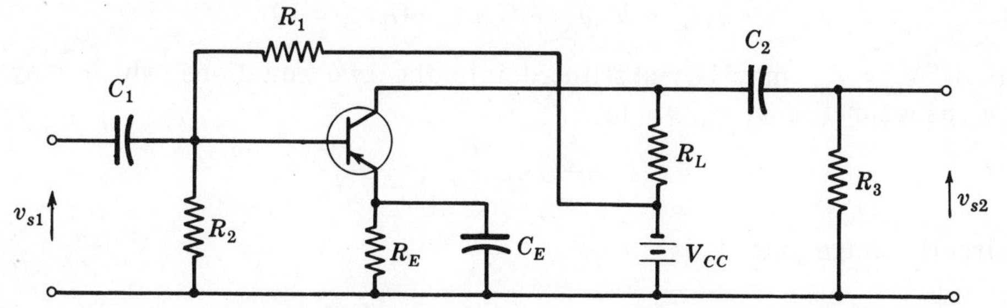

Fig. 6-27. Transistor Amplifier

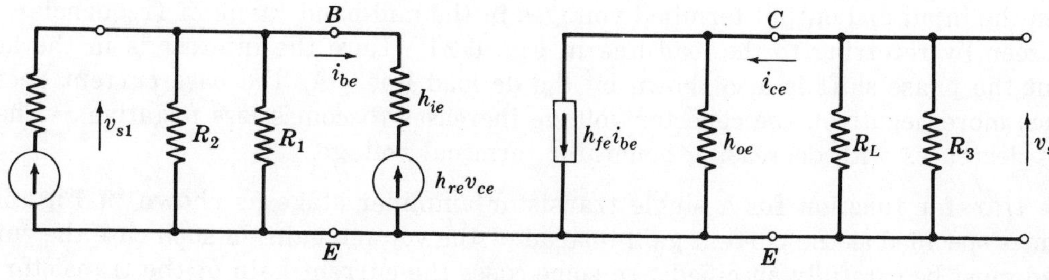

Fig. 6-28. Amplifier Equivalent Circuit

It is assumed that a generator is connected to the input pair of terminals. Since the input signal current is not zero, the internal impedance of the generator must be included in the equivalent circuit. To simplify calculations, Thévenin's theorem may be applied to the circuit to the left of the base and ground terminals of Fig. 6-28. This circuit may then be represented as the effective generator resistance in series with the open circuit terminal voltage as shown in Fig. 6-29 below. The input signal voltage is the generator terminal voltage and denoted v_{s1}. It is necessary to specify whether the magnitude of the generated voltage is constant or the generator terminal voltage is constant. It must also be specified whether the transfer functions are defined in terms of v_{gen} or v_{s1}. This will be discussed further in Chapter 7. When testing a single amplifier in a laboratory, a generator is usually connected to the input terminals of the amplifier and the magnitude of the terminal voltage is maintained at a constant level. This effectively creates a zero impedance generator, since if the terminal voltage is independent of current, then the internal impedance is zero.

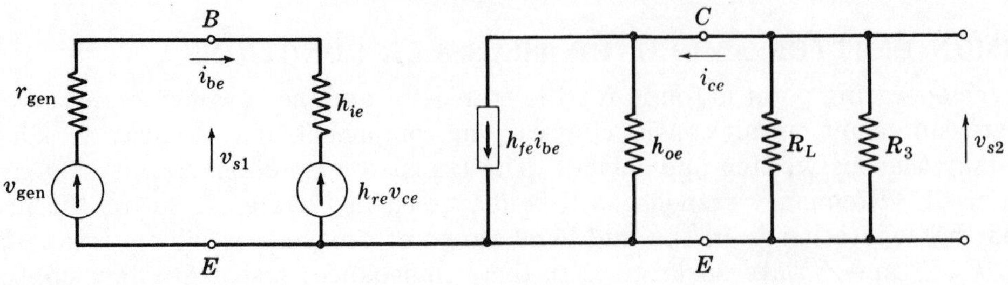

Fig. 6-29. Simplified Amplifier Equivalent Circuit

It is assumed in this example that the magnitude of the terminal voltage is kept constant and that the voltage gain is defined in terms of v_{s1}. The base loop and collector nodal equations are respectively

$$-v_{s1} + h_{ie}i_{be} + h_{re}v_{ce} = 0 \tag{6.30}$$

$$+h_{fe}i_{be} + h_{oe}v_{ce} + G_Lv_{ce} + G_3v_{ce} = 0 \tag{6.31}$$

In this circuit $v_{s2} = v_{ce}$ may be substituted into the two equations which may then be solved for v_{s2} as a function of v_{s1}, giving

$$v_{s2} = \frac{-h_{fe}}{h_{ie}(h_{oe} + G_L + G_3) - h_{fe}h_{re}} v_{s1} \tag{6.32}$$

The open circuit voltage gain A_v is

$$A_v = \frac{v_{s2}}{v_{s1}} = \frac{-h_{fe}}{h_{ie}(h_{oe} + G_L + G_3) - h_{fe}h_{re}} \tag{6.33}$$

where the negative sign indicates that the common emitter amplifier has 180° phase shift between the input and output terminal voltages in the mid-band range of frequencies. This is also seen by referring to the load line in Fig. 6-24. Here the interest is in the ac load line, but the phase shift is also shown by the dc load line. As the base current decreases (becomes more negative), the collector voltage increases (becomes less negative). The base current decreases with decreasing generator terminal voltage.

The transfer function for a single transistor amplifier stage as shown in Fig. 6-27 is sometimes specified as the current gain instead of the voltage gain; in such case the currents involved must be carefully specified. In some cases the current gain of the transistor alone is considered; then the current gain is $A_i = i_{ce}/i_{be}$. The circuit equations may be written for the circuit of Fig. 6-29 and solved for the current gain A_i. When the interest is in the current gain, the terminal voltage at the input is usually not specified. Here an ideal current generator may be assumed at the input and the input loop equation need not be written. This is the dual of the situation where a constant voltage source is applied and the input current is not specified. The collector nodal equation for the circuit of Fig. 6-29 is

$$+h_{fe}i_{be} + (h_{oe} + G_L + G_3)v_{s2} = 0 \tag{6.34}$$

The current i_{ce} may be written as the negative of the conductance external to the collector terminal times the terminal voltage. This is the nodal equation for the non-reference output terminal,

$$i_{ce} = -(G_L + G_3)v_{s2} \tag{6.35}$$

From Equation (*6.34*),
$$v_{s2} = \frac{-h_{fe}}{h_{oe} + G_L + G_3} i_{be} \tag{6.36}$$

which when put into (*6.35*) gives
$$i_{ce} = -(G_L + G_3)\left(\frac{-h_{fe}}{h_{oe} + G_L + G_3}\right) i_{be} \tag{6.37}$$

and thus
$$A_i = \frac{i_{ce}}{i_{be}} = \frac{h_{fe}(G_L + G_3)}{h_{oe} + G_L + G_3} \tag{6.38}$$

The phase relationship may be verified by referring to the load line drawn on the common emitter collector characteristic curves of Fig. 6-24. As the base current decreases (becomes more negative), the collector current also decreases (becomes more negative). Equation (*6.38*) also indicates that the current gain for the transistor can never be greater than h_{fe}. If h_{oe} is very small compared to $G_L + G_3$, then A_i approaches h_{fe}.

In some cases the current gain for an entire amplifier is of interest. As an example, assume the current in the load resistor R_3 is the output current i_{s2}. To be completely correct, the resistors R_1 and R_2 must be included in the input circuit if the current i_{s1} is in the input terminals. If R_1 and R_2 are not large, then $i_{s1} \neq i_{be}$ and the equations for the input circuit must be included. In the present example, assume R_1 and R_2 are so large that the current into the parallel combination is negligibly small compared to i_{be}. Then the collector equation is (*6.36*) with $i_{be} = i_{s1}$. The output current is
$$i_{s2} = -G_3 v_{s2} \tag{6.39}$$

Substituting (*6.36*) into (*6.39*),
$$i_{s2} = -G_3 \frac{-h_{fe}}{h_{oe} + G_L + G_3} i_{be} \tag{6.40}$$

and since $i_{be} = i_{s1}$,
$$A_i = \frac{i_{s2}}{i_{s1}} = \frac{G_3 h_{fe}}{h_{oe} + G_L + G_3} \tag{6.41}$$

which indicates that the current gain can never exceed h_{fe}.

It is sometimes of interest to consider the power gain of a transistor or a transistor amplifier circuit. Again, it is necessary to exercise care in identifying the proper voltages and currents. Only mid-band conditions will be considered at this time so that the circuit elements are resistive. It will still be assumed that the resistors R_1 and R_2 are large. The signal power input to the amplifier is then the signal power input to the transistor. By definition, the average *signal power* applied to the input is
$$P_{s1} = V_{s1} I_{s1} \tag{6.42}$$

where V_{s1} and I_{s1} are the rms values of the signal input voltage and current, respectively. The voltage and current are in phase at the input terminals for the mid-band range of frequencies.

The output signal power of the transistor (power into the circuit to the right of the collector terminal) is
$$P_c = -V_{ce} I_{ce} \tag{6.43}$$

where V_{ce} and I_{ce} are rms values. The negative sign in (*6.43*) is included to account for the fact that the collector current reference is into the collector terminal. In this case, $V_{ce} = V_{s2}$ (rms values). Equations (*6.37*) and (*6.32*) are substituted into (*6.43*), keeping in mind that the ratios of rms values of signals are the same as the ratios of instantaneous values of varying components:

$$P_c = -\left[\frac{-h_{fe}}{h_{ie}(h_{oe}+G_L+G_3) - h_{fe}h_{re}} V_{s1}\right]\left[\frac{h_{fe}(G_L+G_3)}{h_{oe}+G_L+G_3} I_{s1}\right] \qquad (6.44)$$

and thus

$$A_p = \frac{P_c}{P_{s1}} = \frac{h_{fe}^2(G_L+G_3)}{[h_{ie}(h_{oe}+G_L+G_3) - h_{fe}h_{re}](h_{oe}+G_L+G_3)} \qquad (6.45)$$

Equation (*6.44*) states that the average signal power gain is the product of the magnitude of the voltage gain and the magnitude of current gain when appropriate voltages and currents are involved.

The signal power gain for the entire amplifier may be found, assuming the signal output power is the signal power into the resistor R_3. The average signal output power is then

$$P_{s2} = -V_{s2}I_{s2} \qquad (6.46)$$

The negative sign is necessary because of the reference direction for the current I_{s2}. Substituting equations (*6.32*) and (*6.40*) into (*6.46*),

$$P_{s2} = -\left[\frac{-h_{fe}}{h_{ie}(h_{oe}+G_L+G_3) - h_{fe}h_{re}} V_{s1}\right]\left[\frac{G_3h_{fe}}{h_{oe}+G_L+G_3} I_{s1}\right] \qquad (6.47)$$

and the amplifier power gain is

$$A_p = \frac{P_{s2}}{P_{s1}} = \frac{G_3h_{fe}^2}{h_{ie}(h_{oe}+G_L+G_3)^2 - h_{fe}h_{re}(h_{oe}+G_L+G_3)} \qquad (6.48)$$

Again, the average power gain is the magnitude of the appropriate voltage gain times the magnitude of the appropriate current gain.

6.7 COMMON BASE AMPLIFIER TRANSFER FUNCTIONS

A common base transistor amplifier may be considered in the same way as the common emitter amplifier. Examine the transistor circuit of Fig. 6-30. The mid-band equivalent circuit for the amplifier of Fig. 6-30 is shown in Fig. 6-31. The input resistance to the transistor with a common base is small (this will be discussed in Chapter 7), so R_1 will be

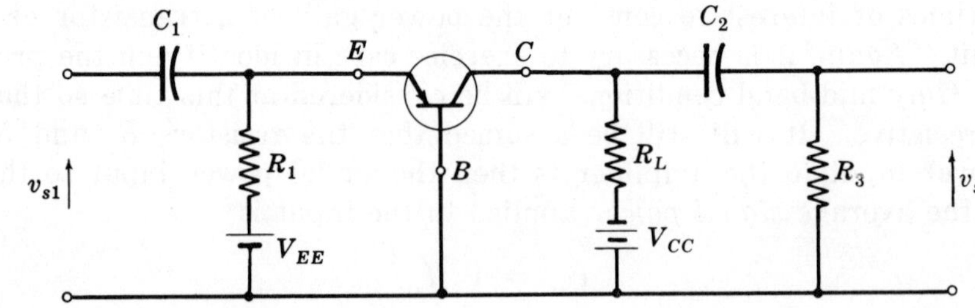

Fig. 6-30. Common Base Transistor Amplifier

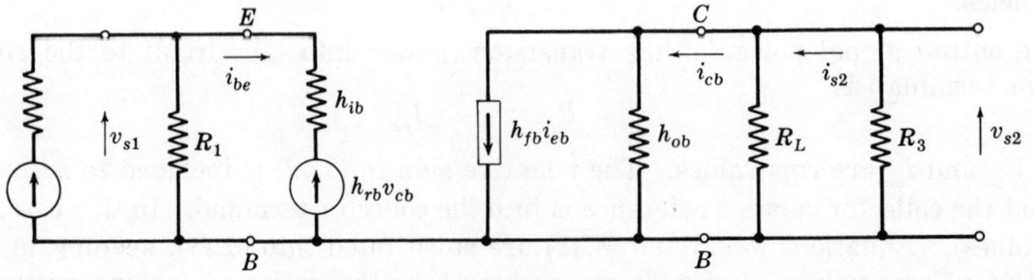

Fig. 6-31. Amplifier Equivalent Circuit

assumed large; hence the signal current in R_1 is negligible compared to i_{eb}. The voltage gain is found as before and it is assumed that the magnitude of v_{s1} is constant and that the transfer function involves v_{s1}. Using $v_{cb} = v_{s2}$, the emitter loop and collector nodal equations are

$$-v_{s1} + h_{ib}i_{eb} + h_{rb}v_{cb} = 0 \tag{6.49}$$

$$+h_{fb}i_{eb} + h_{ob}v_{s2} + G_Lv_{s2} + G_3v_{s2} = 0 \tag{6.50}$$

from which the voltage gain is

$$A_v = \frac{v_{s2}}{v_{s1}} = \frac{-h_{fb}}{h_{ib}(h_{ob} + G_L + G_3) - h_{fb}h_{rb}} \tag{6.51}$$

In common base connections h_{fb} is actually negative, so there is no phase shift between the input and output signal voltages. This is verified by referring to the ac load line of Fig. 6-16 and noting that as i_{EB} increases (becomes less negative) v_{CB} increases.

The current gain may be found in the same way as for the common emitter circuit. Consider the amplifier current gain where the output current i_{s2} is in R_3; then

$$i_{s2} = -G_3v_{s2} \tag{6.52}$$

Solving (*6.50*) for v_{s2} and substituting into (*6.52*),

$$i_{s2} = (-G_3)[-h_{fb}i_{eb}/(h_{ob} + G_L + G_3)] \tag{6.53}$$

and since $i_{eb} = i_{s1}$,

$$A_i = i_{s2}/i_{s1} = G_3h_{fb}/(h_{ob} + G_L + G_3) \tag{6.54}$$

Since h_{fb} is negative, there is a 180° shift in phase between the input and output signal currents. This may also be seen by examining the ac load line of Fig. 6-16 and noting that i_{CB} increases as i_{EB} decreases. Equation (*6.54*) also shows that the magnitude of A_i is always less than that of h_{fb}, which is always less than 1. The magnitude of the current gain of a common base transistor amplifier must always be less than 1.

The power gain of a common base transistor amplifier is positive and may be expressed in terms of the product of the appropriate voltage gain and current gain.

A common base transistor amplifier is analogous to the common (or grounded) grid vacuum tube amplifier. The input impedance of the common base transistor amplifier is very low and the output impedance is high. This will be discussed in Chapter 7.

6.8 EMITTER FOLLOWER AMPLIFIER

An amplifier may also involve a transistor with the collector as the common terminal. This circuit is usually referred to as an emitter follower circuit, probably because it is analogous to the cathode follower vacuum tube circuit. It might be expected that the procedures will be similar to those followed for the common emitter and common base connections, but this is not quite the case. In this circuit the input terminal of the transistor is the base and the output terminal is the emitter. To find the common collector parameters graphically, the static base characteristic curves and the static emitter characteristic curves for the common collector are required. These are not available and the common collector parameters are seldom given. Usually the transistor equivalent circuit is drawn using the common emitter hybrid equivalent circuit. When the equivalent circuit is correctly drawn, it makes no difference which parameters are used.

Consider the simplified emitter follower circuit of Fig. 6-32 which does not show all of the bias circuit. It is assumed that R_E includes all the effective (ac) resistance from the emitter terminal to ground considering the mid-band range of frequencies. Bias resistors in the base circuit are assumed very large and negligible. There are fewer problems involving stability of operating point in this type of circuit. The operating point would be found as before by writing a collector loop equation and a base loop equation based on average values and then solving graphically. The common emitter parameters are usually used, so it is assumed the common emitter characteristic curves are available.

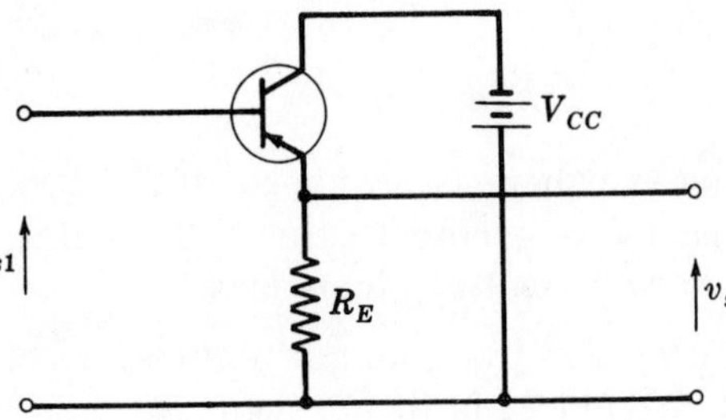

Fig. 6-32. Simplified Emitter Follower Circuit

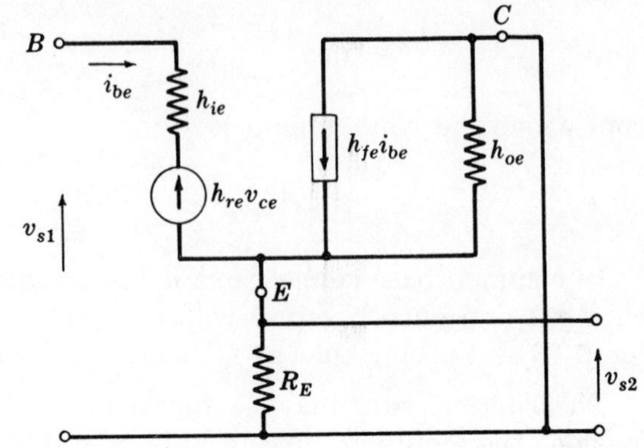

Fig. 6-33. Emitter Follower Equivalent Circuit

The equivalent circuit considering only varying components is drawn by replacing the transistor by its equivalent circuit. Each transistor terminal is marked and the equivalent circuit between the terminals is drawn as in Fig. 6-33. The writing of the equations is simplified if this circuit is redrawn as in Fig. 6-34. It is important that the generator polarities are correctly shown. By inspection of the collector circuit of Fig. 6-34, $v_{ce} = -v_{s2}$. The base loop and emitter nodal equations are

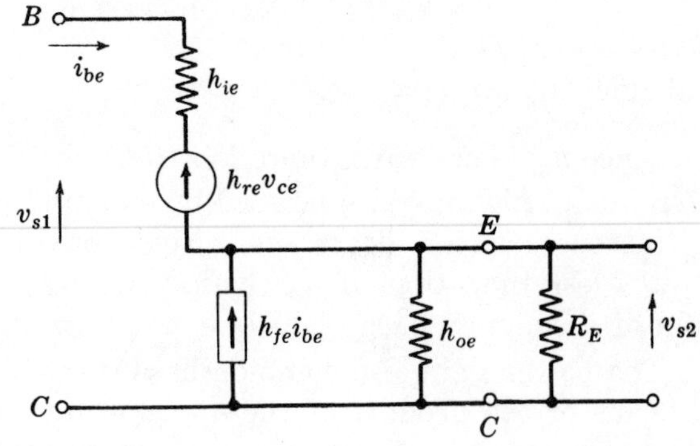

Fig. 6-34. Emitter Follower Equivalent Circuit

$$-v_{s1} + h_{ie}i_{be} + h_{re}(-v_{s2}) + v_{s2} = 0 \tag{6.55}$$

$$-i_{be} - h_{fe}i_{be} + h_{oe}v_{s2} + G_E v_{s2} = 0 \tag{6.56}$$

from which the voltage gain is

$$A_v = \frac{v_{s2}}{v_{s1}} = \frac{h_{fe}+1}{h_{ie}(h_{oe}+G_E) + (h_{fe}+1)(1-h_{re})} \tag{6.57}$$

This may usually be simplified. First, h_{fe} is usually large compared to 1, so $(h_{fe}+1) \cong h_{fe}$. Usually h_{re} is small compared to 1, so $(1-h_{re}) \cong 1$. Making these approximations, the gain is

$$A_v = \frac{h_{fe}}{h_{ie}(h_{oe}+G_E) + h_{fe}} \tag{6.58}$$

Equation (*6.58*) indicates that the input and output signal voltages are in phase for the mid-band range of frequencies. It also shows that the voltage gain is less than 1 because h_{ie}, h_{oe} and G_E are positive. This is analogous to the cathode follower circuit. The input impedance of the emitter follower is not infinite but is higher than for other transistor connections, and the output impedance is low. One of the uses of this circuit is as a type of impedance transformer. Input and output impedances will be discussed in Chapter 7.

The magnitude of the current gain for the emitter follower may be greater than 1. Consider the circuit of Fig. 6-35 showing the amplifier load resistance R_3. The current gain will be defined in terms of i_{s2}, the current in R_3. The emitter nodal equation is

$$-i_{be} - h_{fe}i_{be} + h_{oe}v_{s2} + G_E v_{s2} + G_3 v_{s2} = 0 \tag{6.59}$$

from which

$$v_{s2} = (h_{fe}+1)i_{be}/(h_{oe}+G_E+G_3) \tag{6.60}$$

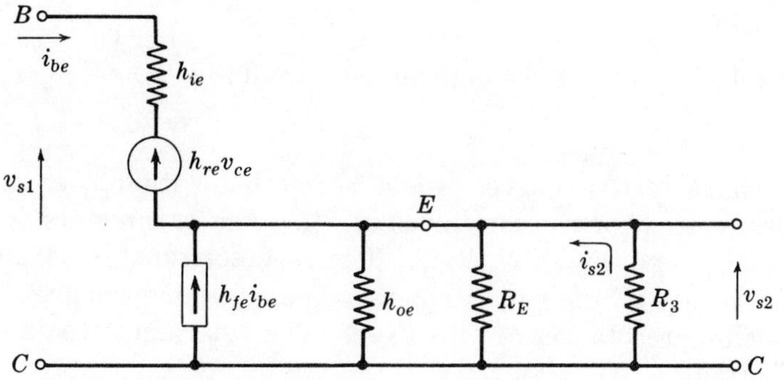

Fig. 6-35. Emitter Follower Circuit

Inspection of Fig. 6-35 shows that

$$i_{s2} = -G_3 v_{s2} \tag{6.61}$$

Substituting equation (6.60) into (6.61) and using $i_{be} = i_{s1}$,

$$A_i = \frac{i_{s2}}{i_{s1}} = \frac{-G_3(h_{fe}+1)}{h_{oe}+G_E+G_3} \tag{6.62}$$

The current gain is negative, indicating a 180° shift in phase, using the established references for the currents. Equation (6.62) shows that if $(h_{oe}+G_E) \ll G_3$, the magnitude of the current gain can approach $(h_{fe}+1) \cong h_{fe}$. Since the current gain can be much larger than 1, the power gain can also be greater than 1.

Formerly when only junction transistors were available, the most convenient way of achieving a high input impedance circuit was to use emitter followers. This resulted in emphasis on emitter follower circuits.

6.9 EQUIVALENT "tee" CIRCUITS

The hybrid equivalent circuit was employed in the previous sections of this chapter. The equivalent "tee" circuit discussed in Section 4.9, page 61, was used quite frequently when transistors first became popular. One reason for this is that the equivalent "tee" circuit more nearly resembles the vacuum tube circuits in which there is a controlled generator only in the output port representation. The two port equivalent circuits for networks frequently involve the z parameters discussed in Section 4.1, page 50. The "tee" equivalent circuit was developed from this.

The static input and output characteristic curves are usually prepared in such a way that the slopes of the input, output, forward transfer and reverse transfer curves at the operating points are parameters. Consider a transistor with the base common, as shown in Fig. 6-36. If the z parameters were to be used, the functional relationship for the collector port would be

$$v_{CB} = f(i_{CB}, i_{EB}) \tag{6.63}$$

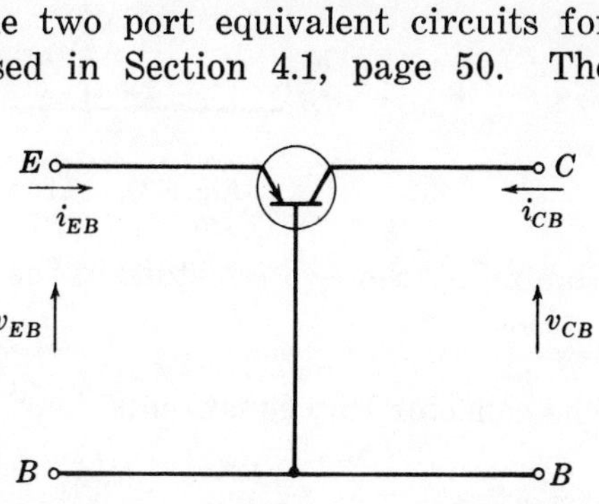

Fig. 6-36. Transistor – Common Base

The static collector characteristic curves would be plotted with v_{CB} as ordinate and i_{CB} as abscissa, and curves plotted for chosen values of i_{EB}. The forward transfer characteristic curves would be plotted with v_{CB} as ordinate and i_{EB} as abscissa, and curves plotted for chosen values of i_{CB}. The two parameters for the output port would be $r_{ob} = \partial v_{CB}/\partial i_{CB}$ and $r_{fb} = \partial v_{CB}/\partial i_{EB}$. These parameters are the slopes of the static collector characteristic curve and the forward transfer characteristic curve, respectively, at the operating point. The voltage generator in the equivalent circuit is current controlled and the parameter r_{fb} is dimensionally a resistance.

The functional relationship for the emitter port would be

$$v_{EB} = f(i_{EB}, i_{CB}) \tag{6.64}$$

The static emitter characteristic curves would be plotted with v_{EB} as ordinate and i_{EB} as abscissa, and curves plotted for chosen values of i_{CB}. The parameters for the emitter port would be $r_{ib} = \partial v_{EB}/\partial i_{EB}$ and $r_{rb} = \partial v_{EB}/\partial i_{CB}$. The reverse transfer parameter r_{rb} is dimensionally a resistance, as are all the r parameters. The r parameters are a special case of the z parameters in which no reactances are involved. The r parameter equivalent circuit is the basis for the "tee" circuit.

The r equivalent circuit for the common base is shown in Fig. 6-37. The "tee" equivalent circuit for the common base transistor is given in Fig. 6-38. The common base circuit was probably the earliest, so this accounts for some of the notation. The "tee" parameters are not defined as partial derivatives (slopes of curves) but are derived from the r parameters. One such derivation was given in Problem 4.6, page 65.

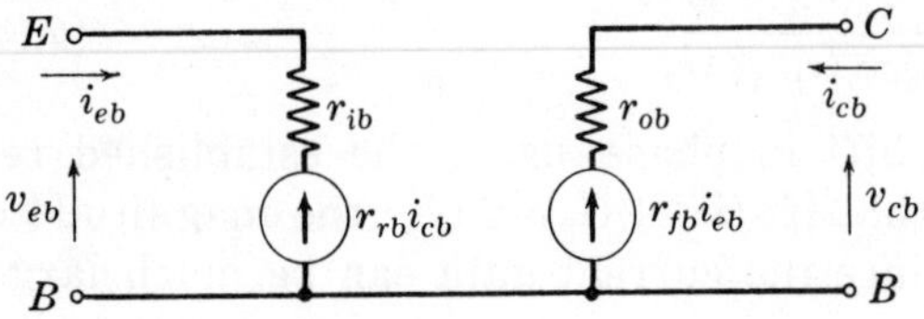

Fig. 6-37. r Parameter Equivalent Circuit – Common Base

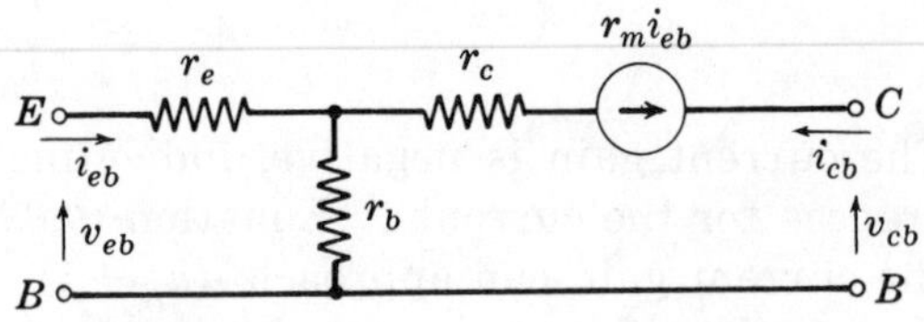

Fig. 6-38. "tee" Parameter Equivalent Circuit – Common Base

Consider an amplifier equivalent circuit using the "tee" parameters as in Fig. 6-39. The voltage gain for this amplifier may be found as before. The emitter loop equation is

$$-v_{s1} + r_e i_{eb} + r_b(i_{cb} + i_{eb}) = 0 \tag{6.65}$$

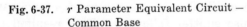

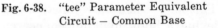

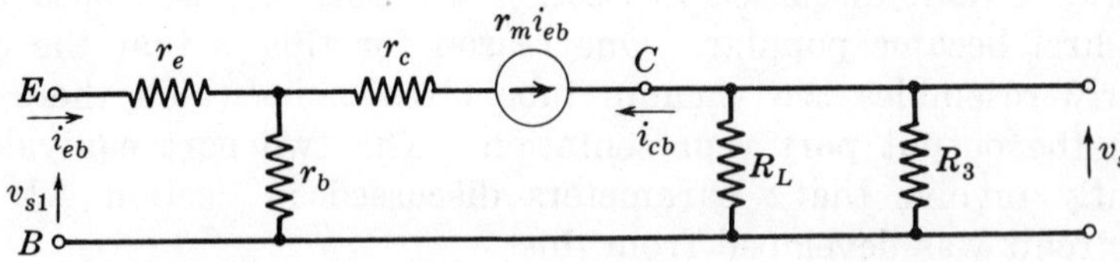

Fig. 6-39. Transistor Equivalent Circuit – Common Base

Combining the two resistors in the collector circuit,

$$R_L^* = R_L R_3/(R_L + R_3) \tag{6.66}$$

The collector loop equation is

$$R_L^* i_{cb} + r_m i_{eb} + r_c i_{cb} + r_b(i_{cb} + i_{eb}) = 0 \tag{6.67}$$

Solving the two equations for i_{cb},

$$i_{cb} = \frac{-(r_m + r_b)v_{s1}}{(r_e + r_b)(r_c + r_b + R_L^*) - r_b(r_m + r_b)} \tag{6.68}$$

From Fig. 6-39 it is seen that

$$v_{s2} = -R_L^* i_{cb} \tag{6.69}$$

which is combined with (6.68) to obtain

$$A_v = \frac{v_{s2}}{v_{s1}} = \frac{R_L^*(r_m + r_b)}{(r_e + r_b)(r_c + r_b + R_L^*) - r_b(r_m + r_b)} \tag{6.70}$$

A variation of the "tee" circuit contains a current generator in the collector circuit as shown in Fig. 6-40. The parameter α can be defined in terms of the parameters of the circuit of Fig. 6-38 by applying Norton's theorem to the portion of the circuit between the collector terminal and the junction of the three resistors which is shown in Fig. 6-41(a). The Norton equivalent circuit for the collector generator is given in Fig. 6-41(b). Both generators in Fig. 6-41 are current controlled. The coefficient of the current generator in Fig. 6-41(b) is defined as $\alpha = r_m/r_c$, a dimensionless current gain. Since both r_m and r_c are positive, the parameter α is positive when the current reference directions of Fig. 6-41 are used.

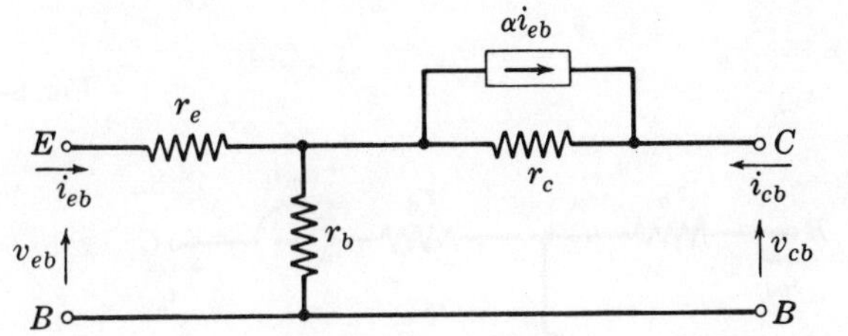

Fig. 6-40. Transistor Equivalent Circuit – Common Base

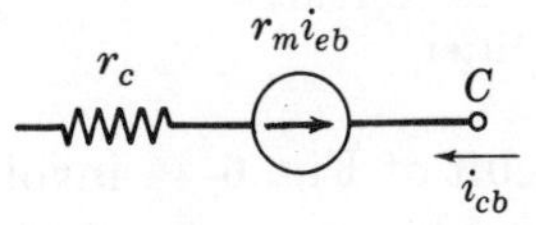

(a) Collector Circuit

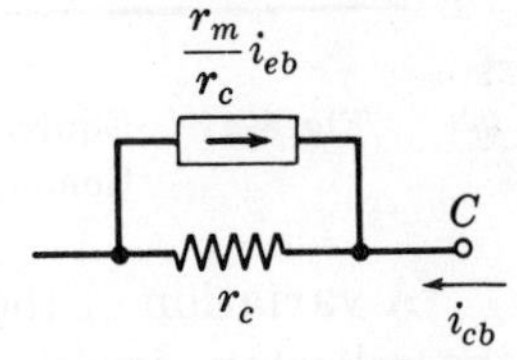

(b) Equivalent Collector Circuit

Fig. 6-41

In most junction transistors, r_m and r_c are large and r_e and r_b are small. If the external load resistance is small compared to r_c, the magnitude of the transistor circuit gain is nearly the same as α. There is some ambiguity in the use of the parameter α because it is occasionally defined in other ways; for example, the common base transistor current gain is sometimes denoted by α.

The equivalent "tee" circuit has also been employed to represent transistors with the emitter common. The common base circuit of Fig. 6-38 has the collector generator controlled by the emitter current and may be redrawn so that the emitter is the common terminal as shown in Fig. 6-42. This involves interchanging the base and emitter terminals and indicating the appropriate voltage and current references. In the circuit of Fig. 6-42 the collector generator is controlled by the emitter current, which is not the current into the nonreference input terminal. The nodal equation for the junction of the resistances in the circuit of Fig. 6-38 is

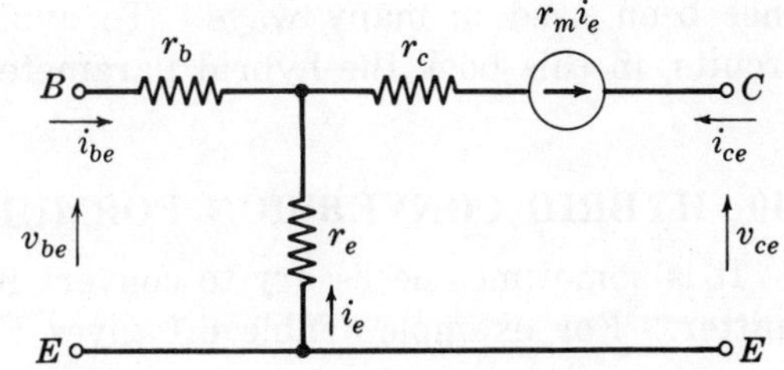

Fig. 6-42. Transistor Equivalent Circuit – Common Emitter

$$-i_e - i_b - i_c = 0 \quad \text{or} \quad i_e = -i_b - i_c \tag{6.71}$$

and the generator voltage in this circuit is then

$$r_m i_e = r_m(-i_b - i_c) = -r_m i_b - r_m i_c \tag{6.72}$$

which indicates that the collector generator may be replaced by two current controlled, voltage generators in series as shown in Fig. 6-43(*a*). The generator controlled by i_c has a voltage drop $-r_m i_c$ when going in the direction of i_c; hence this generator may be represented as a passive element, a resistance $-r_m$. The resistance $-r_m$ may be combined with r_c, giving a single resistor which is denoted $r_d = r_c - r_m$ as shown in Fig. 6-43(*b*). The equivalent "tee" circuit for the common emitter is then drawn as in Fig. 6-44.

Fig. 6-43

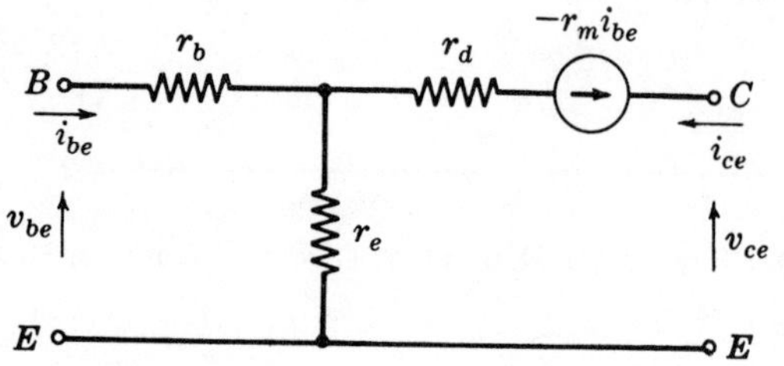

Fig. 6-44. Equivalent "tee" Circuit – Common Emitter

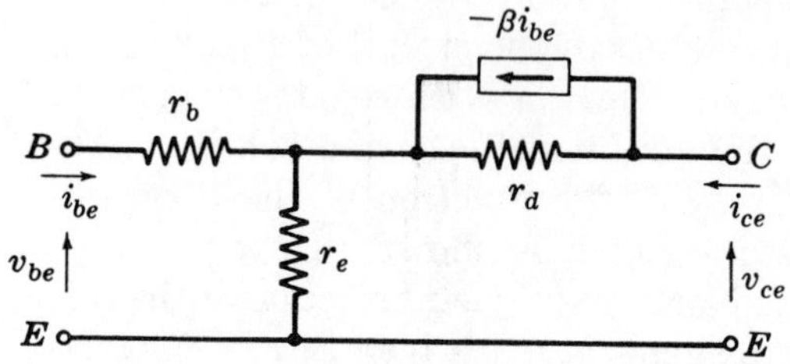

Fig. 6-45. Equivalent "tee" Circuit – Common Emitter

A variation of the circuit of Fig. 6-44 involves a current controlled, current generator in the collector circuit. Using the same current reference as in the circuit of Fig. 6-40, the current generator is

$$-(r_m/r_d)i_{be} \tag{6.73}$$

The coefficient r_m/r_d is sometimes denoted β; then this equivalent circuit is drawn as in Fig. 6-45. The circuits of Fig. 6-44 and 6-45 are employed with many variations, depending on the conventions adopted by the user. Substitutions can be made; for example,

$$\beta = r_m/r_d = r_m/(r_c - r_m) = (r_m/r_c)/(1 - r_m/r_c) = \alpha/(1-\alpha) \tag{6.74}$$

and

$$r_d = r_c - r_m = r_c(1 - r_m/r_c) = r_c(1-\alpha) \tag{6.75}$$

The equivalent "tee" circuits have been discussed because of their historical significance (ref: Bell Telephone System Monograph 1726, 1949). The two parameters α and β have since been used in many ways. To avoid confusion with the parameters of the hybrid circuits, in this book the hybrid parameters have been defined without using α or β.

6.10 HYBRID CONVERSION FORMULAS

It is sometimes necessary to convert hybrid parameters from common base to common emitter. For example, Table 6.1 gives the common base hybrid parameters but not the common emitter parameters, so conversion formulas are useful. The methods are general and can be applied to other conversion problems. This particular set of conversion formulas are derived because they are often needed.

Since the common base parameters are given, the equivalent common base circuit is drawn with the emitter common as shown in Fig. 6-46. This is just a rearrangement of the circuit elements so that the emitter is common. The base loop, collector nodal, base nodal and collector-base loop equations are respectively

$$-v_{be} - h_{rb}v_{cb} + h_{ib}(-i_e) = 0 \qquad (6.76)$$

$$-i_{ce} + h_{fb}i_e + h_{ob}v_{cb} = 0 \qquad (6.77)$$

$$-i_{be} - i_{ce} - i_e = 0 \qquad (6.78)$$

$$-v_{ce} + v_{cb} + v_{be} = 0 \qquad (6.79)$$

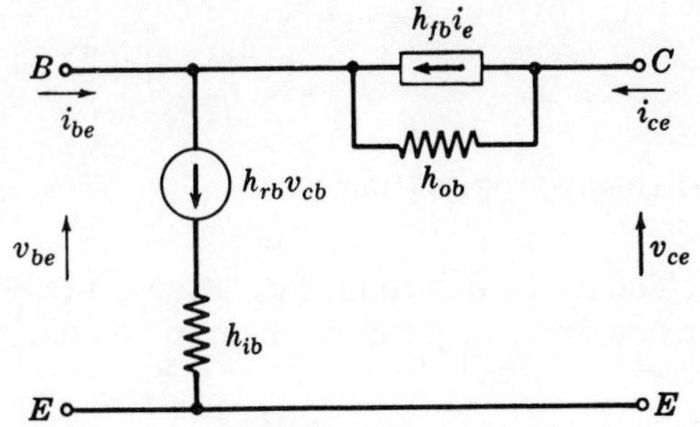

Fig. 6-46. Equivalent Circuit

which may be solved to give equations corresponding to those for the common emitter circuit. The equations for the common emitter equivalent circuit with the common emitter parameters are

$$h_{ie}i_{be} + h_{re}v_{ce} = v_{be} \qquad (6.80)$$

$$h_{fe}i_{be} + h_{oe}v_{ce} = i_{ce} \qquad (6.81)$$

Equations (6.76), (6.77), (6.78) and (6.79) are written as

$$\begin{aligned} -v_{be} - h_{rb}v_{cb} - h_{ib}i_e &= 0 \\ v_{be} + v_{cb} &= v_{ce} \\ -i_e - i_{ce} &= i_{be} \\ +h_{ob}v_{cb} + h_{fb}i_e - i_{ce} &= 0 \end{aligned}$$

These four equations with four dependent variables may be solved for v_{be} and i_{ce}, giving two equations which correspond to (6.80) and (6.81). Usually the number of equations is reduced by substitution, but these four equations are solved to better illustrate the methods. Solving the four equations for v_{be},

$$v_{be} = \frac{h_{ib}i_{be} + (h_{ob}h_{ib} - h_{fb}h_{rb} - h_{rb})v_{ce}}{1 + h_{fb} + h_{ob}h_{ib} - h_{rb} - h_{fb}h_{rb}} \qquad (6.82)$$

Then from (6.82) and (6.80),

$$h_{ie} = \frac{h_{ib}}{1 + h_{fb} + h_{ob}h_{ib} - h_{rb} - h_{fb}h_{rb}} \qquad (6.83)$$

which is frequently approximated (recall h_{fb} is negative) as

$$h_{ie} \cong \frac{h_{ib}}{1 + h_{fb}} \qquad (6.84)$$

$$h_{re} = \frac{h_{ob}h_{ib} - h_{fb}h_{rb} - h_{rb}}{1 + h_{fb} + h_{ob}h_{ib} - h_{rb} - h_{fb}h_{rb}} \qquad (6.85)$$

and the approximation is

$$h_{re} \cong \frac{h_{ob}h_{ib} - h_{fb}h_{rb} - h_{rb}}{1 + h_{fb}} \qquad (6.86)$$

Solving for i_{ce},

$$i_{ce} = \frac{h_{ob}v_{ce} - (h_{fb} + h_{ib}h_{ob} - h_{fb}h_{rb})i_{be}}{1 + h_{fb} + h_{ob}h_{ib} - h_{rb} - h_{fb}h_{rb}} \qquad (6.87)$$

which when combined with (6.81) gives

$$h_{oe} = \frac{h_{ob}}{1 + h_{fb} + h_{ob}h_{ib} - h_{rb} - h_{fb}h_{rb}} \qquad (6.88)$$

and the approximation is

$$h_{oe} \cong \frac{h_{ob}}{1 + h_{fb}} \qquad (6.89)$$

$$h_{fe} = \frac{-(h_{fb} + h_{ib}h_{ob} - h_{fb}h_{rb})}{1 + h_{fb} + h_{ob}h_{ib} - h_{rb} - h_{fb}h_{rb})} \tag{6.90}$$

and the approximation is
$$h_{fe} \cong \frac{-h_{fb}}{1 + h_{fb}} \tag{6.91}$$

The derived formulas give the common emitter parameters in terms of the common base parameters. In practice the approximations are usually used.

Solved Problems

6.1. Given the common emitter transistor equivalent "pi" circuit of Fig. 6-47 and the h parameters for the transistor. Derive the equations for the "pi" circuit parameters in terms of the h parameters.

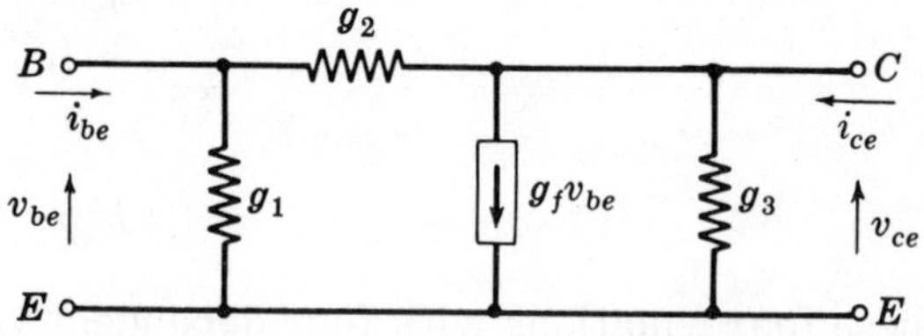

Fig. 6-47

The base nodal and collector nodal are respectively

$$-i_{be} + g_1 v_{be} + g_2(v_{be} - v_{ce}) = 0 \quad \text{or} \quad (g_1 + g_2)v_{be} - g_2 v_{ce} = i_{be} \tag{6.92}$$

$$+g_2(v_{ce} - v_{be}) + g_f v_{be} + g_3 v_{ce} - i_{ce} = 0 \quad \text{or} \quad (g_f - g_2)v_{be} + (g_2 + g_3)v_{ce} = i_{ce} \tag{6.93}$$

The base loop and collector nodal equations for the h parameter equivalent circuit are

$$-v_{be} + h_{ie}i_{be} + h_{re}v_{ce} = 0 \quad \text{or} \quad (1/h_{ie})v_{be} - (h_{re}/h_{ie})v_{ce} = i_{be} \tag{6.94}$$

$$+h_{fe}i_{be} + h_{oe}v_{ce} - i_{ce} = 0 \quad \text{or} \quad (h_{fe}/h_{ie})v_{be} + (h_{oe} - h_{fe}h_{re}/h_{ie})v_{ce} = i_{ce} \tag{6.95}$$

From (*6.92*) and (*6.94*),

$$(g_1 + g_2)v_{be} - g_2 v_{ce} = i_{be} = (1/h_{ie})v_{be} - (h_{re}/h_{ie})v_{ce}$$

From (*6.93*) and (*6.95*),

$$(g_f - g_2)v_{be} + (g_2 + g_3)v_{ce} = i_{ce} = (h_{fe}/h_{ie})v_{be} + (h_{oe} - h_{fe}h_{re}/h_{ie})v_{ce}$$

Since the coefficients of like dependent variables must be equal,

$$g_1 + g_2 = 1/h_{ie}, \quad -g_2 = -h_{re}/h_{ie}, \quad g_f - g_2 = h_{fe}/h_{ie}, \quad g_2 + g_3 = h_{oe} - h_{fe}h_{re}/h_{ie}$$

from which

$$g_2 = \frac{h_{re}}{h_{ie}}, \quad g_1 = \frac{1 - h_{re}}{h_{ie}}, \quad g_f = \frac{h_{fe} + h_{re}}{h_{ie}}, \quad g_3 = h_{oe} - \frac{h_{re}(h_{fe} + 1)}{h_{ie}}$$

6.2. The mid-band equivalent circuit of a common emitter transistor amplifier is given in Fig. 6-48. Transistor parameters are $h_{ie} = 1500\ \Omega$, $h_{re} = 10^{-3}$, $h_{fe} = 100$, $h_{oe} = 50\ \mu\text{mho}$. Find the mid-band current gain $A_i = i_{s2}/i_{s1}$ where i_{s2} is the signal current in the 5 kΩ resistor.

Note that $i_{be} = i_{s1}$ and $v_{ce} = v_{s2}$. The collector nodal equation is

$$+h_{fe}i_{be} + h_{oe}v_{s2} + v_{s2}/10{,}000 + v_{s2}/5000 = 0$$

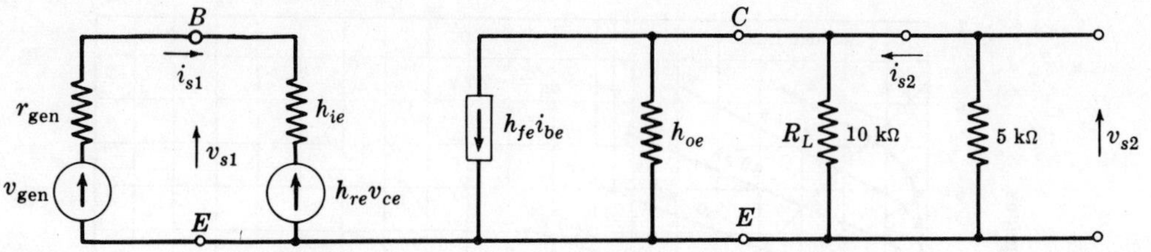

Fig. 6-48

Substituting numerical values,

$$100 i_{s1} + (50 \times 10^{-6} + 100 \times 10^{-6} + 200 \times 10^{-6}) v_{s2} = 0$$

from which $v_{s2} = -(10^8/350) i_{s1}$. Now $i_{s2} = -v_{s2}/5000$ and $A_i = i_{s2}/i_{s1} = 57.2$.

6.3. Given the transistor amplifier circuit of Fig. 6-49, the transistor static base characteristic curves of Fig. 6-50, and the transistor static collector characteristic curves of Fig. 6-51 below. A high resistance voltmeter was used to measure $V_{CE} = -7.7$ v. (*a*) Find I_{CE} and I_{BE}. (*b*) What is the value of R_B? (*c*) What is the "no signal" average collector power dissipation P_C? (*d*) Plot the ac load line on the collector characteristic curves. (*e*) Find the value of h_{fe} at the operating point.

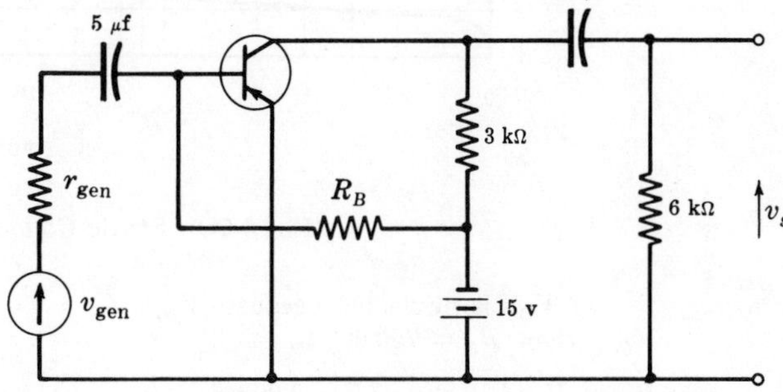

Fig. 6-49

First plot the dc load line on the collector characteristic curves of Fig. 6-51. Then draw the line $V_{CE} = -7.7$ v which intersects the dc load line on the static collector characteristic curve for $i_{BE} = -20\ \mu$A.

(*a*) From the graph of Fig. 6-51, $I_{CE} = -2.45$ ma and $I_{BE} = -20\ \mu$A.

(*b*) V_{BE} is found by locating the operating point on the static base characteristic curves of Fig. 6-50. From this graph, $V_{BE} = -0.165$ v. The base loop equation is

$$+15 + R_B I_{BE} + V_{BE} = 0$$

from which

$$R_B = (-15+0.165)/(-20 \times 10^{-6}) = 741{,}000\ \Omega$$

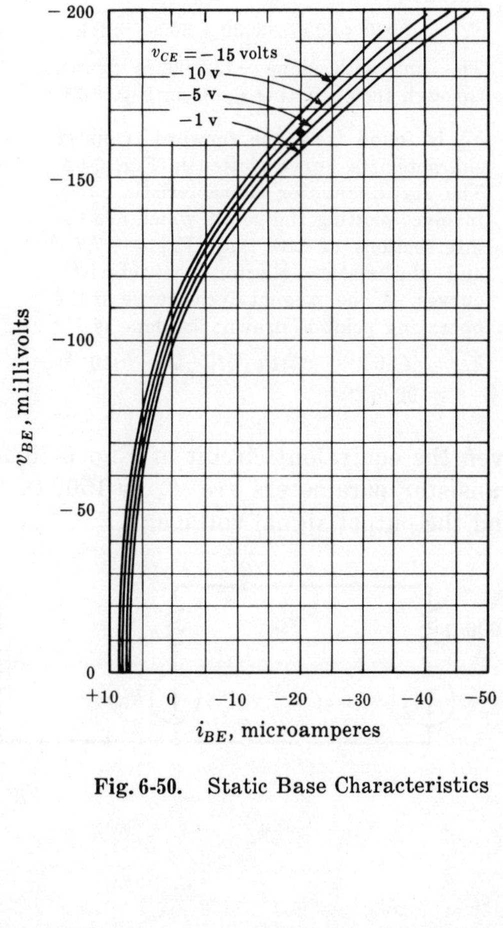

Fig. 6-50. Static Base Characteristics

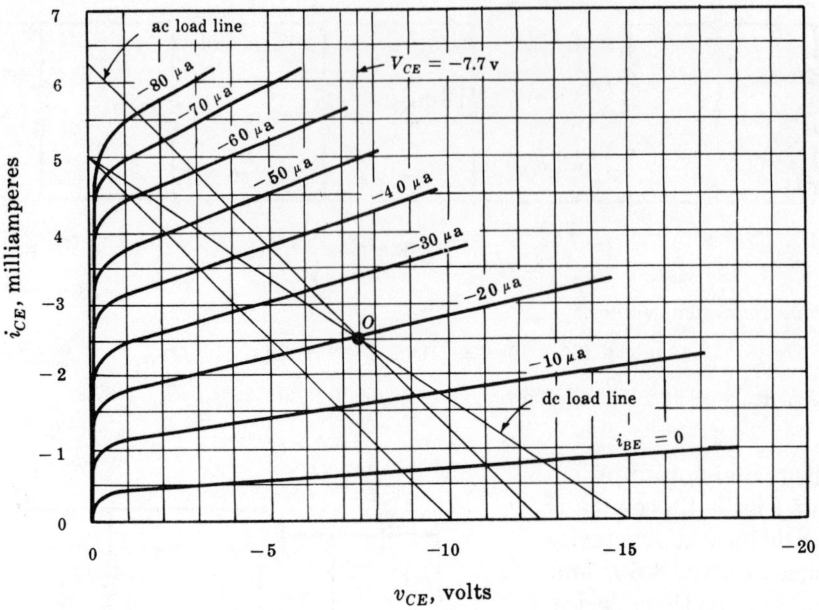

Fig. 6-51. Static Collector Characteristics

If V_{BE} is neglected because $|V_{BE}| \ll V_{CC}$, then $R_B = 750{,}000\ \Omega$.

(c) $P_C = V_{CE} I_{CE} = (-7.7)(-0.00245)$
$= 0.0188$ watts

(d) Combining the 3 kΩ and 6 kΩ resistors in parallel,

$$R_L^* = (3000 \times 6000)/(3000 + 6000) = 2\ \text{k}\Omega$$

The line with slope $-1/2000$ is drawn through the operating point in Fig. 6-51.

(e) h_{fe} is found from the forward transfer characteristic curve plotted in Fig. 6-52. The static transfer characteristic curve involves plotting the points which are the intersections of the line $V_{CE} = -7.7$ v and the static collector characteristic curves. A line tangent to the curve at the operating point is drawn; its slope is

$$h_{fe} = [(-5.2 + 0.67)10^{-3}]/[(-50 - 0)10^{-6}] = 90.6$$

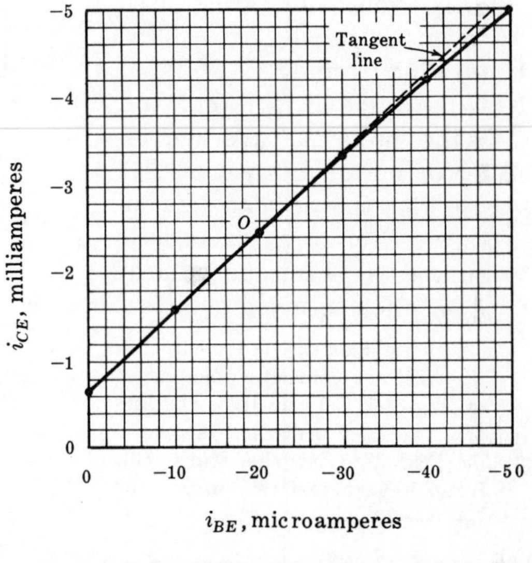

Fig. 6-52. Static Transfer Characteristic Curve

6.4. Given the equivalent circuit of Fig. 6-53 for a common emitter transistor amplifier. Transistor parameters are $h_{ie} = 1500\ \Omega$, $h_{re} = 20 \times 10^{-4}$, $h_{oe} = 40\ \mu$mho, $h_{fe} = 60$. Find the output signal voltage v_{s2}.

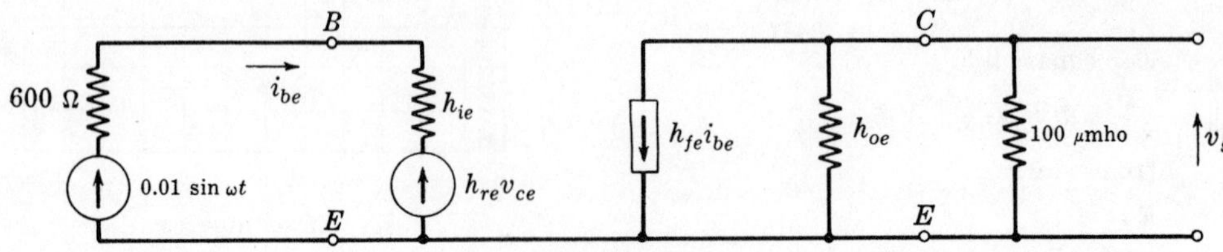

Fig. 6-53

The base loop and collector nodal equations are, with $v_{ce} = v_{s2}$,

$$-0.01 \sin \omega t + 2100 i_{be} + 2000 \times 10^{-6} v_{s2} = 0, \qquad +60 i_{be} + 140 \times 10^{-6} v_{s2} = 0$$

from which $v_{s2} = -3.45 \sin \omega t$.

6.5. Given the common base transistor amplifier circuit of Fig. 6-54. Transistor parameters are $h_{ib} = 30\ \Omega$, $h_{rb} = 250 \times 10^{-6}$, $h_{ob} = 0.35\ \mu\text{mho}$, $h_{fb} = -0.982$. Find the open circuit mid-band voltage gain $A_v = v_{s2}/v_{s1}$, where v_{s1} is the signal voltage at the input terminals of the transistor amplifier.

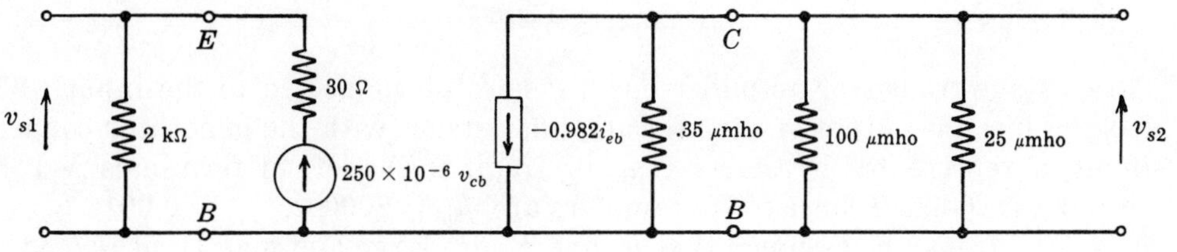

Fig. 6-54

First draw the mid-band equivalent circuit for the amplifier considering varying components, as shown in Fig. 6-55. The 2 kΩ resistor in the emitter circuit may be neglected because the input terminal voltage is involved. Using $v_{cb} = v_{s2}$, the emitter loop and collector nodal equations are respectively

$$-v_{s1} + 30 i_{eb} + (250 \times 10^{-6}) v_{s2} = 0, \qquad +(-0.982) i_{eb} + 125.4 \times 10^{-6} v_{s2} = 0$$

from which $A_v = v_{s2}/v_{s1} = 245$.

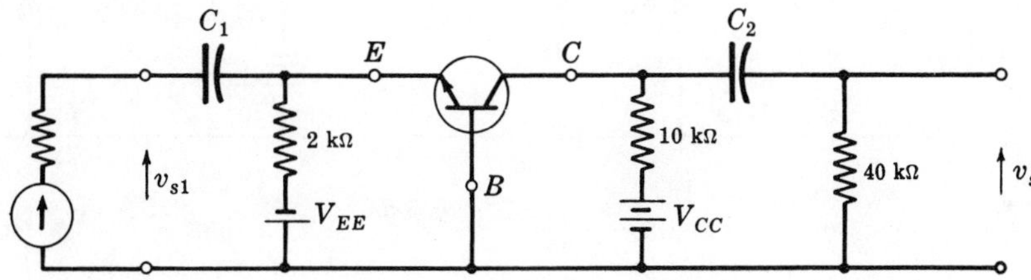

Fig. 6-55

The common base transistor amplifier is analogous to the common (grounded) grid vacuum tube amplifier. There is no phase shift between the input and output signal voltages.

6.6. Given the simplified emitter follower circuit of Fig. 6-56. The bias resistors have been omitted because they are large and negligible. Find the open circuit signal voltage v_{s2}. Transistor parameters are $h_{ie} = 1500\ \Omega$, $h_{re} = 100 \times 10^{-6}$, $h_{oe} = 50\ \mu\text{mho}$, $h_{fe} = 70$.

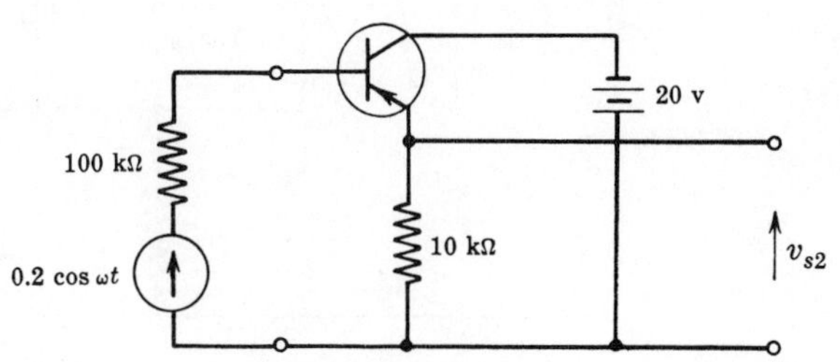

Fig. 6-56

The ac equivalent circuit is shown in Fig. 6-57 below. Note that $v_{s2} = -v_{ce}$.

The base loop and emitter nodal equations are respectively

$$-0.2 \cos \omega t + 101{,}500 i_b + (1 - 1 \times 10^{-4}) v_{s2} = 0, \qquad -71 i_b + 150 \times 10^{-6} v_{s2} = 0$$

Solving, $v_{s2} = 0.165 \cos \omega t$ volts.

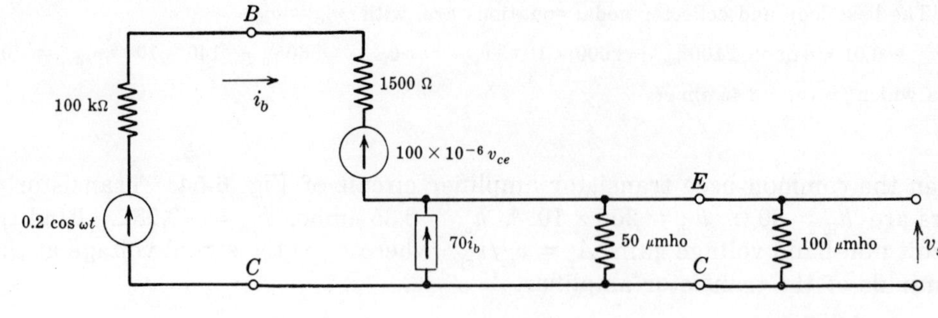

Fig. 6-57

6.7. A two stage transistor amplifier has a generator connected to the input. The circuit looking into the output of the first amplifier stage (with the generator connected to its input) is represented by the equivalent circuit to the left of terminals X-Y in the circuit of Fig. 6-58. Transistor parameters are $h_{ie} = 2000\ \Omega$, $h_{re} = 0.001$, $h_{oe} = 50\ \mu$mho, $h_{fe} = 70$. It may be assumed that R_1 and R_2 are large and may be neglected. Find the overall open circuit, mid-band voltage gain $A_v = v_{s3}/v_{s1}$.

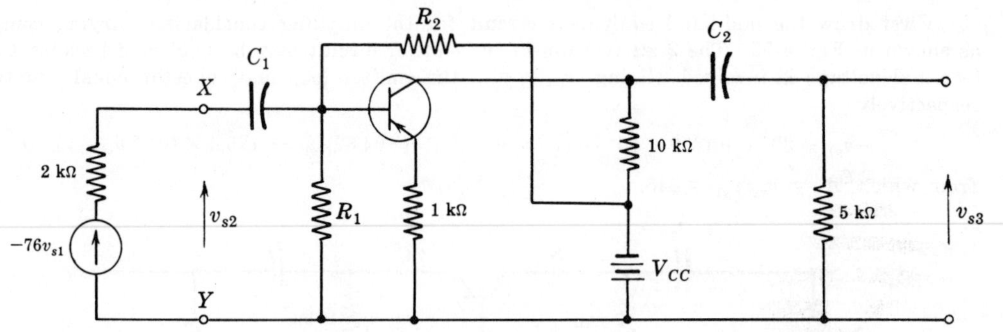

Fig. 6-58

The ac equivalent circuit is shown in Fig. 6-59.

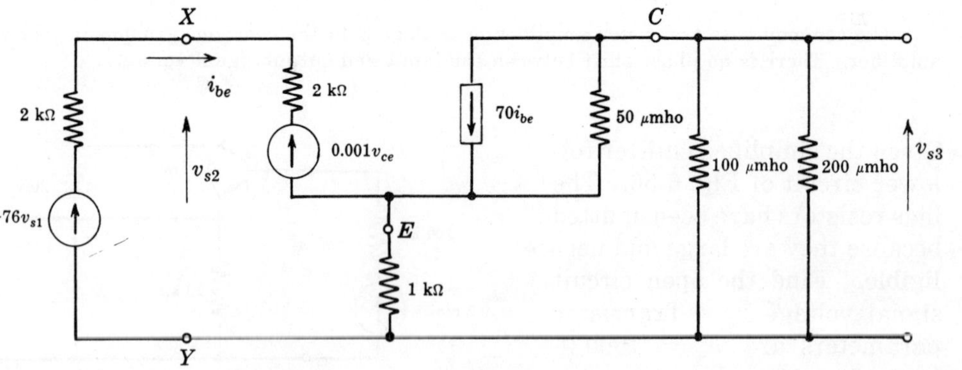

Fig. 6-59

The input loop, emitter nodal and collector nodal equations are respectively

$$-(-76 v_{s1}) + 4000 i_{be} + 0.001 v_{ce} + v_{ey} = 0$$

$$-i_{be} + 10^{-3} v_{ey} - 70 i_{be} - 50 \times 10^{-6} v_{ce} = 0$$

$$+70 i_{be} + 50 \times 10^{-6} v_{ce} + 300 \times 10^{-6} v_{s3} = 0$$

Substitute $v_{ce} = v_{s3} - v_{ey}$ and solve to find $A_v = v_{s3}/v_{s1} = 231$.

When the emitter resistor is unbypassed (as in this problem), it must be included in the equivalent circuit. There is then one more equation for a simple amplifier circuit such as shown in Fig. 6-59.

6.8. In the transistor amplifier circuit of Fig. 6-60, the base resistor R_B is so large that it may be neglected. Transistor parameters are $h_{ie} = 2000\ \Omega$, $h_{re} = 20 \times 10^{-4}$, $h_{oe} = 50\ \mu\text{mho}$, $h_{fe} = 50$. Find the mid-band input impedance (a resistance) where $R_i = v_{s1}/i_{s1}$.

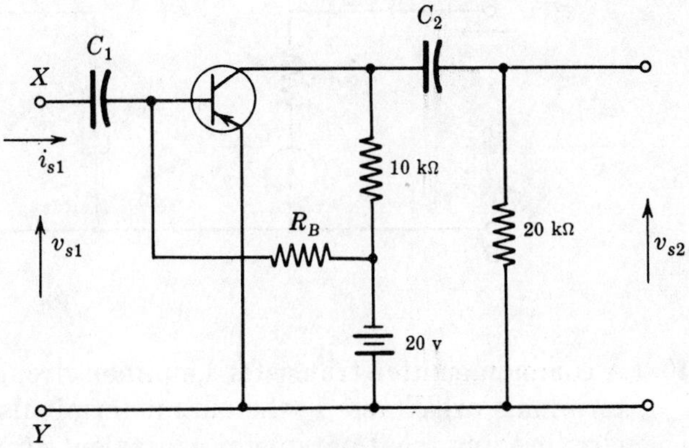

Fig. 6-60

Neglecting R_B, the ac equivalent circuit is drawn in Fig. 6-61. Note that $i_{s1} = i_{be}$ and $v_{s2} = v_{ce}$. The base loop and collector nodal equations are respectively

$$-v_{s1} + 2000 i_{s1} + 0.002 v_{s2} = 0, \qquad 50 i_{s1} + 200 \times 10^{-6} v_{s2} = 0$$

from which $i_{s1} = (200 \times 10^{-6}/0.30) v_{s1}$ and $R_i = v_{s1}/i_{s1} = 1500\ \Omega$. The input resistance is always less than h_{ie}.

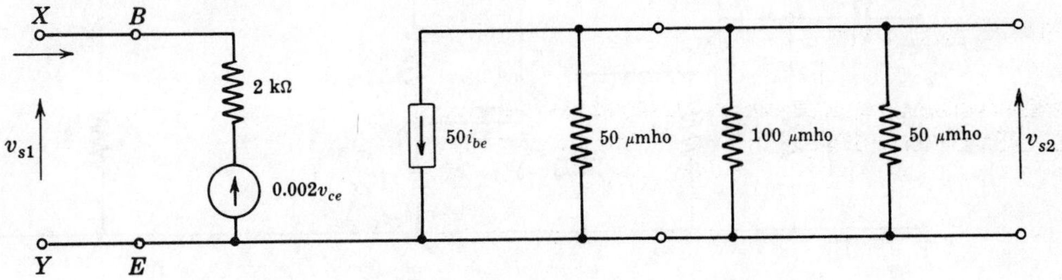

Fig. 6-61

6.9. A common base transistor amplifier circuit is given in Fig. 6-62. The resistor R_X represents the external load resistance, and the emitter resistor R_E is assumed large and negligible. The transfer function is defined as the transfer resistance $r_t = v_{s2}/i_{s1}$ (mid-band frequencies are assumed). Derive the equation for the transfer resistance, using the h parameters for the transistor.

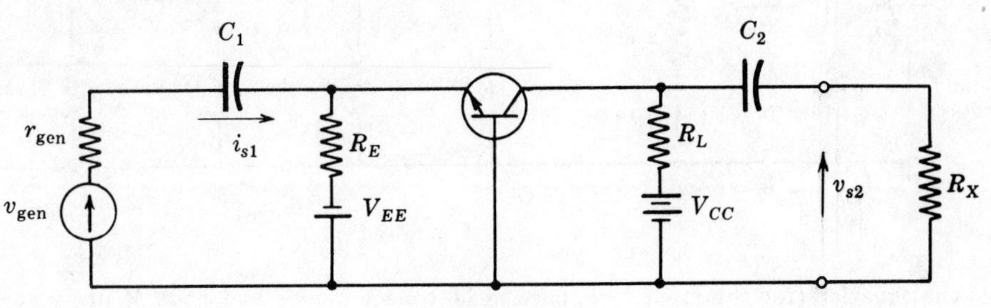

Fig. 6-62

The ac equivalent circuit using the h parameters is shown in Fig. 6-63 below. R_E is negligible. Since the transfer resistance is to be found, it is not necessary to write the emitter loop equation. Substituting $i_{eb} = i_{s1}$, the collector nodal equation is

$$h_{fb}i_{s1} + h_{ob}v_{s2} + G_L v_{s2} + G_X v_{s2} = 0$$

from which $r_t = v_{s2}/i_{s1} = -h_{fb}/(h_{ob} + G_L + G_X)$. Since h_{fb} is negative, the transfer resistance is positive for the common base amplifier circuit.

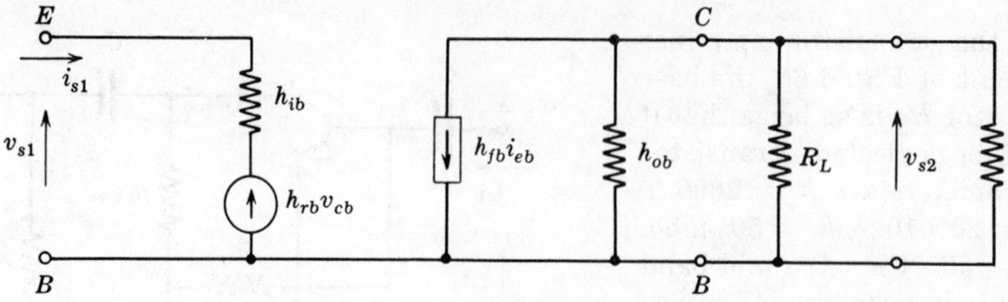

Fig. 6-63

6.10. A common emitter transistor amplifier circuit is given in Fig. 6-64. The circuit input terminals are X and Y; the output terminals are M and N. Represent this amplifier circuit using the g parameter equivalent circuit assuming linear operation and signal frequencies in the mid-band range. Transistor parameters are $h_{ie} = 2000\ \Omega$, $h_{re} = 20 \times 10^{-4}$, $h_{oe} = 50\ \mu$mho, $h_{fe} = 50$.

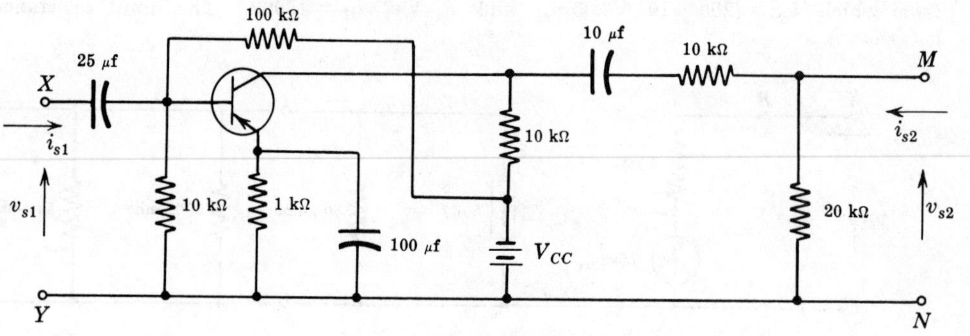

Fig. 6-64

The equivalent circuit for the circuit of Fig. 6-64 is shown in Fig. 6-65. The problem is complicated because the base bias resistors are not neglected but are combined to give the 9.1 kΩ equivalent. The problem is further complicated by having the 10 kΩ resistor between the collector node and the node M. This was done to make the problem less trivial.

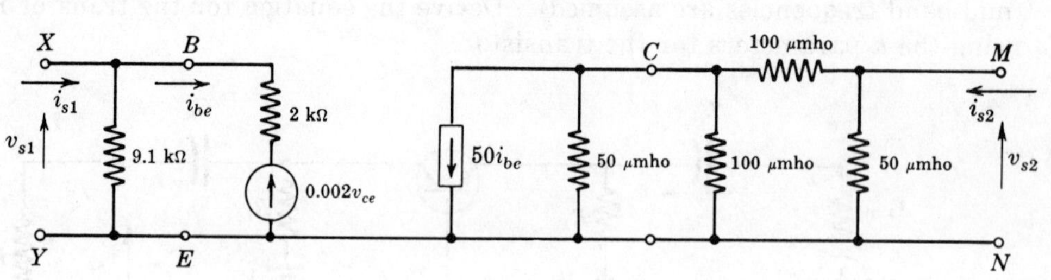

Fig. 6-65

The equations for the input loop, base node, collector node and node M are respectively

$$-v_{s1} + 2000i_{be} + 2000 \times 10^{-6}\, v_{ce} = 0$$

$$-i_{s1} + (1/9100)v_{s1} + i_{be} = 0$$

$$+50i_{be} + 150 \times 10^{-6}\, v_{ce} + 100 \times 10^{-6}\,(v_{ce} - v_{s2}) = 0$$

$$+100 \times 10^{-6}\,(v_{s2} - v_{ce}) + 50 \times 10^{-6}\, v_{s2} - i_{s2} = 0$$

Since the g parameters are to be found, the currents i_{s1} and i_{s2} are the dependent variables; v_{s1} and v_{s2} are the independent variables. The four equations may be written as

$$\begin{aligned} +2000i_{be} + 2000\times 10^{-6}\,v_{ce} &= v_{s1} \\ +9100i_{s1} - 9100i_{be} &= v_{s1} \\ +500{,}000i_{be} + 2.5v_{ce} &= v_{s2} \\ +\ 0.667v_{ce} + 6667i_{s2} &= v_{s2} \end{aligned}$$

Solving for i_{s1} by Cramer's rule,

$$i_{s1} = 0.735\times 10^{-3}\,v_{s1} - 0.5\times 10^{-6}\,v_{s2}$$

When $v_{s2}=0$, $g_i = i_{s1}/v_{s1} = 735$ μmho; when $v_{s1}=0$, $g_r = i_{s1}/v_{s2} = -0.5$ μmho.

Similarly solving for i_{s2},

$$i_{s2} = 100\times 10^{-6}\,v_{s2} + 0.0125v_{s1}$$

When $v_{s2}=0$, $g_f = i_{s2}/v_{s1} = +0.0125$ mho; when $v_{s1}=0$, $g_o = i_{s2}/v_{s2} = 100$ μmho.

The computations are often simpler if the four equations are solved by substitution. The method used illustrates a general approach.

6.11. The equivalent circuit for a common emitter transistor is given in Fig. 6-66. In this case the capacitance from the collector to ground is not small and the impedance is not negligibly large. The input signal generator voltage is 0.01 volts rms at a frequency of 1000 cps. Find the rms value of the output signal voltage and specify the phase angle relative to the input signal generator. Transistor parameters are $h_{ie}=1500\ \Omega$, $h_{re}=5\times 10^{-4}$, $h_{oe}=20\ \mu$mho, $h_{fe}=50$.

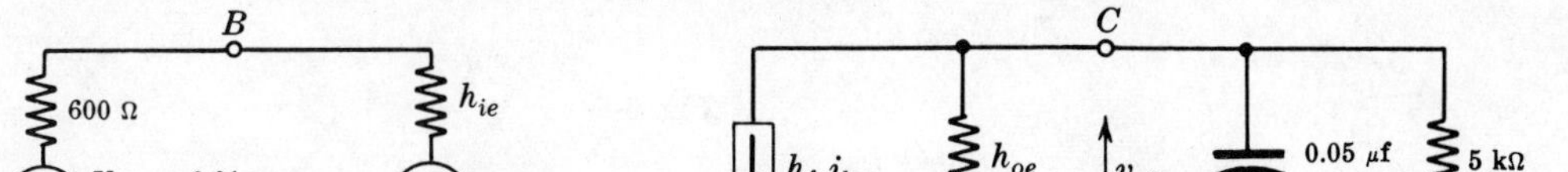

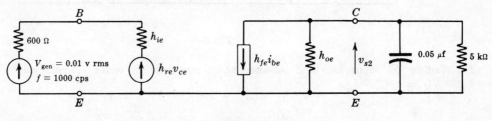

Fig. 6-66

In terms of rms values, the equations for the input loop and collector node are

$$-V_{\text{gen}} + 600I_{be} + 1500I_{be} + 500\times 10^{-6}\,V_{ce} = 0$$

$$+50I_{be} + 20\times 10^{-6}\,V_{s2} + j314\times 10^{-6}\,V_{s2} + 200\times 10^{-6}\,V_{s2} = 0$$

Substitute $V_{ce}=V_{s2}$ and solve for $V_{s2} = \dfrac{-0.5}{0.437 + j0.66} = 0.63\underline{/123.4^\circ}$ volts.

The rms value of the output signal voltage is 0.63 v; hence the open circuit voltage gain at $f=1000$ cps is 63. This is not the mid-band voltage gain. The phase shift at $f=1000$ cps is 123.4°, and in the mid-band range of frequencies it is 180°. Thus the phase shift at $f=1000$ cps is −56.6° relative to the mid-band range of frequencies.

6.12. Given the equivalent circuit for a common base transistor amplifier in Fig. 6-67 below. The transistor is represented by the "tee" equivalent circuit. Find the rms value of the signal output voltage. Transistor parameters are $r_e = 25\ \Omega$, $r_b = 400\ \Omega$, $r_c = 10^6\ \Omega$, $r_m = 0.95\times 10^6\ \Omega$.

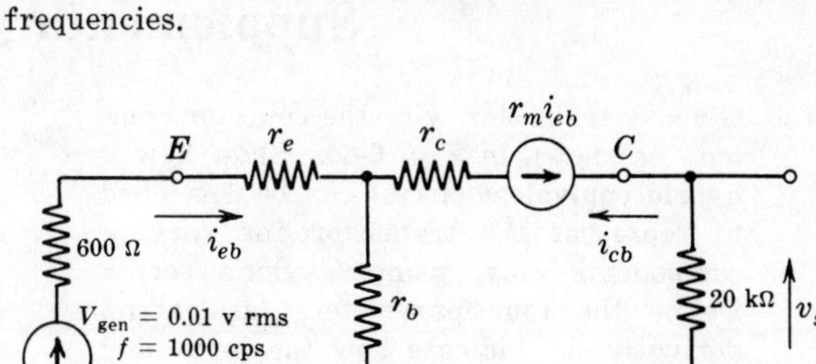

Fig. 6-67

The base loop and collector loop equations in terms of rms values of voltages and currents are

$$-0.01 + 600I_{eb} + 25I_{eb} + 400(I_{eb} + I_{cb}) = 0$$

$$20{,}000I_{cb} + 0.95 \times 10^6 I_{eb} + 10^6 I_{cb} + 400(I_{eb} + I_{cb}) = 0$$

from which $I_{cb} = -14.3 \times 10^{-6}$ amps rms. Now, since $V_{cb} = V_{s2}$, we have $V_{s2} = -20{,}000I_{cb} =$ 0.286 volts rms.

This problem illustrates some of the disadvantages of using the equivalent "tee" circuit to represent the transistor. The collector loop includes r_c and r_m, which are large compared to r_b; hence r_b may be neglected. The external load resistance was chosen as 20 kΩ so that it is not negligible. If the load resistance is much less than 10 kΩ (effective ac resistance), then it also is negligible in the collector loop. The part of the circuit to the left of the collector to base terminals then reduces to a simple constant current generator circuit.

6.13. Given the equivalent circuit for a transistor amplifier as shown in Fig. 6-68 and the transistor parameters. Find the power gain where the power output is that in the 5 kΩ resistor and the input power is that into the base of the transistor. Transistor parameters are $h_{ie} = 1500\ \Omega$, $h_{re} = 5 \times 10^{-4}$, $h_{oe} = 20\ \mu$mho, $h_{fe} = 50$.

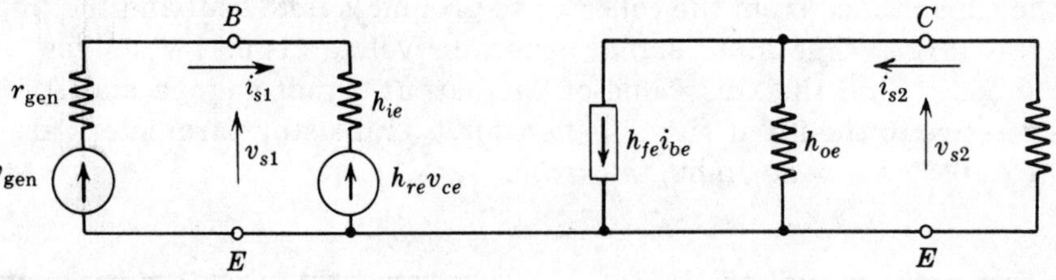

Fig. 6-68

Writing the input loop and collector nodal equations in terms of rms values and using $V_{ce} = V_{s2}$ and $I_{be} = I_{s1}$,

$$-V_{s1} + 1500I_{s1} + 500 \times 10^{-6} V_{s2} = 0 \tag{6.96}$$

$$+50I_{s1} + 20 \times 10^{-6} V_{s2} + 200 \times 10^{-6} V_{s2} = 0 \tag{6.97}$$

from which $V_{s2} = -164V_{s1}$. Then $I_{s2} = -200 \times 10^{-6} V_{s2} = +0.0328V_{s1}$.

For the reference direction of i_{s2}, the average power into the 5 kΩ resistor is $P_o = -V_{s2}I_{s2} =$ $5.38(V_{s1})^2$. From (*6.96*) and (*6.97*), $I_{s1} = 722 \times 10^{-6} V_{s1}$.

The input power $P_I = V_{s1}I_{s1} = 722 \times 10^{-6} (V_{s1})^2$. The power gain $A_p = P_o/P_I = 7450$.

Supplementary Problems

6.14. Given a transistor with the collector common as shown in Fig. 6-69. Show how a hybrid equivalent circuit can be developed to represent the transistor for varying components only, using Taylor's series. Define the four parameters (as partial derivatives). Indicate how the input and output static characteristic curves should be plotted so that the parameters are the slopes of the input, output and transfer characteristic curves.

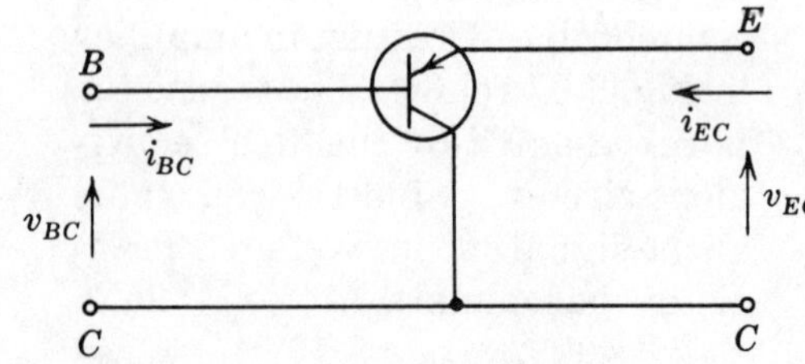

Fig. 6-69

6.15. Given the transistor amplifier circuit of Fig. 6-70 and the static collector and static base characteristic curves of Fig. 6-71 and Fig. 6-72 below. Metering equipment was used to measure $V_{CE} = -10$ v and $I_{BE} = -20\ \mu$A.

(*a*) Find I_{CE}, V_{BE} and V_{CC}.

(*b*) Plot the static forward transfer characteristic curve for the transistor for the given operating point.

(*c*) Find h_{fe} and h_{oe}.

(*d*) Determine the value of the resistor R_1.

(*e*) Plot the ac load line.

(*f*) Plot the dynamic forward transfer characteristic curve based on the dc load line and that based on the ac load line.

(*g*) Estimate the parameters h_{re} and h_{ie}.

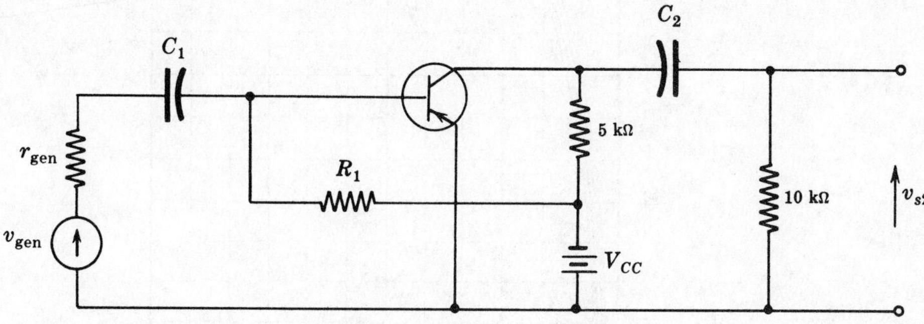

Fig. 6-70

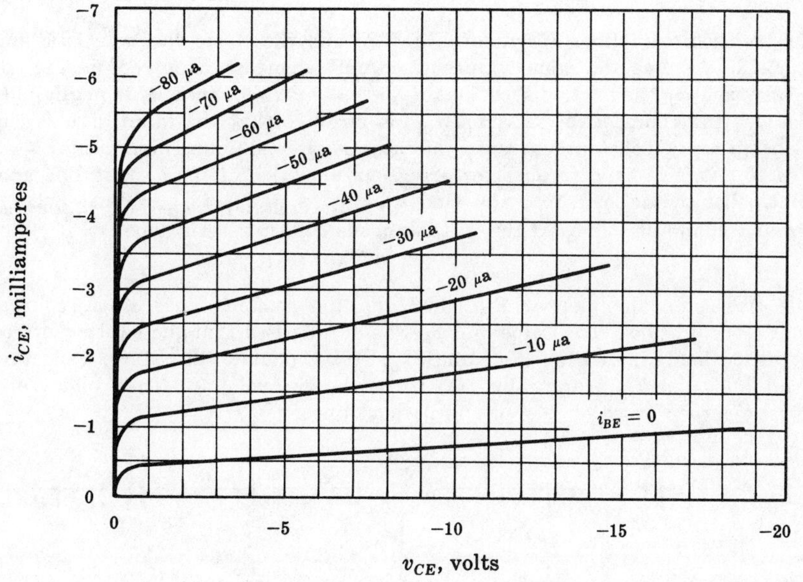

Fig. 6-71. Static Collector Characteristics

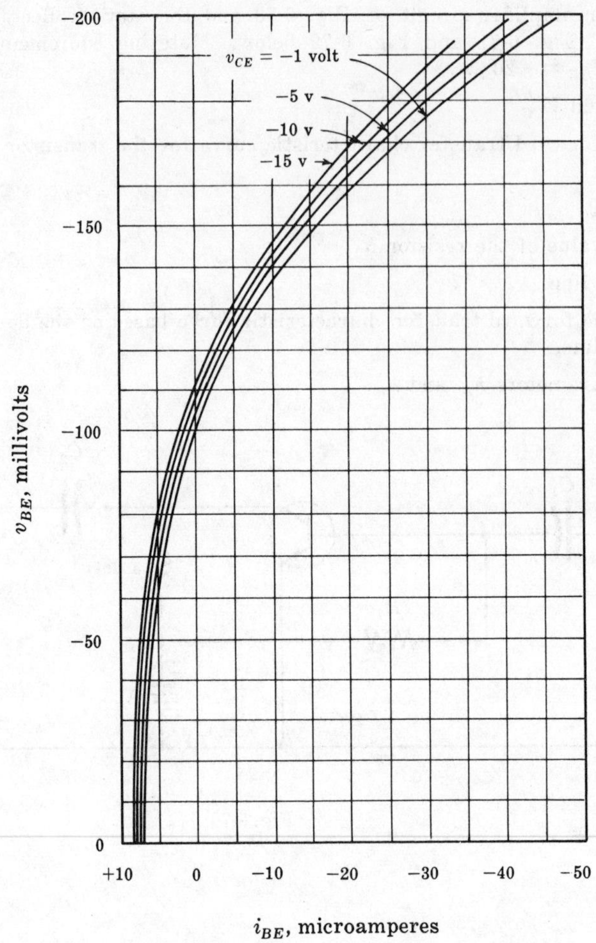

Fig. 6-72. Static Base Characteristics

6.16. Given the transistor amplifier circuit of Fig. 6-70 and the static characteristic curves of Fig. 6-71 and Fig. 6-72. Assume the same values of circuit elements as given in Fig. 6-70, and assume $V_{CC} = -15$ v and R_1 has a value such that $I_{BE} = -20\ \mu$A but that R_1 is negligible in the equivalent circuit for varying components. Assume r_{gen} is large so that the input signal generator is for all practical purposes a constant current generator with maximum magnitude 30 μA and minimum magnitude 10 μA. Mid-band frequencies are assumed. (*a*) Plot the dc load line and locate the operating point. Plot the ac load line. (*b*) Find $v_{CE(max)}$, $v_{CE(min)}$, $i_{CE(max)}$ and $i_{CE(min)}$. Calculate the maximum and minimum values of the instantaneous collector resistance ($r_{C(max)}$ and $r_{C(min)}$).

6.17. Given the transistor amplifier of Fig. 6-73 and the common base static collector characteristic curves of Fig. 6-74 below. The value of V_{EE} was adjusted so that the emitter current $I_{EB} = -3$ ma. (*a*) Plot the dc load line. Find V_{CB} and I_{CB}. (*b*) Calculate the static collector resistance R_{CB}. (*c*) Determine h_{fb} and h_{ob} graphically. (*d*) Draw the ac load line. Then plot the dynamic transfer characteristic curves based on the ac and dc load lines.

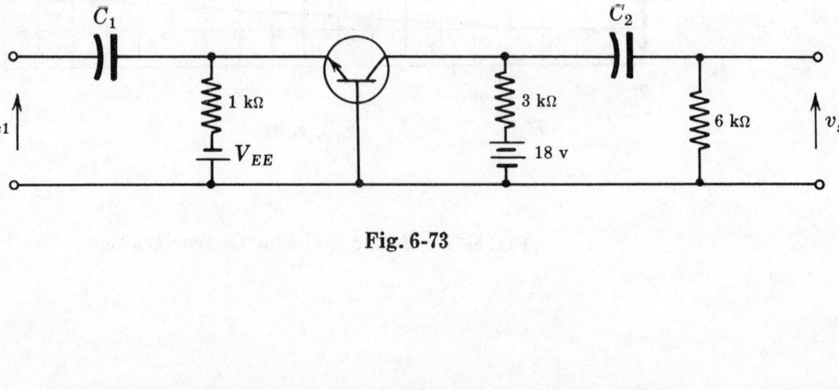

Fig. 6-73

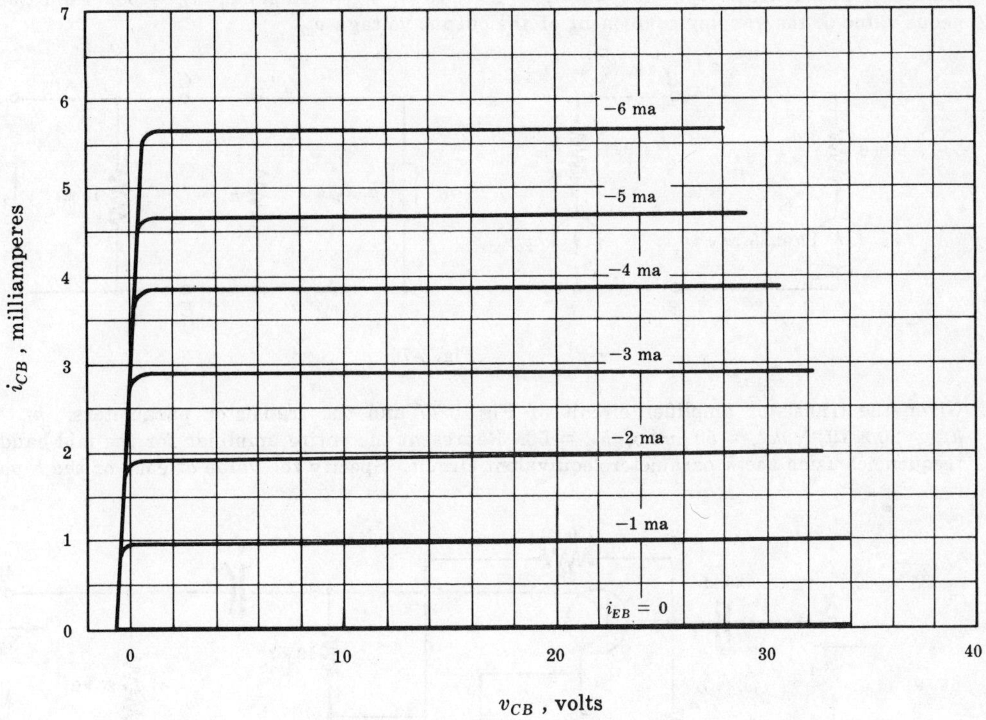

Fig. 6-74. Static Collector Characteristic Curves – Common Base

6.18. Given the transistor circuit of Fig. 6-75 and the static transistor characteristic curves of Fig. 6-71 and 6-72. The milliammeter in the collector circuit reads 3.1 ma. (*a*) Find R_B using engineering approximations. (*b*) Find the following graphically: V_{CE}, V_{BE}, I_{BE}. (*c*) Plot the collector dc and ac load lines. (*d*) Show the correct polarity for the milliammeter in the collector circuit.

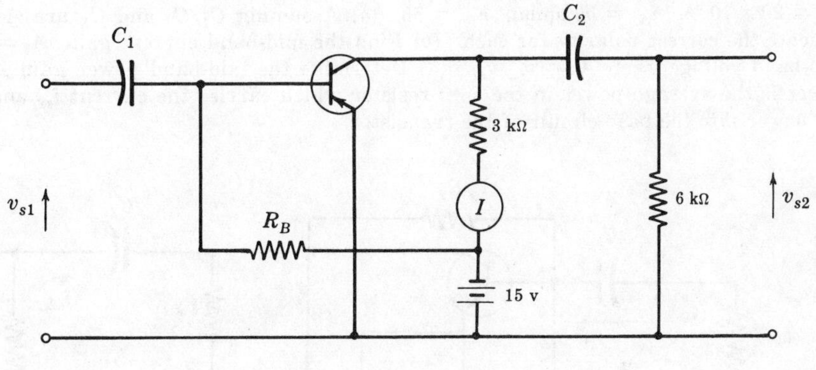

Fig. 6-75

6.19. Given the common base parameters of the 2N117 transistor for the given operating point in Table 6.1, page 131. Using the conversion formulas derived in Section 6.10, page 148, find the common emitter parameters. Then use the approximation formulas to calculate the common emitter parameters and compare results.

6.20. Given the hybrid equivalent circuit for a transistor with the base common. Employing the methods of Section 6.10, page 148, derive the equations for the common collector hybrid equivalent circuit in terms of the common base parameters.

6.21. Given the circuit of Fig. 6-76 which is the equivalent circuit for a transistor amplifier, and the transistor parameters: $h_{ie} = 1500\ \Omega$, $h_{re} = 20 \times 10^{-4}$, $h_{oe} = 40\ \mu$mho, $h_{fe} = 60$. Find the instantaneous value of the varying component of the output voltage v_{s2}.

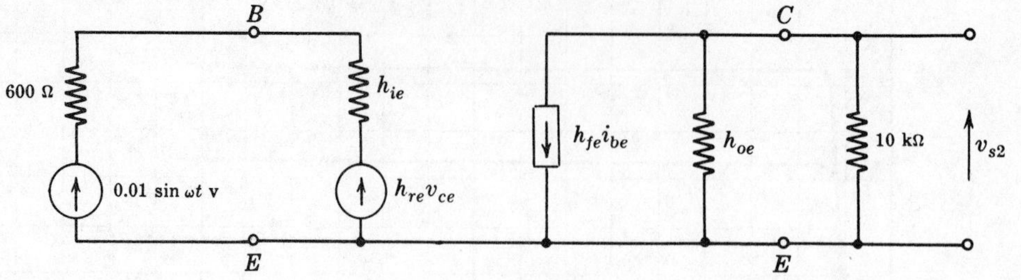

Fig. 6-76

6.22. Given the transistor amplifier circuit of Fig. 6-77 and the transistor parameters: $h_{ie} = 2000\ \Omega$, $h_{re} = 20 \times 10^{-4}$, $h_{oe} = 50\ \mu$mho, $h_{fe} = 50$. Represent the entire amplifier for the mid-band range of frequencies using the h parameters equivalent circuit. Specify the value of each of the h parameters.

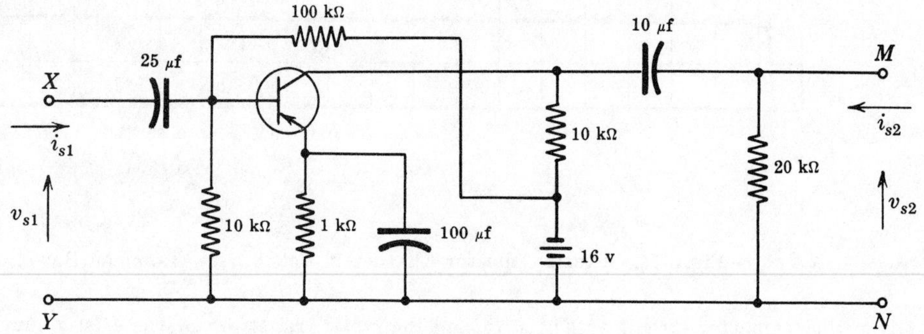

Fig. 6-77

6.23. Given the common emitter amplifier circuit of Fig. 6-78 and the transistor parameters. Assume the base bias resistors R_1 and R_2 are large and may be neglected. Transistor parameters: $h_{ie} = 2000\ \Omega$, $h_{re} = 20 \times 10^{-4}$, $h_{oe} = 50\ \mu$mho, $h_{fe} = 50$. (*a*) Assuming C_1, C_2 and C_E are electrolytic capacitors, indicate the correct polarity for each. (*b*) Find the mid-band current gain $A_i = i_{s2}/i_{s1}$. (*c*) Find the mid-band voltage gain $A_v = v_{s2}/v_{s1}$. (*d*) Determine the mid-band power gain A_p where the output power is the average power in the 5 kΩ resistor which carries the current i_{s2} and the input power is the power into the base circuit of the transistor.

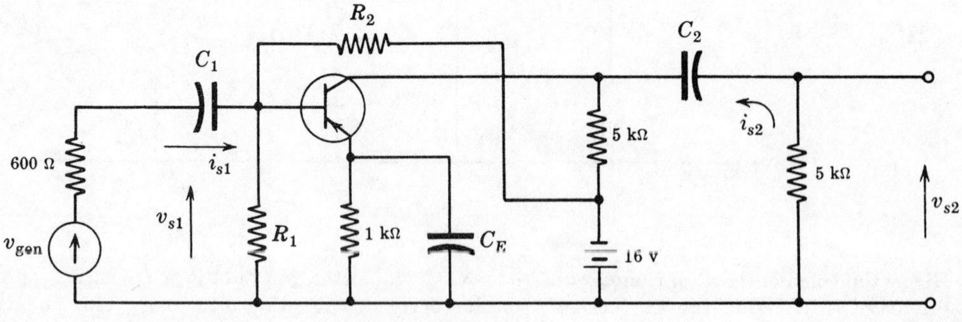

Fig. 6-78

6.24. Given the circuit of Fig. 6-79 and the static transistor characteristic curves of Fig. 6-71 and 6-72. The milliammeter in the collector circuit reads 2.5 ma. (*a*) Indicate the correct polarity for the milliammeter. (*b*) Find V_{CE}, I_{BE}, V_{BE}. (*c*) Determine the value of the resistor R_2 using engineering approximations. Justify each approximation. (*d*) Draw the amplifier equivalent circuit considering varying components and assuming mid-band frequencies.

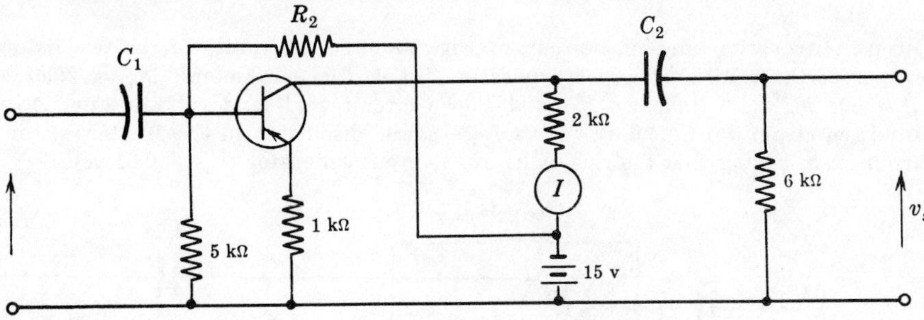

Fig. 6-79

6.25. Given the transistor amplifier circuit of Fig. 6-80 and the transistor parameters. Assume R_B is large and therefore may be neglected. Transistor parameters: $h_{ie} = 1800\ \Omega$, $h_{re} = 20 \times 10^{-4}$, $h_{oe} = 40\ \mu\text{mho}$, $h_{fe} = 60$. (*a*) Show the polarity for the power supply V_{CC}. (*b*) Calculate the open circuit, mid-band voltage gain $A_v = v_{s2}/v_{gen}$. (*c*) Calculate the input resistance (mid-band input impedance) for the amplifier circuit to the right of the terminals X and Y.

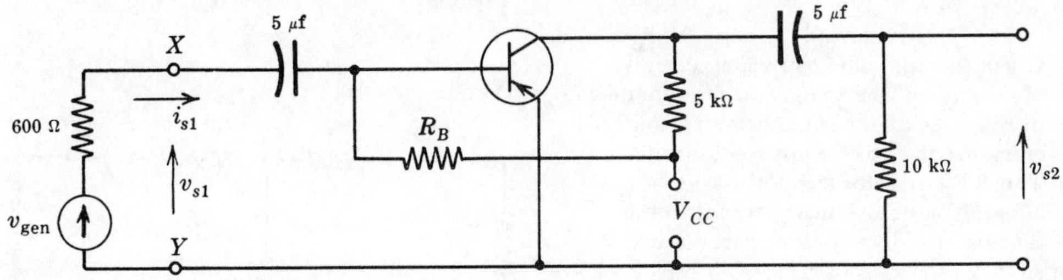

Fig. 6-80

6.26. Given the transistor amplifier circuit of Fig. 6-81 and the transistor parameters: $h_{ib} = 30\ \Omega$, $h_{rb} = 250 \times 10^{-6}$, $h_{ob} = 0.35\ \mu\text{mho}$, $h_{fb} = -0.982$. (*a*) Find the mid-band, open circuit voltage gain $A_v = v_{s2}/v_{s1}$. (*b*) Find the mid-band input impedance for the circuit to the right of the terminals X and Y.

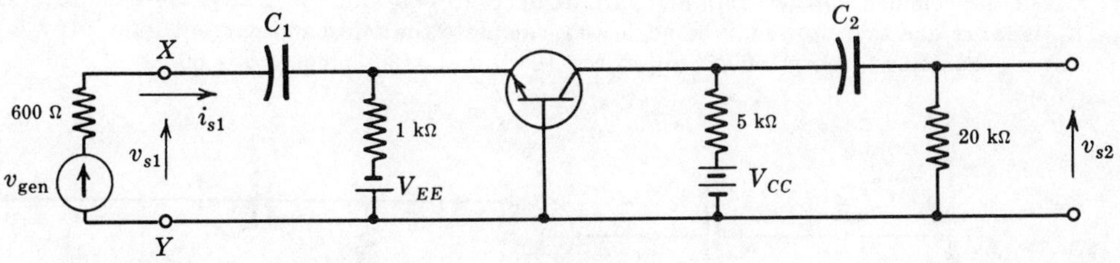

Fig. 6-81

6.27. Given the emitter follower amplifier circuit of Fig. 6-82 and the transistor parameters. Assume the base bias resistor R is large and hence may be neglected. Transistor parameters: $h_{ie} = 1500\ \Omega$, $h_{re} = 1000 \times 10^{-6}$, $h_{oe} = 50\ \mu\text{mho}$, $h_{fe} = 70$. (*a*) Find the mid-band open circuit voltage gain $A_v = v_{s2}/v_{s1}$. (*b*) Find the input resistance of the circuit to the right of the terminals X and Y.

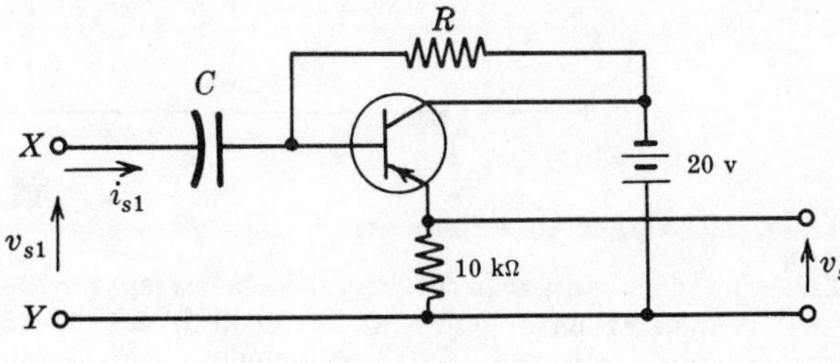

Fig. 6-82

6.28. Given the emitter follower amplifier circuit of Problem 6.27 (Fig. 6-82) and the transistor parameters. Assume the open circuit transfer function is the transfer resistance $r_t = v_{s2}/i_{s1}$. Determine the transfer resistance for the mid-band range of frequencies.

6.29. Given the transistor amplifier circuit of Fig. 6-83 and the values of the transistor parameters. The emitter of the transistor is not bypassed. Transistor parameters: $h_{ie} = 2000\ \Omega$, $h_{re} = 6 \times 10^{-4}$, $h_{fe} = 50$, $h_{oe} = 30\ \mu$mho. (*a*) Find the open circuit, mid-band voltage gain $A_v = v_{s2}/v_{gen}$ for the circuit as given. (*b*) Calculate the average power dissipated in the 10 kΩ resistor across the output terminals assuming that the rms value of the input generator $V_{s1} = 0.01$ volts.

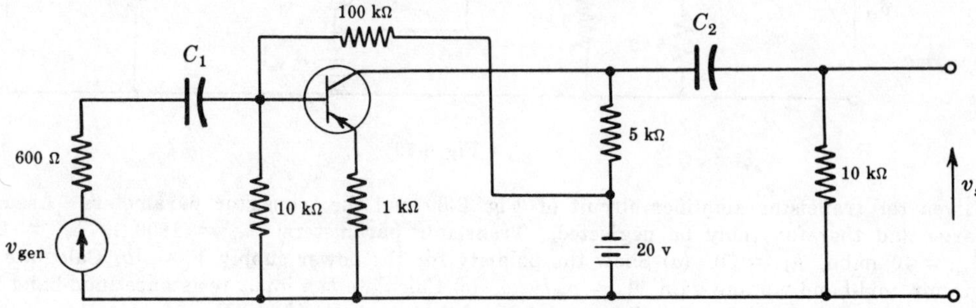

Fig. 6-83

6.30. Given the mid-band equivalent circuit of a common base transistor amplifier in Fig. 6-84 and the transistor parameters for the "tee" equivalent circuit. The 8 kΩ resistor represents all the effective ac resistance external to the transistor. Transistor parameters: $r_e = 25\ \Omega$, $r_b = 400\ \Omega$, $r_c = 10^6\ \Omega$, $r_m = 0.95 \times 10^6\ \Omega$.

(*a*) Calculate the input resistance considering the circuit to the right of the terminals *E* and *B*.

(*b*) Assuming $v_{gen} = 0.02 \sin 2000\pi t$ and $r_{gen} = 600\ \Omega$, calculate v_{s2}.

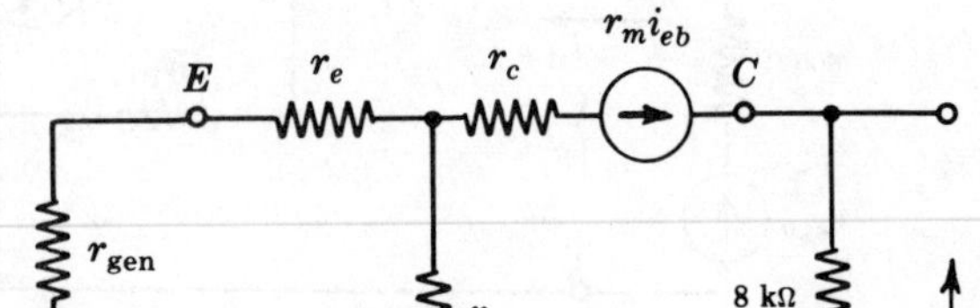

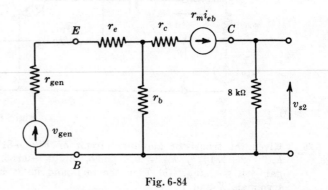

Fig. 6-84

6.31. Given the common emitter amplifier circuit of Fig. 6-85 and the transistor parameters. Assuming R_B is large and therefore may be neglected, calculate the mid-band current gain $A_i = i_{ce}/i_{be}$. Transistor parameters: $h_{ie} = 1500\ \Omega$, $h_{re} = 5 \times 10^{-4}$, $h_{be} = 20\ \mu$mho, $h_{fe} = 50$.

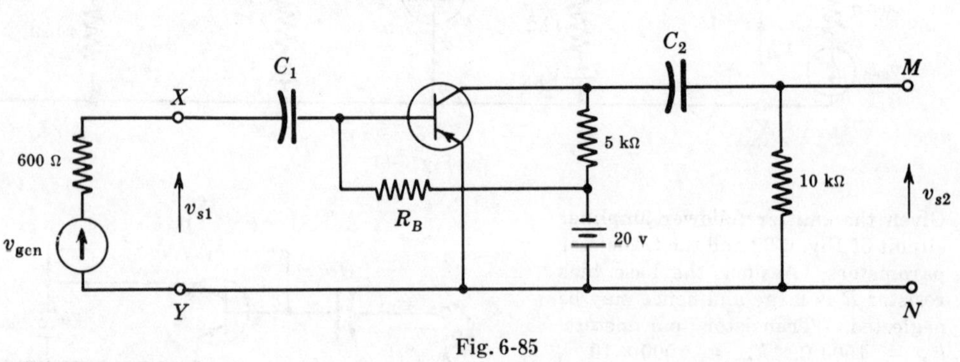

Fig. 6-85

6.32. The circuit diagram of Fig. 6-86 below shows the equivalent circuit for a common emitter amplifier. Transistor parameters: $h_{ie} = 2000\ \Omega$, $h_{re} = 2.5 \times 10^{-3}$, $h_{oe} = 50\ \mu$mho, $h_{fe} = 100$. The voltage v_{ce} was measured as $1.2 \sin 2000\pi t$. Determine the magnitude (amplitude) of the generator voltage v_{gen}.

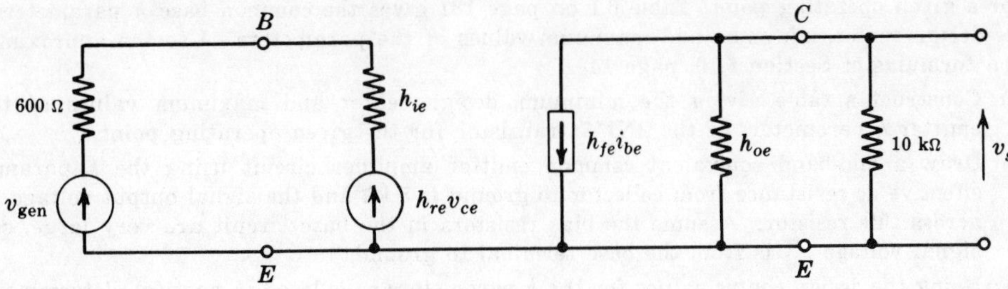

Fig. 6-86

6.33. Given the circuit of Fig. 6-85 and the transistor parameters in Problem 6.31. Assuming R_B is large and mid-band frequencies are involved, represent the amplifier between the terminal pairs *X-Y* and *M-N*, using the hybrid equivalent circuit. Specify the values of the parameters.

6.34. Given the circuit of Fig. 6-82 and the transistor parameters of Problem 6.27. Assuming R is large, represent the amplifier circuit as given, using the r parameter equivalent circuit. Specify the values of the r parameters.

6.35. Given the emitter follower circuit of Fig. 6-87 and the transistor parameters: $h_{ie} = 1800\ \Omega$, $h_{re} = 4 \times 10^{-4}$, $h_{oe} = 40\ \mu$mho, $h_{fe} = 80$. (*a*) Calculate the mid-band current gain $A_i = i_{s2}/i_{s1}$ for the circuit as given. Assume R is large. (*b*) Calculate the mid-band power gain where the output power is into the 10 kΩ resistor carrying i_{s2} and the power input is that to the base circuit of the transistor.

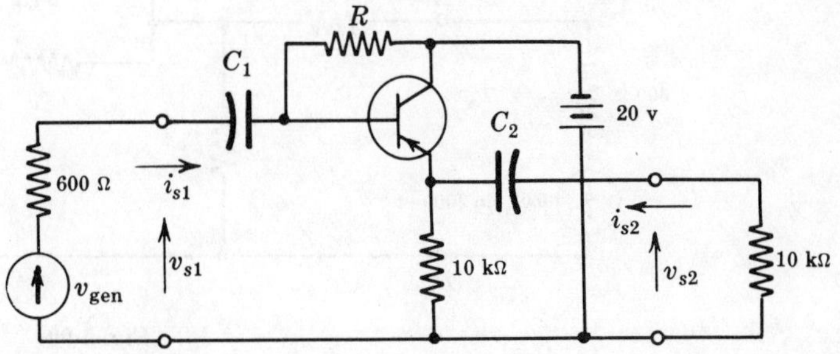

Fig. 6-87

6.36. Given the common emitter amplifier circuit of Fig. 6-88 and the transistor parameters. Assume R_B is large and hence may be neglected. Transistor parameters: $h_{ie} = 2000\ \Omega$, $h_{re} = 20 \times 10^{-4}$, $h_{oe} = 50\ \mu$mho, $h_{fe} = 50$. For the circuit as given, calculate the mid-band forward transfer conductance $g_t = i_{s2}/v_{s1}$.

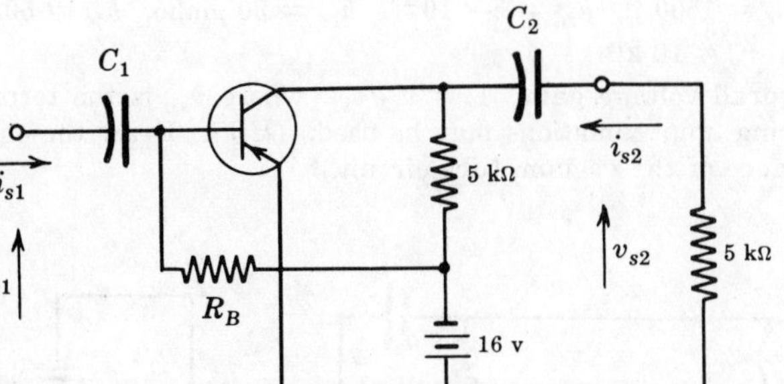

Fig. 6-88

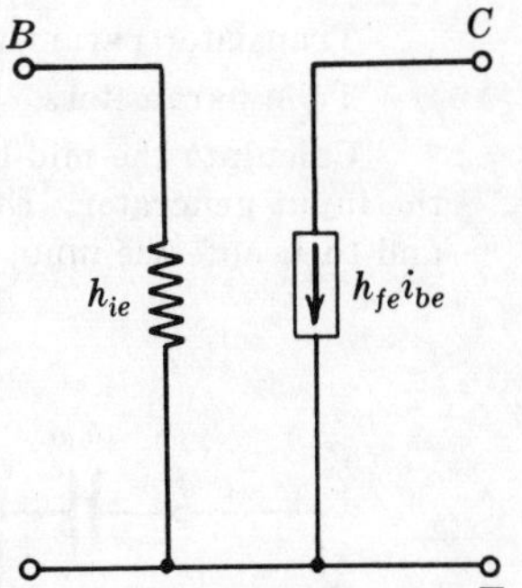

Fig. 6-89

6.37. The equivalent circuit representing a transistor is often simplified to a resistance for the input and a current generator for the output. Consider the common emitter circuit of Fig. 6-80 and the transistor parameters of Problem 6.25. Using the simplified equivalent circuit of Fig. 6-89 for the transistor, calculate the voltage gain $A_v = v_{s2}/v_{gen}$ for the amplifier. Compare this value of A_v with that calculated in Problem 6.25. Does neglecting h_{re} and h_{oe} introduce significant error?

6.38. For a given operating point, Table 6.1 on page 131 gives the common base h parameters. Included are design center, minimum and maximum values of the parameters. Use the approximate conversion formulas of Section 6.10, page 148.

(*a*) Construct a table giving the minimum, design center and maximum values of the common emitter h parameters of the 2N117 transistor for the given operating point.

(*b*) Draw a mid-band equivalent common emitter amplifier circuit using the h parameters. The effective ac resistance from collector to ground is 5 kΩ and the signal output voltage v_{s2} is taken across this resistor. Assume the bias resistors in the base circuit are very large, so the input signal voltage v_{s1} is from the base terminal to ground.

(*c*) Using the design center values for the h parameters calculated in part (*a*), determine the voltage gain $A_v = v_{s2}/v_{s1}$.

(*d*) Using the design center values for h_{re} and h_{oe} and the maximum and minimum values of h_{fe} and h_{ie}, calculate the maximum and minimum values of the voltage gain A_v. This indicates the expected range of the actual gain that might be expected in an amplifier when transistors are selected at random.

6.39. Given the equivalent circuit of Fig. 6-90 for a common emitter amplifier and the transistor parameters. Calculate v_{s2}. Transistor parameters: $r_e = 26\ \Omega$, $r_c = 10^6$, $\alpha = 0.975$, $r_b = 1320\ \Omega$.

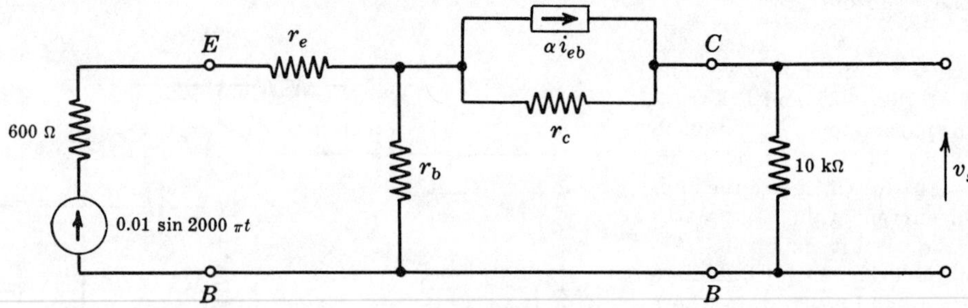

Fig. 6-90

6.40. The connections for the transistor in the circuit of Fig. 6-90 are changed so that the emitter is common. The bias circuits are designed so that the operating point remains the same. The resistors in the base bias circuit are large so that the circuit external to the transistor remains the same as that of Fig. 6-90. (*a*) Calculate the parameters for the equivalent circuit shown in Fig. 6-44 on page 148, i.e. calculate r_b, r_e, r_d and r_m. (*b*) Determine the output voltage v_{s2} for the common emitter circuit. (*c*) Find the value of β for the equivalent circuit of Fig. 6-45, page 148.

6.41. Given the amplifier circuit of Fig. 6-91 and the transistor and tube parameters.

Transistor parameters: $h_{ie} = 1500\ \Omega$, $h_{re} = 5 \times 10^{-4}$, $h_{oe} = 30\ \mu$mho, $h_{fe} = 50$.

Tube parameters: $\mu = 20$, $r_p = 10$ kΩ.

Calculate the mid-band overall voltage gain $A_v = v_{s3}/v_{s1}$ where v_{s1} is the terminal voltage of the input generator. Engineering approximations may be used. (*Hint*: Draw the equivalent circuit and then note the input impedance of the vacuum tube circuit.)

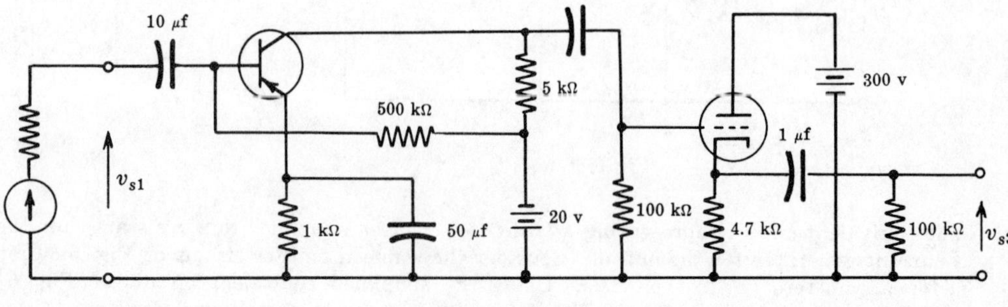

Fig. 6-91

6.42. Given the amplifier circuit of Fig. 6-92 and the parameters for the transistors. Assume the parameters are the same for both transistors. Transistor parameters: $h_{ie} = 1500\ \Omega$, $h_{re} = 5 \times 10^{-4}$, $h_{oe} = 30\ \mu$mho, $h_{fe} = 50$.

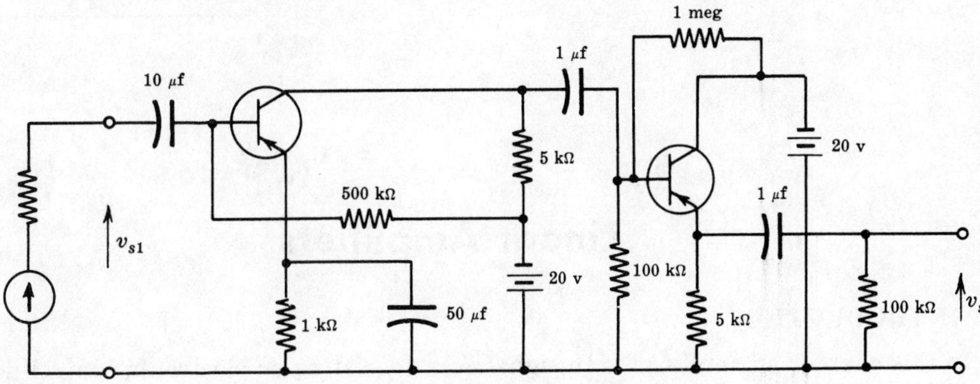

Fig. 6-92

(*a*) Calculate the overall, mid-band voltage gain $A_v = v_{s3}/v_{s1}$ using engineering approximations, where v_{s1} is the terminal voltage of the input signal generator.

(*b*) Assuming no approximations, are there any significant differences between the circuits of Fig. 6-91 and Fig. 6-92 in terms of the calculations for Problems 6.41 and 6.42? (*Hint.* Draw the equivalent circuit for the circuit of Fig. 6-92.)

6.43. Consider a simple transistor amplifier circuit such as given in Fig. 6-75 (Problem 6.18) and the static transistor characteristic curves of Fig. 6-71 and 6-72. Explain how a transistor, acting as an electronic control device, in the given circuit can cause voltage amplification. Illustrate by drawing a simplified equivalent circuit in terms of instantaneous values of voltages and currents, using nonlinear elements when necessary. Demonstrate graphically why the change in instantaneous collector resistance is significant.

6.44. Given the equivalent amplifier circuit of Fig. 6-93 and the transistor parameters. The capacitor in the collector circuit may not be neglected. Transistor parameters: $h_{ie} = 1500\ \Omega$, $h_{re} = 5 \times 10^{-4}$, $h_{oe} = 20\ \mu$mho, $h_{fe} = 50$.

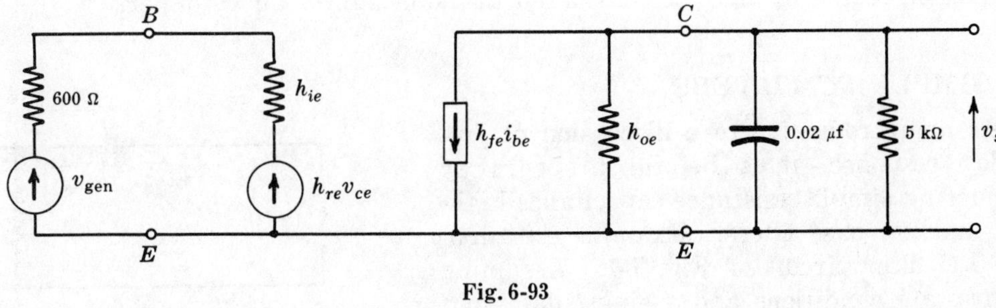

Fig. 6-93

(*a*) If $v_{gen} = 0.01414 \sin 2000\pi t$ volts, calculate the rms value of the output voltage V_{s2} and specify the phase angle relative to v_{gen}.

(*b*) Calculate the voltage gain of the amplifier, $A_v = v_{s2}/v_{be}$, for $f = 1000$ cps. This is complex and the angle should be included.

(*c*) Calculate the current gain $A_i = i_{ce}/i_{be}$ (a complex number) at $f = 1000$ cps.

Chapter 7

Linear Amplifiers

7.1 INTRODUCTION

In this chapter an amplifier is thought of as consisting of electronic control devices and associated circuit elements which together perform a desired function. The power supply may be a part of the amplifier unit or it may be a separate unit, depending on the design. In simple amplifiers there is one pair of input terminals and one pair of output terminals. In general the interest is in the relationship between the signal at the output terminals and that applied at the input terminals. If the output signal is a linear function of the input signal, then the amplifier is a *linear amplifier*. A linear amplifier may be represented using two port equivalent circuits similar to those used to represent individual electronic control devices. This makes the problems appear to be simpler than they actually are.

A linear amplifier may be thought of as a model because there are no truly linear amplifiers. Electronic amplifiers are only approximately linear for given operating conditions. The amount of nonlinearity in an amplifier is one criterion of performance. One way of specifying the nonlinearity of an amplifier for a given signal amplitude is in terms of harmonic distortion.

Another criterion of amplifier performance involves frequency response. The terminal characteristics of an amplifier are not independent of frequency. Moreover, noise is generated within some of the components of an amplifier. Compromises must be made in the actual design. In addition to performance criteria, factors such as space limitations, heat dissipation, cost, etc., must be considered in the design of an amplifier.

7.2 SIMPLE *RC* FILTERS

As an introduction to a discussion of frequency response, it is helpful to begin by considering simple resistance-capacitance high-pass and low-pass filters. Examine the high-pass *RC* filter circuit of Fig. 7-1. Assuming steady-state conditions and a sinusoidal input voltage, the loop equation is

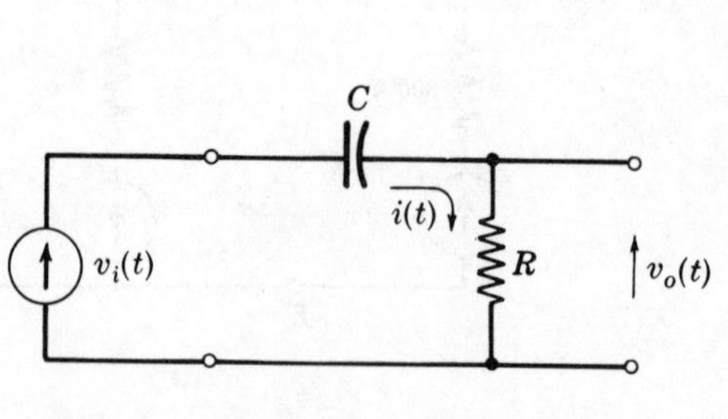

Fig. 7-1. Simple *RC* High-Pass Filter

$$-v_i(t) + Z_C\, i(t) + R\, i(t) = 0 \quad \text{or} \quad i(t) = v_i(t)/(R+Z_C)$$

and the output voltage is

$$v_o(t) = R\, i(t) = R\, v_i(t)/(R+Z_C)$$

The open circuit voltage gain (transfer function) is defined as

$$A_v = v_o(t)/v_i(t) = R/(R+Z_C) = R/(R-j/\omega C) = 1/(1-j/2\pi fRC) \tag{7.1}$$

which shows that A_v is complex and a function of frequency.

When the frequency of $v_i(t)$ is very high so that the imaginary part of the denominator of equation (*7.1*) is small compared to 1, i.e. $1/2\pi RCf \ll 1$, then A_v approaches 1 and the phase shift approaches zero. When the frequency is low so that the imaginary part of the denominator is large compared to 1, i.e. $1/2\pi RCf \gg 1$, then

$$A_v = 1/(-j/2\pi RCf) = j2\pi RCf = 2\pi RCf\,\underline{/90^\circ}$$

Thus for very low frequencies the voltage gain is linearly proportional to the frequency and the phase shift is 90°.

In the range of frequencies where the real and imaginary parts of the denominator of (*7.1*) have the same order of magnitude, the voltage gain is less than 1 and the phase shift is between 0 and 90°. The interest is usually in both the magnitude and phase shift of A_v. For any given values of resistance and capacitance, (*7.1*) may be used to plot curves showing the magnitude and phase shift of A_v as a function of frequency. However, universal curves can be plotted because for any combinations of parameters R and C, the curves have the same shape but are shifted along the frequency axis.

The numerator of (*7.1*) is a real number and the real part of the denominator is independent of frequency. The magnitude of A_v is $1/\sqrt{2} = 0.707$ when the magnitude of the denominator of (*7.1*) is $\sqrt{2}$. The denominator is complex and may be represented as in Fig. 7-2. The real part of the denominator D is unity; hence the magnitude of D is $\sqrt{2}$ when the magnitude of the imaginary part of D is 1. For given values of R and C, this obtains at a frequency f_1 such that $1/2\pi f_1 RC = 1$ or

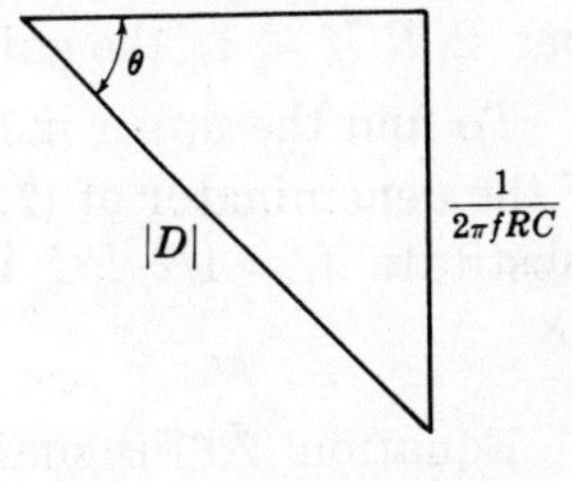

Fig. 7-2. Denominator of Equation (*7.1*)

$$f_1 = 1/2\pi RC \text{ in cps} \quad \text{or} \quad \omega_1 = 1/RC \text{ in radians/sec} \tag{7.2}$$

where f_1 is called the lower half-power frequency. The output power is proportional to the square of the voltage. When the voltage is reduced by a factor of $1/\sqrt{2}$, the power is reduced by a factor of 1/2. Since this is a high-pass filter, there is only one half-power frequency and it is below the pass band (frequency range where the gain is maximum).

Equation (*7.2*) indicates that the lower half-power frequency f_1 for the simple RC high-pass filter is determined by R and C. Substituting (*7.2*) into (*7.1*),

$$A_v = 1/(1 - jf_1/f) \tag{7.3}$$

Equation (*7.3*) may be represented by plotting two universal frequency response curves. One curve is a plot of the magnitude of A_v and the other is a plot of the magnitude of the phase shift of A_v, with both curves plotted as functions of the frequency ratio f_1/f. If the equation for the gain of a simple circuit (one pole) is normalized by dividing A_v by the pass-band gain (call this A_{vo}), then the curves obtained are the same as those of the simple high-pass and low-pass filters. The universal RC filter response curves are shown in Fig. 7-3. Equa-

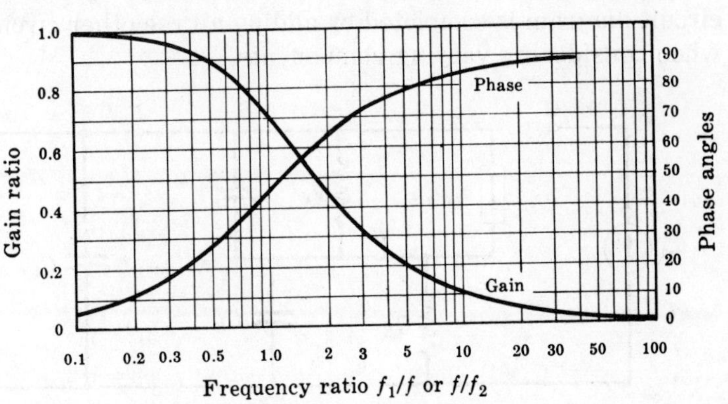

Fig. 7-3. *RC* Filter Response

tion (*7.3*) indicates that for the high-pass RC filter the phase shift is positive and the output voltage leads the input voltage.

A simple resistance-capacitance, low-pass filter is given in Fig. 7-4. Assuming steady-state conditions and a sinusoidal input, the loop equation is

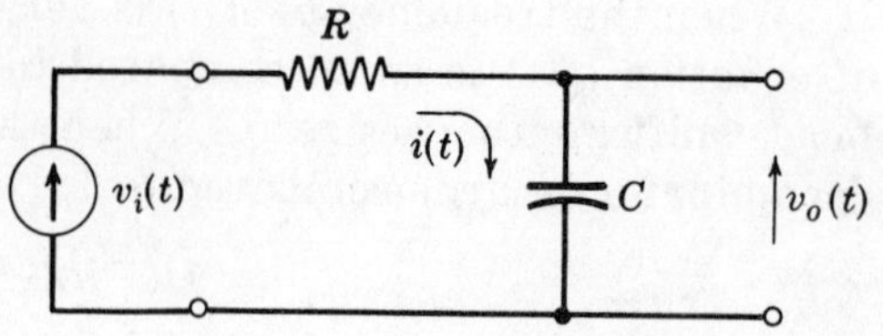

Fig. 7-4. Low Pass Filter

$$-v_i(t) + R\,i(t) + Z_C\,i(t) = 0 \quad \text{or} \quad i(t) = v_i(t)/(R+Z_C)$$

The output voltage $v_o(t)$ and the voltage gain A_v are

$$v_o(t) = Z_C\,i(t) = [Z_C/(R+Z_C)]v_i(t) \tag{7.4}$$

$$A_v = v_o(t)/v_i(t) = Z_C/(R+Z_C) = 1/(1+j2\pi fRC) \tag{7.5}$$

When the frequency is high so that $2\pi RCf \gg 1$, $A_v = 1/j2\pi RCf = -j/2\pi RCf$; the gain then varies inversely with frequency and the phase shift is $-90°$. When the frequency is low so that $2\pi RCf \ll 1$, the gain approaches 1 with no phase shift.

To find the upper half-power frequency f_2 of the low-pass filter, set the imaginary part of the denominator of (*7.5*) equal to 1; then the magnitude of the denominator is $\sqrt{2}$. Now substitute $f_2 = 1/2\pi RC$ into (*7.5*) and obtain

$$A_v = 1/(1+jf/f_2) \tag{7.6}$$

Equation (*7.6*) is similar to (*7.3*) but the frequency ratio is now f/f_2 instead of f_1/f and the phase angle of A_v is negative instead of positive. The magnitude of the phase angle is plotted in Fig. 7-3, so it is necessary to realize that the phase angle is negative for low-pass filters and positive for high-pass filters.

The plotted curves of Fig. 7-3 are universal in that they can be employed to determine the frequency response of simple (one pole) high- and low-pass filters. In this book we are interested largely in the fact that these universal frequency response curves also apply to some amplifier circuits.

7.3 FREQUENCY RESPONSE OF ONE STAGE VACUUM TUBE AMPLIFIERS

A simple vacuum tube amplifier was shown in Fig. 5-21, page 82. The amplifier equivalent circuit considering only varying components is employed in writing the circuit equations. The representation for the vacuum tube (assuming linear operation and only varying components) was developed in Chapter 5. In the present discussion the constant current equivalent circuit will be used for the tube as shown in Fig. 7-5. The equivalent circuit diagram is completed by adding all the other circuit elements which must be included when considering varying components.

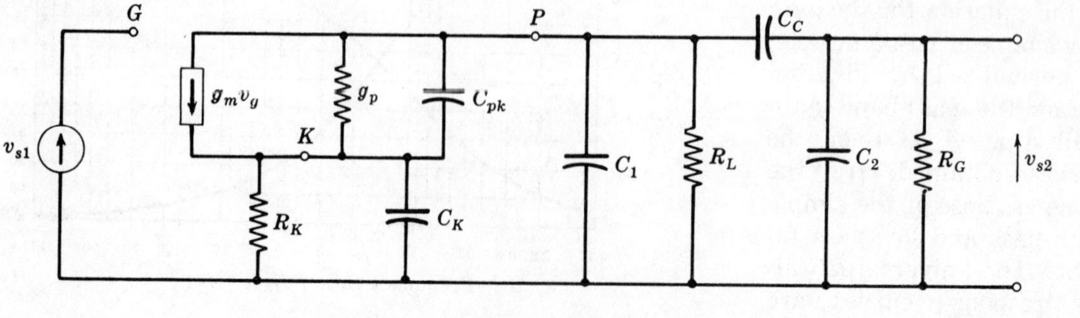

Fig. 7-5. Amplifier Equivalent Circuit

Any physically assembled circuit includes stray shunt capacitances which must be included at high frequencies. Certain of the capacitances are more significant; in the present discussion only those shown in Fig. 7-5 will be considered. In addition, there are interelectrode capacitances between each pair of vacuum tube terminals, but only the capacitance between plate and cathode terminals is considered significant in the present discussion. The circuit of Fig. 7-5 differs from that of Fig. 5-22, page 82, only in the capacitances which are included.

The usual procedure for determining the open circuit voltage gain and the frequency response is to draw the appropriate equivalent circuit, write the circuit equations, and then solve them. Three equivalent circuits are usually drawn: mid-band, low frequency, and high frequency. This simplifies the calculations considerably.

The mid-band range of frequencies is that range where the voltage gain A_v is independent of frequency, hence there are no reactances in the mid-band equivalent circuit. The coupling and bypass capacitances are so large that their impedances are negligibly small in the mid-band range. The shunt capacitances are so small that their impedances are negligibly large. The basis for making comparisons will be shown later. In the mid-band frequency range the capacitors C_K and C_C are assumed to have negligibly small impedances, and all the shunt as well as interelectrode tube capacitances are assumed to have negligibly large impedances. The mid-band equivalent circuit is the circuit of Fig. 5-22, page 82, and is reproduced in Fig. 7-6. The mid-band voltage gain was given by equation (*5.30*), page 82, as

$$A_{vo} = -g_m/(g_p + G_L + G_G)$$

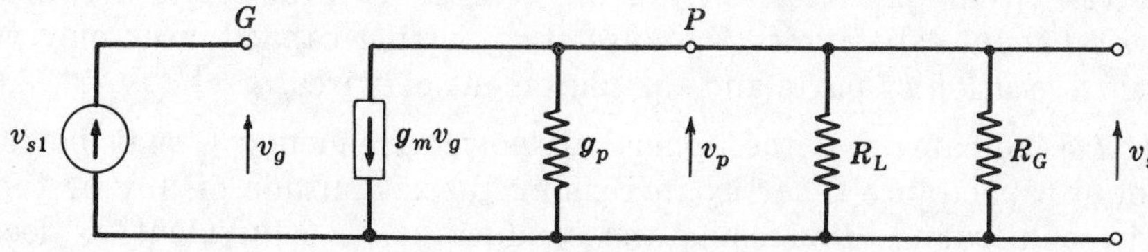

Fig. 7-6. Mid-band Equivalent Circuit

Now examine the high frequency equivalent circuit. Only the shunt and interelectrode capacitances shown in Fig. 7-5 are considered significant. Since the impedance of a capacitor is inversely proportional to frequency, C_K and C_C are also short-circuited in the high frequency equivalent circuit. Then all the shunt capacitances between plate and cathode are in parallel and therefore can be combined to form one effective shunt capacitor C_S as shown in Fig. 7-7. The grid loop and plate nodal equations are

$$-v_{s1} + v_g = 0 \quad \text{or} \quad v_g = v_{s1} \tag{7.7}$$

$$+g_m v_g + g_p v_p + G_L v_p + G_G v_p + j\omega C_S v_p = 0 \tag{7.8}$$

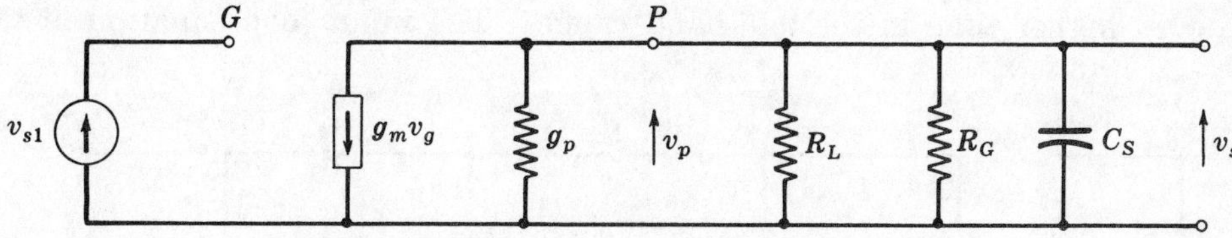

Fig. 7-7. High Frequency Equivalent Circuit

Using $v_g = v_{s1}$ and $v_p = v_{s2}$ and then solving for the high frequency voltage gain,

$$A_v(\text{h.f.}) = v_{s2}/v_{s1} = -g_m/(g_p + G_L + G_G + j\omega C_S) \tag{7.9}$$

The upper half-power frequency f_2 for the amplifier may be determined by using a procedure similar to that employed in finding the half-power frequencies for the simple RC filters. The numerator of (7.9) is real and independent of frequency, and the real part of the denominator is independent of frequency. Then the upper half-power frequency is the frequency at which the real part of the denominator is equal to the imaginary part. For convenience divide the numerator and denominator of (7.9) by the real part of the denominator, giving

$$A_v(\text{h.f.}) = \frac{-g_m/(g_p + G_L + G_G)}{1 + j[2\pi f C_S/(g_p + G_L + G_G)]} \tag{7.10}$$

Set the imaginary part of the denominator equal to the real part and obtain

$$f_2 = (g_p + G_L + G_G)/2\pi C_S \tag{7.11}$$

which when substituted into (7.10) gives

$$A_v(\text{h.f.}) = \frac{-g_m/(g_p + G_L + G_G)}{1 + jf/f_2} \tag{7.12}$$

where the numerator is the mid-band voltage gain. Dividing (7.12) by A_{vo},

$$A_v(\text{h.f.})/A_{vo} = 1/(1 + jf/f_2) \tag{7.13}$$

Equation (7.13) has been normalized, hence the normalized voltage gain of the high frequency circuit is the same as for the simple low-pass filter. The universal response curves of Fig. 7-3 can be used when the upper half-power frequency is known. This can be found for given circuit parameters by using (7.11). In practice it may be difficult to determine the total shunt capacitance C_S, since stray wiring capacitances may vary according to the physical location of parts and the placement of wires.

Equation (7.11) indicates that the upper half-power frequency f_2 may be raised by decreasing the shunt capacitance C_S or by increasing the magnitude of any of the three conductances in the numerator. Increasing the conductance is equivalent to decreasing the resistance. Observe that the numerator of the expression for f_2 is the same as the denominator of the expression for the mid-band gain. Anything which increases the numerator of f_2 decreases the voltage gain correspondingly.

The differential plate conductance g_p is a constant for a given tube and a given operating point. Changing G_G will also affect the low frequency response, as will be shown later. After the shunt capacitance C_S is reduced to a practical minimum, it is necessary to increase G_L (decrease R_L) in order to raise the upper half-power frequency.

The low frequency equivalent circuit is now drawn. It is assumed that C_K can be made so large that the low frequency response will be determined only by the coupling capacitor C_C. The cathode capacitor C_K can be electrolytic, but the coupling capacitor cannot due to the inherent leakage in electrolytic capacitors. The low frequency circuit is drawn in Fig. 7-8. In the range of frequencies below the mid-band the impedances of the shunt capacitances are even higher than in the mid-band range. The input loop equation is the same

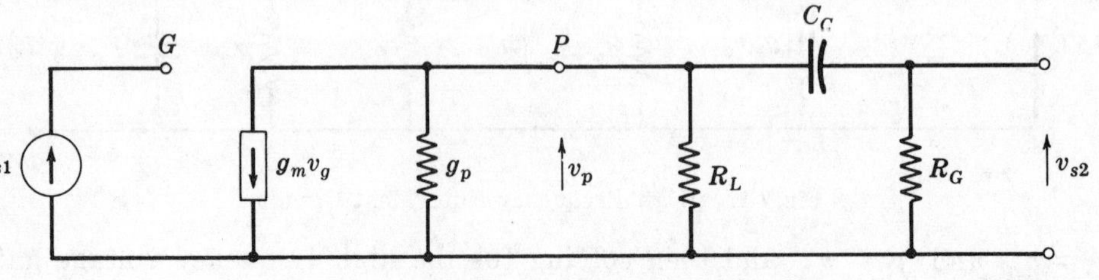

Fig. 7-8. Low Frequency Equivalent Circuit

as (*7.7*). The plate nodal equation is

$$+g_m v_g + g_p v_p + G_L v_p + j\omega C_C(v_p - v_{s2}) = 0 \qquad (7.14)$$

and the nodal equation for the non-reference output node is

$$+j\omega C_C(v_{s2} - v_p) + G_G v_{s2} = 0 \qquad (7.15)$$

Substituting $v_g = v_{s1}$ into (*7.14*) reduces the number of equations to two. Rewriting (*7.14*) and (*7.15*),

$$(g_p + G_L + j\omega C_C)v_p - j\omega C_C v_{s2} = -g_m v_{s1}, \qquad -j\omega C_C v_p + (G_G + j\omega C_C)v_{s2} = 0$$

from which the low frequency voltage gain is

$$A_v(\text{l.f.}) = v_{s2}/v_{s1} = \frac{-j\omega C_C g_m}{G_G(g_p + G_L) + j\omega C_C(G_G + g_p + G_L)}$$

Dividing numerator and denominator of the right side by $j\omega C_C$,

$$A_v(\text{l.f.}) = \frac{-g_m}{(G_G + g_p + G_L) - jG_G(g_p + G_L)/2\pi f C_C}$$

If the numerator and denominator are divided by the real part of the denominator, then the numerator is the mid-band voltage gain,

$$A_v(\text{l.f.}) = \frac{-g_m/(G_G + g_p + G_L)}{1 - jG_G(g_p + G_L)/[2\pi f C_C(G_G + g_p + G_L)]} \qquad (7.16)$$

The lower half-power frequency can be found as before,

$$f_1 = \frac{G_G(g_p + G_L)}{2\pi C_C(G_G + g_p + G_L)} \qquad (7.17)$$

Putting (*7.17*) into (*7.16*) and substituting A_{vo} for the numerator,

$$A_v(\text{l.f.}) = A_{vo}/(1 - jf_1/f)$$

or

$$A_v(\text{l.f.})/A_{vo} = 1/(1 - jf_1/f) \qquad (7.18)$$

whose right side is the same as that of (*7.3*) for the simple *RC* high-pass filter. The universal frequency response curves may be used and f_1 calculated by (*7.17*).

The bandwidth of an amplifier may be defined as the frequency band between the two half-power frequencies. This is denoted as

$$\Delta f = f_2 - f_1$$

At the half-power frequencies the magnitude of the gain is 0.707 times the mid-band value, and the magnitude of the phase shift is 45° with respect to the mid-band value. It is important to emphasize this because there may be 180° phase shift between the input and output in the mid-band range of frequencies. The phase shift indicated by the universal frequency response curves is with respect to the mid-band value. For an amplifier with 180° mid-band phase shift (indicated by a minus sign in the gain equation), the actual overall phase shift at $f = f_2$ is $180° - 45° = 135°$. At $f = f_1$, the overall phase shift is $180° + 45° = 225°$.

The frequency response of the vacuum tube amplifier of Fig. 7-5 has been determined from different equivalent circuits for both the high and low frequency cases. The mid-band equivalent circuit may be obtained from either of these as a limiting case. This is done by letting the frequency become high for the low frequency case and become low for the high frequency case. The main reasons for employing different high and low frequency equivalent circuits is that the solutions to the equations are simplified and the resulting equations have the same form as those for the simple *RC* filters.

When the amplifier equivalent circuit is simple, a general equation for the open-circuit voltage gain may be found. The high frequency equivalent circuit of Fig. 7-7 and the low frequency equivalent circuit of Fig. 7-8 may be combined, yielding a general equivalent circuit for the amplifier. This assumes that the cathode is effectively bypassed, that interelectrode capacitances are negligible, that the power supply is ideal, and all the other assumptions made in drawing the circuits of Fig. 7-7 and Fig. 7-8. The amplifier equivalent circuit is given in Fig. 7-9. The shunt capacitances are shown separately for the parts of the circuit on each side of the coupling capacitor.

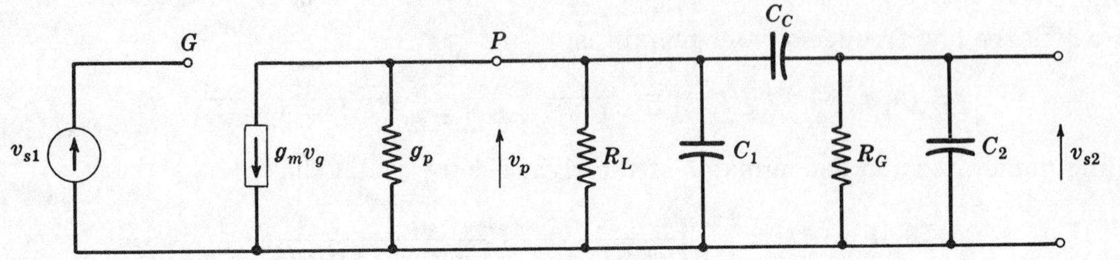

Fig. 7-9. Amplifier Equivalent Circuit

The input loop equation is the same as (*7.7*), so the plate nodal equation is, using $v_g = v_{s1}$,

$$+g_m v_{s1} + g_p v_p + G_L v_p + j\omega C_1 v_p + j\omega C_C(v_p - v_{s2}) = 0 \tag{7.19}$$

The nodal equation for the non-reference output node is

$$j\omega C_C(v_{s2} - v_p) + G_G v_{s2} + j\omega C_2 v_{s2} = 0 \tag{7.20}$$

which when solved with (*7.19*) gives the voltage gain

$$A_v(f) = v_{s2}/v_{s1} = \frac{-j\omega C_C g_m}{[G_G + j\omega(C_2 + C_C)][g_p + G_L + j\omega(C_1 + C_C)] - (j\omega C_C)^2} \tag{7.21}$$

where $A_v(f)$ denotes the voltage gain at any frequency. It might appear that since the coupling capacitor C_C is always large compared to the shunt capacitances C_1 and C_2, equation (*7.21*) should be simplified on this basis. If this is done, the equation will be reduced to the low frequency case. Instead, the real and the imaginary parts of the denominator are written separately and both the numerator and denominator are divided by $j\omega C_C$, giving

$$A_v(f) = \frac{-g_m}{\left[G_G \dfrac{(C_1 + C_C)}{C_C} + (g_p + G_L)\dfrac{(C_2 + C_C)}{C_C}\right] + j\left\{\omega\left[\dfrac{C_C(C_1 + C_2) + C_1 C_2}{C_C}\right] - \left[\dfrac{G_G(g_p + G_L)}{\omega C_C}\right]\right\}} \tag{7.22}$$

which may be simplified. In the real part of the denominator, both C_1 and C_2 are very small compared to C_C and hence $[(C_1 + C_C)/C_C] \cong 1$ and $[(C_2 + C_C)/C_C] \cong 1$. Unless the bandwidth is very small, these approximations cause no error when slide rule accuracy is involved. Considering the first term of the imaginary part of the denominator, it is seen that $C_C(C_1 + C_2)$ will be much larger than $C_1 C_2$ when C_C is large compared to C_1 and C_2. Another way of looking at this is to divide by C_C so that $C_1 + C_2$ is compared to $C_1 C_2/C_C$. The expression within the brackets reduces to $C_1 + C_2$, which is the total shunt capacitance and was denoted C_S in Fig. 7-7. Thus when C_C is large compared to the shunt capacitances, the entire shunt capacitance C_S in the plate circuit may be combined as was done in the high frequency circuit.

Making the simplifications and dividing numerator and denominator by the real part of the denominator, equation (*7.22*) is

$$A_v(f) \;=\; \frac{\dfrac{-g_m}{g_p + G_L + G_G}}{1 + j\left[\dfrac{2\pi f C_S}{g_p + G_L + G_G} - \dfrac{G_G(g_p + G_L)}{2\pi f C_C(g_p + G_L + G_G)}\right]} \tag{7.23}$$

The numerator of (*7.23*) is the mid-band voltage gain. The imaginary part of the denominator includes the equations for f_2 and f_1. Substituting (*7.11*) and (*7.17*) into (*7.23*),

$$\frac{A_v(f)}{A_{vo}} = \frac{1}{1 + j(f/f_2 - f_1/f)} \tag{7.24}$$

Equation (*7.24*) is the general expression of the voltage gain ratio for the vacuum tube amplifier. In most resistance-capacitance coupled amplifiers of the type being considered, f_2 will be large compared to f_1. Then when f is small, the imaginary part of the denominator is dominated by the term f_1/f, giving the low frequency case. There will be a range of frequencies where $f \ll f_2$ and $f \gg f_1$, so that the imaginary part of the denominator is small compared to 1; this is the mid-band case. When the frequency becomes large compared to f_2 so that f/f_2 is not small compared to 1, the high frequency case results because f_1/f is then insignificant.

The gain ratio is often written in the form

$$\frac{A_v(f)}{A_{vo}} = \frac{1}{(1 + jf/f_2)(1 - jf_1/f)} \tag{7.25}$$

or

$$\frac{A_v(f)}{A_{vo}} = \frac{1}{1 + j(f/f_2 - f_1/f) + f_1/f_2} \tag{7.26}$$

Then when $f_1 \ll f_2$, equation (*7.26*) is the same as (*7.24*). Hence (*7.25*) is equivalent to (*7.24*). In the derivation of (*7.24*) it was assumed the coupling capacitor C_C was very large compared to the shunt capacitances C_1 and C_2. For all practical purposes this is the same as assuming $f_2 \gg f_1$.

The procedures for finding the half-power frequencies for other circuit configurations are the same as above whenever there is only one capacitor involved in the equivalent circuit. If more than one capacitor is included in either the high or low frequency equivalent circuit, the equations are not simple.

Consider the low frequency equivalent circuit for a vacuum tube amplifier which includes the cathode capacitor as shown in Fig. 7-10. Since the input current is zero, the generated voltage is the same as the terminal voltage v_{s1}. Denoting the ground node by O, the input loop, plate nodal, cathode nodal and output nodal equations are respectively

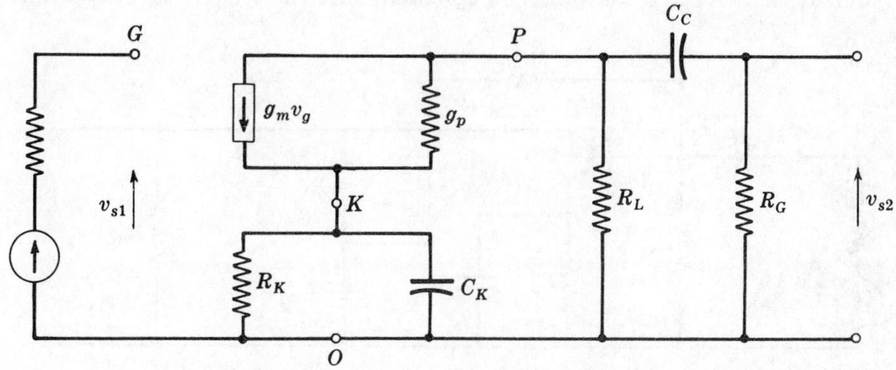

Fig. 7-10. Low Frequency Equivalent Circuit

$$
\begin{aligned}
-v_{s1} + v_{gk} + v_{ko} &= 0 \\
+g_m v_{gk} + g_p v_{pk} + G_L v_{po} + j\omega C_C(v_{po} - v_{s2}) &= 0 \\
-g_m v_{gk} - g_p v_{pk} + G_K v_{ko} + j\omega C_K v_{ko} &= 0 \\
+j\omega C_C(v_{s2} - v_{po}) + G_G v_{s2} &= 0
\end{aligned}
$$

One unknown voltage may be eliminated by writing a loop equation. Eliminating v_{po} by substituting $v_{po} = v_{pk} + v_{ko}$, the four equations become

$$
\begin{aligned}
v_{gk} \qquad\qquad\qquad\qquad + v_{ko} \qquad\qquad &= v_{s1} \\
g_m v_{gk} + (g_p + G_L + j\omega C_C)v_{pk} + (G_L + j\omega C_C)v_{ko} \qquad - j\omega C_C v_{s2} &= 0 \\
-g_m v_{gk} \qquad - g_p v_{pk} + (G_K + j\omega C_K)v_{ko} \qquad &= 0 \\
-j\omega C_C v_{pk} \qquad - j\omega C_C v_{ko} + (G_G + j\omega C_C)v_{s2} &= 0
\end{aligned}
$$

and may be solved for v_{s2} as a function of v_{s1}, from which the equation for the low frequency voltage gain follows. This becomes a rather long and tedious problem and the final equation for the voltage gain contains so many terms that a general equation for the half-power frequency is impossible. Numerical solutions, however, can be obtained. See Problem 7.16.

The two equivalent circuits including two capacitors are examples which show the complexity of the problem and the difficulties in obtaining general solutions for half-power frequencies. Hence approximations are used unless simple circuits are involved.

This section has investigated frequency response of one stage vacuum tube amplifiers. It should be obvious that the discussion may be interpreted on a more general basis. It is equally applicable to any amplifier containing an electronic control device which may be represented by the equivalent circuit of Fig. 7-5. Hence this section may easily be generalized to include amplifiers using any of the types of electronic control devices examined in Chapter 5.

7.4 FREQUENCY RESPONSE OF ONE STAGE TRANSISTOR AMPLIFIERS

Consider the simple transistor amplifier shown in Fig. 7-11. In the mid-band range of frequencies, the capacitors C_1, C_2 and C_E are assumed large, so that their impedances are negligibly small. The shunt, stray and transistor internal capacitances are small, so that their impedances are negligibly large in the mid-band range of frequencies. The mid-band equivalent circuit is given in Fig. 7-12, using the hybrid equivalent circuit for the transistor. It has been assumed that the resistors R_1 and R_2 are combined with the external generator, using Thévenin's theorem. The generator terminal voltage is assumed to be maintained constant by adjustment of the generated voltage. This in effect creates a zero impedance generator (then R_1 and R_2 are completely neglected) but the generator resistance is shown

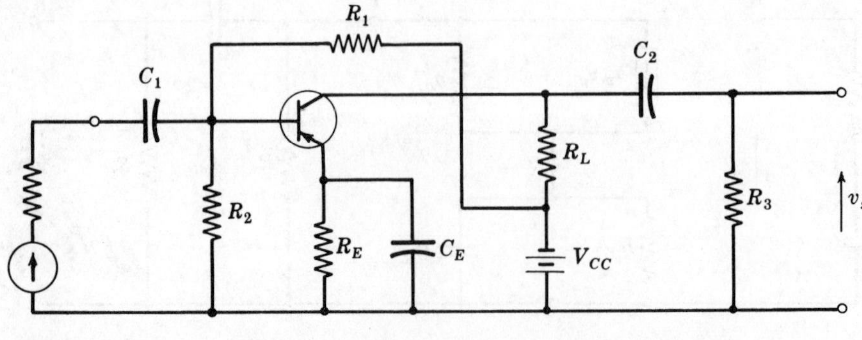

Fig. 7-11. Simple Transistor Amplifier

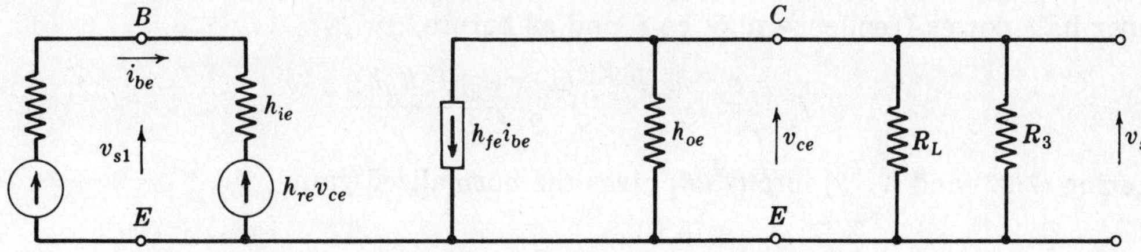

Fig. 7-12. Mid-band Equivalent Circuit

in order to emphasize its existence. The base loop and collector nodal equations are, using $v_{s2} = v_{ce}$,

$$-v_{s1} + h_{ie}i_{be} + h_{re}v_{ce} = 0$$

$$+h_{fe}i_{be} + h_{oe}v_{s2} + G_Lv_{s2} + G_3v_{s2} = 0$$

from which
$$A_{vo} = \frac{v_{s2}}{v_{s1}} = \frac{-h_{fe}}{h_{ie}(h_{oe} + G_L + G_3) - h_{fe}h_{re}} \tag{7.27}$$

This is the mid-band voltage gain which involves a 180° phase shift. In some cases the denominator of (7.27) is simplified if the term $h_{fe}h_{re}$ is small and negligible. This is usually done only in numerical problems where magnitudes can be compared.

The high frequency equivalent circuit may be drawn as it was for the vacuum tube circuit. It will be assumed that the collector capacitance is appreciable and that all the shunt capacitance in the collector circuit may be combined into one effective capacitor C_S as shown in Fig. 7-13. The base loop and collector nodal equations are

$$-v_{s1} + h_{ie}i_{be} + h_{re}v_{ce} = 0$$

$$+h_{fe}i_{be} + h_{oe}v_{s2} + G_Lv_{s2} + G_3v_{s2} + j\omega C_Sv_{s2} = 0$$

from which, using $v_{ce} = v_{s2}$,

$$A_v(\text{h.f.}) = \frac{v_{s2}}{v_{s1}} = \frac{-h_{fe}}{h_{ie}(h_{oe} + G_L + G_3) - h_{fe}h_{re} + j\omega C_Sh_{ie}}$$

Dividing numerator and denominator by the real part of the denominator,

$$A_v(\text{h.f.}) = \frac{\dfrac{-h_{fe}}{h_{ie}(h_{oe} + G_L + G_3) - h_{fe}h_{re}}}{1 + j\left[\dfrac{2\pi C_Sh_{ie}f}{h_{ie}(h_{oe} + G_L + G_3) - h_{fe}h_{re}}\right]} \tag{7.28}$$

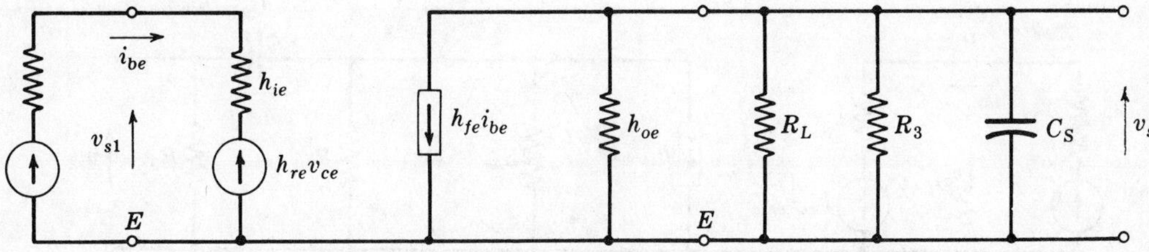

Fig. 7-13. High Frequency Equivalent Circuit

The upper half-power frequency may be found as before,

$$f_2 = \frac{h_{ie}(h_{oe}+G_L+G_3) - h_{fe}h_{re}}{2\pi h_{ie}C_S} \qquad (7.29)$$

Substituting (*7.27*) and (*7.29*) into (*7.28*) gives the normalized gain,

$$\frac{A_v(\text{h.f.})}{A_{vo}} = \frac{1}{1+jf/f_2}$$

which is represented by the universal frequency response curves. In the case of transistor amplifiers, the expression for the upper half-power frequency is different from that for vacuum tube amplifiers.

The low frequency equivalent circuit is displayed in Fig. 7-14. It includes both coupling capacitors C_1 and C_2, but C_E is assumed to be so large that the low frequency response is determined by C_1 and C_2. If C_1 is not so large that its impedance is negligibly small, R_1 and R_2 may need to be included as shown. In some cases R_1 and R_2 are large enough so that they may be neglected. In practice there is little reason to assume that C_1 can be neglected, unless C_2 can also be neglected. However, to simplify the first example, assume C_1 is so large that its impedance is negligible and the low frequency response is determined by C_2 only; then apply Thévenin's theorem to the part of the circuit of Fig. 7-14 to the left of the base terminal. This results in the circuit shown in Fig. 7-15. The signal voltage v_{s1} is the terminal voltage of the input generator. The base loop, collector nodal and non-reference output nodal equations are respectively

$$-v_{s1} + h_{ie}i_{be} + h_{re}v_{ce} = 0$$

$$+h_{fe}i_{be} + h_{oe}v_{ce} + G_Lv_{ce} + j\omega C_2(v_{ce} - v_{s2}) = 0$$

$$+j\omega C_2(v_{s2} - v_{ce}) + G_3v_{s2} = 0$$

The three equations are solved for v_{s2}. Dividing the numerator and the denominator by $j\omega C_2$, the voltage gain is

$$A_v(\text{l.f.}) = \frac{v_{s2}}{v_{s1}} = \frac{-h_{fe}}{h_{ie}(G_3+h_{oe}+G_L) - h_{fe}h_{re} - jG_3[h_{ie}(h_{oe}+G_L) - h_{fe}h_{re}]/\omega C_2}$$

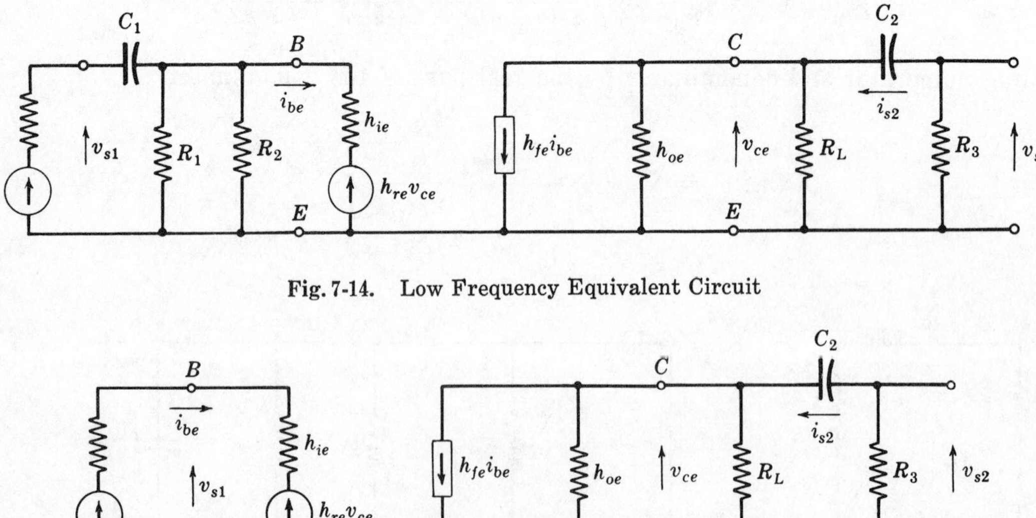

Fig. 7-14. Low Frequency Equivalent Circuit

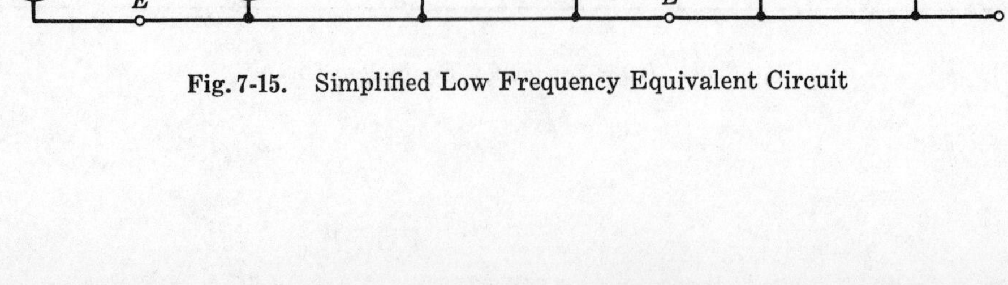

Fig. 7-15. Simplified Low Frequency Equivalent Circuit

Dividing the numerator and denominator by the real part of the denominator,

$$A_v(\text{l.f.}) \quad = \quad \frac{A_{vo}}{1 - j\,\dfrac{G_3[h_{ie}(h_{oe}+G_L) - h_{fe}h_{re}]}{2\pi C_2 f[h_{ie}(G_3+h_{oe}+G_L) - h_{fe}h_{re}]}} \tag{7.30}$$

The lower half-power frequency is found by equating the real and imaginary parts of the denominator:

$$f_1 \quad = \quad \frac{G_3[h_{ie}(h_{oe}+G_L) - h_{fe}h_{re}]}{2\pi C_2[h_{ie}(G_3+h_{oe}+G_L) - h_{fe}h_{re}]} \tag{7.31}$$

Substituting Equation (*7.31*) into (*7.30*),

$$\frac{A_v(\text{l.f.})}{A_{vo}} \quad = \quad \frac{1}{1 - jf_1/f}$$

When the circuit of Fig. 7-15 is considered, the normalized low frequency gain equation is the same as that of the simple *RC* high-pass filter. The expression for f_1 is not as simple as that for the vacuum tube circuit. Each of the derived formulas for the upper and lower half-power frequencies may be considered to have the form $f = 1/2\pi CR_{\text{eff}}$.

Examine equation (*7.31*) for the low frequency case where, for example,

$$R_{\text{eff}} \quad = \quad \frac{h_{ie}(G_3+h_{oe}+G_L) - h_{fe}h_{re}}{G_3[h_{ie}(h_{oe}+G_L) - h_{fe}h_{re}]}$$

Assuming $h_{fe}h_{re}$ is small in both numerator and denominator, then for the transistor amplifier,

$$R_{\text{eff}} \quad \cong \quad \frac{h_{ie}(G_3+h_{oe}+G_L)}{h_{ie}G_3(h_{oe}+G_L)} \quad = \quad \frac{G_3+h_{oe}+G_L}{G_3(h_{oe}+G_L)} \tag{7.32}$$

It is seen by inspecting the circuit of Fig. 7-15 that $h_{oe}+G_L$ is the conductance to the left of the coupling capacitor (assuming h_{re} small). The conductor G_3 is to the right of the coupling capacitor. The series combination of the conductance to the left and to the right of the coupling capacitor is the reciprocal of R_{eff} in (*7.32*). Hence when h_{re} is negligibly small (then the generator $h_{fe}i_{be}$ is independent) the effective resistance R_{eff} for the lower half-power frequency f_1 is the series resistance on both sides of the coupling capacitor. This could also be shown in other ways. In practice, the approximate half-power frequencies are often calculated by using the appropriate approximate values of R_{eff}.

In (*7.29*) for the upper half-power frequency f_2, the effective resistance is

$$R_{\text{eff}} \quad = \quad \frac{h_{ie}}{h_{ie}(h_{oe}+G_L+G_3) - h_{fe}h_{re}}$$

If h_{re} is small so that $h_{fe}h_{re}$ is small, this becomes

$$R_{\text{eff}} \quad \cong \quad \frac{1}{h_{oe}+G_L+G_3}$$

Thus when h_{re} is small, the effective resistance is approximately the parallel combination of the three shunt resistances in the collector circuit of Fig. 7-13. This is the value of R_{eff} used in many practical problems.

In practice, approximations are usually used for R_{eff} because of the difficulty in making the calculations for other than simple (one pole) circuits. However, the purpose of the present discussion is to develop concepts and methods of analysis.

As an illustration, refer to the circuit of Fig. 7-14. Since C_1 is usually as appreciable as C_2, C_1 is included in the low frequency equivalent circuit shown in Fig. 7-16 below. The base loop, collector nodal and output nodal equations are

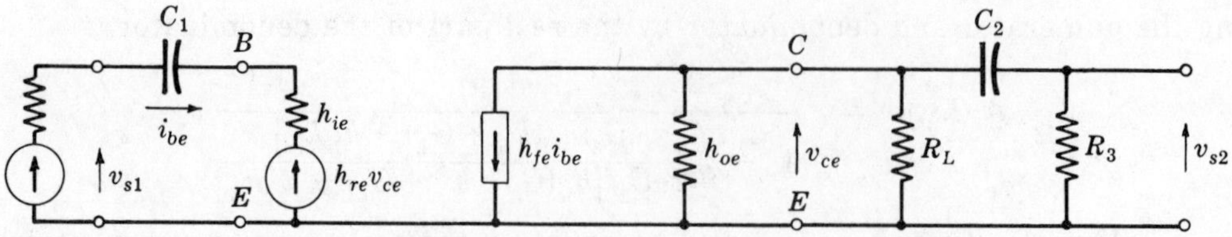

Fig. 7-16. Simplified Low Frequency Equivalent Circuit

$$-v_{s1} - ji_{be}/\omega C_1 + h_{ie}i_{be} + h_{re}v_{ce} = 0$$

$$+h_{fe}i_{be} + h_{oe}v_{ce} + G_L v_{ce} + j\omega C_2(v_{ce} - v_{s2}) = 0$$

$$j\omega C_2(v_{s2} - v_{ce}) + G_3 v_{s2} = 0$$

Solving these equations for v_{s2}/v_{s1} and dividing by $j\omega C_2$,

$$v_{s2}/v_{s1} = \frac{-h_{fe}}{h_{ie}(G_3 + G_L + h_{oe}) - h_{fe}h_{re} - \dfrac{G_3(h_{oe}+G_L)}{\omega^2 C_1 C_2} - \dfrac{j}{\omega}\left\{\dfrac{G_3 + h_{oe} + G_L}{C_1} + \dfrac{G_3[h_{ie}(h_{oe}+G_L) - h_{fe}h_{re}]}{C_2}\right\}} \tag{7.33}$$

The numerator is real and independent of frequency, but the real part of the denominator is not independent of frequency. Setting the real part of the denominator equal to the imaginary part will not give the correct equation for f_1. Instead, it is necessary to set the magnitude of the denominator equal to $\sqrt{2}$ times the magnitude of the denominator for the mid-band frequency range. Since this is the low frequency circuit, the mid-band value can be found by examining the limit as the frequency becomes large. Then

$$\frac{v_{s2}}{v_{s1}} = \frac{-h_{fe}}{h_{ie}(G_3 + G_L + h_{oe}) - h_{fe}h_{re}}$$

The lower half-power frequency is the frequency for which the magnitude of the denominator of (7.33) is $\sqrt{2}[h_{ie}(G_3 + G_L + h_{oe}) - h_{fe}h_{re}]$. The magnitude of the denominator can be found by taking the square root of the sum of the squares of the real and imaginary parts. This can be worked out when a numerical problem is involved, but a general equation for f_1 is difficult to derive unless (7.33) can be simplified. This example illustrates how complicated the equations become when there are two capacitors in the equivalent circuit.* See Problem 7.7, page 210.

In the most general cases, the half-power frequencies are found by solving the equations of the high or low frequency equivalent circuit for the high or low frequency transfer function (voltage gain in this case). The half-power frequency is that frequency at which the magnitude of the high or low frequency transfer function is $1/\sqrt{2}$ times the mid-band transfer function. Calculations are simplified if the transfer function can be written so that the numerator is independent of frequency. Then for simple circuits the real part of the denominator is also independent of frequency, so that the calculations are further simplified as previously illustrated.

The general transistor amplifier circuit including only the collector coupling capacitor C_2 and the shunt capacitances can be solved. The general equation for the normalized voltage gain $A_v(f)/A_{vo}$ can be simplified so that it is identical with (7.25). See Problem 7.20.

*Then the transfer function involves two poles. There are two energy storing elements included in the equivalent circuit. This differs from the equivalent circuit of Fig. 7-9 which essentially involves two capacitors C_C and C_S, and where the two poles are so far apart that the equation for the transfer function can be simplified.

The frequency response of amplifiers has been defined in terms of the open circuit voltage gain. In the case of vacuum tube amplifiers (the grounded grid is an exception) the signal input current is zero, so that the amplifier transfer function is either the voltage gain or the transfer admittance. The transfer admittance is used only in special cases. In transistor amplifiers the transfer function is frequently expressed as the current gain for a finite terminating admittance. If the short circuit current gain is considered as the transfer function, then the problem is usually easier to solve but there may be significant error.

In many problems in system analysis, the elements of the system are assumed completely isolated from each other. This is necessary in many cases; otherwise the equations become so complicated that they are difficult to solve (and sometimes solutions are impossible). However, in this book a special effort is made to emphasize the general cases even though in practice approximations are made on the basis of engineering judgment.

When the transfer function of a transistor amplifier is specified as the current gain, the input and output currents must be identified. Referring to the low frequency equivalent circuit of Fig. 7-14, let the output current i_{s2} be that in the resistor R_3 (directed toward the collector); this resistor is usually the load external to this amplifier stage. Let the input current i_{s1} be the same as the base current i_{be}; this assumes any bias resistors are large. In this case it is not necessary to write an input loop equation. Using $i_{be} = i_{s1}$, the equations for the collector node and non-reference output node are respectively

$$h_{fe}i_{s1} + h_{oe}v_{ce} + G_L v_{ce} + j\omega C_2(v_{ce} - v_{s2}) = 0$$

$$+j\omega C_2(v_{s2} - V_{ce}) + G_3 v_{s2} = 0$$

from which

$$v_{s2} = \frac{-j\omega C_2 h_{fe} i_{s1}}{G_3(h_{oe} + G_L) + j\omega C_2(h_{oe} + G_L + G_3)} \tag{7.34}$$

By inspection of Fig. 7-15,

$$i_{s2} = -G_3 v_{s2} \tag{7.35}$$

Substituting (*7.34*) into (*7.35*) and dividing numerator and denominator by $j\omega C_2$, the low frequency current gain is

$$A_i(\text{l.f.}) = i_{s2}/i_{s1} = \frac{G_3 h_{fe}}{(h_{oe} + G_L + G_3) - jG_3(h_{oe} + G_L)/2\pi f C_2}$$

Dividing numerator and denominator by the real part of the denominator,

$$A_i(\text{l.f.}) = \frac{\dfrac{G_3 h_{fe}}{h_{oe} + G_L + G_3}}{1 - j\left[\dfrac{G_3(h_{oe} + G_L)}{2\pi f C_2(h_{oe} + G_L + G_3)}\right]} \tag{7.36}$$

If f is so large that the imaginary part of the denominator is small compared to 1, the mid-band current gain is obtained. This is the numerator of (*7.36*). In Chapter 6 the mid-band current gain for the same circuit was derived [equation (*6.41*), page 141]; the result agrees with (*7.36*).

The lower half-power frequency f_1 is now defined as the frequency at which the current gain is $1/\sqrt{2}$ times the mid-band value. This may be found by equating the real and imaginary parts of the denominator of (*7.36*) and solving to obtain

$$f_1 = \frac{G_3(h_{oe} + G_L)}{2\pi C_2(h_{oe} + G_L + G_3)} \tag{7.37}$$

Writing (*7.36*) in terms of the gain ratio and using (*7.37*), the gain ratio becomes

$$\frac{A_i(\text{l.f.})}{A_{io}} = \frac{1}{1 - jf_1/f}$$

Since this is the same as that for the high-pass filter, the universal response curves may be used.

The lower half-power frequency f_1 as determined from the current gain and given by (*7.37*) is slightly different from the equation for f_1 found from the voltage gain and given by (*7.31*). This is to be expected. Even though the same equivalent circuit was used in writing the equations, the voltage gain and current gain involve two different problems.

It is not appropriate to consider all the possible transfer functions and to derive the corresponding equations for the half-power frequencies. The methods outlined can be used to derive equations for the high and the low frequency transfer impedances and transfer admittances. The equations for the half-power frequencies may be derived from these. It is to be expected that when the appropriate substitutions are made, the normalized transfer functions will involve equations containing $1/(1+jf/f_2)$ or $1/(1-jf_1/f)$.

7.5 GAIN-BANDWIDTH PRODUCT

One type of linear amplifier is the video amplifier. This amplifier has an upper half-power frequency f_2 which is usually of the order of megacycles per second and a lower half-power frequency f_1 which is of the order of a few cycles per second. Here $f_2 \gg f_1$ and hence the bandwidth is for all practical purposes $\Delta f = f_2$. If a simple amplifier of the type shown in Section 7.3 is used as an example (this is an uncompensated video amplifier),

$$\Delta f = f_2 = \frac{g_p + G_L + G_G}{2\pi C_S}$$

The product of the magnitude of the mid-band voltage gain and the bandwidth is the *gain-bandwidth product,*

$$GBW = |A_{vo}|\,\Delta f = \left(\frac{+g_m}{g_p + G_L + G_G}\right)\left(\frac{g_p + G_L + G_G}{2\pi C_S}\right) = \frac{+g_m}{2\pi C_S} \qquad (7.38)$$

which indicates that the bandwidth can be increased only by sacrificing the mid-band voltage gain.

The gain-bandwidth product for the simple transistor amplifier may be found by using the same procedure as for the vacuum tube amplifier. It is again assumed that $f_1 \ll f_2$ and $\Delta f = f_2$. From equations (*7.27*) and (*7.29*) we obtain

$$GBW = |A_{vo}|\,\Delta f = \left[\frac{h_{fe}}{h_{ie}(h_{oe} + G_L + G_3) - h_{fe}h_{re}}\right]\left[\frac{h_{ie}(h_{oe} + G_L + G_3) - h_{fe}h_{re}}{2\pi h_{ie} C_S}\right]$$

$$= \frac{h_{fe}}{2\pi h_{ie} C_S} \qquad (7.39)$$

which is similar to (*7.38*) for the vacuum tube circuit, except for the parameter h_{ie} in the denominator. In the transistor amplifier, the gain-bandwidth product is increased only by increasing h_{fe} or by decreasing either h_{ie} or C_S.

A discussion of video amplifier design problems is not appropriate at this time. It seems advisable, however, to point out the simple relationships given by (*7.38*) and (*7.39*) because of their importance.

7.6 DECIBEL

The frequency response and gain of amplifiers are often specified in relative terms using logarithmic units. One advantage of a logarithmic unit is that the calculation of the overall gain of cascaded amplifier stages involves addition rather than multiplication.

Historically, a logarithmic unit of relative power called the *bel* was defined by

$$\text{b} = \log_{10} P_2/P_1$$

which gives in bels the power level of P_2 relative to the power level P_1.

The bel is a rather large unit and seldom used. A more convenient unit of relative power, the *decibel,* is defined by

$$\text{db} = 10 \log_{10} P_2/P_1 \tag{7.40}$$

Consider a complete amplifier as shown in Fig. 7-17 where P_2 is the output power (delivered by the amplifier into an external load) and P_1 is the power input to the amplifier. The power gain of the amplifier in db is given by equation (*7.40*). The power gain in db is positive, negative or zero according as the power ratio P_2/P_1 is respectively $>1, <1, = 1$.

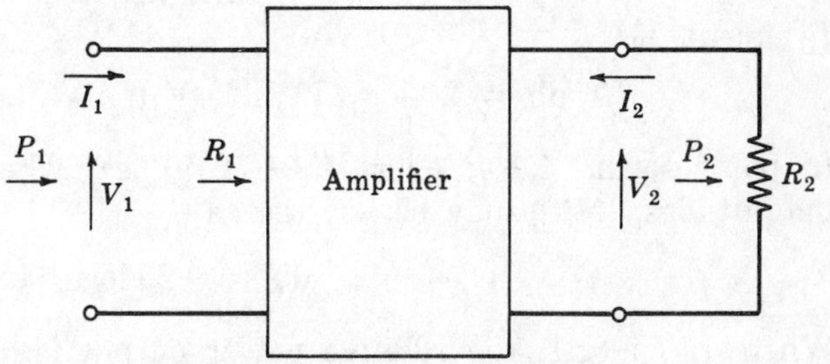

Fig. 7-17. Amplifier Block Diagram

In some problems such as those involving passive filters where the output power P_2 is never greater than the input power P_1, the attenuation (or power loss) is defined by

$$\text{db attenuation} = 10 \log_{10} P_1/P_2$$

By including this definition, the power ratio can be written so that it is never less than 1. It follows that the attenuation in db is the negative of the power gain in db.

In most practical situations the power is not measured but rather the rms value of the voltage is measured. In the mid-band range where the parameters are all resistances, the output power P_2 and input power P_1 are $P_2 = V_2^2/R_2$ and $P_1 = V_1^2/R_1$ where V_2 and V_1 are rms values. Substituting into (*7.40*),

$$\text{db} = 10 \log_{10} (V_2^2/R_2)/(V_1^2/R_1)$$

When $R_2 = R_1$,
$$\text{db} = 10 \log_{10} V_2^2/V_1^2 = 20 \log_{10} V_2/V_1 \tag{7.41}$$

Note that (*7.41*) gives the relative power in db only when $R_1 = R_2$. Because voltages are usually more easily measured, it is common practice to use (*7.41*) even though $R_1 \neq R_2$. In such case (*7.41*) does not give the power gain in db, so this is often called "db voltage gain" or by some other name to distinguish it from the true relative power in db.

When considering amplifier frequency response problems, the output power at any frequency not in the mid-band frequency range may be compared to the output power in the mid-band frequency range. Usually a reference frequency within the mid-band range is specified. The relative output power is then in db if the amplifier output is terminated in a pure resistance of constant magnitude. When frequency response measurements are made, the magnitude of the input voltage V_1 is kept constant as the frequency of the input signal is varied. At any frequency, the power output in db relative to that at the reference frequency f_o is given by

$$\text{db} = 10 \log_{10} P_2(f)/P_2(f_o) = 20 \log_{10} V_2(f)/V_2(f_o)$$

The frequency response curve for an amplifier is often plotted on semilog graph paper, taking the relative power output at the reference frequency as zero db.

Consider the normalized high frequency voltage gain vs. frequency ratio for a simple amplifier as given by

$$A_v(\text{h.f.})/A_{vo} = 1/(1+jf/f_2) \qquad (7.13)$$

Since f_o is in the mid-band range of frequencies, $V_2(f_o) = A_{vo}V_1$. At high frequencies,

$$V_2(\text{h.f.}) = A_v(\text{h.f.})\,V_1 = \left(\frac{1}{1+jf/f_2}\right)A_{vo}V_1$$

The ratio of the output voltage at any high frequency to the reference output voltage for constant input V_1 is

$$V_2(\text{h.f.})/V_2(f_o) = 1/(1+jf/f_2)$$

In db this is

$$\text{db} = 20\log_{10} V_2(\text{h.f.})/V_2(f_o) = 20[\log_{10} 1 - \log_{10}(1+jf/f_2)] \qquad (7.42)$$

In polar form, $1+jf/f_2 = [1+(f/f_2)^2]^{1/2}\,e^{j\theta}$ where $\theta = \tan^{-1} f/f_2$. Considering only the magnitudes, (7.42) may be written as

$$\text{db} = -20\log_{10}[1+(f/f_2)^2]^{1/2} \qquad (7.43)$$

When $(f/f_2)^2 \ll 1$, the relative power output is db $\cong -20\log_{10} 1 = 0$. If $(f/f_2)^2 \ll 1$, the frequency f is either in the mid-band range or below the mid-band range (in which case the equations do not apply). When $(f/f_2)^2 \gg 1$, the relative power output is

$$\text{db} \cong -20\log_{10} f/f_2 = -20(\log_{10} f - \log_{10} f_2) \qquad (7.44)$$

Since f_2 is a constant for a given circuit, the term $\log_{10} f$ determines the frequency response. If f is doubled, the output decreases by 6 db. This may be seen by assuming a particular frequency f_a (f_a must be so large that $(f_a/f_2)^2 \gg 1$). Doubling the frequency f_a means that $f = 2f_a$; substituting into (7.44),

$$\text{db} = -20(\log_{10} 2f_a - \log_{10} f_2) = -20\log_{10} 2 - 20(\log_{10} f_a - \log_{10} f_2) \qquad (7.45)$$

The last term of (7.45) is the power output in db at the frequency f_a (relative to the mid-band reference frequency f_o). The other term gives the output at the frequency $2f_a$ relative to f_a. Thus when the frequency is doubled [for $(f/f_2)^2 \gg 1$] the output decreases by approximately 6 db, since

$$-20\log_{10} 2 = -20(0.301) = -6.02 \text{ db}$$

In the frequency range where $(f/f_2)^2 \gg 1$, the attenuation is said to be 6 db per octave (frequency doubled). It is easily shown that in this frequency range, if the frequency is increased by a factor of ten, the gain is decreased by 20 db. Thus the attenuation is 6 db per octave or 20 db per decade.

A high-pass filter may be treated in the same way. Consider the normalized low frequency voltage gain vs. frequency ratio for a simple amplifier as given by

$$A_v(\text{l.f.})/A_{vo} = 1/(1-jf_1/f) \qquad (7.18)$$

In this case the magnitude of the denominator is $[1+(f_1/f)^2]^{1/2}$. The relative power output at any frequency below the mid-band range may be found in the same way as for the low-pass filter. The equation equivalent to (7.43) is

$$\text{db} = -20\log_{10}[1+(f_1/f)^2]^{1/2}$$

When the frequency is high so that $(f_1/f)^2 \ll 1$, the relative power in db = 0. When the frequency is low so that $(f_1/f)^2 \gg 1$,

$$\text{db} \;=\; 20 \log_{10} f_1/f \;=\; -20 \log_{10} f_1 + 20 \log_{10} f$$

Doubling the frequency in this equation causes the gain to increase by 6 db. Thus the attenuation is −6 db per octave (or −20 db per decade). However, with respect to mid-band frequencies the attenuation is 6 db per octave as the frequency decreases in the range where $(f_1/f)^2 \gg 1$.

An amplifier transfer function may be given as the current gain A_i. The input power can then be written as $P_1 = I_1^2 R_1$ and the output power as $P_2 = I_2^2 R_2$ (I_1 and I_2 must be rms values). Substituting in (*7.40*),

$$\text{db} \;=\; 10 \log_{10} I_2^2 R_2 / I_1^2 R_1$$

When $R_1 = R_2$, this becomes

$$\text{db} \;=\; 20 \log_{10} I_2/I_1 \tag{7.46}$$

which has the same form as (*7.41*). If the magnitude of the input current I_1 is kept constant, then the output current I_2 as a function of frequency is a measure of the amplifier frequency response. The output at any frequency may be specified in db relative to the mid-band value in the same way as was done considering the voltage gain as the transfer function. Since (*7.46*) and (*7.41*) have the same form, it is only necessary to substitute the corresponding I for V. As was previously shown, the equations for the half-power frequencies when based on the voltage gain are usually not the same as when based on the current gain.

The decibel is a measure of relative power. There are several reference power levels used. The one most frequently employed by engineers is called the dbm and is one milliwatt, and is particularly important because it is the basis for calibration of the db scales on many laboratory voltmeters. Since the voltmeter measures the rms value of the varying voltage (or usually the true rms value only if a sine wave is applied), the voltage is an indication of the power if the resistance is specified. Most meters are calibrated in terms of a 1 milliwatt power level in a 600 ohm resistance (600 Ω is a standard line impedance). The power level in dbm is

$$\text{dbm} \;=\; 10 \log_{10} P/0.001$$

where P is in watts. However, since voltages are usually measured, it is more convenient to specify the dbm in terms of the voltage ratio. The rms value of the voltage across a 600 ohm resistor when 1 milliwatt is dissipated is $V = \sqrt{0.6} = 0.775$ volts. Hence the 0 dbm mark is at 0.775 volts on the voltmeter scale. The power level in dbm when 600 ohms resistance is used is

$$\text{dbm} \;=\; 20 \log_{10} V/0.775$$

where V is the rms value. A voltmeter with a dbm scale reads the absolute power in dbm only when the voltmeter is connected across 600 ohms and when the scale reading is the rms value of the voltage. When a constant resistance other than 600 ohms is used, the *relative power* in db can be read directly on the meter (assuming the meter reads the rms value). This scale is quite convenient in making frequency response measurements.

7.7 TRANSISTOR HIGH FREQUENCY RESPONSE

In the previous sections the high frequency response of amplifiers was assumed to be a function of the shunt capacitance only. The tube and transistor parameters were assumed constant and independent of frequency unless the signal frequency becomes very high. In the case of transistors, the parameters may change at frequencies low enough to affect the frequency response of an untuned amplifier. This was especially true for the earlier types of transistors. Information on frequency dependence is not included in the static transistor characteristic curves.

It is possible to develop physical models which account for frequency effects, but these become rather complicated and are beyond the scope of this book. In developing simple circuit models which account for frequency effects, it is usually assumed that only the forward transfer parameter is affected; hence the magnitude of this parameter and phase relationships (with respect to low frequencies) are of interest.

Consider a transistor with the base common. Assume that the forward transfer parameter h_{fb} is frequency dependent and may be written as

$$h_{fb}(f) = |h_{fb}(f)|\, e^{j\theta(f)}$$

The cutoff frequency $f_{\alpha b}$ (also denoted by f_{co}) is defined as the frequency at which the magnitude of h_{fb} is $1/\sqrt{2}$ times the low frequency value. In the low frequency range, h_{fb} is constant and independent of frequency. An expression for $h_{fb}(f)$ which is a good approximation in the frequency range below $f_{\alpha b}$ is

$$h_{fb}(f) = h_{fb}/(1 + jf/f_{\alpha b}) \qquad (7.47)$$

This approximation becomes poorer as the frequency increases above $f_{\alpha b}$.

The value of the cutoff frequency is usually given on data sheets furnished by the manufacturer of the transistor. Table 6.1, page 131, gives the value of $f_{\alpha b}$ as 4 mc for the 2N117 transistor for the conditions specified. The notation for the cutoff frequency was chosen to agree with that of Table 6.1. The first subscript α is probably used because the cutoff frequency was first defined for the parameter α of the "tee" circuit of Fig. 6-36, page 145. The second subscript b indicates that the base is common.

The cutoff frequency for the common emitter forward transfer parameter h_{fe} is not the same as that for the common base forward transfer parameter. This may be easily shown by considering the approximate conversion equation (6.91), page 150, for the parameters,

$$h_{fe} \cong \frac{-h_{fb}}{1 + h_{fb}} \qquad (6.91)$$

Substituting (7.47) into (6.91),

$$h_{fe}(f) \cong \frac{-h_{fb}(f)}{1 + h_{fb}(f)} = \frac{-h_{fb}}{(1 + h_{fb}) + jf/f_{\alpha b}}$$

Since the numerator is real and independent of frequency and the real part of the denominator is independent of frequency, the real part of the denominator may be set equal to the imaginary part to find the half-power frequency denoted as $f_{\beta e}$:

$$1 + h_{fb} = f_{\beta e}/f_{\alpha b} \qquad \text{or} \qquad f_{\beta e} = f_{\alpha b}\,(1 + h_{fb})$$

Since h_{fb} is negative and slightly less than 1, $f_{\beta e} < f_{\alpha b}$.

There is ambiguity in the notation, because as originally defined α and β refer to the equivalent "tee" circuits, but the notation frequently used involves $f_{\alpha b}$ and $f_{\beta e}$ for the half-power frequencies considering h_{fb} and h_{fe}, respectively. The differences in magnitude between h_{fb} and α and between β and h_{fe} are small, so the approximations are usually good. Sometimes α and β are defined so that they are the same as h_{fb} and h_{fe}, but then there is ambiguity with the equivalent "tee" parameters.

Another problem involved in the high frequency hybrid equivalent circuit is the internal transistor capacitance. A *P-N* junction involves a capacitance because of the charge distributed in the regions on each side of the junction. In addition there is stray capacitance, including the leads and the capacitance of the transistor. Frequently the effective capacitance across the collector to base junction is specified for given operating conditions. This is denoted as C_{ob} where o indicates output capacitance and b indicates common base. The internal transistor capacitance for the common emitter (between the collector and emitter) is C_{oe} and may not be specified. An equation relating C_{oe} and C_{ob} is desired. See Problem 7.3, page 205.

The approximate equation for C_{oe} is $C_{oe} \cong C_{ob}/(1+h_{fb})$. Since h_{fb} is negative and slightly less than 1, C_{oe} is much larger than C_{ob}. For example, consider the data in Table 6.1, page 131, where $C_{ob} = 7\ \mu\mu\text{f}$ and $h_{fb} = -0.925$ (design center value). Then $C_{oe} = 7/(1-0.925)\ \mu\mu\text{f} = 93.3\ \mu\mu\text{f}$.

7.8 AMPLIFIER INPUT ADMITTANCE

Another consideration at high frequencies is the input admittance of an amplifier stage, since it terminates any previous amplifier stage. Historically, the input admittance of vacuum tube triode amplifiers was of special interest because of problems in tuned amplifiers. The vacuum tube triode amplifier is the simplest example which illustrates how the input admittance to an amplifier may change at high frequencies depending on the plate load admittance. For this reason it will be studied first.

Consider the high frequency equivalent circuit for a triode amplifier which includes the interelectrode capacitances between the terminals of the vacuum tube, as shown in Fig. 7-18. Stray wiring capacitances are included. The external stray capacitances may be neglected if they are small compared to the internal interelectrode capacitances. The input admittance to the vacuum tube triode is defined as $y_i = i_g/v_g$.

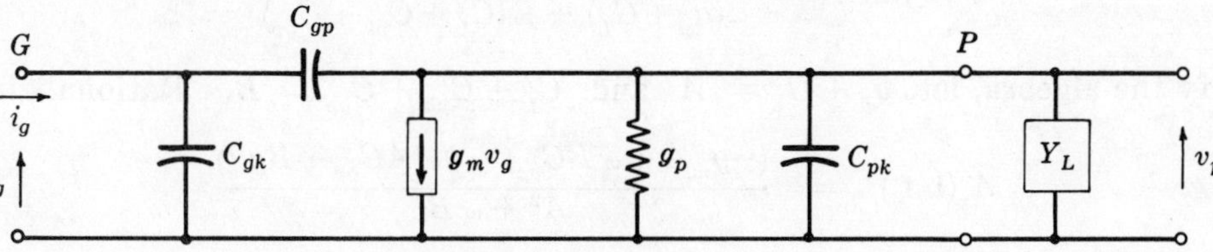

Fig. 7-18. High Frequency Triode Equivalent Circuit

The plate nodal equation is

$$+j\omega C_{gp}(v_p - v_g) + g_m v_g + g_p v_p + j\omega C_{pk} v_p + Y_L v_p = 0 \tag{7.48}$$

from which the high frequency voltage gain is

$$A_v(\text{h.f.}) = v_p/v_g = (-g_m + j\omega C_{gp})/[g_p + Y_L + j\omega(C_{pk} + C_{gp})] \tag{7.49}$$

The grid nodal equation is

$$-i_g + j\omega C_{gk} v_g + j\omega C_{gp}(v_g - v_p) = 0$$

and this is solved with (7.48) to obtain

$$v_g = \frac{[g_p + Y_L + j\omega(C_{gp} + C_{pk})]i_g}{j\omega(C_{gk} + C_{gp})[g_p + Y_L + j\omega(C_{gp} + C_{pk})] + j\omega C_{gp}(g_m - j\omega C_{gp})}$$

from which the input admittance is

$$y_i = i_g/v_g = j\omega C_{gk} + j\omega C_{gp} + \frac{j\omega C_{gp}(g_m - j\omega C_{gp})}{g_p + Y_L + j\omega(C_{gp} + C_{pk})}$$

and, using (7.49), is simplified to

$$y_i = j\omega\{C_{gk} + C_{gp}[1 - A_v(\text{h.f.})]\} \tag{7.50}$$

which indicates that the input admittance involves more than just the two capacitors connected to the grid terminal.

To further investigate the input admittance, consider three separate cases.

Case I. Let $Y_L = G_L$, a pure conductance. Substituting into Equation (*7.49*) gives the expression for the voltage gain,

$$A_v(\text{h.f.}) = \frac{-g_m + j\omega C_{gp}}{g_p + G_L + j\omega(C_{pk} + C_{gp})} \tag{7.51}$$

Assuming $g_m \gg C_{gp}$, then $-g_m + j\omega C_{gp} \cong -g_m$. If $(g_p + G_L) \gg \omega(C_{pk} + C_{gp})$, equation (*7.51*) reduces to $A_v = -g_m/(g_p + G_L)$. For this case,

$$y_i = j\omega\left[C_{gk} + C_{gp}\left(1 + \frac{g_m}{g_p + G_L}\right)\right] \tag{7.52}$$

Hence the input admittance is a pure capacitance but the magnitude of the capacitance is the grid to cathode capacitance, added to the capacitance from grid to plate times (1 plus the magnitude of the mid-band voltage gain).

Equation (*7.52*) indicates that at high frequencies the previous amplifier stage is terminated by a capacitance which is much larger than just the stray and interelectrode capacitances. Hence the effective capacitance must be considered in any frequency response problem. For this case the effective capacitance from grid to cathode is

$$C_{\text{eff}} = C_{gk} + C_{gp}[1 + |A_v(\text{h.f.})|]$$

Case II. Let $Y_L = G_L + j\omega C_L$, a combination of a conductor and a capacitor in parallel. Now

$$A_v(\text{h.f.}) = \frac{-g_m + j\omega C_{gp}}{(g_p + G_L) + j\omega(C_L + C_{pk} + C_{gp})} \tag{7.53}$$

To simplify the algebra, let $g_p + G_L = A$ and $C_L + C_{pk} + C_{gp} = B$. Rationalizing (*7.53*),

$$A_v(\text{h.f.}) = \frac{(-g_m A + \omega^2 BC_{gp}) + j\omega(AC_{gp} + Bg_m)}{A^2 + \omega^2 B^2} \tag{7.54}$$

Since $g_m A \gg \omega^2 BC_{gp}$, equation (*7.54*) is of the form

$$A_v(\text{h.f.}) = -k_1 + jk_2$$

and when substituted into (*7.50*) gives

$$y_i = \omega k_2 C_{gp} + j\omega[C_{gk} + C_{gp}(1 + k_1)]$$

which shows that when the plate load is a conductor (resistor) and a capacitor in parallel, the input admittance involves a conductance and capacitive susceptance. The magnitudes can easily be worked out for numerical problems.

Case III. Let $Y_L = G_L - j/\omega L_L$, a combination of a conductor and an inductor in parallel. The equation for the gain is now

$$A_v(\text{h.f.}) = \frac{-g_m + j\omega C_{gp}}{(g_p + G_L) - j[1/\omega L_L - \omega(C_{pk} + C_{gp})]} \tag{7.55}$$

Assume $1/\omega L_L \gg \omega(C_{pk} + C_{gp})$ and let $A = g_p + G_L$ and $1/\omega L_L - \omega(C_{pk} + C_{gp}) = 1/\omega D$; then after rationalizing, (*7.55*) is

$$A_v(\text{h.f.}) = \frac{-g_m A - C_{gp}/D - j(g_m/\omega D - \omega AC_{gp})}{A^2 + 1/\omega^2 D^2}$$

When $g_m/\omega D \gg \omega AC_{gp}$, the equation for the high frequency voltage gain has the form

$$A_v(\text{h.f.}) = -k_3 - jk_4$$

which when substituted into (*7.50*) yields

$$y_i = -\omega k_4 C_{gp} + j\omega[C_{gk} + C_{gp}(1 + k_3)]$$

Here the input admittance involves a negative conductance in shunt with a capacitive susceptance.

Case I is important when high frequency response is being considered. This example shows that in multistage amplifier circuits it is necessary to consider the total effective shunt capacitance from plate to ground. The input of the vacuum tube triode is not an open circuit at high frequencies when the grid to plate capacitance is not negligible, but involves a capacitance which is a function of the voltage gain.

Case II is important because there is always some capacitance from plate to ground. Hence the input admittance always involves a positive conductance in an ordinary resistance-capacitance coupled amplifier. This conductance will usually be quite low.

Case III is most likely to occur in tuned amplifiers. In tuned amplifiers at high frequencies a negative conductance can cause oscillation. If a resonant circuit in the plate circuit is tuned so that it is inductive, then the input admittance has a negative real part and hence a negative conductance exists in the input circuit.

Historically, the effects of the plate impedance and voltage gain on the input admittance of a vacuum tube triode amplifier were studied by John M. Miller, hence the name Miller effect is often used.

The Miller effect also exists in a transistor amplifier. It is, however, less pronounced because of the lower impedances of the input circuit. Consider the high frequency transistor amplifier circuit of Fig. 7-19. This circuit is obtained by adding the capacitances between the transistor terminals to the mid-band equivalent circuit. The current i_{be} is that in the branch containing h_{ie}, hence $i_{be} \neq i_{s1}$.

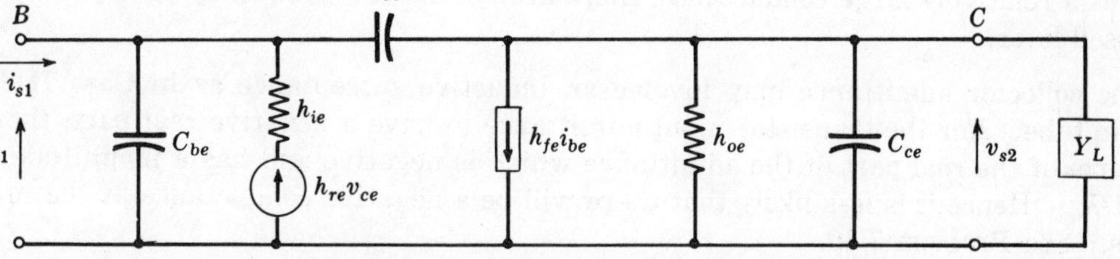

Fig. 7-19. High Frequency Transistor Equivalent Circuit

The input admittance is defined as $y_i = i_{s1}/v_{s1}$ and is found by solving the circuit equations. The base nodal, base loop and collector nodal equations are (using $v_{ce} = v_{s2}$)

$$-i_{s1} + j\omega C_{be} v_{s1} + i_{be} + j\omega C_{bc}(v_{s1} - v_{s2}) = 0 \tag{7.56}$$

$$-v_{s1} + h_{ie} i_{be} + h_{re} v_{s2} = 0 \tag{7.57}$$

$$j\omega C_{bc}(v_{s2} - v_{s1}) + h_{fe} i_{be} + h_{oe} v_{s2} + j\omega C_{ce} v_{s2} + Y_L v_{s2} = 0 \tag{7.58}$$

which yield

$$y_i = i_{s1}/v_{s1} = j\omega(C_{be} + C_{bc}) + \frac{\omega^2 h_{ie} C_{bc}^2 + h_{oe} + Y_L + j\omega[C_{ce} + C_{bc}(h_{fe} + 1 - h_{re})]}{h_{ie}[h_{oe} + Y_L + j\omega(C_{ce} + C_{bc})] - h_{fe} h_{re}} \tag{7.59}$$

The high frequency voltage gain is found by solving (*7.57*) and (*7.58*) for v_{s2} as a function of v_{s1}. Then

$$A_v(\text{h.f.}) = \frac{j\omega C_{bc} h_{ie} - h_{fe}}{h_{ie}[h_{oe} + Y_L + j\omega(C_{ce} + C_{bc})] - h_{fe} h_{re}} \tag{7.60}$$

Now rewrite (*7.59*) as

$$y_i = j\omega(C_{be} + C_{bc}) + \frac{j\omega C_{bc}[-(-h_{fe} + j\omega h_{ie} C_{bc})]}{h_{ie}[h_{oe} + Y_L + j\omega(C_{bc} + C_{ce})] - h_{fe}h_{re}}$$

$$+ \frac{h_{oe} + Y_L + j\omega[C_{ce} + C_{bc}(1 - h_{re})]}{h_{ie}[h_{oe} + Y_L + j\omega(C_{bc} + C_{ce})] - h_{fe}h_{re}}$$

and into this equation put (*7.60*) and obtain

$$y_i = j\omega\{C_{be} + C_{bc}[1 - A_v(\text{h.f.})]\} + \frac{h_{oe} + Y_L + j\omega[C_{ce} + C_{bc}(1 - h_{re})]}{h_{ie}[h_{oe} + Y_L + j\omega(C_{bc} + C_{ce})] - h_{fe}h_{re}} \qquad (7.61)$$

which indicates that the equation for the transistor amplifier input admittance is not as simple as it was for the vacuum tube triode amplifier. Equation (*7.61*) may be simplified. The parameter h_{re} is usually very small compared to 1, so that $1 - h_{re} \cong 1$ in the numerator. If h_{re} is small so that the term $h_{fe}h_{re}$ in the denominator is small, then (*7.61*) becomes

$$y_i = j\omega\{C_{be} + C_{bc}[1 - A_v(\text{h.f.})]\} + (1/h_{ie})$$

which has the same form as (*7.50*) but for the term $1/h_{ie}$ which is a conductance. When h_{re} is small as assumed above, then h_{ie} is the mid-band input impedance of the transistor amplifier [see equation (*7.80*)].

Regardless of the collector admittance, there is never a situation which corresponds to Case I for the vacuum tube triode amplifier. Because of the conductance $1/h_{ie}$, the input admittance will never approach a pure capacitive susceptance as in vacuum tube Case I.

When the collector admittance includes a capacitive susceptance, there will be additional real terms in the equation for the input admittance. However, the term $1/h_{ie}$ will usually dominate so that the additional terms are negligible. Since the input admittance always includes a relatively large conductance, there are special low frequency problems as will be discussed later.

The collector admittance may involve an inductive susceptance as in Case III for the vacuum tube. For the transistor input admittance to have a negative real part, there must be a term of the real part of the admittance which is negative and has a magnitude greater than $1/h_{ie}$. Hence it is less likely that there will be a negative conductance at the input terminals. See Problem 7.10.

It might seem appropriate to consider also vacuum tube tetrode and pentode circuits. However, the effect of the grid to plate capacitance is reduced in the tetrode and pentode tubes. These tubes are specially designed to isolate the grid and plate circuits by adding the screen grid (grid No. 2) and operating it (considering varying components) at ground potential. The effective capacitance between the grid and plate is also much lower. Hence there is no need to give special consideration to amplifier circuits with tetrode and pentode tubes.

7.9 AMPLIFIER TRANSIENT RESPONSE

This section may be omitted without any loss of continuity. It is included for readers who are familiar with Laplace transforms and would like to use them in the analysis of the circuits.

In the previous sections of this chapter the steady-state response of amplifiers to sinusoidal excitation was determined. In general, the complete amplifier response for any arbitrary input signal would be desired. Then the steady-state response to sinusoidal excitation is a special case.

The procedures for drawing the equivalent amplifier circuits are the same as outlined in previous sections. When energy storing elements are included in the equivalent circuits, the general circuit equations are differential equations rather than algebraic equations. The amplifier response then involves the solution of a set of simultaneous differential equations for the particular input excitation. Solving a set of simultaneous differential equations by classical methods can become very tedious, so transform techniques are usually used.

Electrical engineers usually write the circuit equations directly in terms of Laplace transforms by defining operational impedances and admittances. The operational impedance of a passive circuit element is the ratio of the Laplace transform of the voltage to the Laplace transform of the current (sometimes denoted as s-domain voltages, currents, impedances and admittances). Since the equivalent circuits involve varying components only, initial values are usually zero.

As an example, consider the low frequency amplifier circuit of Fig. 7-8, page 172. The input loop equation indicates that $v_g = v_{s1}$. Using Laplace transforms, the plate nodal equation in terms of the complex variable s is

$$+g_m V_{s1}(s) + (g_p + G_L)V_p(s) + sC_C[V_p(s) - V_{s2}(s)] = 0$$

and the non-reference output nodal equation is

$$sC_C[V_{s2}(s) - V_p(s)] + G_G V_{s2}(s) = 0$$

These two equations are solved to give the transfer function which here is the *low frequency voltage gain*:

$$A_v(s) = \frac{-g_m C_C s}{G_G(g_p + G_L) + sC_C(g_p + G_L + G_G)}$$

Dividing numerator and denominator by $C_C(g_p + G_L + G_G)$,

$$A_v(s) = \left(\frac{-g_m}{g_p + G_L + G_G}\right)\left\{\frac{s}{s + \left[\dfrac{G_G(g_p + G_L)}{C_C(g_p + G_L + G_G)}\right]}\right\}$$

The first term (multiplier or scale factor) of the right side of this equation is the mid-band voltage gain A_{vo} which is independent of s, so the transfer function may be normalized to give

$$\frac{A_v(s)}{A_{vo}} = \frac{s}{s + \left[\dfrac{G_G(g_p + G_L)}{C_C(g_p + G_L + G_G)}\right]} \qquad (7.62)$$

The transfer function has one zero and one pole. The pole is on the negative real axis at $s = s_1 = \sigma_1 = \dfrac{-G_G(g_p + G_L)}{C_C(g_p + G_L + G_G)}$. The zero is at the origin $(s = 0)$. This zero is to be expected since the equivalent circuit includes a series coupling capacitor.

The transfer function is completely specified by the scale factor (mid-band gain) and the location of all its poles and zeros. A pole-zero plot of (*7.62*) is shown in Fig. 7-20. The half-power frequency is that at which $\omega_1 = |\sigma_1|$. Substituting the value of $|\sigma_1|$ and solving for f_1, the lower half-power frequency is

$$f_1 = \frac{G_G(g_p + G_L)}{2\pi C_C(g_p + G_L + G_G)}$$

which is identical with (*7.17*), page 173.

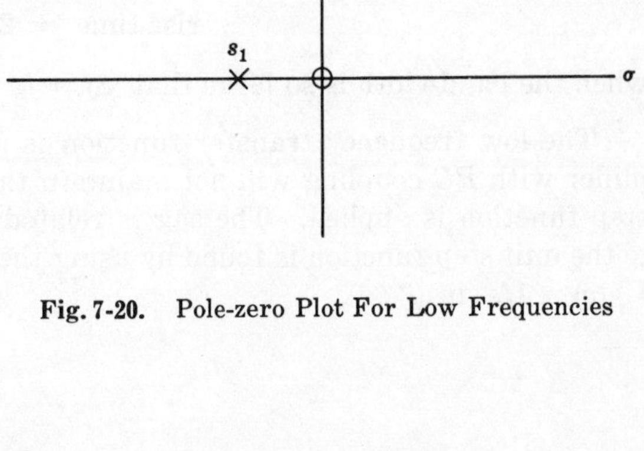

Fig. 7-20. Pole-zero Plot For Low Frequencies

The normalized steady-state transfer function may be found from (*7.62*) by letting $s = j\omega$:

$$\frac{A_v(j\omega)}{A_{vo}} = \frac{j\omega}{j\omega + \left[\dfrac{G_G(g_p + G_L)}{C_C(g_p + G_L + G_G)}\right]} = \frac{1}{1 - jf_1/f}$$

which is the same as (*7.18*), page 173.

The high frequency equivalent circuit of Fig. 7-7 may be analyzed in a similar manner; this is left as an exercise for the reader. The *high frequency transfer function* is

$$A_v(s) = \left(\frac{-g_m}{g_p + G_L + G_G}\right)\left(\frac{(g_p + G_L + G_G)/C_S}{s + [(g_p + G_L + G_G)/C_S]}\right) \tag{7.63}$$

and has one pole and no zero. There are no zeros because there are no series coupling capacitors; recall, however, that this equivalent circuit does not obtain for low frequencies. The pole is located at $s = s_2 = \sigma_2 = -(g_p + G_L + G_G)/C_S$. Equation (*7.63*) was given in the above form so that the normalized transfer function could be written by inspection. Since C_S is usually small, the pole s_2 is far to the left of s_1.

The upper half-power frequency is that at which $\omega_2 = |\sigma_2|$. Substituting for $|\sigma_2|$ and solving for f_2,

$$f_2 = (g_p + G_L + G_G)/2\pi C_S$$

which is the same as equation (*7.11*), page 172.

In transient response problems, the response to the unit step function $U(t)$ is often studied. The Laplace transform of $U(t)$ is $1/s$, which is substituted for $V_{s1}(s)$ to find the output voltage. The high frequency transfer function may be employed to find the rise time of the amplifier. Using (*7.63*), the Laplace transform of the output voltage is

$$V_{s2}(s) = A_v\left(\frac{a_2}{s + a_2}\right)\frac{1}{s} = A_v\left(\frac{a_2}{s(s + a_2)}\right)$$

where $a_2 = -\sigma_2 = (g_p + G_L + G_G)/C_S$. Taking the inverse Laplace transform of $V_{s2}(s)$, the output voltage as a function of time is

$$v_{s2}(t) = A_v(1 - e^{-a_2 t}) \tag{7.64}$$

The output voltage is zero until the unit step function is applied and then it increases exponentially until the final value of A_v is reached.

The rise time of an amplifier may be defined in several ways. A definition frequently used is that the *rise time* is the time required for the step response to rise from 10% to 90% of the final value. Using (*7.64*), the times t_1 and t_2 when $v_{s2}(t)$ is respectively 10% and 90% of the final value are given by

$$0.1 = 1 - e^{-a_2 t_1} \quad \text{and} \quad 0.9 = 1 - e^{-a_2 t_2}$$

from which $t_1 = 0.11/a_2$, $t_2 = 2.30/a_2$, and the rise time $= t_2 - t_1 = 2.19/a_2$. In terms of the upper half-power frequency, $a_2 = 2\pi f_2 = \omega_2$. Hence

$$\text{rise time} \cong 2.2/2\pi f_2 = 2.2/\omega_2$$

when the bandwidth is so large that $\Delta f \cong f_2$ (assuming $f_2 \gg f_1$).

The low frequency transfer function is used to determine what is called *sag*. An amplifier with *RC* coupling will not maintain the jump A_v in the output voltage when the unit step function is applied. The sag is related to the low frequency response. The response to the unit step function is found by using the low frequency transfer function. Substituting $V_{s1}(s) = 1/s$ in (*7.62*),

$$V_{s2}(s) \;=\; A_v\left\{\frac{1}{s+\left[\dfrac{G_G(g_p+G_L)}{C_C(g_p+G_L+G_G)}\right]}\right\}$$

Taking the inverse Laplace transform of this equation,

$$v_{s2}(t) \;=\; A_v e^{-a_1 t} \qquad (7.65)$$

where $a_1 = \dfrac{G_G(g_p+G_L)}{(g_p+G_L+G_G)C_C} = 2\pi f_1 = \omega_1$. Equation (*7.65*) indicates that there is an instantaneous jump of A_v in the output as $t=0$. This is not true, as was shown by (*7.64*); but it is known that the low frequency equivalent circuit neglects this effect. Equation (*7.65*) does correctly show that the output will eventually decay and approach zero. In terms of the usual definition of time constant, the time constant using (*7.65*) is $T = 1/a_1 = 1/\omega_1 = 1/2\pi f_1$.

The sag is usually defined in terms of an applied pulse rather than a step function. If the pulse duration is d seconds, then the output voltage at the end of the pulse is

$$v_{s2}(d) \;=\; A_v e^{-a_1 d}$$

The sag is defined as the difference between the initial and final value of the output pulse. In this case,

$$\text{sag} \;=\; A_v(1-e^{-a_1 d}) \;=\; A_v(1-e^{-\omega_1 d}) \qquad (7.66)$$

When the sag is small (in other words, the amplifier time constant is long compared to the pulse duration), the output decrease is approximately linear rather than exponential. The exponential of (*7.66*) may be expanded in a series, giving

$$\text{sag} \;=\; A_v\left\{1 - \left[1-\omega_1 d + \frac{(\omega_1 d)^2}{2!} - \frac{(\omega_1 d)^3}{3!} + \cdots\right]\right\}$$

When the sag is small, the first two terms of the series give a good approximation which is frequently used: sag $\cong A_v\omega_1 d$.

The amplifier equivalent circuit may include both the coupling capacitor C_C and the shunt capacitors C_1 and C_2 as shown in Fig. 7-9. The circuit equations may be written and solved, giving the following equation for the voltage gain:

$$A_v(s) \;=\; \frac{-g_m C_C s}{[C_1C_2 + C_C(C_1+C_2)]s^2 \;+\; [(C_1+C_C)G_G + (C_2+C_C)(g_p+G_L)]s \;+\; G_G(g_p+G_L)}$$

This may be simplified somewhat, since C_C is large compared to C_1 and C_2. The general transfer function has one zero and two poles. The zero is at $s=0$ because of the series capacitor. Unless it is known that the two poles are widely separated, considerable effort may be expended in attempting to find a solution for the quadratic function which is the denominator of the above equation. Care must be exercised in making simplifications, to avoid reducing the transfer function to the low frequency case. In the simplified transfer function one of the two poles should be approximately the same as the pole of the high frequency transfer function [equation (*7.63*)], and the other pole approximately the same as that of the low frequency transfer function [equation (*7.62*)].

7.10 TWO PORT AMPLIFIER REPRESENTATIONS

Only single stages of amplification have been considered in previous sections of this chapter. When several stages are cascaded, the number of equations becomes large and in general the computations become more involved. In vacuum tube circuits the grid is represented as an open circuit for the mid-band and low frequencies. Hence there is no loading on the previous stage unless the grounded grid circuit is under consideration. When several amplifier stages are connected in cascade, a two port representation may be convenient to use for individual stages and combinations of stages. In vacuum tube circuits the equivalent circuit representations are simpler because the inputs are infinite impedances. Any of the two port circuits may be used, so the most convenient one should be chosen.

Examine the mid-band equivalent vacuum tube amplifier circuit of Fig. 7-21. A two port representation which has a voltage controlled generator for the output port must be used, since the input current $i_{s1} = 0$. Let the output port be represented by a constant voltage generator which is voltage controlled. The input loop and plate nodal equations are respectively

$$-v_{s1} + v_g = 0 \quad \text{or} \quad v_g = v_{s1}$$

$$+g_m v_g + g_p v_{s2} + G_L v_{s2} + G_G v_{s2} - i_{s2} = 0$$

from which

$$-v_{s2} + \frac{-g_m}{g_p + G_L + G_G} v_{s1} + \frac{1}{g_p + G_L + G_G} i_{s2} = 0 \qquad (7.67)$$

Considering this equation term by term, the mid-band equivalent circuit for the output port is drawn as in Fig. 7-22. The input port is represented as an infinite impedance since $i_{s1} = 0$. When the frequency of the exciting signal is in the mid-band range and operation is restricted to the linear range, then the circuit of Fig. 7-22 is equivalent to that of Fig. 7-21 considering the terminal pairs.

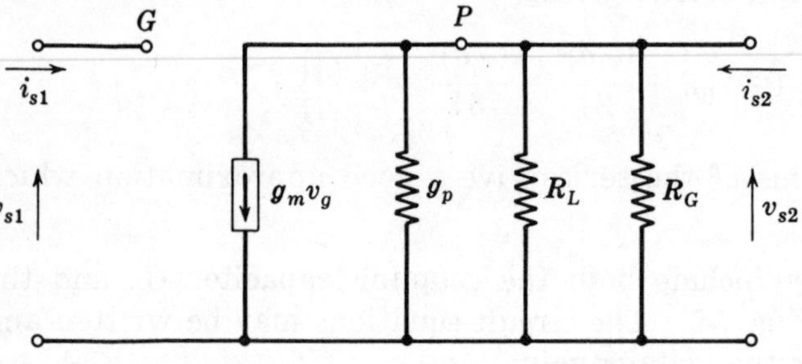

Fig. 7-21. Equivalent Circuit

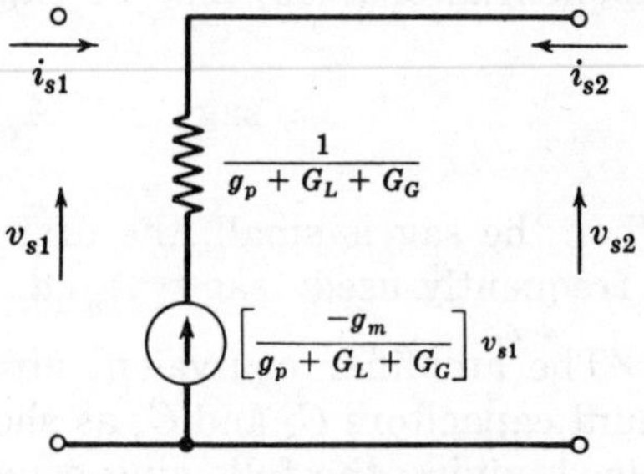

Fig. 7-22. Two Port Representation

Since the circuit of Fig. 7-21 is rather simple and involves only independent generators, the two port representation of Fig. 7-22 may seem trivial. In Fig. 7-21 the voltage v_g is independent of i_{s1}, v_{s2} and i_{s2}; thus the current generator is independent. For this case the equivalent circuit of Fig. 7-22 could easily be obtained by applying Thévenin's theorem to the circuit of Fig. 7-21. However, this is not usually the case. In general the controlled generators are dependent.

Examine the mid-band transistor amplifier circuit of Fig. 7-23 below. The input bias resistors are assumed large to avoid making the problem unnecessarily complicated. A two port representation involving voltage generators which are voltage controlled is developed by writing the circuit equations and then solving them. Noting that $i_{be} = i_{s1}$ and $v_{ce} = v_{s2}$, the input loop equation is

$$-v_{s1} + h_{ie} i_{s1} + h_{re} v_{s2} = 0 \qquad (7.68)$$

Let $h_{ie} = r_i$ and $h_{re} = \mu_r$, so that the two port parameters are identified as in Chapter 4. Equation (7.68) is then

$$r_i i_{s1} + \mu_r v_{s2} = v_{s1} \qquad (7.69)$$

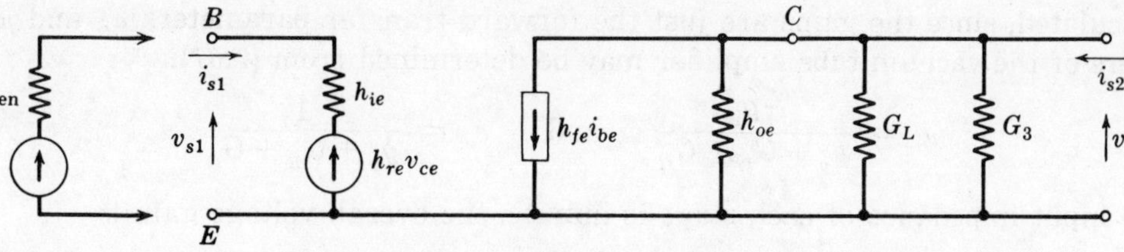

Fig. 7-23. Transistor Amplifier

The collector nodal equation is

$$h_{fe}i_{s1} + (h_{oe} + G_L + G_3)v_{s2} - i_{s2} = 0 \tag{7.70}$$

Solving (*7.68*) for i_{s1} and substituting into (*7.70*), we obtain

$$-v_{s2} + \left[\frac{h_{ie}}{h_{ie}(h_{oe} + G_L + G_3) - h_{fe}h_{re}}\right]i_{s2} + \left[\frac{-h_{fe}}{h_{ie}(h_{oe} + G_L + G_3) - h_{fe}h_{re}}\right]v_{s1} = 0 \tag{7.71}$$

Let
$$r_o = \frac{h_{ie}}{h_{ie}(h_{oe} + G_L + G_3) - h_{fe}h_{re}}, \qquad \mu_f = \frac{-h_{fe}}{h_{ie}(h_{oe} + G_L + G_3) - h_{fe}h_{re}} \tag{7.72}$$

Equation (*7.71*) now becomes

$$-v_{s2} + r_o i_{s2} + \mu_f v_{s1} = 0$$

Using this equation and (*7.69*), the equivalent circuit for the amplifier of Fig. 7-24 is drawn as in Fig. 7-24.

If the output port in Fig. 7-24 is unterminated ($i_{s2} = 0$), then the open circuit mid-band voltage gain $A_v = \mu_f$. In this case equation (*7.72*) for the parameter μ_f is the same as (*6.33*), page 140, for the open circuit voltage gain. The parameter r_o is the output impedance for the two port equivalent circuit of Fig. 7-24; it should not be confused with the output impedance of an amplifier which also is often denoted as r_o.

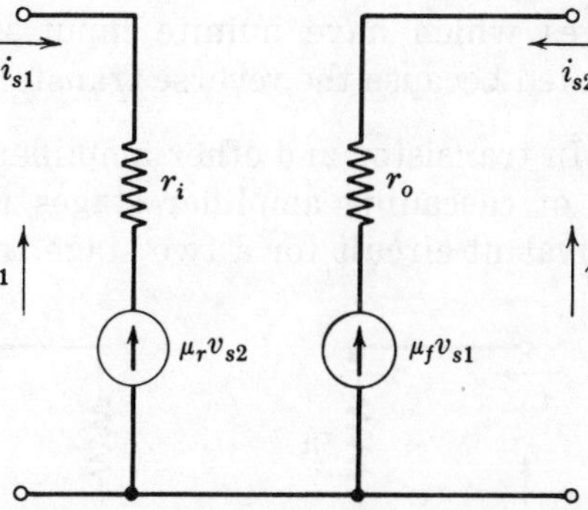

Fig. 7-24. Transistor Amplifier Representation

There are notational problems which have not been satisfactorily solved. Two of the parameters of the two port representation are usually denoted as r_o and r_i whether the generators are voltage or current controlled. The same notation is usually also used for one port representations. There is also ambiguity between the parameters for two port representations for the control device, a single stage amplifier, several stages of amplification, or an entire amplifier. There is no consistent notation which distinguishes the parameters for these different representations. In addition, not all authors use the same notation.

A linear amplifier stage may be represented using any of the two port equivalent circuit representations. The two port parameters are determined by drawing the equivalent circuit for the amplifier stage, writing the pertinent circuit equations and then solving them. The two previous examples illustrate the procedures.

When two amplifier stages are cascaded, the problems are much simpler if the inputs of the two port representations are open circuits. This is the case for vacuum tube amplifiers when the grid is negative. Consider the equivalent circuit of Fig. 7-25 below which shows two such amplifier stages cascaded. The open circuit voltage gains for each stage are easily calculated, since the gains are just the forward transfer parameters μ_{f1} and μ_{f2}. The parameters of the vacuum tube amplifier may be determined from (7.67):

$$\mu_f = \frac{-g_m}{g_p + G_L + G_G} \qquad \text{and} \qquad r_o = \frac{1}{g_p + G_L + G_G}$$

Since the input impedance of each stage is infinite, the overall voltage gain is

$$A_v = A_{v1} A_{v2} = \mu_{f1} \mu_{f2} \tag{7.73}$$

where $A_{v1} = v_{s2}/v_{s1}$ and $A_{v2} = v_{s3}/v_{s2}$.

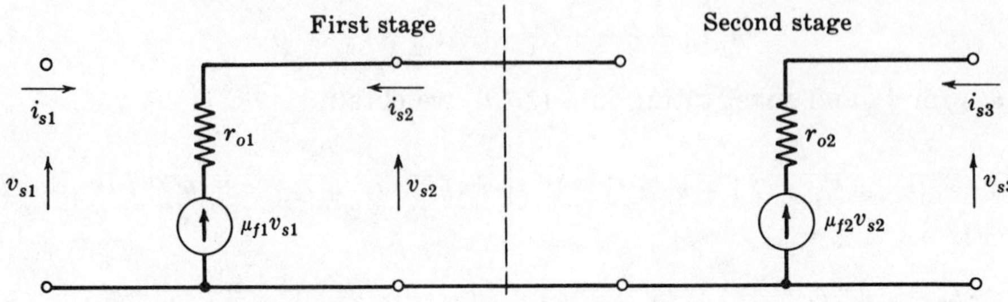

Fig. 7-25. Equivalent Circuit, Two Stage Amplifier

In vacuum tube amplifiers, the overall voltage gain is found by multiplying the voltage gains for the individual stages. When the gains are expressed in "db voltage gain", the gains for the individual stages are added. This is a correct procedure for any amplifier stages which have infinite input impedances. In such cases the stages are considered isolated because the reverse transfer function is zero.

In transistor and other amplifiers where the input port is not an open circuit, the problem of cascading amplifier stages is not as simple. Examine Fig. 7-26 which shows the equivalent circuit for a two stage transistor amplifier.

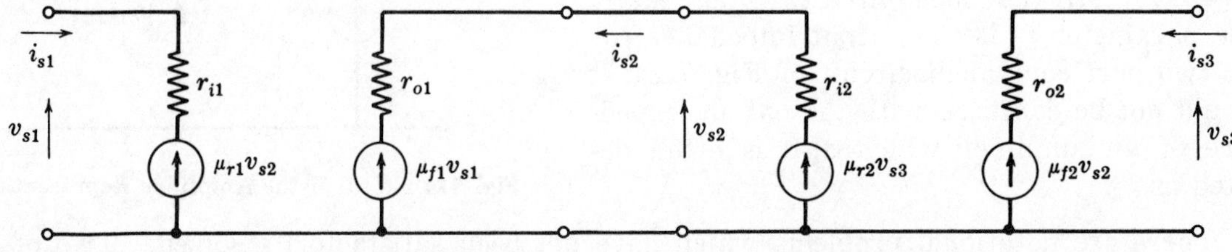

Fig. 7-26. General Equivalent Circuit, Two Stage Amplifiers

The overall open circuit voltage gain ($i_{s3} = 0$) is defined as $A_v = v_{s3}/v_{s1}$. The input loop equation is

$$-v_{s1} + r_{i1}i_{s1} + \mu_{r1}v_{s2} = 0 \tag{7.74}$$

and the interior loop equation in the direction of i_{s2} is

$$-\mu_{r2}v_{s3} + r_{i2}i_{s2} + r_{o1}i_{s2} + \mu_{f1}v_{s1} = 0 \tag{7.75}$$

Since $i_{s3} = 0$, the output loop equation is

$$-v_{s3} + \mu_{f2}v_{s2} = 0 \qquad \text{or} \qquad v_{s2} = v_{s3}/\mu_{f2} \tag{7.76}$$

We need one additional loop equation which includes v_{s2}, so the loop equation including the output of the first stage is

$$-v_{s2} + r_{o1}i_{s2} + \mu_{f1}v_{s1} = 0 \tag{7.77}$$

Equation (*7.76*) is now substituted into the other equations to eliminate the variable v_{s2} and then (*7.74*), (*7.75*) and (*7.77*) are solved to yield the overall voltage gain,

$$A_v = v_{s3}/v_{s1} = \frac{r_{i1}r_{i2}\mu_{f1}\mu_{f2}}{r_{i1}(r_{i2}+r_{o1}) - r_{i1}r_{o1}\mu_{f2}\mu_{r2}}$$

which should reduce to the product of the forward transfer parameters in the limit as the input resistances become large and the output resistances become negligibly small. If $r_{i2} \gg r_{o1}$, then $r_{i2}+r_{o1} \cong r_{i2}$ and

$$A_v = \frac{r_{i2}\mu_{f1}\mu_{f2}}{r_{i2} - r_{o1}\mu_{f2}\mu_{r2}} = \frac{\mu_{f1}\mu_{f2}}{1-(r_{o1}/r_{i2})\mu_{f2}\mu_{r2}}$$

If $r_{o1} \ll r_{i2}$ and $\mu_{f2}\mu_{r2}$ is small (μ_{r2} is usually small), then

$$A_v = \mu_{f1}\mu_{f2} \tag{7.78}$$

Any two port representations may be employed to represent amplifier stages. The desired transfer functions may be found by solving the circuit equations. The voltage gain is most frequently the overall transfer function for the amplifier, so the equivalent circuit of Fig. 7-26 was used. The forward transfer parameters then correspond to the open circuit voltage gains which are frequently specified for individual stages. When the input resistances r_i are large compared to the output resistances r_o (in this case r_i and r_o are two port parameters), then (*7.78*) is a good approximation.

Since there are two stages, the overall transfer function is positive if each individual stage gain has a minus sign. This is true for common emitter and common cathode circuits. The methods of analysis apply equally well to other amplifier configurations, but these are most common and hence used most frequently in illustrations.

In practice the number of simultaneous equations becomes rather large when there are many stages in a complete amplifier. This tends to make the amount of work prohibitive if each stage is represented as in Fig. 7-26 and all equations are solved simultaneously. Because of this, terminating impedances are often included in calculating gain for the amplifier stage. Instead of employing a true two port representation, a model is used which assumes the terminating impedances are known. This means that the input impedance (or admittance) must be known for each stage when terminated; the transfer function is then calculated for the specific terminations. These are usually good approximations in practice; however, the approximations which are made should be recognized.

Two port representations are emphasized in this book because they are general and the basis for representing all types of circuits, including for example passive filters. It is anticipated that the importance of two port models will increase because they will provide representations for integrated amplifier modules.

Any number of linear amplifier stages may be represented using a single two port model, or the desired transfer functions may be determined for specific terminations. Other linear circuits may be included in the amplifier as, for example, passive filters or other circuits which can be represented using two port equivalent circuits. As the circuit equations become more complicated, it is usually necessary to make more approximations in order to simplify the calculations.

7.11 ONE PORT AMPLIFIER REPRESENTATIONS

If either the input or output port of an amplifier is terminated, then a one port representation (equivalent circuit) may be found by looking into the unterminated port. This concept is often discussed in terms of particular problems such as determining the input impedance, output impedance, or transfer function. These are special cases. In general, one port equivalent circuits are involved and the concepts of input and output impedance are defined in terms of one port equivalent circuit representations.

Consider the two port representation of Fig. 7-22, page 194, with an external generator connected to the input port as shown in Fig. 7-27. This is the equivalent circuit for the vacuum tube amplifier of Fig. 7-21. Since the input port is terminated, a one port model may be developed for the output port. Thévenin's theorem could be applied to the circuit of Fig. 7-27, since the generators are both independent. This will usually not be the case. Hence all examples will be solved by general methods.

The input loop equation is $-v_{gen} + r_{gen} i_{s1} + v_{s1} = 0$. Since $i_{s1} = 0$, $v_{gen} = v_{s1}$.

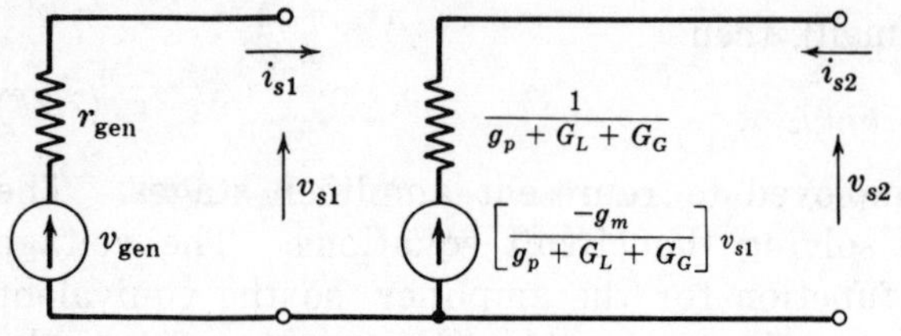

Fig. 7-27. Equivalent Circuit

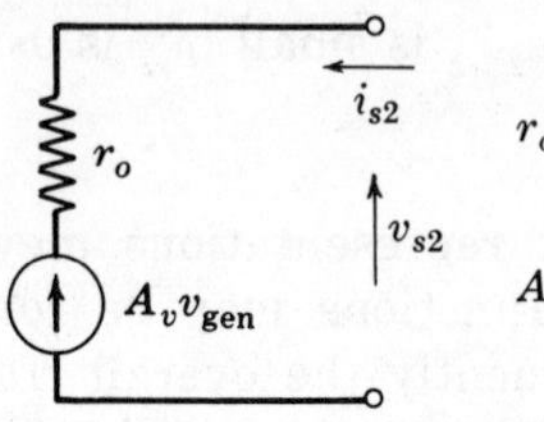

$$r_o = \frac{1}{g_p + G_L + G_G}$$

$$A_v = \frac{-g_m}{g_p + G_L + G_G}$$

Fig. 7-28. Equivalent Circuit

The output loop equation is the same as *(7.67)* with $v_{gen} = v_{s1}$. Therefore looking into the output port of Fig. 7-27, the circuit may be represented as in Fig. 7-28. The resistance r_o shown in the one port model for the output port is the so-called "output resistance". The coefficient of the voltage controlled generator is the mid-band open circuit voltage gain as derived in Chapter 5. If $i_{s2} = 0$ (open circuit), the terminal voltage $v_{s2} = A_v v_{gen}$; if $i_{s2} \neq 0$, then $v_{s2} \neq A_v v_{gen}$. The output resistance is an important concept. It is significant in amplifier gain calculations, unless the output is unterminated or the output resistance is small compared to the external terminating impedance. In this simple example both r_o and A_v are independent of the external input generator resistance because $i_{s1} = 0$. In general this is not the case.

If the output port is terminated (usually by a passive element), then a one port representation may be found for the input port. Consider an external resistance R_X connected to the output port of Fig. 7-22, as shown in Fig. 7-29. This is a trivial problem, since $i_{s1} = 0$ regardless of the termination on the output port; hence the input port is always represented as an infinite impedance for the mid-band and low frequencies. This is not true at high frequencies, since then the input involves an appreciable capacitance as was shown in Section 7.8. Fig. 7-29 does emphasize the fact that vacuum tube circuits are rather simple special cases when the input is to the control grid (and is negative). There is no loading or reverse coupling in the mid-band and lower frequency ranges when the grid is negative with respect to the cathode.

Consider the circuit of Fig. 7-24 with an external generator connected to the input port as shown in Fig. 7-30. A one port model for the output port involves writing the equations and solving.

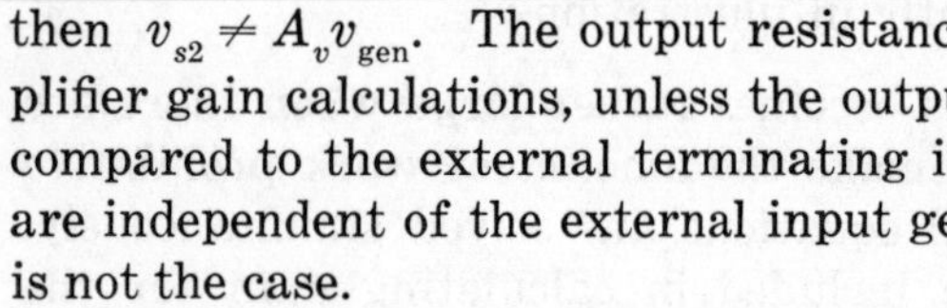

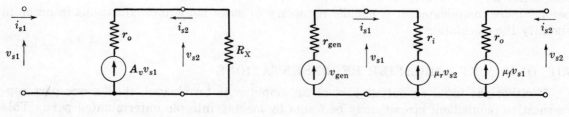

Fig. 7-29. Equivalent Circuit

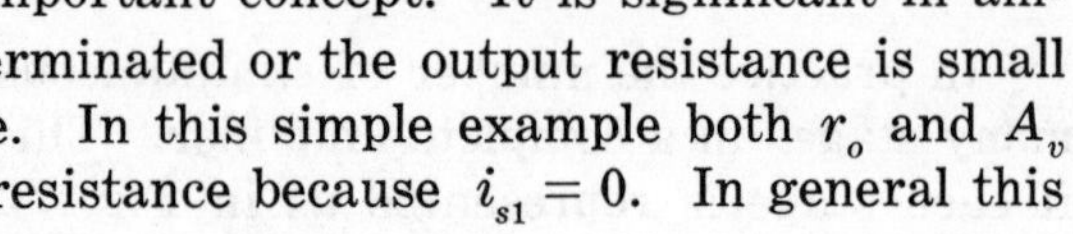

Fig. 7-30. Equivalent Circuit

The equations for the entire input loop, for the input port loop and for the output loop are respectively

$$-v_{gen} + r_{gen}i_{s1} + r_i i_{s1} + \mu_r v_{s2} = 0, \quad -v_{s1} + r_i i_{s1} + \mu_r v_{s2} = 0, \quad -v_{s2} + r_o i_{s2} + \mu_f v_{s1} = 0$$

from which
$$-v_{s2} + \frac{\mu_f r_i v_{gen} + r_o(r_{gen} + r_i)i_{s2}}{r_{gen} + r_i - \mu_f \mu_r r_{gen}} = 0$$

This equation may be considered term by term, giving a representation looking into the output port. The coefficient of i_{s2} is the output resistance r_o and the coefficient of v_{gen} is the open circuit voltage gain (defined with respect to v_{gen}). If the generator terminal voltage v_{s1} is constant, then effectively r_{gen} is zero and a different output impedance is obtained. The term involving v_{gen} is represented as a controlled generator because it is independent of v_{s2} and i_{s2}.

The equivalent one port representation is shown in Fig. 7-31. The circuit of Fig. 7-31 is equivalent to that of Fig. 7-30 with the input generator connected. The equivalence means that when looking into the output port with the input port terminated as shown, no experiment can distinguish one circuit from the other. This assumes linear operation and frequencies in the mid-band range. There may be a difference in the circuits internally. No power is dissipated in the circuit of Fig. 7-31 unless the output is terminated; this is not true for the circuit of Fig. 7-30.

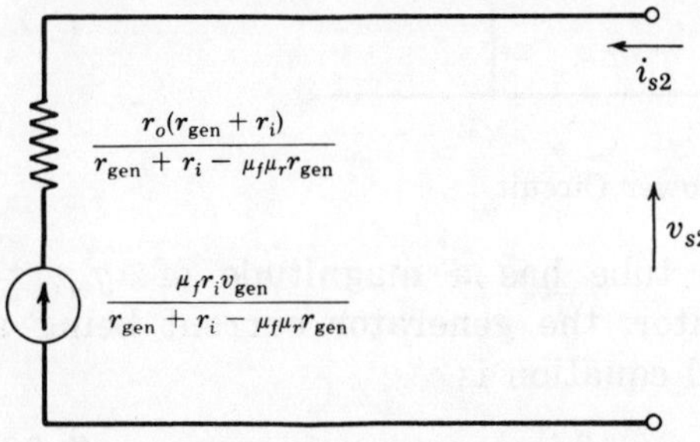

Fig. 7-31. One Port Representation

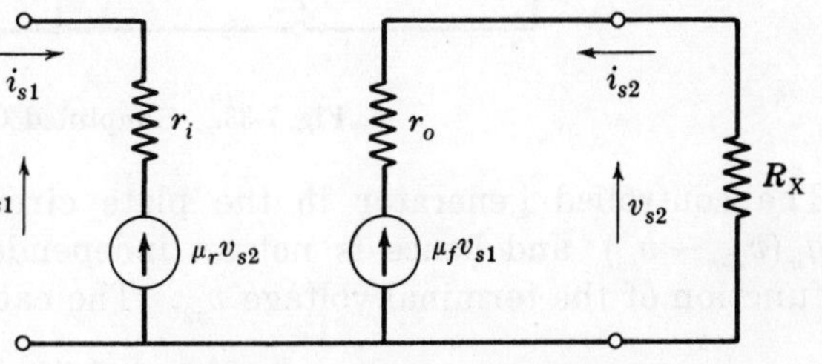

Fig. 7-32. Equivalent Circuit

Consider the circuit of Fig. 7-24, page 195, with a load resistance R_X connected across the output terminals as shown in Fig. 7-32. A one port model looking into the input port involves a passive element because there is no generator external to the output terminals. The input loop and output loop equations are respectively

$$-v_{s1} + r_i i_{s1} + \mu_r v_{s2} = 0, \quad -v_{s2} + r_o i_{s2} + \mu_f v_{s1} = 0$$

Substitute $i_{s2} = -v_{s2}/R_X$ and solve to obtain

$$R_i = \frac{v_{s1}}{i_{s1}} = \frac{r_i(1 + r_o/R_X)}{(1 + r_o/R_X) - \mu_f\mu_r} \tag{7.79}$$

which is not in the form usually used, since it is in terms of the parameters of the two port representation.

The external resistance R_X may be added to the transistor equivalent circuit of Fig. 7-23, page 195. Consider the circuit looking into the input port. Using $v_{ce} = v_{s2}$, the input loop equation is

$$-v_{s1} + h_{ie}i_{s1} + h_{re}v_{s2} = 0$$

Using $i_{be} = i_{s1}$, the collector nodal equation is

$$+h_{fe}i_{s1} + (h_{oe} + G_L + G_3 + G_X)v_{s2} = 0$$

From the two equations, we obtain

$$R_i \;=\; \frac{v_{s1}}{i_{s1}} \;=\; h_{ie} - \frac{h_{fe}h_{re}}{h_{oe} + G_L + G_3 + G_X} \tag{7.80}$$

which shows that the input resistance is less than h_{ie}, since each quantity in the last term is positive. In practice, problems are sometimes simplified by assuming that the input resistance of a common emitter transistor amplifier is h_{ie}. This is often done in calculating the overall gain of a number of cascaded amplifier stages.

Equations (7.79) and (7.80) are the same, as may be shown by substituting the equations for r_i, r_o, μ_f and μ_r from Section 7.10 into (7.79).

Another example involving dependent generators is the simplified cathode follower circuit of Fig. 7-33. Here the circuit is drawn so that the equations may be more conveniently written by inspection. The grid current $i_{s1} = 0$, hence $v_{gen} = v_{s1}$. The grid loop equation is

$$-v_{gen} + r_{gen}(0) + v_g + v_{s2} \;=\; 0 \quad \text{or} \quad v_g \;=\; v_{gen} - v_{s2} \tag{7.81}$$

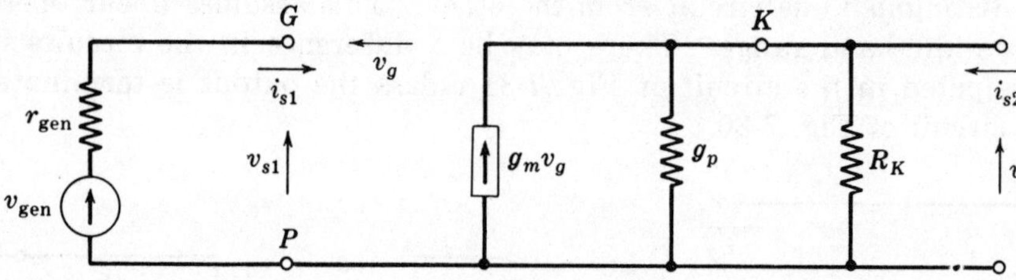

Fig. 7-33. Simplified Cathode Follower Circuit

The controlled generator in the plate circuit of the tube has a magnitude of $g_m v_g = g_m(v_{gen} - v_{s2})$ and hence is not an independent generator, the generator current being a function of the terminal voltage v_{s2}. The cathode nodal equation is

$$-g_m v_g + g_p v_{s2} + G_K v_{s2} - i_{s2} \;=\; 0 \tag{7.82}$$

Solving (7.81) and (7.82) for v_{s2} as a function of v_{gen} and i_{s2},

$$-v_{s2} + \left(\frac{1}{g_p + G_K + g_m}\right) i_{s2} + \left(\frac{g_m}{g_p + G_K + g_m}\right) v_{gen} \;=\; 0 \tag{7.83}$$

The coefficient of i_{s2} is the output resistance r_o, and the coefficient of v_{gen} is the open circuit voltage gain. Equation (7.83) may be represented as in Fig. 7-34. This is the model for the output port of a cathode follower. Since the input current is zero, the output port equivalent circuit is independent of the input generator resistance r_{gen}.

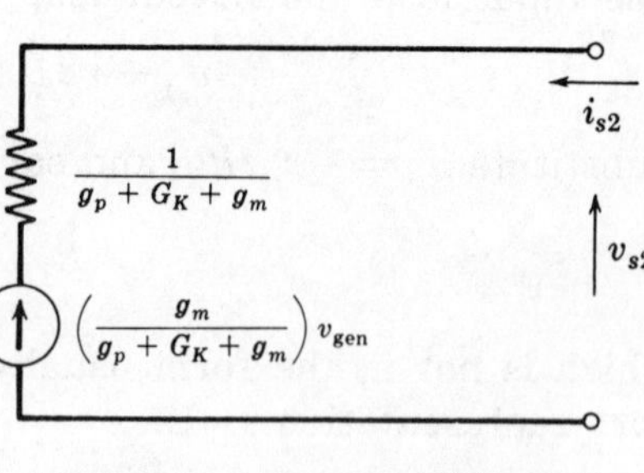

Fig. 7-34. Cathode Follower Output Circuit

The equation for the output impedance of a cathode follower is often given in another form which may be derived from (7.83). Using $\mu = g_m r_p$,

$$r_o \;=\; \frac{1}{g_p + G_K + g_m} \;=\; \frac{1}{1/r_p + 1/R_k + g_m} \;=\; \frac{r_p R_K}{r_p + (\mu + 1)R_K}$$

which shows that the effective output resistance r_o is quite different from the parallel combination of the resistors in the cathode circuit (unless $\mu \ll 1$, which is never the case).

An approximate relation often used in practice when g_p and G_K are small compared to g_m is $r_o \cong 1/g_m$.

The vacuum tube cathode followers and the transistor emitter followers are used very frequently to isolate circuits. The cathode follower has a very high input resistance, and an output resistance in the order of hundreds of ohms. The emitter follower input resistance is not as high as that of cathode followers, but is high compared to other junction transistor circuits. The output impedances are also low.

Consider the simplified emitter follower equivalent circuit of Fig. 7-35 and assume the resistor R_E includes all the external load resistance. Using $i_{be} = i_{s1}$ and $v_{s2} = -v_{ce}$, the input loop and emitter nodal equations are

$$-v_{s1} + h_{ie}i_{s1} + h_{re}(-v_{s2}) + v_{s2} = 0 \tag{7.84}$$

$$-i_{s1} - h_{fe}i_{s1} + (h_{oe} + G_E)v_{s2} - i_{s2} = 0 \tag{7.85}$$

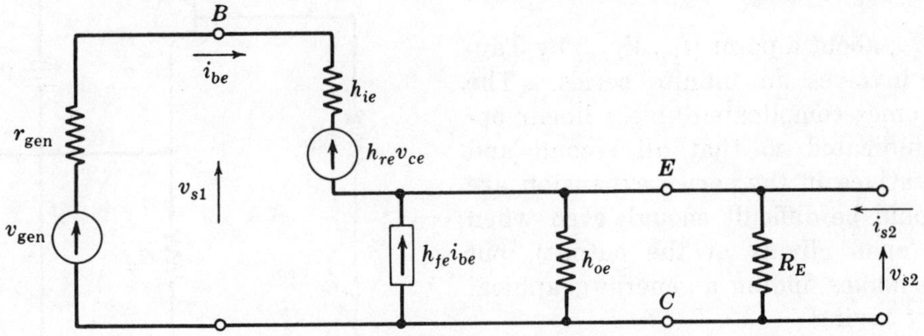

Fig. 7-35. Emitter Follower Equivalent Circuit

The representation for the input port of the emitter follower of Fig. 7-35 is found by setting $i_{s2} = 0$ (R_E includes the load resistance) and solving the above equations for the input resistance

$$R_i = v_{s1}/i_{s1} = h_{ie} + \frac{(h_{fe}+1)(1-h_{re})}{h_{oe}+G_E}$$

which may be simplified by using approximations which are usually reasonable. Usually: $h_{fe} \gg 1$, so $h_{fe}+1 \cong h_{fe}$; $h_{oe} \ll G_E$, so $h_{oe}+G_E \cong G_E$; $h_{re} \ll 1$, so $1-h_{re} \cong 1$. Then $R_i \cong h_{ie} + h_{fe}/G_E = h_{ie} + h_{fe}R_E$; and since usually $h_{fe}R_E \gg h_{ie}$, we have

$$R_i \cong h_{fe}R_E$$

which is a reasonably good approximation in most problems and is often used by designers.

The representation for the output port is found by including the external generator of Fig. 7-35 in the input loop. The input loop equation is now

$$-v_{gen} + (r_{gen} + h_{ie})i_{s1} - h_{re}v_{s2} + v_{s2} = 0$$

Solving this equation with (7.85) for v_{s2} as a function of v_{gen} and i_{s2},

$$-v_{s2} + \frac{(r_{gen}+h_{ie})i_{s2} + (h_{fe}+1)v_{gen}}{(r_{gen}+h_{ie})(h_{oe}+G_E) + (h_{fe}+1)(1-h_{re})} = 0 \tag{7.86}$$

where the coefficient of i_{s2} is the output resistance r_o and the coefficient of v_{gen} is the open circuit voltage gain A_v. The equivalent circuit could be drawn, but there should be no problem in visualizing the circuit which can be drawn by considering (7.86) term by term. When the output is an open circuit ($i_{s2} = 0$), v_{s2}/v_{gen} is the coefficient of v_{gen}. When $v_{gen} = 0$, the voltage across the output resistor is v_{s2} and hence $r_o = v_{s2}/i_{s2}$. Equation (*7.86*) can usually be simplified by using several approximations.

The one and two port amplifier representations are usually found from equivalent amplifier circuits so that only varying components are involved. However, the one and two port representations may be developed in terms of instantaneous values of voltages and currents as shown in Chapter 4. The parameters are then defined as partial derivatives rather than in terms of the parameters of an equivalent circuit. Any equations for the same parameters are equal to each other, assuming the same amplifier circuit is involved. Since all of Chapter 4 was devoted to a general analysis of two port equivalent circuits, it seems redundant to repeat the discussion here. However, it may be appropriate to study one example for emphasis.

Consider the emitter follower circuit of Fig. 7-36 involving instantaneous voltages and currents. A representation for the output port may be developed by writing the functional relationship

$$v_{S2} = f(i_{S2}, v_{GEN}) \qquad (7.87)$$

Expanding v_{S2} about a point (I_{S2}, V_{GEN}) by Taylor's series involves an infinite series. The problem becomes complicated unless linear operation is indicated so that all second and higher derivatives in the series expansion are zero. It would be difficult enough even when i_{S2} is zero (open circuit at the output), but including i_{S2} makes finding a general graphical solution very trying.

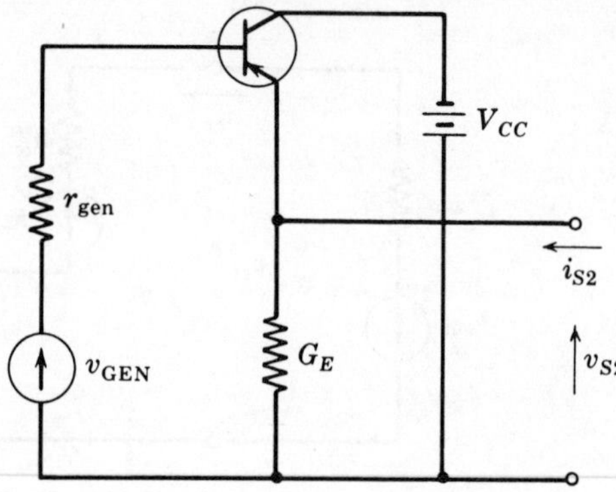

Fig. 7-36. Emitter Follower Circuit

Assuming linear operation, the series expansion for (*7.87*) is

$$v_{S2} = f(I_{S2}, V_{GEN}) + (i_{S2} - I_{S2})\frac{\partial v_{S2}}{\partial i_{S2}} + (v_{GEN} - V_{GEN})\frac{\partial v_{S2}}{\partial v_{GEN}}$$

The term $V_{S2} = f(I_{S2}, V_{GEN})$ is the average value of the output voltage. If superposition is applied and only varying components are considered,

$$v_{s2} = (\partial v_{S2}/\partial i_{S2})i_{s2} + (\partial v_{S2}/\partial v_{GEN})v_{gen}$$

which is the same as (*7.86*). Hence for this emitter follower,

$$r_o = \partial v_{S2}/\partial i_{S2} = \frac{r_{gen} + h_{ie}}{(r_{gen} + h_{ie})(h_{oe} + G_E) + (h_{fe} + 1)(1 - h_{re})}$$

$$A_v = \partial v_{S2}/\partial v_{GEN} = \frac{h_{fe} + 1}{(r_{gen} + h_{ie})(h_{oe} + G_E) + (h_{fe} + 1)(1 - h_{re})}$$

This example emphasizes the fact that the parameters of the one and two port circuit representations are in general defined as partial derivatives. However, it is usually more convenient to employ equations which are derived in terms of the parameters of an equivalent circuit. In earlier chapters it was shown that when linear operation is involved, the partial derivative is the same as the ratio of the varying components; for example, $A_v = \partial v_{S2}/\partial v_{GEN} = v_{s2}/v_{gen}$.

One port representations may contain current generators which are voltage or current controlled. If current controlled, current generators are used, the one port representation will include two parameters. The forward transfer parameter is the short circuit current gain A_i (by definition, $A_i = \partial i_{S2}/\partial i_{S1}$); the other parameter will be the output conductance

g_o, which is the reciprocal of the output resistance r_o. There are very many possible representations for the same amplifier; the examples that were given illustrate some general methods.

As a matter of convenience, problems are not always solved in the most general way. Open circuit voltage gains were calculated for a number of amplifier circuits in Chapters 5 and 6. The equations are simpler when the output port has a specific termination. Hence it is common practice to calculate the transfer functions for either an open circuit, a short circuit, or a specific termination.

The output impedance (or admittance) is also frequently calculated by a particular procedure. It was shown that when the external input generated voltage is zero ($v_{gen} = 0$), then $r_o = v_{s2}/i_{s2}$. Hence the equivalent circuit is often drawn showing the external generator as a resistance r_{gen} (the generated voltage is zero, so there is no generator). Then the circuit equations are written to include the output current. The output resistance is the ratio of the output voltage v_{s2} to the output current i_{s2}. See Problem 7.11, page 213.

7.12 CASCADED AMPLIFIER STAGES

When each amplifier stage is linear, the equivalent circuits for the individual stages can be connected together. The overall mid-band transfer function can be found by using the methods of Section 7.10. In this chapter the individual stages have been assumed linear, which ignores many problems. The circuits used in finding the frequency response are linear but include energy storing elements.

It is necessary to consider the overall frequency response of a number of cascaded stages. In general, the complete amplifier equivalent circuit could be drawn, including the capacitances which are significant in determining the frequency response. The simple examples of Sections 7.3 and 7.4 demonstrate that if there are many stages, it is convenient to consider the low frequency equivalent circuit separately from the high frequency circuit.

In low or high frequency two port representations for an amplifier stage, the parameters are no longer real numbers. The input and output parameters are complex admittances or impedances. The transfer functions are also complex. See Problems 7.12 and 7.15. An amplifier containing several stages can become quite complicated. In practice, approximations are usually taken.

Consider an amplifier consisting of two stages which have identical upper and lower half-power frequencies. If the stages are isolated so that the overall gain is the product of the stage gains, then the low frequency voltage gain is

$$A_v(\text{l.f.}) = A_{v1}(\text{l.f.})\,A_{v2}(\text{l.f.}) = A_{v1}A_{v2}\left(\frac{1}{1-jf_1/f}\right)\left(\frac{1}{1-jf_1/f}\right) \qquad (7.88)$$

The overall gain ratio is

$$\frac{A_v(\text{l.f.})}{A_{v1}A_{v2}} = \frac{1}{1-(f_1/f)^2 - j2f_1/f}$$

Since the numerator is a real number and independent of frequency, the half-power frequency of the amplifier is the frequency at which the denominator is $\sqrt{2}$ times the mid-band value. In this case the mid-band magnitude is 1. It will be necessary to distinguish the half-power frequency of a single stage from the amplifier half-power frequency. Let the overall amplifier lower half-power frequency be denoted f_1^*. Then f_1^* is found by solving

$$\{[1-(f_1/f_1^*)^2]^2 + (2f_1/f_1^*)^2\}^{1/2} = 2^{1/2}$$

to obtain $\quad (f_1/f_1^*)^2 = -1+\sqrt{2} = 0.414 \quad$ or $\quad f_1^* = f_1/0.644$

When the high frequency circuit is considered, equation (*7.88*) will be the same but the term in the denominator is $1 + jf/f_2$. Making this substitution, the amplifier upper half-power frequency f_2^* is found to be

$$f_2^* = 0.644\,f_2 \tag{7.89}$$

When there are n identical stages, it is more convenient to derive the equations for the amplifier half-power frequencies by using the equations which express the relative gain in db. For n identical stages, equation (*7.43*), page 184, may be written as

$$\text{db} = -20\log_{10}[1 + (f/f_2)^2]^{n/2} = -10\log_{10}[1 + (f/f_2)^2]^n$$

At the amplifier upper half-power frequency f_2^*, the relative gain is -3 db. Then

$$-3 = -10\log_{10}[1 + (f_2^*/f_2)^2]^n \quad \text{or} \quad 10^{0.3/n} = 1 + (f_2^*/f_2)^2$$

from which

$$f_2^* = f_2\sqrt{10^{0.3/n} - 1} \tag{7.89a}$$

When $n = 2$, $\quad f_2^* = f_2\sqrt{10^{0.15} - 1} = f_2\sqrt{1.414 - 1} = 0.644\,f_2$

which is the same as equation (*7.89*).

By similar analysis it can be shown that for n identical stages the amplifier lower half-power frequency f_1^* is

$$f_1^* = \frac{f_1}{\sqrt{10^{0.3/n} - 1}} \tag{7.90}$$

In more general cases the individual stages may have different half-power frequencies. Then the derivation of general equations becomes complicated; numerical problems are more easily solved than the general equations.

If the amplifier stages are not isolated, the two port equivalent circuits must be used. The two port parameters are complex for the high and low frequency circuits. In the case of transistor amplifiers, the forward transfer parameter may also be complex and frequency dependent. The problems then are usually very difficult to solve unless approximations are used.

Solved Problems

7.1. Given a simple amplifier circuit which can be represented as a high-pass filter for frequencies below the mid-band range. Assume that the normalized low frequency voltage gain is

$$\frac{A_v(\text{l.f.})}{A_{vo}} = \frac{1}{1 - jf_1/f} \tag{7.91}$$

Plot the gain ratio vs. frequency ratio curve on semilog paper, showing the gain ratio (magnitude only) in decibels. Then draw the low frequency asymptote (this is the extension of the straight line for $f_1/f \gg 1$). Show that the mid-band asymptote and low frequency asymptote intersect at $f_1/f = 1$. Asymptotes to frequency response curves are often used in practice. The half-power frequency is often called the corner frequency or break frequency because it is the intersection of the two asymptotes.

The gain ratio in db $= 20 \log \dfrac{1}{1 - jf_1/f}$. In terms of magnitudes only,

$$\text{db} = 20\{\log 1 - \log [1 + (f_1/f)^2]^{1/2}\} = -10 \log [1 + (f_1/f)^2]$$

The curve is plotted, as shown in Fig. 7-37, on a point by point basis. For example, let $f_1/f = 5$. Then the gain ratio is

$$\text{db} = -10 \log [1 + 25]$$
$$= -10(1.415) = -14.15$$

Hence when $f = f_1/5$, the output is 14.15 db below the mid-band value of zero. The asymptotes intersect at $f_1/f = 1$.

This can also be shown analytically by considering the two asymptotes of (*7.91*):

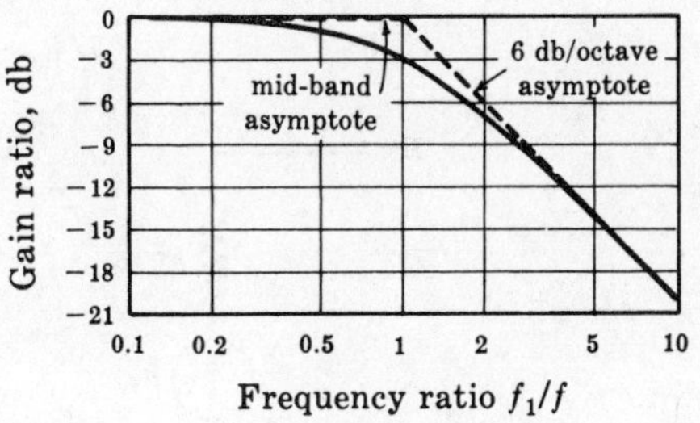

Fig. 7-37

mid-band asymptote: $f_1/f \ll 1$, then $A_v(\text{l.f.}) = A_{vo}$

low frequency asymptote: $f_1/f \gg 1$, then $A_v(\text{l.f.}) = j(f/f_1)A_{vo}$

At the point of intersection of the two asymptotes, the magnitudes of the two gain equations are the same: $A_{vo} = A_{vo}(f/f_1)$ or $f = f_1$. Hence the two asymptotes intersect at $f = f_1$.

The asymptotes give the poorest approximation to the true frequency response curve at $f = f_1$ where the true response curve is 3 db below the asymptotes (intersection point).

7.2. Given the equivalent "tee" circuit for a transistor with the base common as in Fig. 6-40, page 147, and the same circuit with the emitter common as in Fig. 6-45, page 148. Take $\alpha(f) = \dfrac{\alpha}{1 + jf/f_{\alpha co}}$ where $f_{\alpha co}$ is the cutoff frequency, i.e. the frequency at which $|\alpha(f)|$ is $\alpha/\sqrt{2}$. Derive the equation for the cutoff frequency $f_{\beta co}$, the frequency at which $|\beta(f)|$ is $\beta/\sqrt{2}$ in terms of $f_{\alpha co}$.

By equation (*6.74*), page 148, $\beta = \alpha/(1 - \alpha)$. Substituting $\alpha(f)$ into equation (*6.74*),

$$\beta(f) = \frac{\alpha(f)}{1 - \alpha(f)} = \frac{\alpha/(1 + jf/f_{\alpha co})}{1 - [\alpha/(1 + jf/f_{\alpha co})]} = \frac{\alpha}{(1 - \alpha) + jf/f_{\alpha co}}$$
$$= \frac{\alpha/(1 - \alpha)}{1 + jf/[f_{\alpha co}(1 - \alpha)]} = \frac{\beta}{1 + jf/[f_{\alpha co}(1 - \alpha)]}$$

The cutoff frequency for $\beta(f)$ is the frequency at which $|\beta(f)| = \beta/\sqrt{2}$. Since the numerator is real and independent of frequency and since the real part of the denominator is independent of frequency, set the real part of the denominator equal to the imaginary part and solve for the frequency.

$$1 = f_{\beta co}/[f_{\alpha co}(1 - \alpha)] \quad \text{or} \quad f_{\beta co} = f_{\alpha co}(1 - \alpha)$$

7.3. Derive the equation of the effective output capacitance C_{oe} for the common emitter in terms of the output capacitance C_{ob} and other common base parameters.

Consider the circuit diagram of Fig. 6-46, page 149. Add the capacitance C_{ob} between the base and collector terminals as shown in the following diagram:

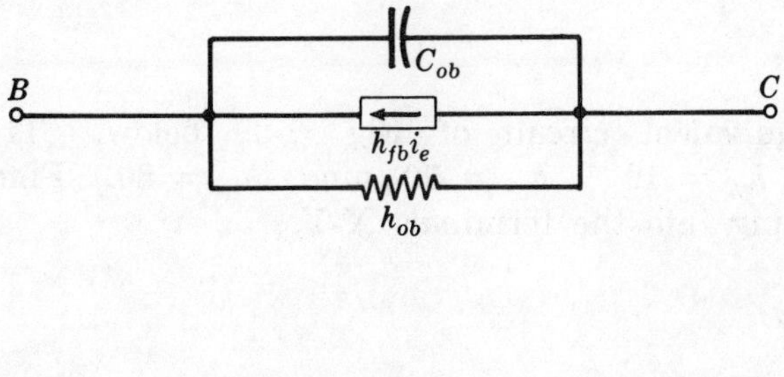

Equations (6.76), (6.78) and (6.79), page 149, remain the same. Equation (6.77) is now

$$-i_{ce} + h_{fb}i_e + (h_{ob} + j\omega C_{ob})v_{cb} = 0 \qquad (6.77a)$$

The common emitter equivalent circuit for the transistor now includes the capacitance C_{oe} as shown in Fig. 7-38. Equation (6.80) remains the same but (6.81) is now

$$h_{fe}i_{be} + (h_{oe} + j\omega C_{oe})v_{ce} = i_{ce} \qquad (6.81a)$$

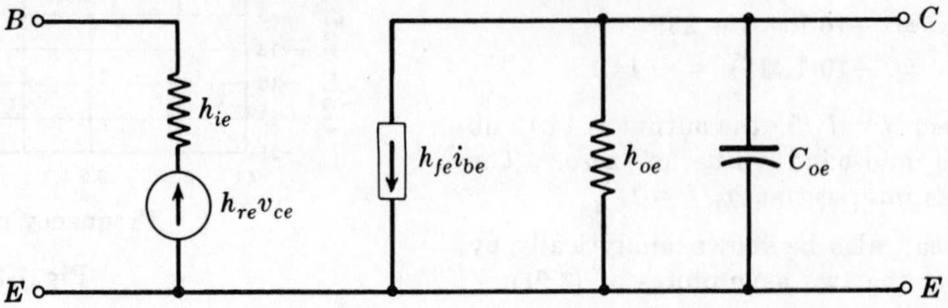

Fig. 7-38

It might be tempting to guess that since

$$h_{oe} = \frac{h_{ob}}{1 + h_{fb} + h_{ob}h_{ib} - h_{rb} - h_{fb}h_{rb}} \qquad (6.88)$$

the equation for C_{oe} would be

$$C_{oe} = \frac{C_{ob}}{1 + h_{fb} + h_{ob}h_{ib} - h_{rb} - h_{fb}h_{rb}} \qquad (7.92)$$

This guess would be based on the fact that the coefficient of v_{cb} in (6.77a) includes the capacitive susceptance of C_{ob}.

Rewriting equations (6.76), (6.77a), (6.78) and (6.79),

$$\begin{aligned} -v_{be} - \quad h_{rb}v_{cb} - h_{ib}i_e &= 0 \\ (h_{ob} + j\omega C_{ob})v_{cb} + h_{fb}i_e - i_{ce} &= 0 \\ -i_e - i_{ce} &= i_{be} \\ v_{be} + \quad v_{cb} &= v_{ce} \end{aligned}$$

from which $$i_{ce} = \frac{+(h_{ob} + j\omega C_{ob})v_{ce} - [h_{fb} + h_{ib}(h_{ob} + j\omega C_{ob}) - h_{fb}h_{rb}]i_{be}}{1 + h_{fb} + h_{ib}(h_{ob} + j\omega C_{ob}) - h_{rb} - h_{fb}h_{rb}} \qquad (7.93)$$

Since the coefficients of v_{ce} in (7.93) and (6.81a) are equal,

$$\frac{h_{ob} + j\omega C_{ob}}{1 + h_{fb} + h_{ib}(h_{ob} + j\omega C_{ob}) - h_{rb} - h_{fb}h_{rb}} = h_{oe} + j\omega C_{oe} \qquad (7.94)$$

To solve this equation in a completely general way would require that the denominator be rationalized; this becomes a tedious task. The imaginary part of the denominator ($j\omega h_{ib}C_{ob}$) is small compared to the real part and may be neglected. Then C_{oe} is

$$C_{oe} \cong \frac{C_{ob}}{1 + h_{fb} + h_{ib}h_{ob} - h_{rb} - h_{fb}h_{rb}} \qquad (7.95)$$

Using approximations, this becomes

$$C_{oe} \cong \frac{C_{ob}}{1 + h_{fb}}$$

Equation (7.95) shows that the intuitive estimate for C_{oe} as given by (7.92) is sufficiently accurate, because this involves only dropping the imaginary part of the denominator of (7.94) which is very small.

7.4. Given the equivalent circuit of Fig. 7-39 below. Transistor parameters are $h_{ie} = 2000\ \Omega$, $h_{re} = 10^{-3}$, $h_{oe} = 50\ \mu$mho, $h_{fe} = 80$. Find the mid-band input admittance looking into the terminals X-Y.

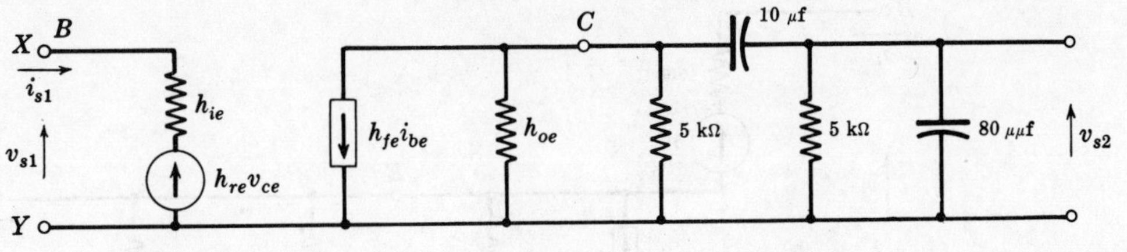

Fig. 7-39

Writing the input loop equation and the collector nodal equation for the mid-band range, noting that $i_{be} = i_{s1}$ and $v_{ce} = v_{s2}$ (due to the mid-band assumption),

$$2000 i_{s1} + 1000 \times 10^{-6} v_{s2} = v_{s1}, \qquad 80 i_{s1} + 450 \times 10^{-6} v_{s2} = 0$$

from which $i_{s1} = 549 \times 10^{-6} v_{s1}$.

The input admittance $y_i = i_{s1}/v_{s1} = 549$ μmho. The input resistance $R_i = 10^6/549 = 1825\ \Omega$.

7.5. Given the simplified diagram, Fig. 7-40, for an emitter follower (common collector) amplifier. Transistor parameters are $h_{ie} = 1500\ \Omega$, $h_{re} = 10^{-4}$, $h_{oe} = 50$ μmho, $h_{fe} = 80$.

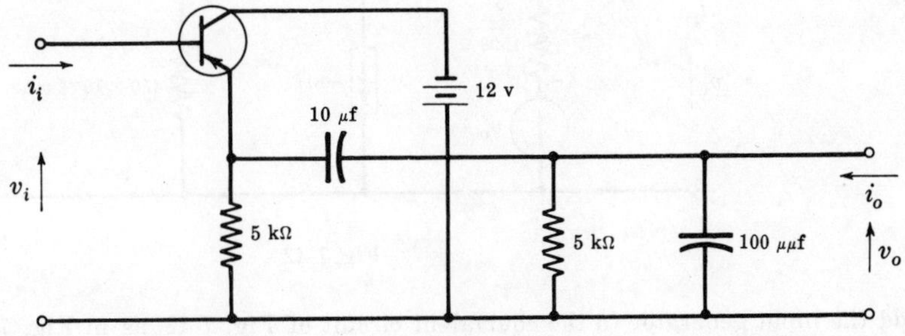

Fig. 7-40

(a) Represent the transistor for the mid-band range of frequencies using the hybrid two port representation (*h* parameters). Determine the parameters and draw the equivalent circuit.

(b) Add the following generator at the input terminals of the above circuit. Then develop a one port representation (equivalent circuit) looking into the output terminals. Specify the parameters.

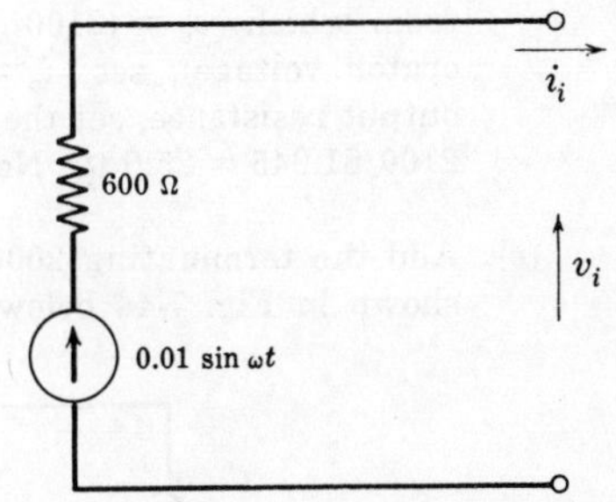

(c) Add a 2000 Ω resistor across the output terminals. Develop a one port representation for the input pair of terminals with the generator removed. Specify the parameters.

(a) First draw the mid-band equivalent circuit. The equations are more easily written by inspection if the circuit is displayed as in Fig. 7-41 below.

It is important that the polarities of the controlled generator are correctly shown. Note that $v_o = -v_{ce}$ and $i_i = i_b$. The input loop and emitter nodal equations are

$$1500 i_i + (1 - 10^{-4}) v_o = v_i \tag{7.96}$$

$$-81 i_i + 450 \times 10^{-6} v_o = i_o \tag{7.97}$$

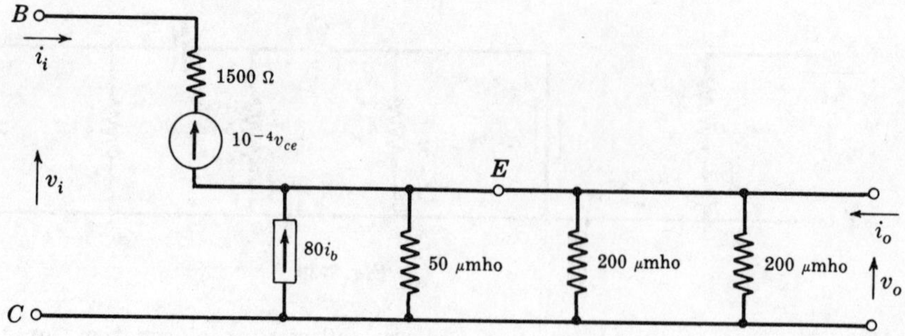

Fig. 7-41

Here the solution is rather simple. It was shown in Chapter 4 that the functional relationship for the input and output ports is written as above so that the h parameters are the slopes of the input, output, forward transfer, and reverse transfer characteristic curves.

To find h_i, set $v_o = 0$ in (*7.96*); then $h_i = v_i/i_i = 1500\ \Omega$. To find h_r, put $i_i = 0$ in (*7.96*); then $h_r = v_i/v_o = 1 - 10^{-4} \cong 1.0$. To find h_f, let $v_o = 0$ in (*7.97*); then $h_f = i_o/i_i = -81$. To find h_o, set $i_i = 0$ in (*7.97*); then $h_o = i_o/v_o = 450 \times 10^{-6}$ mho. The equivalent circuit is now drawn as in Fig. 7-42.

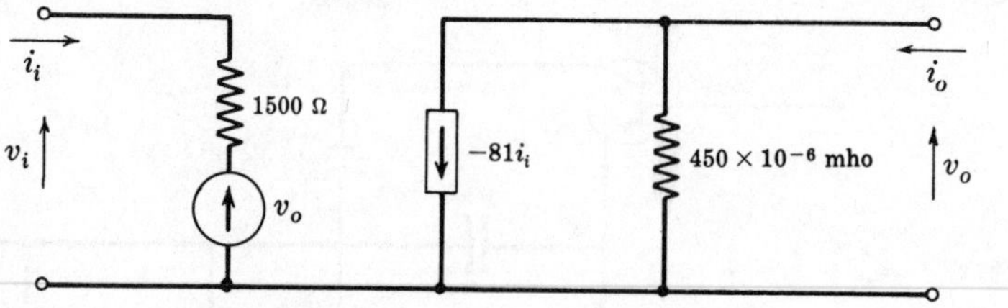

Fig. 7-42

(*b*) Add the input generator to the equivalent circuit of Fig. 7-42, as in Fig. 7-43.

The input loop and non-reference output nodal equations are

$$2100 i_i + v_o = 0.01 \sin \omega t, \qquad -81 i_i + 450 \times 10^{-6}\, v_o = i_o$$

from which $v_o = (2100 i_o + 0.81 \sin \omega t)/81.945$. To find the open circuit terminal voltage (generator voltage), set $i_o = 0$ and obtain v_o (open circuit) $= 0.00988 \sin \omega t$ volts. To find the output resistance, set the external input generator voltage equal to zero and obtain $r_o = v_o/i_o = 2100/81.945 = 25.6\ \Omega$. Note that r_o is quite different from the reciprocal of h_o which is 2220 Ω.

(*c*) Add the terminating 2000 Ω resistor across the output terminals of the circuit of Fig. 7-42, as shown in Fig. 7-44 below.

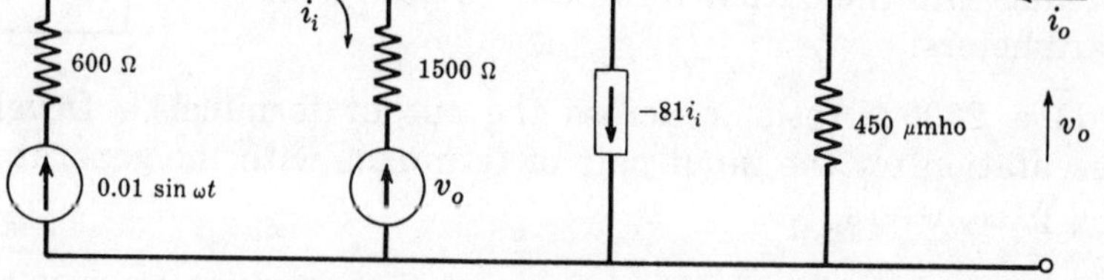

Fig. 7-43

The input loop and output nodal equations are respectively

$$-v_i + 1500 i_i + v_o = 0, \qquad -81 i_i + 450 \times 10^{-6}\, v_o + 500 \times 10^{-6}\, v_o = 0$$

from which $R_i = v_i/i_i = 86{,}800\ \Omega$. Since there is no generator external to the output terminals, the input is represented as a passive element: the input resistance R_i.

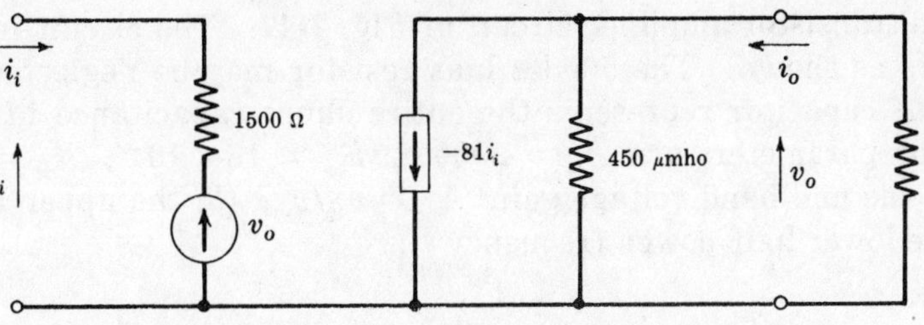

Fig. 7-44

7.6. Given the common grid amplifier circuit of Fig. 7-45. The tube parameters are $\mu = 30$, $r_p = 8000\ \Omega$. Assume the input terminal voltage v_{s1} has constant magnitude. Find (*a*) the mid-band voltage gain $A_{vo} = v_{s2}/v_{s1}$ and (*b*) the lower half-power frequency f_1.

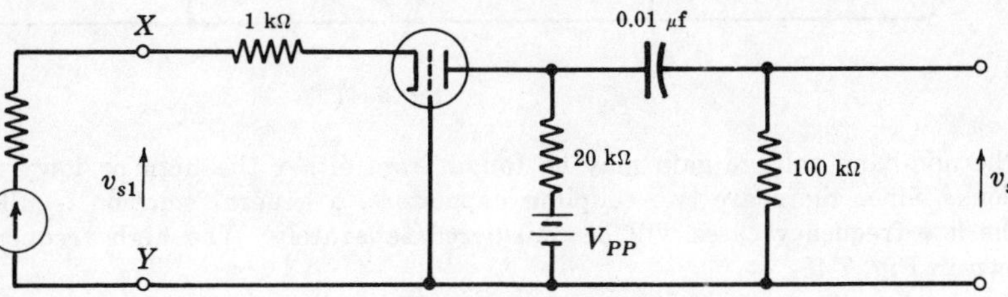

Fig. 7-45

In this problem the solution to part (*a*) may be found from the solution to part (*b*). The low frequency equivalent circuit is drawn in Fig. 7-46.

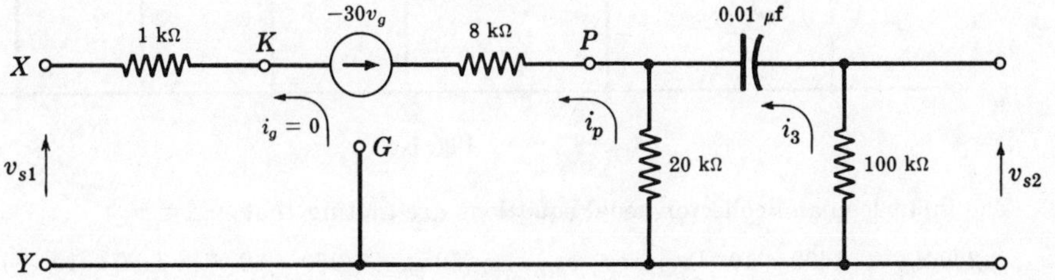

Fig. 7-46

The grid-cathode loop equation, the plate-cathode loop equation and the equation for the current loop i_3 are respectively

$$+v_g + 1000i_p + v_{s1} = 0$$

$$20{,}000(i_p - i_3) + 8000i_p - 30v_g + 1000i_p + v_{s1} = 0$$

$$100{,}000i_3 - ji_3/\omega(0.01 \times 10^{-6}) + 20{,}000(i_3 - i_p) = 0$$

from which $$i_3 = \frac{-620v_{s1}}{6.68 \times 10^6 - j59/2\pi f(0.01 \times 10^{-6})}$$

and, since $v_{s2} = -100{,}000i_3$,

$$A_v(\text{l.f.}) = v_{s2}/v_{s1} = 62/(6.68 - j59/0.02\pi f) \tag{7.98}$$

(*a*) A_{vo} is found by letting f become large in (*7.98*): $A_{vo} = 62/6.68 = 9.28$.

(*b*) Since the numerator of (*7.98*) is real and independent of frequency and the real part of the denominator is independent of frequency, the lower half-power frequency is found by setting the real part of the denominator equal to the imaginary part: $6.68 = 59/0.02\pi f_1$ or $f_1 = 140.8$ cps.

7.7. Given the transistor amplifier circuit of Fig. 7-47. The circuit includes both coupling capacitors as shown. The 500 kΩ bias resistor may be neglected in the calculations. The 50 $\mu\mu$f capacitor represents the entire shunt capacitance in the collector circuit. Transistor parameters are $h_{ie} = 2000\ \Omega$, $h_{re} = 16 \times 10^{-4}$, $h_{oe} = 50\ \mu$mho, $h_{fe} = 50$. Find (*a*) the mid-band voltage gain $A_{vo} = v_{s2}/v_{s1}$, (*b*) the upper half-power frequency f_2, (*c*) the lower half-power frequency f_1.

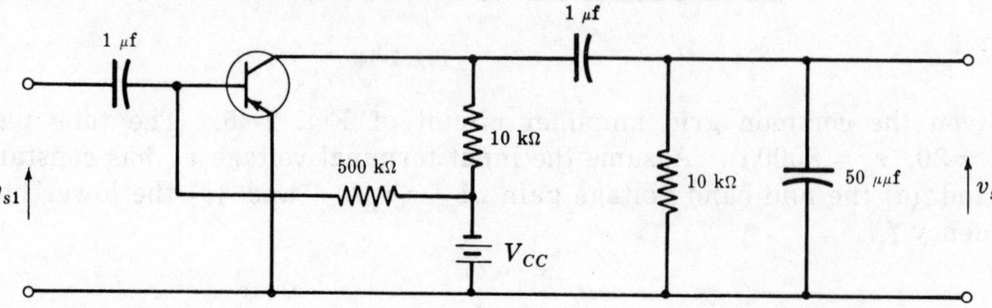

Fig. 7-47

The mid-band voltage gain may be found from either the high or low frequency voltage gain equations. Since there are two coupling capacitors, a general solution is difficult; hence the high and the low frequency cases will be considered separately. The high frequency equivalent circuit is shown in Fig. 7-48.

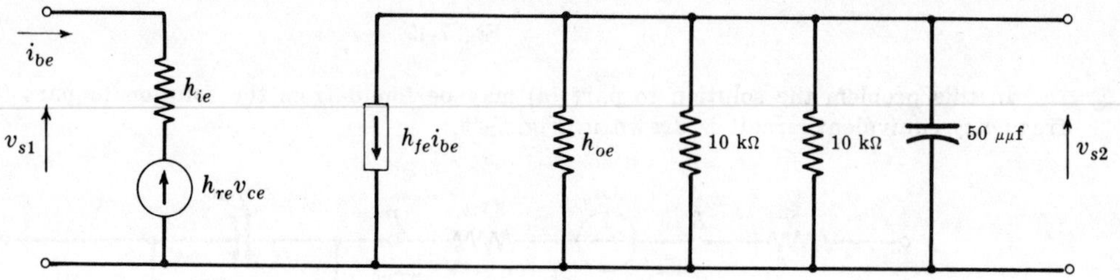

Fig. 7-48

The input loop and collector nodal equations are (noting that $v_{ce} = v_{s2}$)

$$2000 i_{be} + 1600 \times 10^{-6} v_{s2} = v_{s1}, \qquad 50 i_{be} + (250 \times 10^{-6} + j\omega 50 \times 10^{-12}) v_{s2} = 0$$

from which

$$v_{s2} = \frac{-50 v_{s1}}{0.42 + j(0.2\pi f \times 10^{-6})} \tag{7.99}$$

(*a*) A_{vo} is found from (*7.99*) by letting f approach zero: $A_{vo} = v_{s2}/v_{s1} = -50/0.42 = -119$.

(*b*) The upper half-power frequency is found by setting the real part of the denominator of (*7.99*) equal to the imaginary part: $0.2\pi f_2 \times 10^{-6} = 0.42$ or $f_2 = 668{,}000$ cps.

(*c*) The low frequency equivalent circuit is shown in Fig. 7-49.

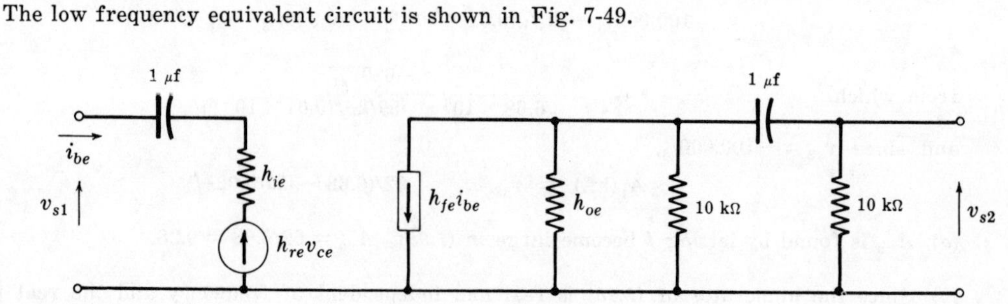

Fig. 7-49

The equations are simplified if the impedances and admittances of the capacitors are written as Z_C and Y_C; both capacitors are the same in this problem. The input loop, collector nodal, and output nodal equations are

$$-v_{s1} + Z_C i_{be} + 2000 i_{be} + 1600 \times 10^{-6} v_{ce} = 0$$

$$50 i_{be} + (150 \times 10^{-6}) v_{ce} + Y_C(v_{ce} - v_{s2}) = 0$$

$$Y_C(v_{s2} - v_{ce}) + 100 \times 10^{-6} v_{s2} = 0$$

from which

$$A_v(\text{l.f.}) = v_{s2}/v_{s1} = \frac{-50}{0.42 + (272 \times 10^{-6})/Y_C + (150 \times 10^{-10})/Y_C^2}$$

Letting $Y_C = j\omega(1 \times 10^{-6})$, the denominator is $D = 0.42 - 380/f^2 - j43.3/f$. At the lower half-power frequency, the magnitude of the denominator is $\sqrt{2}$ times the mid-band value of the denominator: $D(\text{mid-band}) = \sqrt{2}(0.42) = 0.594$. Setting $|D| = 0.594$, the equation may be simplified to

$$0.176 f_1^4 - 1556 f_1^2 - 144{,}400 = 0$$

from which $f_1^2 = 4420 \pm 4510$. The real solution is $f_1 = \sqrt{8930} = 94.6$ cps.

7.8. Given the common emitter transistor amplifier equivalent circuit of Fig. 7-50. The entire shunt capacitance in the collector circuit is C_S. Assume h_{fe} is frequency dependent and can be approximated as

$$h_{fe}(f) = \frac{h_{fe}}{1 + jf/f_{\beta e}} \qquad (7.100)$$

Investigate the problem of deriving the equation for the upper half-power frequency f_2 for the amplifier circuit. The voltage gain is defined as $A_v = v_{s2}/v_{gen}$.

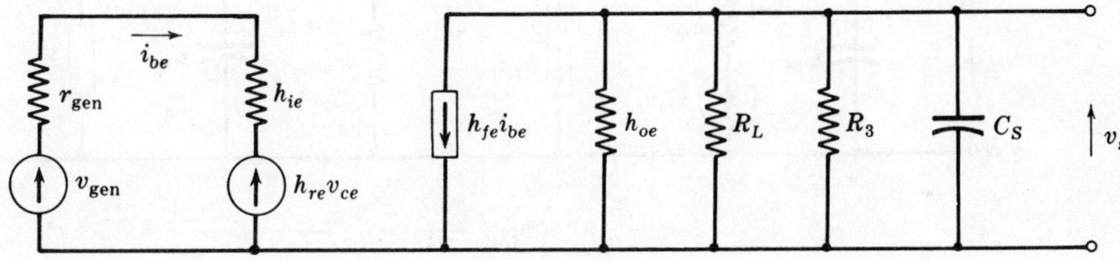

Fig. 7-50

The base loop and collector nodal equations are, using $v_{ce} = v_{s2}$,

$$-v_{gen} + r_{gen} i_{be} + h_{ie} i_{be} + h_{re} v_{s2}, \qquad h_{fe}(f)\, i_{be} + (h_{oe} + G_L + G_3 + j\omega C_S) v_{s2} = 0$$

from which the high frequency voltage gain is

$$A_v(\text{h.f.}) = v_{s2}/v_{gen} = \frac{-h_{fe}(f)}{(r_{gen} + h_{ie})(h_{oe} + G_L + G_3 + j\omega C_S) - h_{re} h_{fe}(f)} \qquad (7.101)$$

Substituting (7.100) into (7.101), we obtain

$$A_v(\text{h.f.}) = \frac{-h_{fe}}{(r_{gen} + h_{ie})(h_{oe} + G_L + G_3 - 2\pi C_S f^2/f_{\beta e}) - h_{fe}h_{re} + jf[(r_{gen} + h_{ie})(h_{oe} + G_L + G_3 + 2\pi C_S f_{\beta e})]/f_{\beta e}} \qquad (7.102)$$

whose numerator is real and independent of frequency. The real part of the denominator is a function of f^2, so the upper half-power frequency is found by setting the magnitude of the denominator equal to $\sqrt{2}$ times the mid-band value. The mid-band value is found by letting f approach zero, giving

$$D(\text{mid-band}) = (r_{gen} + h_{ie})(h_{oe} + G_L + G_3) - h_{fe}h_{re} \qquad (7.103)$$

Equations (7.102) and (7.103) show that it is impossible to derive a general equation for the half-power frequency. Solutions may be worked out for numerical problems as shown in Problem 7.9.

7.9. Given the circuit of Problem 7.8 and the following values for the circuit parameters:

$$r_{gen} = 600\ \Omega \qquad C_S = 100\ \mu\mu f \qquad h_{fe} = 70$$
$$R_L = 5\ k\Omega \qquad f_{\beta e} = 300{,}000 \text{ cps} \qquad h_{oe} = 50\ \mu\text{mho}$$
$$R_3 = 5\ k\Omega \qquad h_{ie} = 1400\ \Omega \qquad h_{re} = 10^{-3}$$

Find the upper half-power frequency for the amplifier of Problem 7.8.

Substituting the given numerical values, (*7.102*) becomes

$$A_v(\text{h.f.}) = \frac{-70}{0.83 - 4.19 \times 10^{-12} f^2 + jf(4.26 \times 10^{-6})} \qquad (7.104)$$

Let $f \to 0$; then the mid-band denominator is 0.83, and $\sqrt{2}(0.83) = 1.175$. Setting the magnitude of the denominator of (*7.104*) equal to 1.175 and squaring both sides of the equation, we obtain

$$17.6 \times 10^{-24} f_2^4 + 11.20 \times 10^{-12} f_2^2 + 0.689 = 1.38$$

or $f_2^4 + 0.637 \times 10^{12} f_2^2 - 0.0393 \times 10^{24} = 0$ from which $f_2^2 = 0.057 \times 10^{12}$ and $f_2 = 238{,}000$ cps.

7.10. Given the high frequency equivalent transistor amplifier circuit of Fig. 7-51. The capacitors shown between the transistor terminals represent the entire capacitance, including that external to the transistor. Derive the general equation for the input admittance $y_i = i_{s1}/v_{s1}$. Under what conditions is the real part of the input admittance negative?

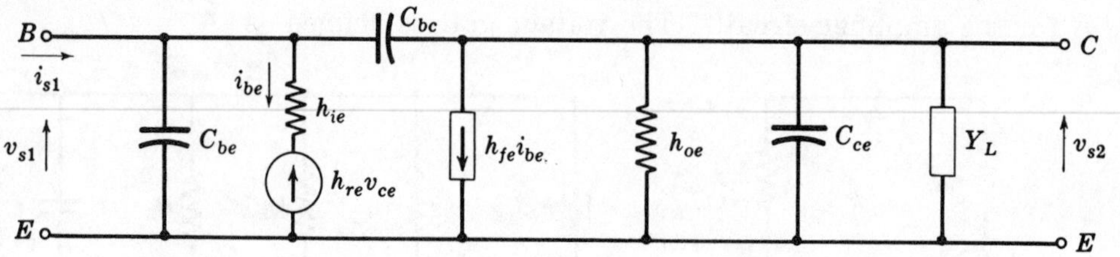

Fig. 7-51

Since there are several branches in the input loop, the base circuit of the transistor is redrawn as in Fig. 7-52.

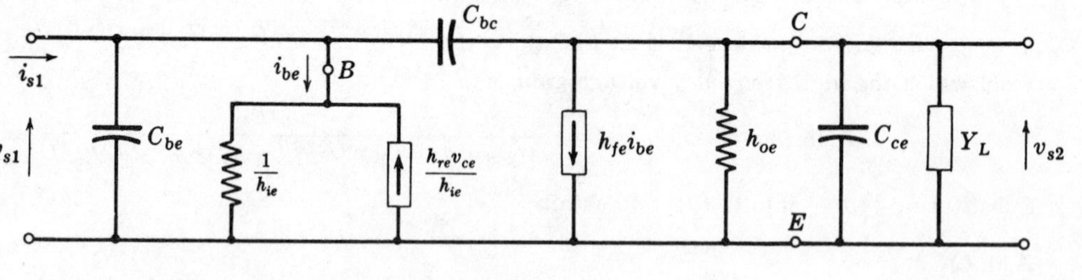

Fig. 7-52

Using $v_{ce} = v_{s2}$, the base nodal equation is

$$-i_{s1} + Y_{be}v_{s1} + Y_{bc}(v_{s1} - v_{s2}) + (1/h_{ie})v_{s1} - (h_{re}/h_{ie})v_{s2} = 0 \qquad (7.105)$$

The collector nodal equation is

$$Y_{bc}(v_{s2} - v_{s1}) + h_{fe}i_{be} + (h_{oe} + Y_{ce} + Y_L)v_{s2} = 0 \qquad (7.106)$$

where $Y_{be} = j\omega C_{be}$, $Y_{bc} = j\omega C_{bc}$, and $Y_{ce} = j\omega C_{ce}$. To eliminate the unknown i_{be}, write a nodal equation for the base terminal:

$$i_{be} = (1/h_{ie})v_{s1} - (h_{re}/h_{ie})v_{s2} \tag{7.107}$$

Substituting (*7.107*) into (*7.106*) and rewriting (*7.105*) and (*7.106*),

$$(Y_{be} + Y_{bc} + 1/h_{ie})v_{s1} - (Y_{bc} + h_{re}/h_{ie})v_{s2} = i_{s1}$$

$$(h_{fe}/h_{ie} - Y_{bc})v_{s1} + (h_{oe} + Y_{bc} + Y_{ce} + Y_L - h_{fe}h_{re}/h_{ie})v_{s2} = 0$$

Solving for v_{s1} as a function of i_{s1}, the input admittance is

$$y_i = 1/h_{ie} + j\omega(C_{be} + C_{bc}) + \frac{(h_{fe}h_{re}/h_{ie}^2 + \omega^2 C_{bc}^2) + j\omega C_{bc}(h_{fe} - h_{re})/h_{ie}}{(h_{oe} - h_{fe}h_{re}/h_{ie}) + j\omega(C_{bc} + C_{ce}) + Y_L} \tag{7.108}$$

It was shown in Section 7.8 that the input admittance may have a negative real part if the collector load is inductive. Consider the case where $Y_L = G_L - jB_L$. Substituting this into (*7.108*), the denominator of the last term becomes

$$[h_{oe} + G_L - h_{fe}h_{re}/h_{ie}] + j[\omega(C_{bc} + C_{ce}) - B_L]$$

To simplify the algebra, let $G^* = h_{oe} + G_L - h_{fe}h_{re}/h_{ie}$ and $B^* = B_L - \omega(C_{bc} + C_{ce})$. Approximations will be made to simplify (*7.108*). Since $h_{fe} \gg h_{re}$, then $h_{fe} - h_{re} \cong h_{fe}$. Since the interest is only in the real part of the input admittance,

$$\text{Re}(y_i) = \frac{1}{h_{ie}} + \frac{G^*(h_{fe}h_{re}/h_{ie}^2 + \omega^2 C_{bc}^2)}{G^{*2} + B^{*2}} - \frac{\omega C_{bc}B^* h_{fe}}{h_{ie}(G^{*2} + B^{*2})}$$

When $\text{Re}(y_i) < 0$, then

$$\frac{\omega C_{bc}B^* h_{fe}}{h_{ie}(G^{*2} + B^{*2})} > \frac{1}{h_{ie}} + G^*\left(\frac{h_{fe}h_{re}/h_{ie}^2 + \omega^2 C_{bc}^2}{G^{*2} + B^{*2}}\right)$$

which may be written as

$$\omega C_{bc}B^* h_{fe} > G^*(G^* + h_{fe}h_{re}/h_{ie} + \omega^2 h_{ie}C_{bc}^2) + B^{*2}$$

For the inequality to hold, the input admittance of the transistor amplifier must have a negative real part. Hence there is a negative conductance (or resistance) looking into the input of the amplifier.

7.11. Given the grounded grid amplifier circuit of Fig. 7-53, find the mid-band output impedance of the amplifier looking into the terminals *A-B*. The magnitude of v_{gen} is constant. Tube parameters are $\mu = 20$, $r_p = 10{,}000\ \Omega$.

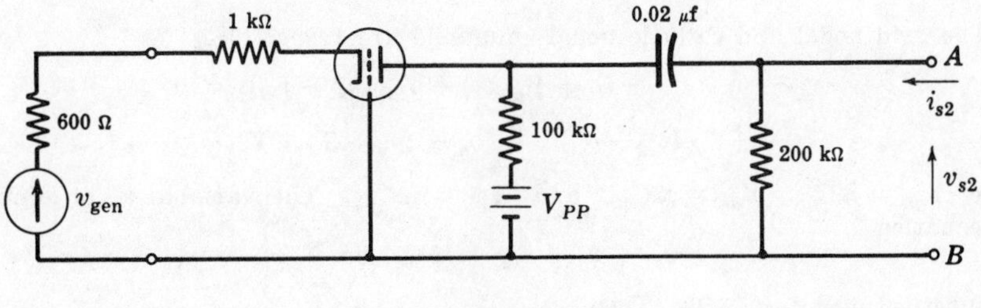

Fig. 7-53

The output impedance is often found by letting the external input generator voltage be zero and then solving for $r_o = v_{s2}/i_{s2}$. Since the magnitude of v_{gen} is constant, the equivalent circuit is drawn as shown in Fig. 7-54.

Combining the resistors in the plate circuit, $R_L^* = 66{,}700\ \Omega$.

The plate-cathode, grid-cathode and output loop equations are respectively

$$66{,}700(i_p - i_{s2}) + 10{,}000 i_p - 20 v_g + 1600 i_p = 0, \qquad v_g + 1600 i_p = 0, \qquad -v_{s2} + 66{,}700(i_{s2} - i_p) = 0$$

from which the output impedance $r_o = v_{s2}/i_{s2} = 26{,}360\ \Omega$.

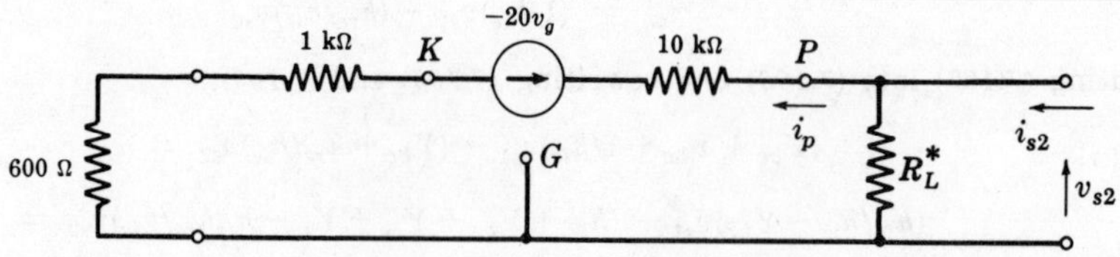

Fig. 7-54

In some cases the terminal voltage of the input generator is kept constant. Then the external generator effectively has zero internal resistance ($r_{gen} = 0$). The output impedance is then different. The 600 Ω resistor would be shorted in the circuit of Fig. 7-54. The plate-cathode, grid-cathode and output loop equations are now respectively

$$77{,}700 i_p - 66{,}700 i_{s2} - 20 v_g = 0, \qquad 1000 i_p + v_g = 0, \qquad 66{,}700(i_{s2} - i_p) = v_{s2}$$

from which the output impedance is now $r_o = v_{s2}/i_{s2} = 21{,}200\ \Omega$.

Comparing the two results, the output impedance changes significantly if the input generator terminal voltage is kept constant or if the generated voltage is constant in this problem.

7.12. The high frequency equivalent circuit for a cathode follower is given in Fig. 7-55. The conductor G_K is assumed to include all the conductance in the cathode circuit. The capacitances shown in the circuit diagram include all the capacitance between the terminal pairs. The cathode follower is to be represented for high frequencies using a two port equivalent circuit containing voltage controlled, current generators as in the circuit of Fig. 4-9, page 57. Derive the equations for the two port parameters in terms of the parameters of the circuit of Fig. 7-55.

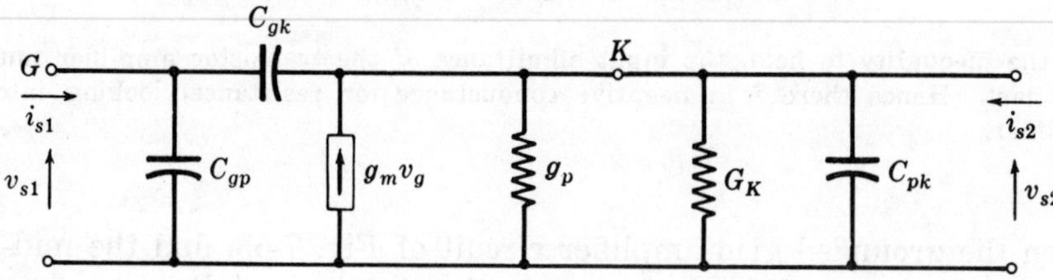

Fig. 7-55

The grid nodal and cathode nodal equations are respectively

$$-i_{s1} + Y_{gp} v_{s1} + Y_{gk}(v_{s1} - v_{s2}) = 0 \tag{7.109}$$

$$+Y_{gk}(v_{s2} - v_{s1}) - g_m v_g + (g_p + G_K + Y_{pk}) v_{s2} - i_{s2} = 0 \tag{7.110}$$

where $Y_{gp} = j\omega C_{gp}$, $Y_{gk} = j\omega C_{gk}$, and $Y_{pk} = j\omega C_{pk}$. The variable v_g is eliminated by writing a loop equation

$$-v_{s1} + v_g + v_{s2} = 0 \qquad \text{or} \qquad v_g = v_{s1} - v_{s2} \tag{7.111}$$

and substituting into (*7.110*). Then

$$(Y_{gp} + Y_{gk}) v_{s1} - Y_{gk} v_{s2} = i_{s1} \tag{7.112}$$

$$-(g_m + Y_{gk}) v_{s1} + (g_p + G_k + g_m + Y_{pk} + Y_{gk}) v_{s2} = i_{s2} \tag{7.113}$$

Equations (*7.112*) and (*7.113*) are in the desired form and may be compared with equations (*4.17*) and (*4.16*), page 57, respectively. In (*7.112*), the coefficient of v_{s1} is y_i and the coefficient of v_{s2} is y_r:

$$y_i = j\omega(C_{gp} + C_{gk}), \qquad y_r = -j\omega C_{gk}$$

In (*7.113*), the coefficient of v_{s1} is y_f and that of v_{s2} is y_o:

$$y_f = -(g_m + j\omega C_{gk}), \qquad y_o = g_p + G_K + g_m + j\omega(C_{pk} + C_{gk})$$

The two port parameters are complex and functions of frequency outside the mid-band range.

7.13. Given an amplifier consisting of two cascaded stages which are effectively isolated from each other. The lower half-power frequency of the first stage is 60 cps and that of the second stage is 20 cps. Sketch an approximate low frequency response curve for the amplifier and find the relative output at a frequency of 5 cps.

It is assumed the gain ratio for the individual stages may be written

$$A_v(\text{l.f.})/A_{vo} = 1/(1 - jf_1/f)$$

The low frequency response is most conveniently expressed in decibels and plotted on semilog graph paper. The low frequency asymptote for the first stage may be drawn with a 6 db per octave slope intersecting the zero db line at 60 cps as shown in Fig. 7-56. For the second stage, the asymptote intersects the zero db line at 20 cps. The two asymptotes are combined by drawing a line with slope 12 db per octave through the intersection point of the first stage asymptote and the $f = 20$ cps line as shown in Fig. 7-56. This approximation to the true response curve is very good at low frequencies and is the poorest approximation in the regions near the half-power frequencies. The approximation is improved if a correcting curve is sketched. At the break frequencies, the true response is 3 db below the asymptote. At one octave above and below the break frequencies, the true response curves are one db below the asymptotic approximation. Points may be located as in Fig. 7-56 and a smooth curve sketched through the points. The overall attenuation at any frequency is the sum of the attenuations of the first and second stages when the attenuation is expressed in decibels.

The relative response at $f = 5$ cps is found by using the response curve and is -33.5 db.

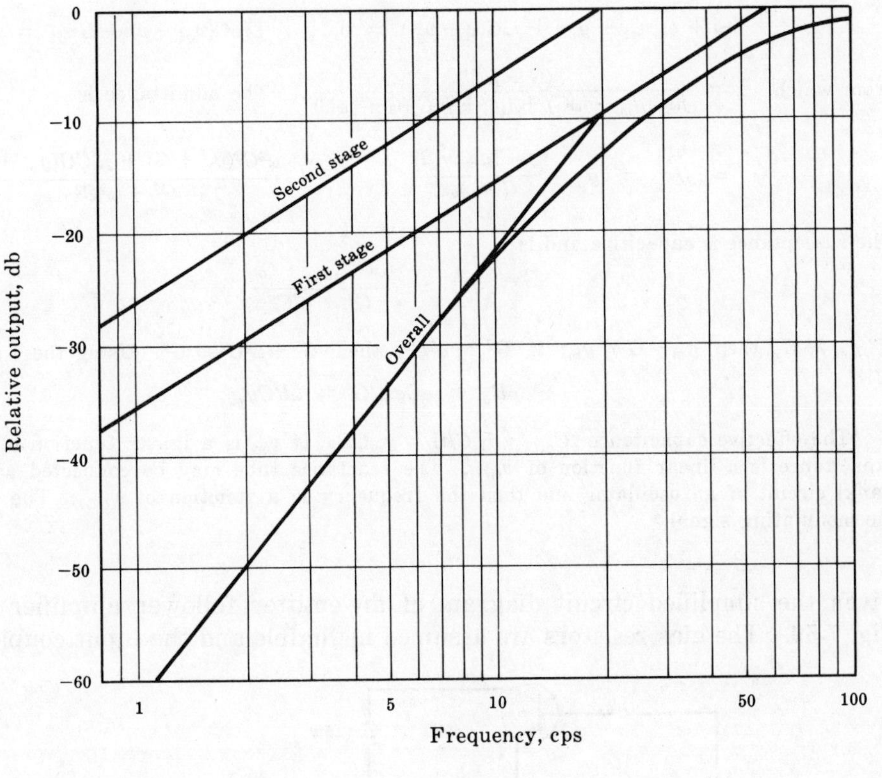

Fig. 7-56. Amplifier Low Frequency Response

7.14. Given the "reactance tube" circuit of Fig. 7-57. The circuit at the terminals A and B has the characteristics of a capacitor. Over a limited range of operation, the magnitude of the capacitance is approximately proportional to instantaneous voltage v_{GG1}. A pentode tube is usually used because pentode transfer characteristics have the desired features. The screen grid voltage is assumed constant. The inductor rfc (radio frequency choke) may be assumed to have infinite impedance. The equation for the capacitive susceptance looking into the terminals A and B is often given as $B_c \cong \omega g_m RC$. Derive this equation, showing the approximations involved.

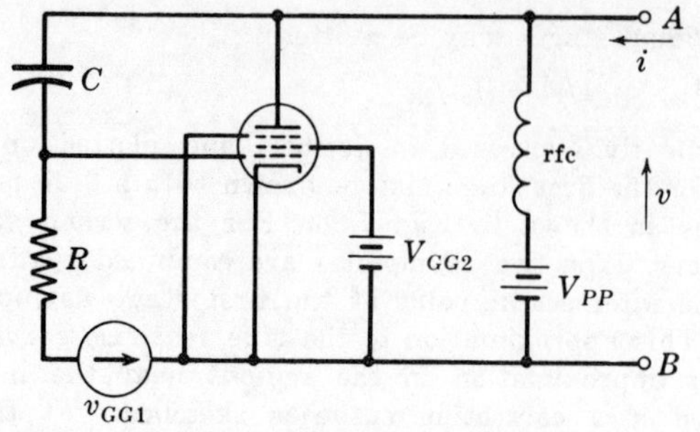

Fig. 7-57

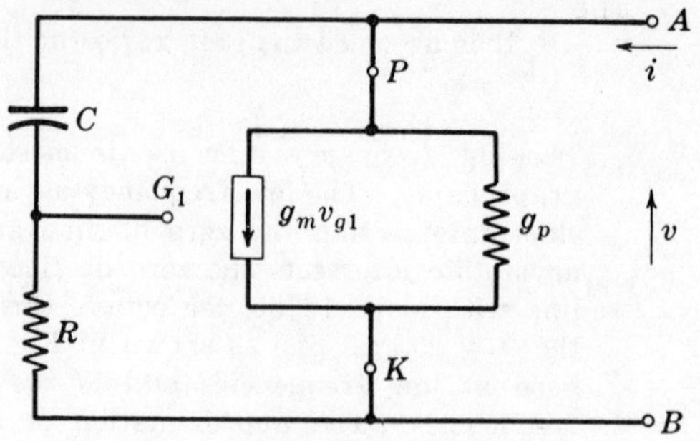

Fig. 7-58

The equivalent circuit for the reactance tube is given in Fig. 7-58. Since the suppressor grid and screen grid voltages are constant, the equivalent circuit for the pentode reduces to that in Fig. 7-58. As v_{GG1} changes, the value of g_m changes. The circuit of Fig. 7-58 (which is for a particular value of v_{GG1}) appears as a fixed capacitor. The equivalent circuit applies only when g_m is constant.

Noting that $v = v_p$, the plate nodal and grid nodal equations are

$$-i + g_m v_{g1} + g_p v + j\omega C(v - v_{g1}) = 0, \qquad +j\omega C(v_{g1} - v) + G v_{g1} = 0$$

from which $v = \dfrac{(G + j\omega C)i}{j\omega C(g_m - j\omega C) + (g_p + j\omega C)(G + j\omega C)}$. The admittance is

$$y = i/v = g_p + \frac{j\omega C(g_m + G)}{G + j\omega C} = g_p + \frac{\omega^2 C^2(g_m + G) + j\omega CG(g_m + G)}{G^2 + \omega^2 C^2}$$

The susceptance is capacitive and is

$$B_c = \frac{\omega CG(g_m + G)}{G^2 + \omega^2 C^2}$$

If $g_m \gg G$, then $g_m + G \cong g_m$; if $G^2 \gg \omega^2 C^2$, then $G^2 + \omega^2 C^2 \cong G^2$. Using these approximations,

$$B_c \cong \omega C g_m / G = \omega RC g_m$$

The effective capacitance $C_{\text{eff}} \cong g_m C/G = g_m CR$. If g_m is a linear function of v_{GG1}, then the capacitance is a linear function of v_{GG1}. The reactance tube may be connected across the tuned (tank) circuit of an oscillator and then the frequency is a function of v_{GG1}. The voltage v_{GG1} is the modulating signal.

7.15. Given the simplified circuit diagram of an emitter follower amplifier as shown in Fig. 7-59. The bias resistors are assumed negligible and the input coupling capacitor

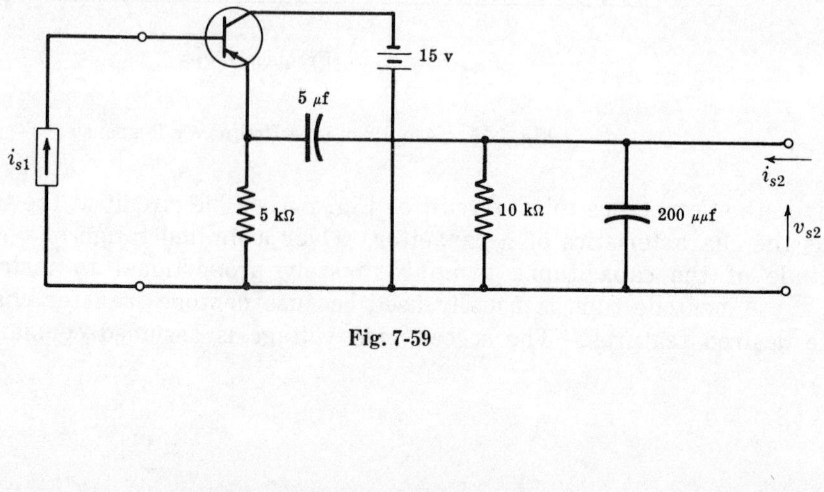

Fig. 7-59

may be neglected because the input generator is a constant current generator. The open-circuit transfer function is the transfer resistance $r_t = v_{s2}/i_{s1}$. Find the mid-band transfer resistance and the upper and lower half-power frequencies. The half-power frequencies are the frequencies at which the transfer resistance is $1/\sqrt{2}$ times the mid-band value. Transistor parameters are $h_{ie} = 2000\ \Omega$, $h_{re} = 10^{-3}$, $h_{oe} = 50\ \mu$mho, $h_{fe} = 100$.

The low frequency equivalent circuit is shown in Fig. 7-60. Note that $i_{s1} = i_b$. The emitter nodal equation is

$$-i_{s1} - 100i_{s1} + 250 \times 10^{-6}\, v_{ec} + Y_C(v_{ec} - v_{s2}) = 0$$

where $Y_C = j\omega(5 \times 10^{-6})$; and the output nodal equation is (with $i_{s2} = 0$)

$$Y_C(v_{s2} - v_{ec}) + 100 \times 10^{-6}\, v_{s2} = 0$$

Then $v_{s2} = \dfrac{101 Y_C i_{s1}}{0.025 \times 10^{-6} + 350 \times 10^{-6}\, Y_C}$ and the low frequency transfer resistance is

$$r_t(\text{l.f.}) = v_{s2}/i_{s1} = \frac{101}{350 \times 10^{-6} - j(0.025 \times 10^{-6})/(2\pi f\, 5 \times 10^{-6})} \qquad (7.114)$$

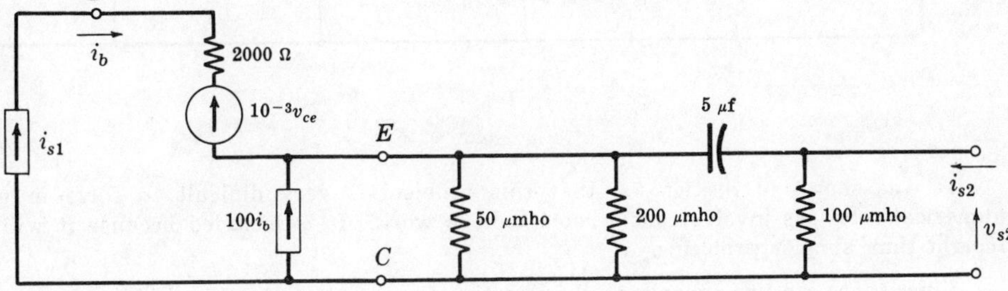

Fig. 7-60

The mid-band transfer resistance is found by letting f become large, giving

$$r_t = 101/(350 \times 10^{-6}) = 289{,}000\ \Omega$$

The lower half-power frequency is found from (7.114) by setting the real part of the denominator equal to the imaginary part:

$$(2\pi\, 5 \times 10^{-6}) f_1 = 0.025/350 \qquad \text{or} \qquad f_1 = 2.27 \text{ cps}$$

Since the input generator is a constant current generator, the base circuit has no effect on the transfer resistance. The high frequency equivalent circuit is drawn as shown in Fig. 7-61. The base circuit need not be included since the interest is in the transfer resistance.

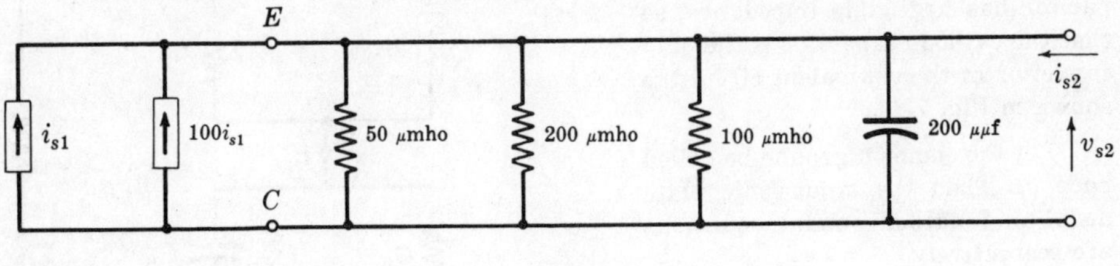

Fig. 7-61

The emitter nodal equation is

$$-i_{s1} - 100i_{s1} + (350 \times 10^{-6} + j\omega C)v_{s2} = 0$$

The high frequency transfer resistance is

$$r_t(\text{h.f.}) = v_{s2}/i_{s1} = \frac{101}{350 \times 10^{-6} + j\omega 200 \times 10^{-12}}$$

The mid-band transfer resistance is found from this equation by letting $f \to 0$; this is the same value as found from the equation for the low frequency transfer resistance. For the upper half-power frequency,

$$(2\pi\, 200 \times 10^{-12}) f_2 \;=\; 350 \times 10^{-6} \qquad \text{or} \qquad f_2 \;=\; 278{,}000 \text{ cps}$$

7.16. Given the low frequency equivalent circuit of Fig. 7-62. This is a common cathode amplifier in which the cathode capacitor is included. Derive the equation for the low frequency open circuit voltage gain. Then find the lower half-power frequency. Tube parameters are $g_m = 2500\ \mu$mho, $r_p = 7000\ \Omega$.

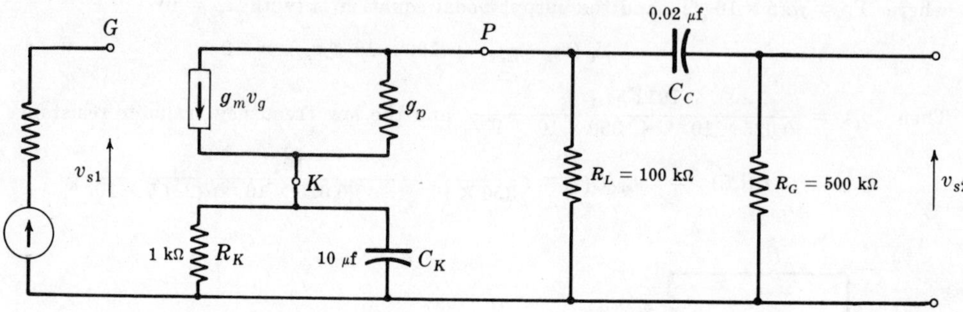

Fig. 7-62

It was shown in Section 7.3 that this problem is very difficult to solve in general. Even numerical solutions involve considerable tedious work. It is included because it will provide more insight than simpler problems.

Since there are two capacitors, the equation for the open circuit voltage gain cannot be reduced to a rational fraction in which the numerator is real and independent of frequency and in which the real part of the denominator is independent of frequency. The real part of the denominator will be a function of f^2 (the transfer function has two poles).

One approach is to approximate the low frequency response curve by two asymptotes. It is anticipated that there are two break frequencies*. One of these may be found by assuming that the cathode is effectively bypassed. The equation for this break frequency (the half-power frequency for the equivalent circuit with the cathode grounded) may be found by substituting into *(7.17)*, page 173, giving

$$f_1 \;=\; \frac{G_G(g_p + G_L)}{2\pi C_C(G_G + g_p + G_L)} \;=\; \frac{(2 \times 10^{-6})(143 + 10)10^{-6}}{2\pi(0.02 \times 10^{-6})(2 + 143 + 10)10^{-6}} \;=\; 15.7 \text{ cps}$$

The other break frequency is obtained by assuming the coupling capacitor has negligible impedance, so that the cathode capacitor is the only capacitor in the equivalent circuit as shown in Fig. 7-63.

Let the common ground be called node *o*. Then the input loop, plate nodal and cathode nodal equations are respectively

$$-v_{s1} + v_{gk} + v_{ko} \;=\; 0$$

$$+g_m v_{gk} + g_p v_{pk} + (G_L + G_G) v_{s2} \;=\; 0$$

$$-g_m v_{gk} - g_p v_{pk} + (G_K + j\omega C_K) v_{ko} = 0$$

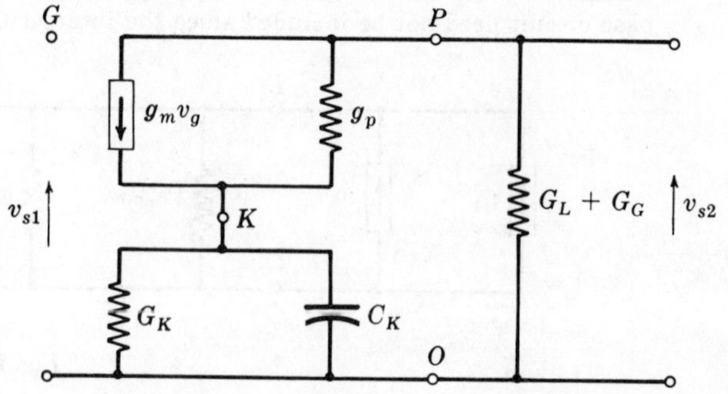

Fig. 7-63

*The break frequencies are not necessarily half-power frequencies unless there is only one pole. If there are two widely separated break frequencies below the mid-band range, then the higher is approximately the lower half-power frequency.

One unknown voltage is eliminated by substituting $v_{pk} = v_{s2} - v_{ko}$, and then the equations are solved for

$$v_{s2} = \frac{-g_m(G_K + j\omega C_K)v_{s1}}{(G_K + j\omega C_K)(G_L + G_G + g_p) + (g_m + g_p)(G_L + G_G)}$$

which may be simplified by dividing numerator and denominator by $(G_K + j\omega C_K)(G_L + G_G + g_p)$. The equation for the low frequency voltage gain becomes

$$A_v(\text{l.f.}) = \frac{-g_m/(G_L + G_G + g_p)}{1 + (g_m + g_p)(G_L + G_G)/[(G_K + j\omega C_K)(G_L + G_G + g_p)]} \qquad (7.115)$$

The numerator is the mid-band gain A_{vo}, so both sides of (*7.115*) may be divided by A_{vo}. To find a general equation for the break frequency (half-power frequency for the circuit of Fig. 7-63), the right term of the denominator must be rationalized. Then the frequency is found at which the magnitude of the denominator is equal to $\sqrt{2}$ (the mid-band value is 1). A general solution becomes very tedious. Substituting numerical values, the denominator of (*7.115*) is

$$D = 1 + \frac{(2643 \times 10^{-6})(12 \times 10^{-6})}{[1000 \times 10^{-6} + j2\pi f(10 \times 10^{-6})](155 \times 10^{-6})}$$

or

$$D = 1 + \frac{3.25}{15.86 + jf} \qquad (7.116)$$

Rationalizing, $D = \dfrac{f^2 + 303.5 - j3.25f}{f^2 + 252}$. Letting $|D| = \sqrt{2}$, we obtain $f_1^4 + 390f_1^2 + 35{,}000 = 0$ from which $f_1^2 = -195 \pm 54.8$ whose solutions are imaginary numbers. This might suggest that there is a mistake in the calculations, but further checking shows that the equations are correct within slide rule accuracy.

Inspection of Fig. 7-63 shows that even as the frequency approaches zero, the magnitude of the gain does not go to zero but has a value somewhat lower than the mid-band value. This lower value may be found by letting f approach zero in (*7.116*); then the denominator becomes $D = 1 + 3.25/15.86 = 1.205$. Hence at $f = 0$, the gain ratio is $A_v(\text{l.f.})/A_{vo} = 1/1.205 = 0.829$. Thus the gain ratio for the circuit of Fig. 7-63 does not go below 0.829 even as the frequency approaches zero. This is why there is no real solution for the half-power frequency. Hence it is not practical to seek an approximate solution by using asymptotes to the true frequency response curve unless the effect of the cathode circuit is neglected.

The equations for the low frequency circuit of Fig. 7-62 were given in Section 7.3 and may be solved for v_{s2} as a function of v_{s1}. Then the equation for the low frequency gain may be found and this equation normalized, giving the gain ratio

$$\frac{A_v(\text{l.f.})}{A_{vo}} = \frac{1}{1 + \dfrac{G_G(g_p + G_L)}{Y_C(g_p + G_L + G_G)} + \dfrac{(g_p + g_m)(G_L + G_G)}{(G_K + Y_K)(g_p + G_L + G_G)} + \dfrac{G_G G_L(g_p + g_m)}{Y_C(G_K + Y_K)(g_p + G_L + G_G)}} \qquad (7.117)$$

where $Y_C = j\omega C_C$ and $Y_K = j\omega C_K$.

If the cathode admittance $G_K + Y_K$ approaches infinity, then the gain ratio equation is the same as given in Section 7.3 for the amplifier with the bypassed cathode. A general solution to (*7.117*) for the lower half-power frequency is impossible. Even a numerical solution to this problem becomes quite tedious.

This problem was included to show how difficult some problems can become even though they appear to be relatively simple. Other bypassing problems are similar, but the fact that the cathode circuit is a part of both the plate loop and grid loop makes this particular bypassing problem much more complicated. For example, a low-pass *R-C* filter in the plate power supply circuit is not included in the grid loop.

In practice, the half-power frequency for the given circuit would probably be calculated using approximations and then the exact value determined experimentally.

7.17. Given a line amplifier designed to operate into a 600 Ω load. In the mid-band range of frequencies, the input impedance is resistive and is 600 Ω. The gain of the amplifier is 50 db. The amplifier is operating with a 600 Ω resistor as a load, and the amplifier input power is −40 dbm at a frequency in the mid-band range. What is the rms value of the output voltage v_{s2}?

Here the amplifier input and load resistances are equal. The average output power P_o is the average input power in dbm added to the amplifier power gain in db: $P_o = -40 + 50 = 10$ dbm. The output power in milliwatts is found from

$$10 \text{ dbm} = 10 \log P_o/1, \qquad \log P_o = 1, \qquad P_o = 10 \text{ milliwatts}$$

Then $V_o^2/600 = P_o = 0.01$ watts, $V_o^2 = 6$, and $V_o = 2.45$ volts rms.

7.18. Given the grounded grid amplifier circuit of Fig. 7-64. Find the mid-band input impedance R_i looking into the terminals X-Y. Assume there is provision for the plate current in the circuit external to the input terminals. Tube parameters are $g_m = 2000\ \mu$mho, $r_p = 10$ kΩ.

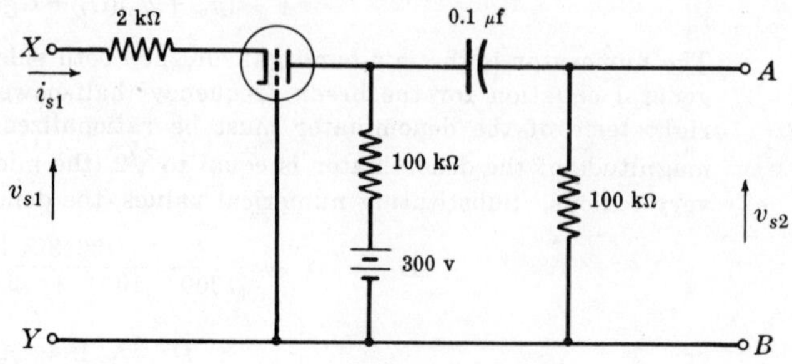

Fig. 7-64

Since it is most convenient to use the constant voltage generator for the tube, the parameter μ is calculated: $\mu = g_m/g_p = (2000 \times 10^{-6})/(100 \times 10^{-6}) = 20$. The parallel combination of the two resistors in the plate circuit is $R_L^* = 50{,}000\ \Omega$. The mid-band equivalent circuit is shown in Fig. 7-65.

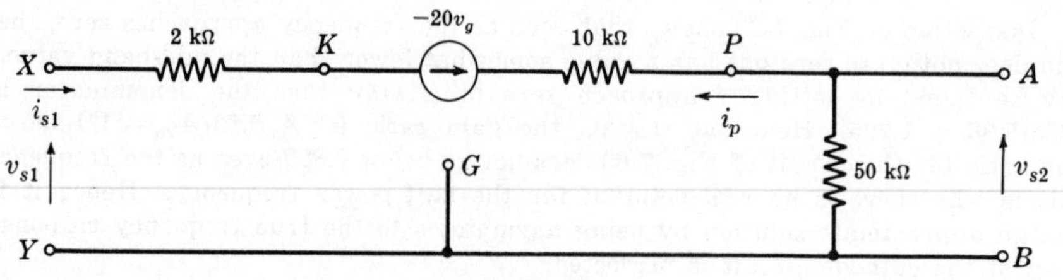

Fig. 7-65

Noting that $i_p = -i_{s1}$, the plate-cathode and grid-cathode loop equations are

$$-50{,}000 i_{s1} - 10{,}000 i_{s1} - 20 v_g - 2000 i_{s1} + v_{s1} = 0, \qquad +v_g - 2000 i_{s1} + v_{s1} = 0$$

from which $i_{s1} = (21/102{,}000)v_{s1}$, and the mid-band input impedance $R_i = v_{s1}/i_{s1} = 4860\ \Omega$.

7.19. An amplifier is being designed. Because of the gain specifications, it is necessary to use 3 stages. The half-power frequencies for each of the 3 stages will be approximately the same. The amplifier lower half-power frequency should not be above 20 cps and the upper half-power frequency should not be below 20,000 cps. Determine the limits on the half-power frequencies for the individual stages.

Equation (*7.89a*), page 204, is solved for the upper half-power frequency for each stage:

$$f_2 = \frac{f_2^*}{\sqrt{10^{0.3/n} - 1}} = \frac{20{,}000}{\sqrt{10^{0.1} - 1}} = \frac{20{,}000}{0.51} = 39{,}200 \text{ cps}$$

Equation (*7.90*) is solved for the lower half-power frequency for each stage:

$$f_1 = \sqrt{10^{0.3/3} - 1}\, f_1^* = 0.51(20) = 10.2 \text{ cps}$$

Supplementary Problems

7.20. Given the equivalent circuit of Fig. 7-66. This is a transistor amplifier circuit which includes the shunt capacitance and the coupling capacitor in the collector circuit. Derive the general equation for the open circuit voltage gain $A_v(f) = v_{s2}/v_{s1}$. Show that the equation for the gain ratio can be simplified to

$$A_v(f)/A_{vo} = 1/[1 + j(f/f_2 - f_1/f)]$$

and state the approximations involved. Also, give the equations for the half-power frequencies f_1 and f_2.

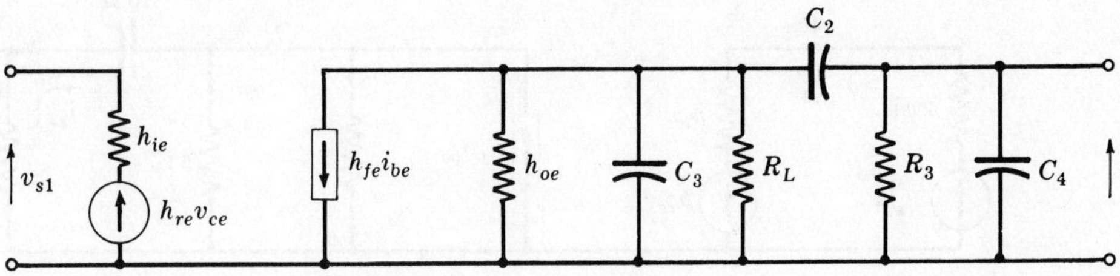

Fig. 7-66

7.21. Given the transistor amplifier equivalent circuit of Fig. 7-67. Transistor parameters are $h_{ie} = 1400\ \Omega$, $h_{re} = 10^{-3}$, $h_{oe} = 50\ \mu$mho, $h_{fe} = 100$. The total shunt capacitance in the collector circuit is $C_S = 100\ \mu\mu$f. The magnitude of v_{gen} is kept constant as the frequency is varied. Find (*a*) the mid-band voltage gain $A_{vo} = v_{s2}/v_{gen}$, (*b*) the lower half-power frequency f_1, (*c*) the upper half-power frequency f_2, (*d*) the bandwidth when it is defined as the band between the half-power frequencies.

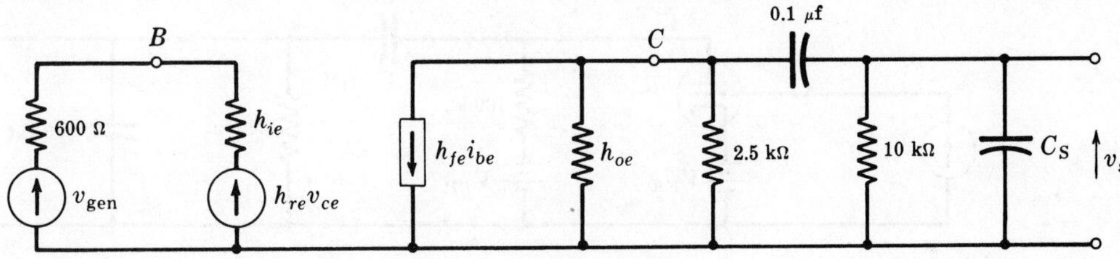

Fig. 7-67

7.22. Given the high frequency equivalent circuit of Fig. 7-68. This is the equivalent circuit for a cathode follower amplifier. All capacitances are assumed negligible, except C_S which is the total cathode to ground capacitance. Tube parameters are $\mu = 20$, $r_p = 10\ \text{k}\Omega$.

(*a*) Find the upper half-power frequency f_2; this is the frequency at which $A_v(\text{h.f.}) = A_{vo}/\sqrt{2}$.

(*b*) Find the mid-band voltage gain $A_{vo} = v_{s2}/v_{gen}$.

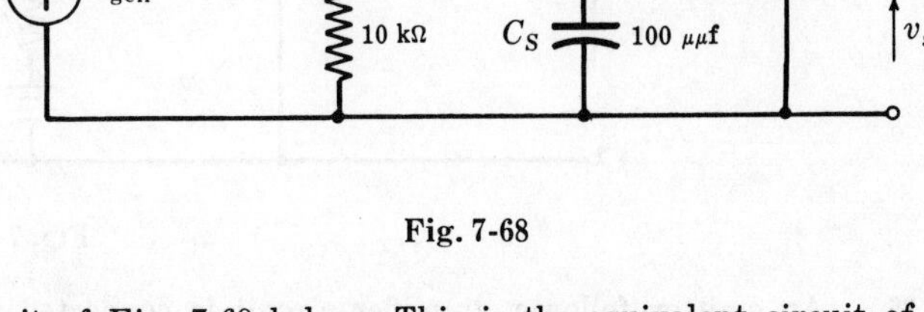

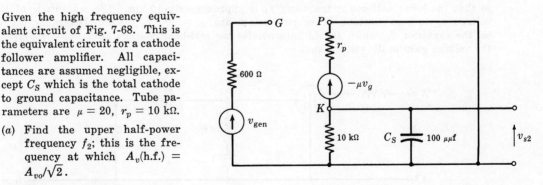

Fig. 7-68

7.23. Given the low frequency equivalent circuit of Fig. 7-69 below. This is the equivalent circuit of a common emitter amplifier in which the input circuit is direct coupled. The low frequency response is determined by the collector coupling capacitor only. Transistor parameters are $h_{ie} = 2000\ \Omega$, $h_{re} = 10^{-3}$, $h_{oe} = 50\ \mu$mho, $h_{fe} = 80$. For the circuit as given, find:

(*a*) The mid-band current gain $A_i = i_{s2}/i_{s1}$.

(*b*) The lower half-power frequency f_1 when f_1 is the frequency at which $A_i(\text{l.f.}) = A_i/\sqrt{2}$.

(c) The mid-band voltage gain $A_v = v_{s2}/v_{gen}$.
(d) The lower half-power frequency f_1 when f_1 is the frequency at which $A_v(\text{l.f.}) = A_v/\sqrt{2}$.
(e) The mid-band transfer conductance $y_t = i_{s2}/v_{gen}$.
(f) The lower half-power frequency f_1 when f_1 is the frequency where $y_t(\text{l.f.}) = y_t/\sqrt{2}$.
(g) The mid-band transfer resistance $r_t = v_{s2}/i_{s1}$.
(h) The lower half-power frequency f_1 when f_1 is the frequency where $r_t(\text{l.f.}) = r_t/\sqrt{2}$.
(i) Explain any differences in answers among parts (b), (d), (f), and (h).

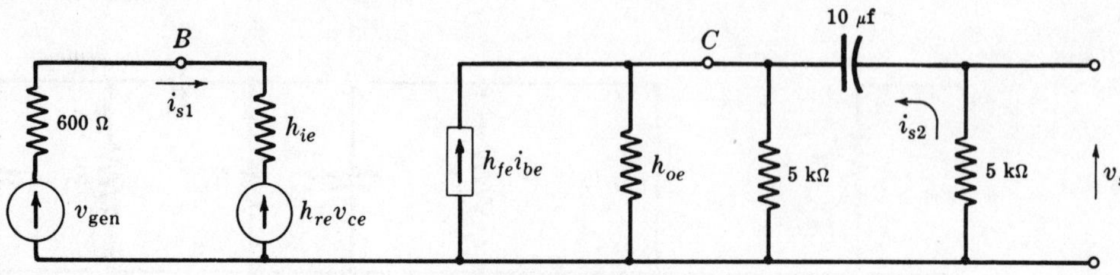

Fig. 7-69

7.24. Given the circuit of Fig. 7-70. The tube is operating on the linear part of the range. It may be assumed that the only significant capacitances are those shown. Tube parameters are $\mu = 60$, $r_p = 50$ kΩ. Find (a) the mid-band voltage gain $A_{vo} = v_{s2}/v_{s1}$, (b) the lower half-power frequency f_1 when f_1 is the frequency at which $A_v(\text{l.f.}) = A_{vo}/\sqrt{2}$, (c) the upper half-power frequency f_2 when f_2 is the frequency at which $A_v(\text{h.f.}) = A_{vo}/\sqrt{2}$.

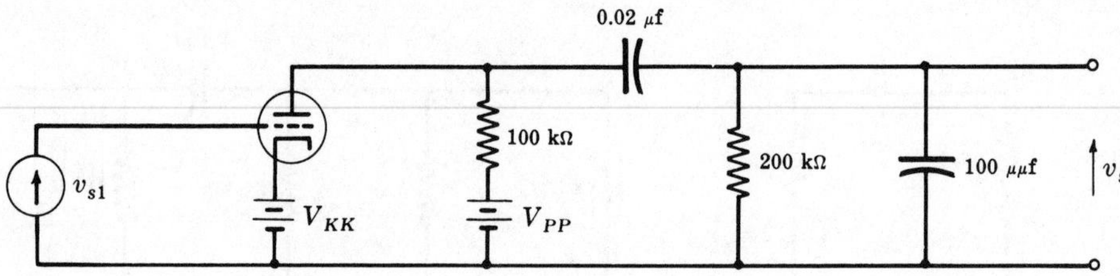

Fig. 7-70

7.25. Given the grounded grid amplifier circuit of Fig. 7-71. The coupling capacitor C_C is to be chosen so that the lower half-power frequency f_1 is approximately 10 cps. The magnitude of the input voltage v_{s1} is kept constant. Tube parameters are $\mu = 30$, $r_p = 8$ kΩ. Determine the value of (a) the capacitor C_C which should be used, (b) the mid-band voltage gain $A_v = v_{s2}/v_{s1}$. Express the voltage gain in db voltage gain.

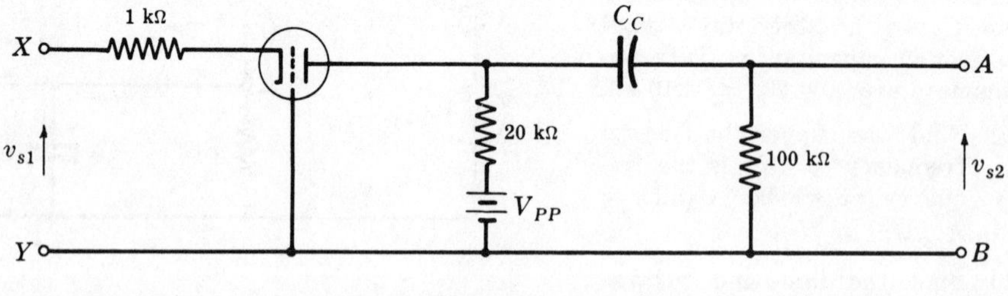

Fig. 7-71

7.26. An emitter follower amplifier circuit is connected to a 10 kΩ load resistor as in Fig. 7-72 below. The input circuit is direct coupled with a bias source in the base circuit as shown. The magnitude of v_{gen} is kept constant. The entire shunt capacitance in the emitter circuit is 100 μμf, and the shunt capacitance external to the amplifier is negligible. Transistor parameters are $h_{ie} = 1400$ Ω, $h_{re} = 10^{-4}$, $h_{oe} = 50$ μmho, $h_{fe} = 80$. Find (a) the mid-band transfer conductance $y_t = i_{s2}/v_{gen}$, (b) the upper and the lower half-power frequencies when they are the frequencies at which $y_t(f) = y_t/\sqrt{2}$.

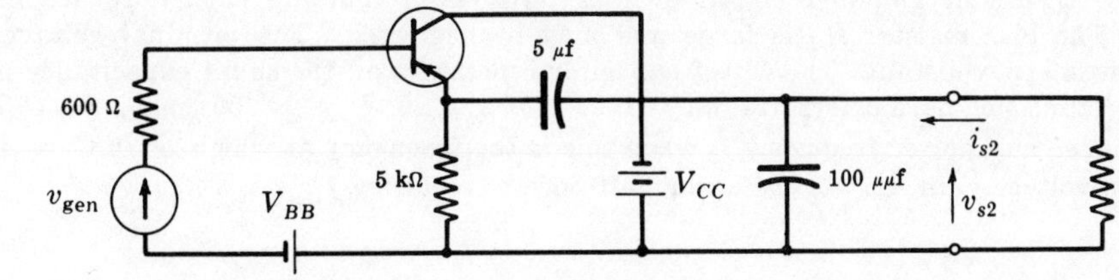

Fig. 7-72

7.27. A three stage transistor amplifier has stages which have identical half-power frequencies. The half-power frequencies for each stage are: $f_1 = 10$ cps, $f_2 = 25{,}000$ cps. (*a*) Determine the amplifier half-power frequencies. (*b*) What are the slopes of the low frequency and high frequency asymptotes of the amplifier response curve?

7.28. A two stage transistor amplifier has stages which do not have identical half-power frequencies. The half-power frequencies for the first stage are: $f_1 = 30$ cps, $f_2 = 18{,}000$ cps. The second stage half-power frequencies are: $f_1 = 10$ cps, $f_2 = 12{,}000$ cps. The gain ratio vs. frequency ratio equations for the individual stages are the same as those for the *R-C* high and low-pass filters. Find the amplifier half-power frequencies.

7.29. Given the high frequency equivalent circuit of Fig. 7-73. The capacitor C_S includes all the shunt capacitance in the collector circuit. Assume the shunt capacitance external to the output terminals is negligible. The output current i_{s2} is the current in the conductor G_3 as shown. (*a*) Derive the equations for the high frequency current gain $A_i(\text{h.f.}) = i_{s2}(\text{h.f.})/i_{s1}(\text{h.f.})$. (*b*) From the equations for $A_i(\text{h.f.})$ from part (*a*), derive the equations for the mid-band current gain A_i. (*c*) Derive the equations for the upper half-power frequency f_2 when f_2 is the frequency at which $A_i(\text{h.f.}) = A_i/\sqrt{2}$. (*d*) Assuming $f_2 \gg f_1$ so that $\Delta f \cong f_2$, derive the equation for the gain-bandwidth product defined as $\text{GBW} = |A_i|\,\Delta f$. Compare this equation with (*7.39*), page 182.

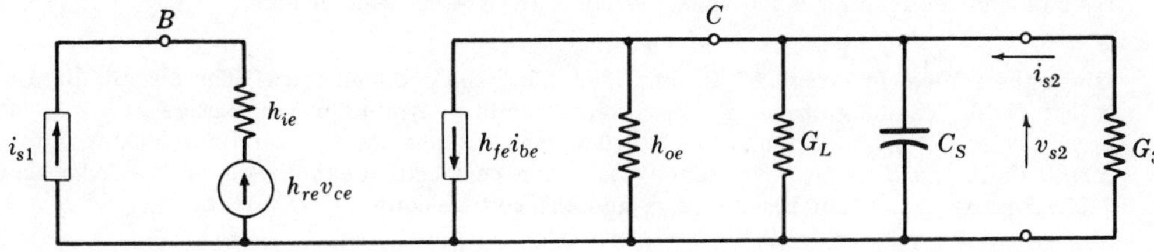

Fig. 7-73

7.30. Given the pentode amplifier circuit of Fig. 7-74. It may be assumed that the cathode and screen grid are effectively bypassed at any frequency of interest in this problem. The 30 μμf capacitor is the entire shunt capacitance in the plate circuit, and other interelectrode capacitances are negligible. The tube parameters are $g_m = 6200$ μmho, $r_p = 0.6$ megohms. (*a*) Given $R_L = 100$ kΩ, find the mid-band voltage gain $A_{vo} = v_{s2}/v_{s1}$ and the upper half-power frequency f_2. (*b*) Given $R_L = 2$ kΩ, find A_{vo} and f_2. (*c*) Calculate the gain-bandwidth product for this amplifier.

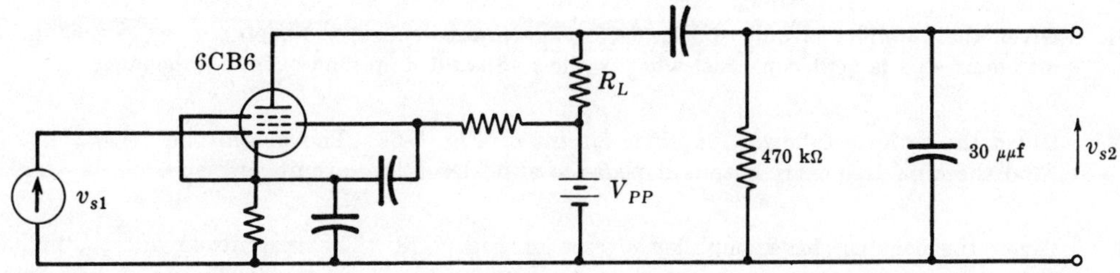

Fig. 7-74

7.31. Given the transistor amplifier circuit of Fig. 7-75. Both coupling capacitors are included in the circuit. The bias resistor R_B is large and may be neglected. The terminal voltage of the input generator is kept constant. The 40 $\mu\mu$f capacitor represents all the shunt capacitance in the collector circuit. Transistor parameters are $h_{ie} = 1500\ \Omega$, $h_{re} = 10^{-3}$, $h_{oe} = 50\ \mu$mho, $h_{fe} = 70$. Determine (a) the upper half-power frequency f_2 when this is the frequency at which $A_v(\text{h.f.}) = A_v/\sqrt{2}$, (b) the mid-band voltage gain A_v, (c) the lower half-power frequency f_1.

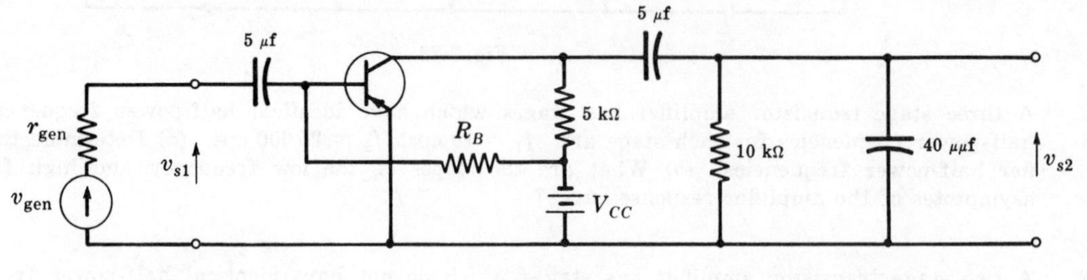

Fig. 7-75

7.32. An amplifier has a 2000 Ω resistor terminating the output. A voltmeter across the 2000 Ω resistor reads +24 on the dbm scale. What is the output power in watts?

7.33. An amplifier has an input resistance of 100 kΩ (mid-band input impedance). The amplifier output is terminated with a 600 Ω resistor. The input voltage is $V_{s1} = 0.1$ volts rms, and the output voltage $V_{s2} = 90$ volts rms. What is the amplifier gain in decibels (not the "db voltage gain")?

7.34. An amplifier with an input resistance 1000 Ω has an input current $I_{s1} = 0.002$ amperes rms. The amplifier output is terminated with a 600 Ω resistor. The amplifier transfer resistance is 40,000 Ω when terminated with 600 Ω. What is the amplifier gain in decibels?

7.35. A 600 Ω resistor is connected to a source of energy. A true rms voltmeter is connected across the resistor and the reading is 9.6 volts. What is the power level in dbm?

7.36. Given the equivalent circuit of an amplifier using the g parameters. The circuit diagram is given in Fig. 7-76. The magnitude of v_{gen} is kept constant. Amplifier parameters are $g_f = 0.0125$ mho, $g_o = 100\ \mu$mho, $g_i = 735\ \mu$mho, $g_r = -0.5\ \mu$mho. Represent the amplifier looking into the output terminals A and B using a constant voltage one port equivalent circuit of the form shown in Fig. 7-77. Specify the output resistance r_o and the voltage gain $A_v = v_o/v_{gen}$.

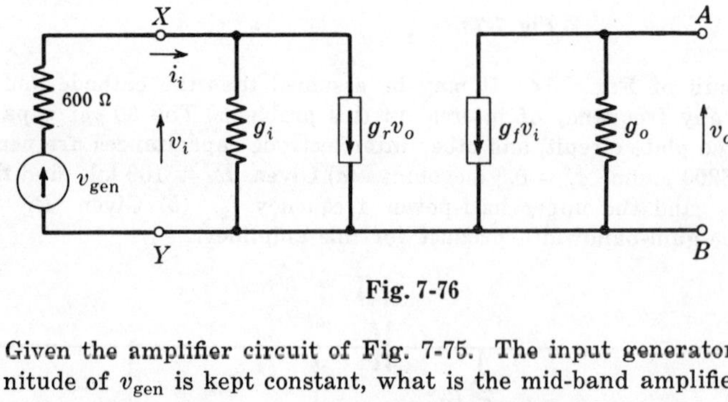

Fig. 7-76

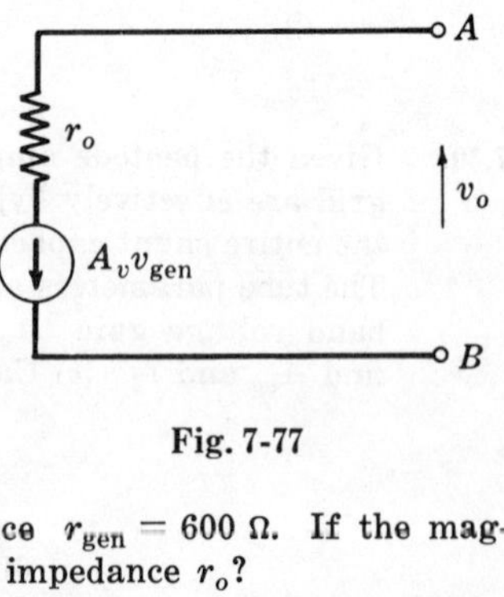

Fig. 7-77

7.37. Given the amplifier circuit of Fig. 7-75. The input generator resistance $r_{gen} = 600\ \Omega$. If the magnitude of v_{gen} is kept constant, what is the mid-band amplifier output impedance r_o?

7.38. Given the cathode follower amplifier circuit of Fig. 7-68. The magnitude of v_{gen} is kept constant. Find the high frequency output impedance $z_o(\text{h.f.})$ for the circuit as given.

7.39. Given the common base amplifier circuit of Fig. 7-78. The magnitude of v_{gen} is kept constant. Transistor parameters are $h_{ib} = 30\ \Omega$, $h_{rb} = 250 \times 10^{-6}$, $h_{fb} = -0.982$, $h_{ob} = 0.35\ \mu$mho.

(a) For the circuit as given, find the mid-band input resistance $R_i = v_{s1}/i_{s1}$.

(*b*) For the given circuit with v_{gen} constant, find the parameters r_o and A_v for a one port representation looking into the terminals A and B. This representation is the type shown in Fig. 7-77.

(*c*) Can the mid-band input resistance R_i ever exceed h_{ib}?

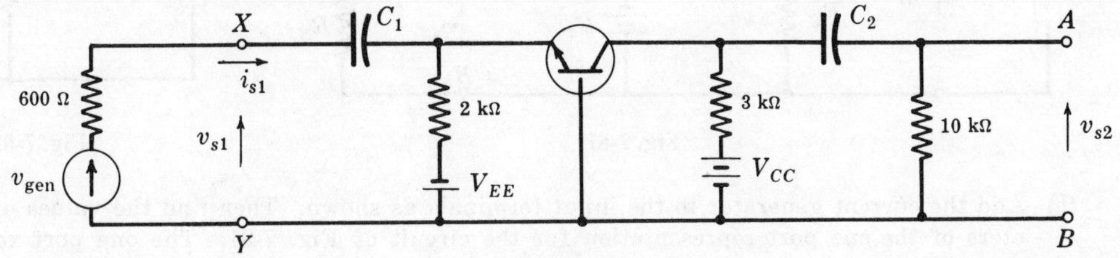

Fig. 7-78

7.40. Given the emitter follower amplifier circuit of Fig. 7-79. The transistor is to be represented by using the h parameters. The input of the emitter follower is connected to a transistor amplifier stage which is represented as shown in Fig. 7-79 to the left of the terminals X and Y. The magnitude of v_{gen} is kept constant. Derive the equations for the one port parameters looking into the terminals A and B where a representation of the form given in Fig. 7-77 is to be used. The equations are to be in terms of the h parameters of the transistor of Fig. 7-79 and the circuit parameters of Fig. 7-79.

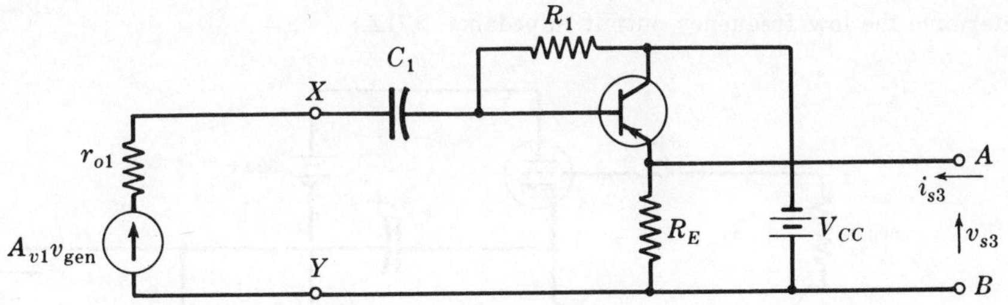

Fig. 7-79

7.41. Given the equivalent transistor amplifier circuit of Fig. 7-80. The transistor parameters are independent of frequency in the range of interest. Transistor parameters are $h_{ie} = 2000\ \Omega$, $h_{re} = 2.5 \times 10^{-3}$, $h_{oe} = 50\ \mu\text{mho}$, $h_{fe} = 100$.

(*a*) The output voltage v_{s2} is 1.2 sin ωt where ω is in the mid-band range. Find v_{gen} and specify the phase angle using v_{s2} as a reference.

(*b*) The upper half-power frequency f_2 was measured experimentally as 105,000 cps. Assuming that only the collector shunt capacitance is significant, find the value of the effective collector shunt capacitance.

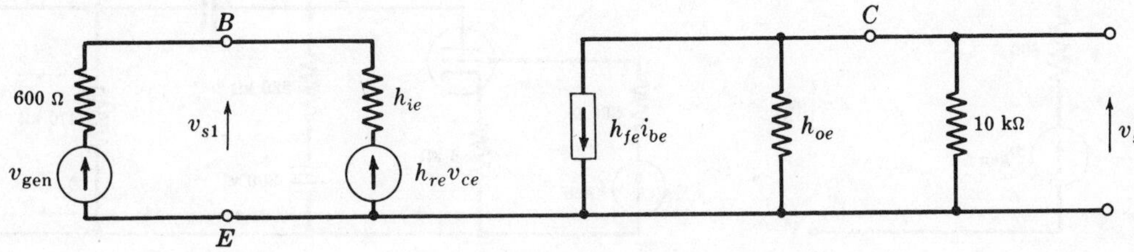

Fig. 7-80

7.42. Given the simplified emitter follower circuit of Fig. 7-81 below. Assume the bias resistors are large and negligible. The effective mid-band resistance in the emitter circuit of the amplifier is R_E, and R_X is the external load resistance. The amplifier input terminals are X and Y, and the output terminals are A and B. Transistor parameters are $h_{ie} = 2000\ \Omega$, $h_{re} = 20 \times 10^{-4}$, $h_{oe} = 50\ \mu\text{mho}$, $h_{fe} = 50$.

(*a*) Represent the amplifier using the h parameter two port equivalent circuit.

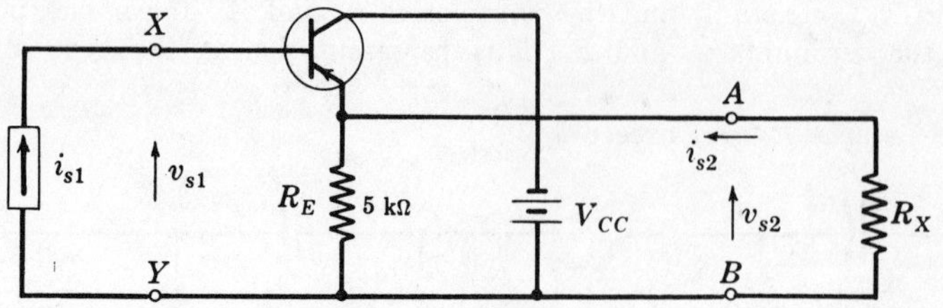

Fig. 7-81

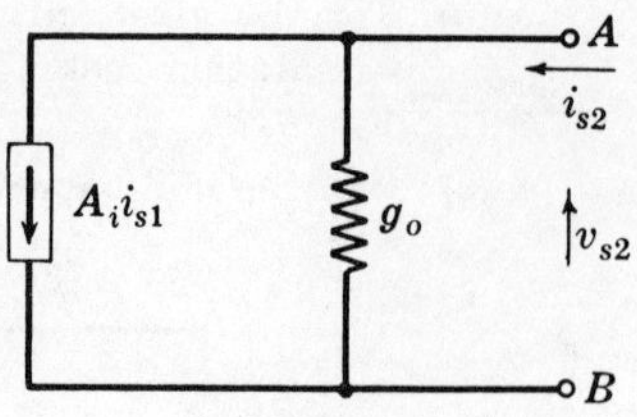

Fig. 7-82

(*b*) Add the current generator to the input terminals as shown. Then find the values of the parameters of the one port representation for the circuit of Fig. 7-81. The one port representation should be of the form shown in Fig. 7-82.

(*c*) If R_X is 10 kΩ, find the input conductance $G_i = i_{s1}/v_{s1}$ looking into the terminals X and Y.

7.43. Given the cathode follower amplifier circuit of Fig. 7-83. The value of v_{gen} is kept constant. Tube parameters are $\mu = 20$, $r_p = 10$ kΩ.

(*a*) Calculate the mid-band output impedance r_o.

(*b*) Find the lower half-power frequency f_1 when f_1 is the frequency where $A_v(\text{l.f.}) = A_v/\sqrt{2}$.

(*c*) Determine the mid-band voltage gain $A_v = v_{s2}/v_{gen}$.

(*d*) Determine the low frequency output impedance $z_o(\text{l.f.})$

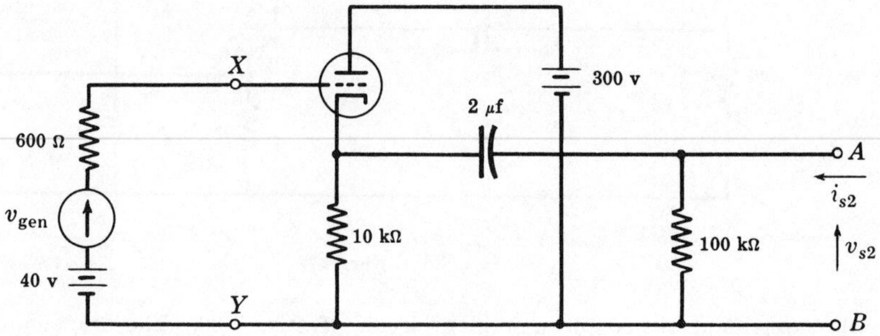

Fig. 7-83

7.44. Given the grounded grid amplifier circuit of Fig. 7-71. Add the generator of Fig. 7-84 to the input of the grounded grid amplifier. Find the mid-band output impedance r_o of the amplifier using the parameters given in Fig. 7-71.

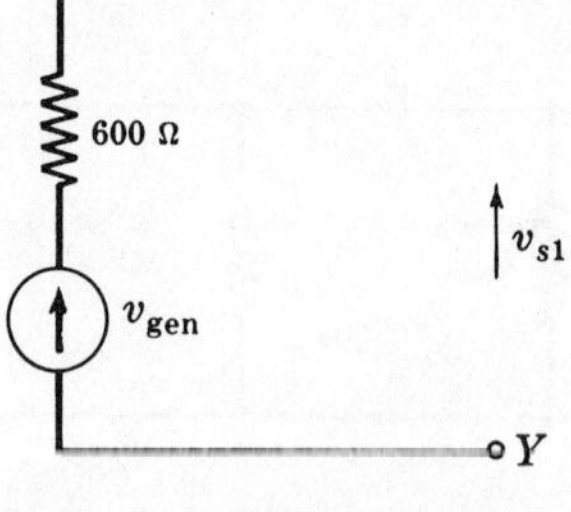

Fig. 7-84

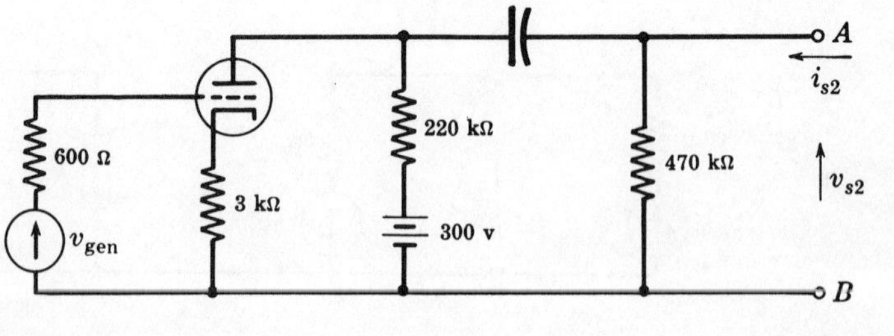

Fig. 7-85

7.45. Given the amplifier circuit of Fig. 7-85. The magnitude of v_{gen} is kept constant. Tube parameters are $\mu = 70$, $r_p = 60$ kΩ.

(*a*) Find the mid-band output impedance r_o for the circuit as given looking into the terminals *A-B*.

(*b*) Add a 20 μf capacitor from cathode to ground. Determine the mid-band output impedance for the circuit with the cathode capacitor. Compare the answer with that of part (*a*).

7.46. Given the transistor amplifier equivalent circuit of Fig. 7-86. The magnitude of v_{gen} is kept constant. Transistor parameters are $r_e = 26\ \Omega$, $r_b = 1320\ \Omega$, $r_c = 10^6\ \Omega$, $\alpha = 0.975$. Find (a) the mid-band output impedance r_o, (b) the mid-band voltage gain $A_v = v_{s2}/v_{gen}$.

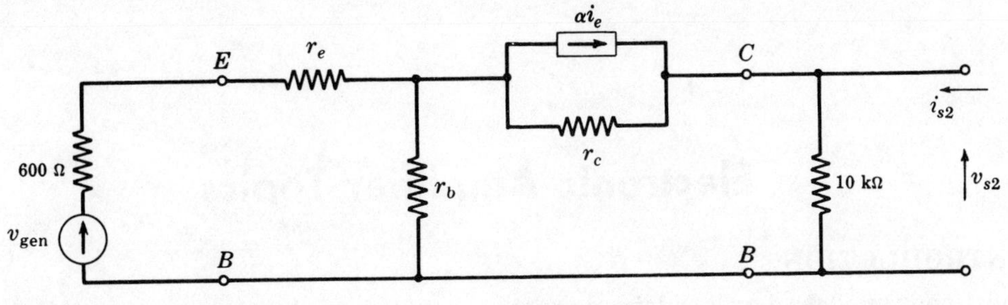

Fig. 7-86

7.47. Given the two stage amplifier circuit of Fig. 7-87. Tube parameters are $\mu_1 = 100$, $r_{p1} = 70$ kΩ, $\mu_2 = 20$, $r_{p2} = 10$ kΩ. The two stages are direct coupled as shown, so there are no coupling capacitors. There is a capacitor in the circuit external to the cathode of T_2, so the average value of the output voltage is no problem.

(a) Find the mid-band voltage gain $A_v = v_{s3}/v_{gen}$.

(b) Find the mid-band output impedance r_o looking into the terminals A and B.

(c) Suggest a reason why it may be an advantage to include T_2 even though the voltage gain of the second stage is less than one.

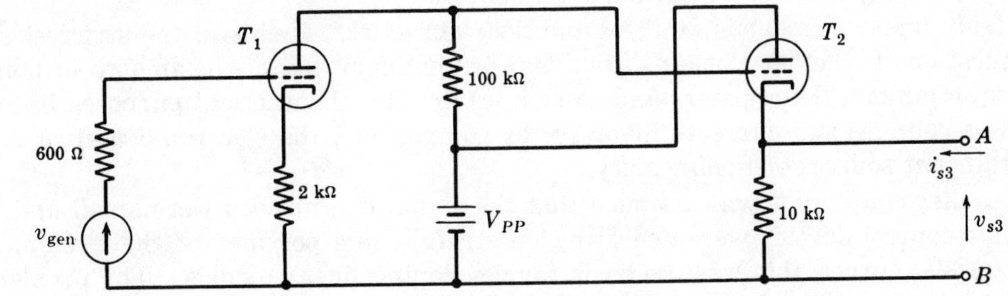

Fig. 7-87

7.48. An amplifier consists of three cascaded stages. The db voltage gain of each stage is as follows: stage 1, 18 db; stage 2, 20 db; stage 3, 14 db. What is the overall voltage gain $A_v = v_o/v_i$?

Chapter 8

Electronic Amplifier Topics

8.1 INTRODUCTION

The linear models developed in previous chapters are useful representations for amplifiers, but there are important considerations not included in these models. It is necessary to have performance criteria for amplifiers. Several amplifier topics must be introduced and briefly discussed as a basis for stating amplifier "goodness of performance" criteria.

When a sine wave is applied to a circuit consisting of linear elements, all of the steady-state voltages and currents are sinusoidal. However, if any elements are nonlinear, then this is no longer true. In a circuit consisting only of linear resistors, the voltages and currents are independent of frequency. If the circuit includes inductors or capacitors, then the voltages and currents are in general not independent of frequency; they are independent only for a limited band of frequencies (mid-band).

A truly linear element is an ideal and does not exist. Hence in the strictest sense it is not a question of whether elements are linear or nonlinear, but of the amount of nonlinearity. In most amplifiers, the resistors and capacitors are for all practical purposes linear for the ranges of voltages and currents involved; by comparison, the electronic control devices are the significant sources of nonlinearity.

In earlier chapters it was assumed that the signal magnitudes were small and hence the electronic control devices were operating on straight line portions of the dynamic transfer characteristic curves; this was the basis for developing linear models. The problem of nonlinearity usually becomes more significant in amplifier stages where the signal levels are large. One of the topics discussed in this chapter is amplitude distortion, which is one way of specifying nonlinearity in amplifiers.

Electrical noise is generated in any resistive circuit element when the temperature is above absolute zero. Additional noise sources exist in electronic control devices. Hence the topic of amplifier noise needs to be considered. The problem of noise is usually most significant in the stages where the signal levels are small, usually the first stage.

Noise, distortion, and frequency response are problems which cannot be eliminated, so compromises must be made in meeting desired amplifier performance specifications. This leads to the important topic of feedback. By properly applying feedback, the overall performance of an amplifier may be improved by sacrificing gain, as will be shown.

8.2 AMPLIFIER CLASSIFICATION

There are many categories and subcategories of electronic amplifiers. It is very difficult to provide a comprehensive scheme of amplifier classification. When there were only vacuum tube amplifiers, they were often classified as either voltage or power amplifiers. Amplifiers are still roughly divided into two categories.

One category is that which includes voltage amplifiers. Amplifiers of this category are designed to increase the magnitude of the signal whether it is a voltage or a current. Obviously, voltage and current signal levels are not independent quantities in general, but it may be more convenient to use either a voltage or a current amplifier model.

The other general category includes power amplifiers. These are designed to provide the signal power necessary to drive recorders, reproducers and other output devices. There are special problems with amplifiers of this type. Since a significant amount of power is usually involved, efficiency is a consideration. The signal levels are usually high, so nonlinear distortion is a problem. It is necessary to effectively couple the high power electronic control device to the load. A detailed discussion of special power amplifier problems is beyond the scope of this book.

It is difficult to establish a firm basis for classifying every amplifier into one of the two general categories. The last stage of one amplifier may be a power amplifier in the context of the design of this amplifier unit. However, there may be a separate amplifier connected to the output of the first amplifier. In the context of the design of the second amplifier, the output stage of the first amplifier may be considered as a "driver" amplifier stage.

Amplifiers are also classified as tuned or untuned amplifiers. Tuned amplifiers include tuned resonant circuits and usually operate in the frequency range above several hundred kilocycles per second. These amplifiers are usually rather narrow-band and are purposely designed to be frequency selective. Untuned amplifiers are usually broad-band. For example, audio amplifiers are designed for voice or musical signals. Another example of an untuned amplifier is a video amplifier which may have a bandwidth in the order of megacycles per second.

Audio amplifiers are sometimes classified according to the type of interstage coupling employed. Transformer coupling is sometimes used in input and output stages because of impedance matching considerations. Transformer coupling may also be used as interstage coupling, especially in the input of a push-pull amplifier stage (see Section 8.3). Resistance and capacitance networks of the types used in the amplifier circuits of earlier chapters are employed most frequently as interstage coupling in audio and instrument amplifiers.

Amplifier stages are divided into four categories according to the location of the operating points on the dynamic transfer characteristic curves for the electronic control devices. The definitions of the four classes of operation were formulated for vacuum tube amplifiers but may be generalized to include transistor amplifiers.

A Class-A amplifier stage is biased so that the magnitude of the plate or collector current is greater than zero at all times and the signal may swing an equal amount above or below the operating point. Untuned, single-ended (unbalanced with respect to ground) amplifier stages usually operate as Class-A amplifiers. The amplifier stages of earlier chapters were all Class-A.

A Class-B amplifier stage has the operating point such that the average plate or collector current is nearly zero (cutoff). The plate or collector current increases in magnitude as the input signal increases in magnitude. In order to obtain approximately linear operation, untuned Class-B amplifiers are push-pull.

A Class-AB amplifier has the operating point between that for Class-A and that for Class-B operation. The magnitude of the plate or collector current must be above cutoff for appreciably more than half but less than the entire cycle of the input signal voltage or current. Untuned Class-AB amplifier stages are also push-pull.

A Class-C amplifier has the operating point below the cutoff value. The magnitude of the plate or collector current is greater than zero for appreciably less than half of a cycle of the input signal voltage. Class-C amplifiers are usually tuned amplifiers.

Amplifiers are also classified according to the intended application. Some are designed for very low input signal levels and hence are called preamplifiers. The name for an amplifier category usually indicates the intended service, e.g., line amplifier, isolation amplifier, monitor amplifier, distribution amplifier, etc.

8.3 UNTUNED POWER AMPLIFIERS

The power (or large signal) amplifier stage is designed to provide the required amplifier signal power output. The signal levels are the highest in this stage, so signal distortion is an important consideration in power amplifier stages. Unless operation is truly linear, the equivalent circuit for an electronic control device is only an approximation. For this reason, graphical techniques and approximations are used in the analysis of power amplifiers.

A simple single-ended Class-A power amplifier circuit is shown in Fig. 8-1. Power amplifier tubes are especially designed for higher signal levels and include triodes, tetrodes, pentodes and beam power tubes. A beam power tube is shown in the circuit of Fig. 8-1 because such tubes are especially designed for power amplifier applications.

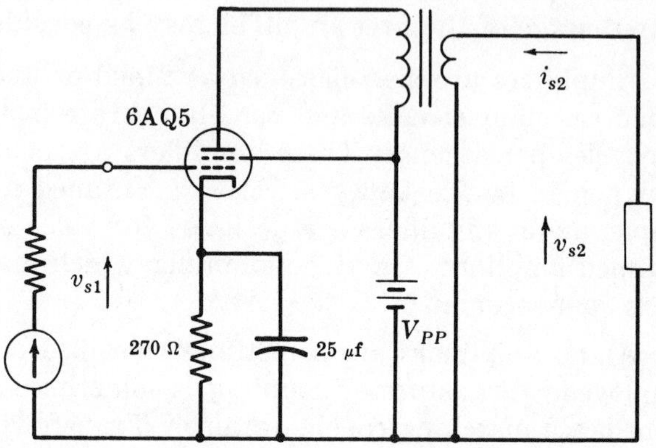

Fig. 8-1. Single-ended Class-A Power Amplifier

Transformer coupling is usually employed to couple the output amplifier stage to the load. Since a transformer involves magnetic fields and magnetic materials, it introduces many problems which are beyond the scope of this book. In the present discussion, the ideal transformer model will be used. The ideal transformer has no resistance in the windings and dissipates no energy.

Consider the ideal transformer of Fig. 8-2 with a resistive load connected to the secondary (output) terminals. Since this is an ideal transformer, the voltage induced per turn is the same in each winding. The primary winding has N_p turns, the secondary has N_s turns, and the turns ratio is

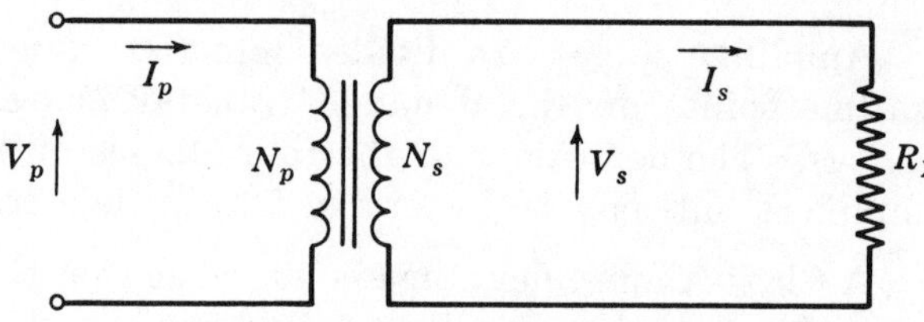

Fig. 8-2. Ideal Transformer

$$a = N_p/N_s = V_p/V_s$$

where V_p and V_s are the rms values of the primary and secondary voltages, respectively. The power dissipated in the load R_X is $P_s = V_sI_s$. Since the transformer is ideal, there is no power loss in it and it is resistive looking into the primary terminals (resistive load connected to the secondary). Hence the power input to the primary is equal to the power dissipated in R_X; thus $P_p = V_pI_p = V_sI_s = P_s$ and

$$a^2 = (V_p/V_s)(I_s/I_p) = (V_p/I_p)/(V_s/I_s) \quad \text{or} \quad a = (Z_p/Z_s)^{1/2} \tag{8.1}$$

The interest is in the resistance looking into the primary terminals,

$$R_X^* = \frac{V_p}{I_p} = \frac{V_sI_s/I_p}{I_p}\cdot\frac{I_s}{I_s} = \frac{I_s^2}{I_p^2}\cdot\frac{V_s}{I_s} = a^2R_X \tag{8.2}$$

For an ideal transformer the effective primary resistance is the square of the turns ratio times the load resistance. For a given load R_X, the output transformer is designed to provide the desired load resistance for the particular electronic control device.

The power amplifier stage converts the power supplied by the plate or collector power supply to signal power. Efficiency is a consideration, especially when the signal power is large. The plate or collector efficiency is defined as

$$\eta = \frac{\text{signal power output}}{\text{dc power input}} \times 100\%$$

It is possible to develop limits for the efficiency of a stage based on circuit models. For a Class-A stage, the upper limit to the efficiency is 50% and the actual efficiency is much lower. For this reason, high power stages are usually not Class-A. See Problem 8.1, page 253.

Since a power amplifier stage is designed to provide significant signal output power, it seems intuitively appropriate to use the power gain as the transfer function. This is a problem in vacuum tube amplifiers where the input signal power is negligible unless the grid is positive[1]. Then the power gain is not a useful transfer function. The transfer function in this case is the power sensitivity. Since power is proportional to the square of the rms value of the voltage, the power sensitivity of a vacuum tube amplifier stage is defined in terms of the square of the rms value of the grid signal voltage:

$$\text{Power sensitivity} = \frac{\text{signal power output}}{V_g^2}$$

A push-pull (balanced with respect to ground) Class-A amplifier[2] stage is shown in Fig. 8-3. Two tubes are coupled to the same load through a single output transformer. The plate winding is center tapped so that the instantaneous plate currents of the two tubes are in opposite directions. Hence the magnetic fields of the two primary half-windings are opposed. This is an advantage in terms of core saturation effects in the design of the transformer, since the average values of the plate currents are high.

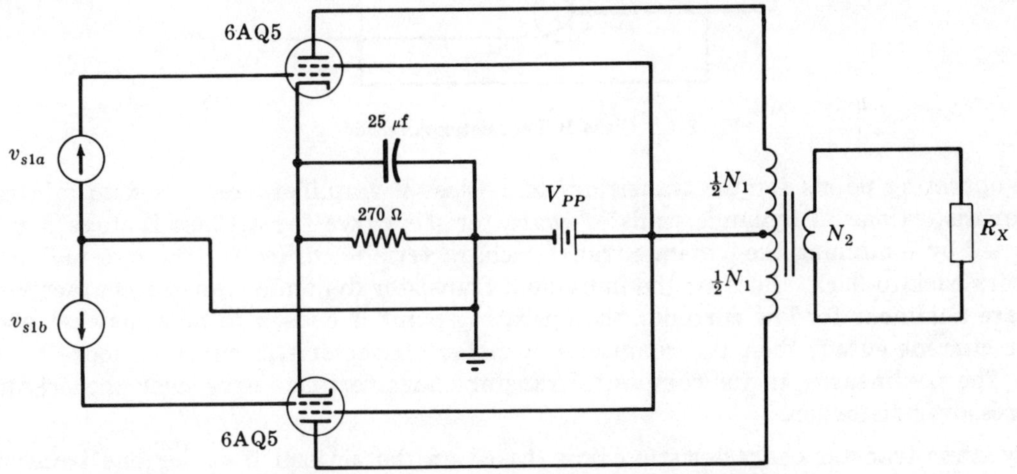

Fig. 8-3. Push-pull Class-A Power Amplifier

The two halves of a push-pull amplifier should be as nearly balanced as possible. Ideally the output transformer is perfectly balanced, the two tubes have identical characteristics, and the grid signal voltages are equal in magnitude and 180° out of phase. Interstage transformer coupling is sometimes used because the secondary of the interstage transformer can be center tapped so that grid signals of equal magnitude and opposite phase are provided. In most amplifiers an R-C coupled phase inverter stage is used. See Problem 5.4, page 100.

[1]In vacuum tube power amplifier stages, a subscript 1 indicates the grid is never positive and a 2 indicates that it is positive on peaks. For example, a Class-AB_1 stage draws no grid current and a Class-AB_2 stage does.

[2]Vacuum tube circuits are used as examples of Class-A power amplifier stages because transistor power amplifiers usually operate as Class-B.

The name push-pull was probably chosen for this type of amplifier because the signal voltages and currents in the two tubes (or transistors) are out of phase with respect to each other. When the current in one tube is above the average value, the other is below. There are amplifiers in which all stages are push-pull and the amplifier input is balanced with respect to ground.

A push-pull Class-B transistor amplifier circuit is shown in Fig. 8-4. The general principles of operation are the same for vacuum tube or transistor Class-B push-pull amplifier stages. The operating point is near the value for plate or collector current cutoff. As the input signal swings positive, one transistor (or tube) conducts, with the collector (or plate) current proportional to the signal magnitude. The other transistor (or tube) conducts as the signal swings negative. Since the no-signal collector (or plate) current is near cutoff, the efficiency is higher than for Class-A amplifiers. The maximum efficiency of Class-B amplifiers is $\pi/4 = 78.5\%$.

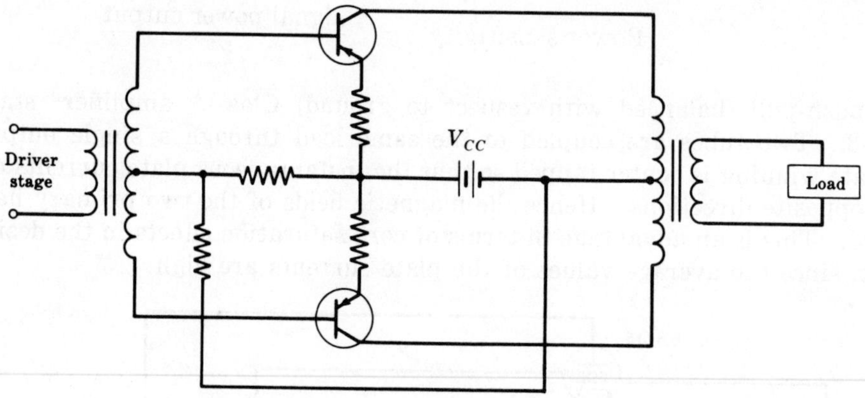

Fig. 8-4. Class-B Transistor Amplifier

The operating points for the transistors of a Class-B amplifier are chosen to minimize nonlinear distortion. A dynamic transfer characteristic curve for a Class-B stage may be constructed by combining the dynamic transfer characteristic curves for the two individual transistors back-to-back. Because the individual transistor dynamic transfer characteristic curves are nonlinear for low currents, the operating point is chosen to be somewhat above collector current cutoff; then the composite transfer characteristic curve is more nearly linear. The nonlinearity in the composite transfer characteristic curve near the origin is called crossover distortion.

A dynamic transfer characteristic curve (based on the ac load line) for one transistor is given in Fig. 8-5 below. When a transformer is in the collector circuit, the dc resistance is low (zero for the ideal transformer) and hence the dc load line on the collector characteristic curves is almost vertical. The slope of the ac load line is determined by the effective resistance R_X^*. The operating point is sometimes chosen by extending the straight line portion of the dynamic transfer characteristic curve and taking the intersection with the v_{BE} axis as the operating point.

The composite dynamic transfer characteristic curve is drawn from the two individual dynamic transfer characteristic curves after inverting one of them. The two characteristic curves are positioned so that the operating points of the two transistors are aligned as shown in Fig. 8-6 below. The composite curve is plotted point by point. For each value of v_{BE}, the corresponding values of i_{CE1} and i_{CE2} are added algebraically to give one point of the composite transfer characteristic curve. Since the crossover distortion would be appreciable if the two individual curves were aligned at $v_{BE} = 0$, the operating point is not at the origin of the individual dynamic transfer characteristic curves.

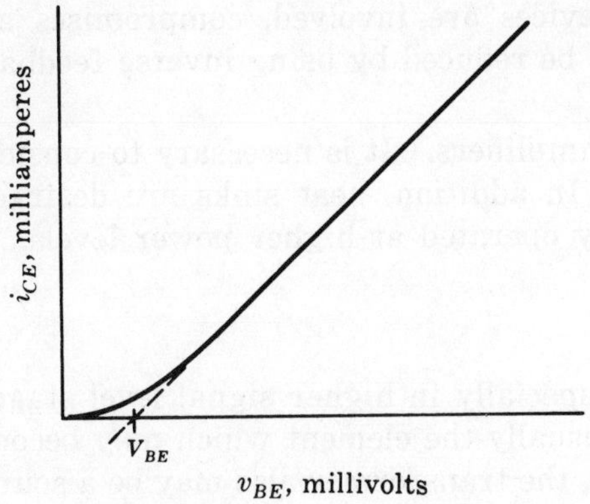

Fig. 8-5. Dynamic Transfer Characteristic Curve

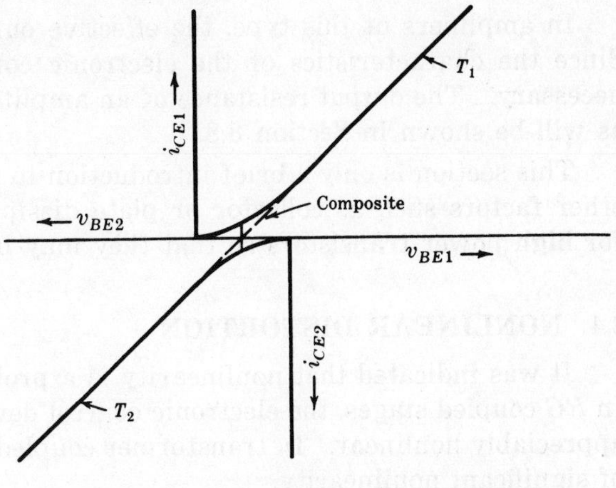

Fig. 8-6. Composite Dynamic Transfer Characteristic Curve

Since transistors may be either *NPN* or *PNP*, a complementary symmetric Class-B transistor push-pull amplifier using an *NPN* and a *PNP* transistor is possible because the corresponding voltages and currents are of opposite polarity. A simplified circuit diagram is given in Fig. 8-7. In this circuit a phase inverter is not needed because a positive excursion of the input signal voltage causes the *NPN* transistor collector current to increase in magnitude and drives the *PNP* transistor below cutoff. A negative swing of the input signal voltage causes the reverse of this to occur, so the transistor inputs may be driven in parallel.

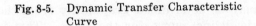

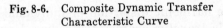

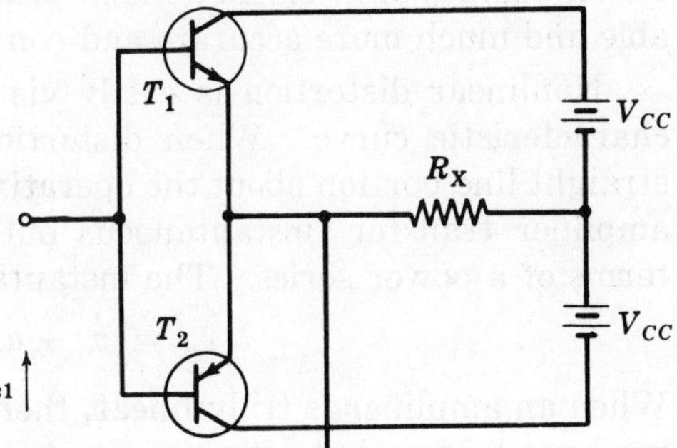

Fig. 8-7. Complementary Symmetric Push-pull Amplifier

The signal currents add in the load resistance R_X, and the average values of the collector currents are not included in the load currents. An output transformer is needed only for impedance matching. Sound reproducer (speaker) impedances are often the values desired as load impedances, and when this is the case they may be connected directly in the circuit as the load R_X.

Since efficiency and signal power output are of major importance in power amplifiers, consideration must be given to impedance matching. It would seem appropriate to apply the maximum power transfer theorem; however, this is not advisable. There is an optimum value of the effective load resistance R_X^* which provides the maximum average output signal power into the rated load with the nonlinear distortion less than a specified limit. This value of R_X^* may be quite different from that chosen for maximum power transfer. For maximum power transfer, the effective load resistance R_X^* would be the value of the output resistance of the tube or transistor circuit.

If an amplifier is designed for sound reproduction, then its effective output resistance is an important consideration. The sound reproducer (speaker) is an energy storing device; hence sound reproducer damping must be considered[1]. The dynamics of sound reproducers is beyond the scope of this book.

[1]In vacuum tube amplifiers before feedback was used, many preferred triodes because of the effective damping even though the power sensitivity of triode stages is low. Pentode amplifiers have high power sensitivity but damping is a problem because of the high output resistance.

In amplifiers of this type, the effective output resistance is made as small as possible. Since the characteristics of the electronic control devices are involved, compromises are necessary. The output resistance of an amplifier may be reduced by using inverse feedback as will be shown in Section 8.8.

This section is only a brief introduction to power amplifiers. It is necessary to consider other factors such as collector or plate dissipation. In addition, heat sinks are desirable for high power transistors so that they may be safely operated at higher power levels.

8.4 NONLINEAR DISTORTION

It was indicated that nonlinearity is a problem, especially in higher signal level stages. In *RC* coupled stages, the electronic control device is usually the element which may become appreciably nonlinear. In transformer coupled stages, the transformer also may be a source of significant nonlinearity.

The harmonic distortion of individual stages may be determined graphically from the dynamic transfer characteristic (based on the ac load line) curve for the stage. This is rather tedious and seldom done in practice. Instruments for analyzing distortion are available and much more accurate and convenient to use.

Nonlinear distortion is easily visualized by considering the overall mid-band transfer characteristic curve. When distortion obtains, the operating range extends beyond the straight line portion about the operating point. It is then convenient to represent the overall amplifier transfer (instantaneous output vs. instantaneous input) characteristic curve in terms of a power series. The instantaneous output voltage may be written as

$$v_O = a_0 + a_1 v_I + a_2 v_I^2 + a_3 v_I^3 + \cdots + a_n v_I^n \tag{8.3}$$

When an amplifier is truly linear, then all the coefficients a_i are zero except for a_1 and a_0; a_0 need not be zero but often is.

Nonlinear distortion measurements usually involve sinusoidal test signals applied to the amplifier input. Assuming the input signal is a pure sine wave[1] in the mid-band frequency range, the input voltage may be written $v_I = V \sin \omega t$. Substituting for v_I in equation (*8.3*), the instantaneous output voltage is

$$v_O = a_0 + a_1 V \sin \omega t + a_2 (V \sin \omega t)^2 + a_3 (V \sin \omega t)^3 + \cdots$$

Assuming the coefficients a_i are zero for $i \geqq 4$, and using the identities $\sin^2 \omega t = \frac{1}{2} - \frac{1}{2}\cos 2\omega t$ and $\sin^3 \omega t = \frac{3}{4}\sin \omega t - \frac{1}{4}\sin 3\omega t$,

$$v_O = a_0 + \tfrac{1}{2}a_2 V^2 + (a_1 V + \tfrac{3}{4}a_3 V^3)\sin \omega t - \tfrac{1}{2}a_2 V^2 \cos 2\omega t - \tfrac{1}{4}a_3 V^3 \sin 3\omega t \tag{8.4}$$

Note that the output signal includes harmonics of the signal input. Whenever there is nonlinear distortion, the output contains harmonics of the input signal; hence this is also called harmonic distortion.

Equation (*8.4*) may be written as

$$v_O = b_0 + b_1 \sin(\omega t + \phi_1) + b_2 \sin(2\omega t + \phi_2) + b_3 \sin(3\omega t + \phi_3)$$

where the coefficients of the harmonic terms indicate the amount of harmonic distortion. A wave analyzer may be tuned to the frequency of each harmonic and then the magnitude of each harmonic measured, giving the value of each b_i. The magnitudes of the harmonic frequency terms are usually compared with the magnitude of the fundamental frequency (input signal frequency) and may be expressed in per cent.

Many distortion analyzers measure the total harmonic distortion and compare its magnitude with that of the fundamental frequency. The per cent harmonic distortion

[1]In practice this is not exactly true; however, function generators with very low distortion are usually used.

(or distortion factor) is defined by

$$D = \frac{\sqrt{\sum_{i=2}^{n} b_i^2}}{b_1} \times 100\% \tag{8.5}$$

This is sometimes given in terms of the square root of the sum of the squares of the rms voltages (or currents) of each individual harmonic frequency divided by the rms voltage (or current) of the fundamental. Since sine waves are involved, the definition in terms of amplitudes is equivalent.

Unless a tuned wave analyzer is used, the magnitude (or rms value) of the fundamental frequency is not conveniently measured. If the harmonic distortion as defined by equation (*8.5*) is less than 10%, then the error is small if the distortion is defined by

$$D = \frac{\sqrt{\sum_{i=2}^{n} b_i^2}}{\sqrt{\sum_{i=1}^{n} b_i^2}} \times 100\%$$

In this equation the harmonics are not compared to the fundamental alone but to the signal which includes the harmonics. This defines the measurements made by many distortion analyzers. See Problem 8.4, page 254.

When an individual single-ended stage is considered, the collector (or plate) current may be written as a power series. When appreciable nonlinearity exists, operation is beyond the straight line portion of the dynamic transfer characteristic curve. Since the dynamic transfer characteristic curve (based on the ac load line) involves one independent variable only, the function $i_C = f(i_B)$ may be expanded about the no-signal operating point. The instantaneous collector current may be written as

$$i_C = I_C + a_1 i_b + a_2 i_b^2 + a_3 i_b^3 + \cdots + a_n i_b^n$$

where $I_C = f(I_B)$. When the base signal current is sinusoidal, $i_b = I \sin \omega t$; substituting into the above equation and assuming $a_i = 0$ for $i \geq 3$,

$$i_C = I_C + a_1 I \sin \omega t + a_2 (I \sin \omega t)^2 = I_C + \tfrac{1}{2} a_2 I^2 + a_1 I \sin \omega t - \tfrac{1}{2} a_2 I^2 \cos 2\omega t$$

When there is second harmonic distortion, the average value (operating point) of the collector current changes. The average value of the collector current with signal is

$$I_{CQ} = I_C + \tfrac{1}{2} a_2 I^2$$

The amount of second harmonic distortion may be determined by the shift in the average value of the collector current when there is only second and third harmonic distortion (third harmonic distortion does not change the average value). Second harmonic distortion is a problem in single-ended Class-A amplifier stages.

One advantage of push-pull amplifier stages is that the even order harmonic frequency components cancel each other when there is perfect balance. Consider the simplified circuit for a push-pull amplifier stage as given in Fig. 8-8. Transistors are shown but the results are general. It is not necessary to draw an equivalent circuit (this assumes linear operation). It is assumed that the collector current for each transistor may be written in terms of a power series. Since the output transformer is balanced, there are $N_1/2$ turns of the primary winding in series with each collector terminal.

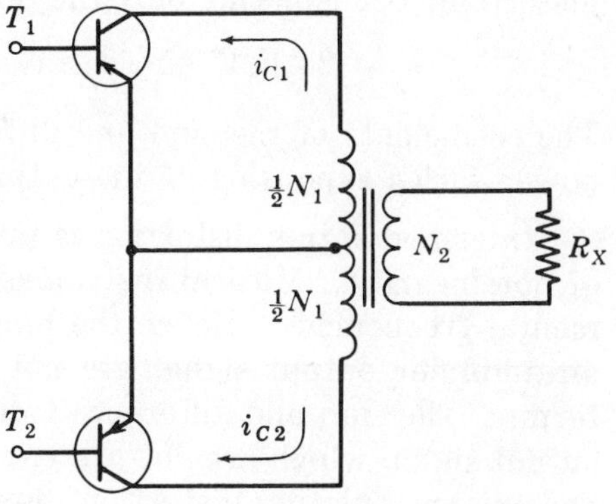

Fig. 8-8. Simplified Push-pull Amplifier

It is convenient to think in terms of the net magnetic potential rise in the entire primary winding. The two collector currents are in opposite directions in each half of the primary winding, so the net magnetic potential rise in the primary is

$$\mathcal{F}_1 = (N_1/2)i_{C1} - (N_1/2)i_{C2}$$

Writing each collector current as a power series,

$$\begin{aligned}\mathcal{F}_1 &= (N_1/2)\,[I_{C1} + a_1 i_{b1} + a_2(i_{b1})^2 + \cdots + a_n(i_{b1})^n] \\ &\quad - (N_1/2)\,[I_{C2} + a_1 i_{b2} + a_2(i_{b2})^2 + \cdots + a_n(i_{b2})^n] \qquad (8.6)\end{aligned}$$

If I_{C1} and I_{C2} are equal in magnitude, then $(N_1/2)I_{C1} - (N_1/2)I_{C2} = 0$ which is important in terms of transformer nonlinearity. The varying components of the collector currents may be considered by using (recall $i_{b1} = -i_{b2}$)

$$i_{b1} = I \sin \omega t \quad \text{and} \quad i_{b2} = -I \sin \omega t$$

For convenience assume $a_i = 0$ for $i \geq 3$. Since the no-signal average values cancel, *(8.6)* becomes (assuming perfect balance),

$$\mathcal{F}_1 = N_1 a_1 I \sin \omega t$$

which indicates that the second harmonic frequency terms cancel and that the net primary magnetic potential at the second harmonic frequency is zero. It may be easily shown that this is the case for all even order harmonic frequencies.

The fact that even order harmonics are cancelled in a balanced push-pull stage is one of the reasons that high power amplifier stages are push-pull. It is possible to obtain significantly higher power output for a specified amount of harmonic distortion.

In a push-pull stage that is balanced, it is not necessary to bypass a common emitter (or cathode) resistor to eliminate signal degeneration (inverse feedback; see Section 8.6). See Problem 8.2, page 253.

Nonlinear distortion may also be characterized in terms of intermodulation distortion. An input signal consisting of the linear sum of two sine waves whose frequencies are quite different may be applied to the amplifier input. The input signal is then, where $\omega_2 \gg \omega_1$,

$$v_I = V_1 \sin \omega_1 t + V_2 \sin \omega_2 t \qquad (8.7)$$

Substituting this into *(8.3)* and assuming $a_i = 0$ for $i \geq 3$, the output signal is

$$v_O = a_0 + a_1 V_1 \sin \omega_1 t + a_1 V_2 \sin \omega_2 t + a_2 V_1^2 \sin^2 \omega_1 t + 2a_2 V_1 V_2 \sin \omega_2 t \sin \omega_1 t + a_2 V_2^2 \sin^2 \omega_2 t$$

When there is nonlinear distortion, the output contains not only the input signal frequencies but also harmonics of these frequencies and terms involving sum and difference frequencies. The harmonic components of the output signal are usually not included in intermodulation distortion. Considering only the term involving both ω_1 and ω_2,

$$2a_2 V_1 V_2 \sin \omega_2 t \sin \omega_1 t = a_2 V_1 V_2 [\cos (\omega_2 - \omega_1)t - \cos (\omega_2 + \omega_1)t]$$

The coefficients of the sum and difference frequency terms include the coefficient a_2 of the power series expansion of the output signal.

Intermodulation distortion is preferred by some audio amplifier designers as a measure of nonlinearity. Musical instruments generate signals containing harmonics of the fundamental frequencies. Hence the harmonic frequency terms of a power series expansion of an amplifier output signal are not as objectionable as the sum and difference frequency terms. The sum and difference frequency terms cause frequencies to exist in the amplifier output signal which are not present in complex musical input signals. Harmonic distortion and intermodulation distortion, however, are not independent but are just different ways of characterizing nonlinear distortion.

8.5 NOISE

In this book noise will be considered as any undesired signals. One category of undesired signals includes power supply ripple, hum, and other such unwanted signals which are inherent in amplifiers. This type of noise may be considered as "man-made." The source of a man-made noise signal may be known and its magnitude may be reduced, but it usually cannot be eliminated entirely; hence design compromises are made.

Another category includes the random types of noise which are relatively broad-band. Thermal noise is an example and is inherent in any conductor; it provides an absolute limit for the noise level of an amplifier. Since the present interest is only in amplifier performance criteria, the discussion of noise will be limited accordingly. The topic of noise is very extensive and requires background in probability and statistics for even an elementary treatment.

Thermal noise is the result of random motion of charges in conductors. Consider a resistor R which is in thermal equilibrium at an absolute temperature T. The thermal noise generated in the resistor may be represented as the voltage generator of Fig. 8-9. Since thermal noise is random, instantaneous values of the equivalent generator are not known. When the resistor is in thermal equilibrium, then the mean-squared value of the noise generator is specified.

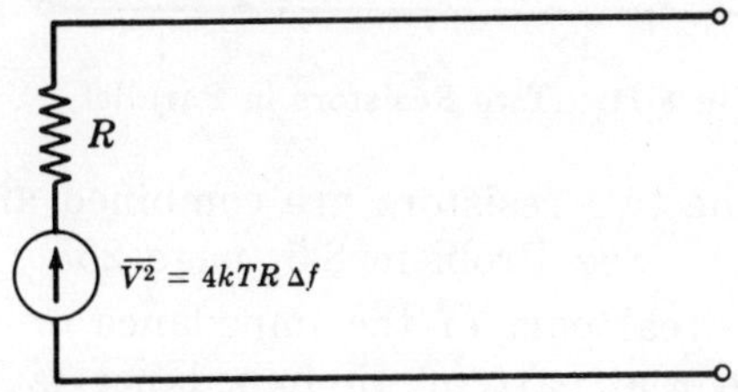

k = 1.38×10^{-23} joule/degree, Boltzmann's constant
T = absolute temperature in degrees Kelvin
R = resistance in ohms
Δf = bandwidth in cps
$\overline{V^2}$ = mean-squared voltage[1]

Fig. 8-9. Thermal Noise Model

The circuit of Fig. 8-9 includes no capacitance and no limitation to the frequency bandwidth; this type of generated noise is called *white noise* and has a flat spectrum extending from zero to infinity. This is of course a model, because a white noise source involves infinite energy. It is convenient in some problems to consider a resistor as a white noise source and the bandwidth as being limited by the circuits connected to the resistor.

In some problems it is more convenient to use a current generator model for thermal noise in a resistor. Norton's theorem may be applied to the circuit of Fig. 8-9, giving the circuit of Fig. 8-10. The representation which is most convenient is usually used.

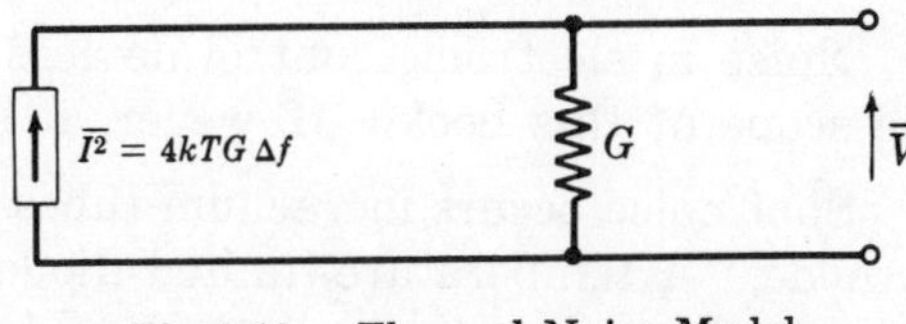

Fig. 8-10. Thermal Noise Model

Since every resistor is a noise source, the noise free resistors shown in all earlier circuit diagrams are models which are realizable only at absolute zero of temperature. In practice, thermal noise in resistors is a problem only in low signal level amplifier stages; the noise free resistor model is a good approximation in most circuits. Thermal noise is present in

---

[1]The mean-squared voltage is defined as

$$\overline{V^2} = \frac{1}{T}\int_0^T v^2(t)\,dt$$

where $v(t)$ is the instantaneous value of the voltage. The square root of V^2 is the root-mean-square (rms) value. The bar is placed over the mean-squared value to indicate that a relatively long period of time T is assumed. If T is small, the mean-squared value may not be the same for each short interval selected. Hence $\overline{V^2}$ is an "average" mean-squared voltage.

all physical resistors at temperatures above absolute zero. The thermal noise models of this chapter assume ideal resistors. There is additional noise generated in physical resistors which depends on their construction. Carbon resistors are rarely used in circuits where resistor noise is a problem; instead, wire wound or other special types of resistors are used.

Noise generators are not added linearly. Noise powers are added, hence the generators are specified in terms of mean-squared values. For example, consider two resistors R_1 and R_2 connected in parallel as in Fig. 8-11. Both are at the same temperature and each is represented using the model of Fig. 8-10. The two conductors G_1 and G_2 may be combined in parallel, giving $G = G_1 + G_2$. Assuming the two noise generators are independent, the noise powers may be added; hence the mean-squared values of the two generators are added. The mean-squared value of an equivalent single generator is $\overline{I^2} = \overline{I_1^2} + \overline{I_2^2}$, and the total noise power is

$$\overline{I^2}R = \overline{I^2}/G = \overline{V^2}/R = \overline{V^2}G$$

from which the mean-squared voltage is $\overline{V^2} = \overline{I^2}/G^2$. In terms of the individual generators,

$$\overline{V^2} = \frac{\overline{I_1^2} + \overline{I_2^2}}{(G_1 + G_2)^2} = \frac{4kT(G_1 + G_2)\,\Delta f}{(G_1 + G_2)^2} = \frac{4kT\,\Delta f}{G_1 + G_2}$$

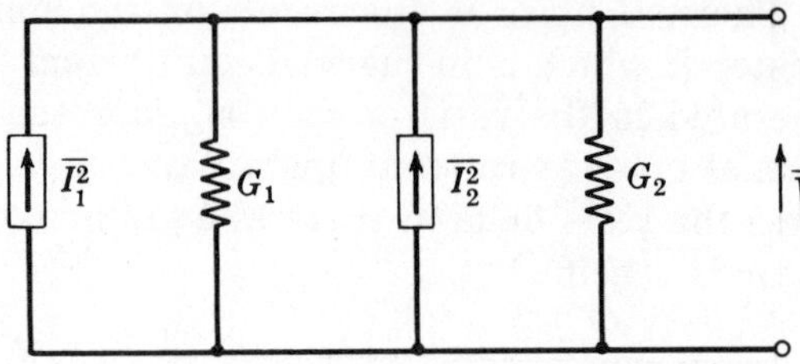

Fig. 8-11. Two Resistors in Parallel

which indicates that the same result is obtained when the two resistors are combined and considered as an effective single resistor $R = 1/(G_1 + G_2)$. See Problem 8.5, page 255.

When reactive elements are included in circuits, the real part of the impedance or of the admittance is used in noise calculations. Since the real part of an impedance is in general a function of frequency, the equation for the voltage noise source may be written

$$\overline{V^2} = 4kT \int_{f_1}^{f_2} \text{Re}\,(Z)\, df \tag{8.8}$$

For the current source, the noise equation is

$$\overline{I^2} = 4kT \int_{f_1}^{f_2} \text{Re}\,(Y)\, df \tag{8.9}$$

See Problem 8.6.

Noise in electronic control devices is a much more complicated problem and is beyond the scope of this book. However, it seems appropriate to mention certain useful results.

Shot noise occurs in vacuum tubes because of randomness in electron emission from the cathode. A temperature-limited diode (operation in the temperature-limited region rather than in the space-charge-limited region) is a simple example which illustrates shot noise. The mean-squared value of the noise component of the plate current is

$$\overline{I_p^2} = \Gamma^2\, 2eI_P\,\Delta f$$

where Γ = a constant depending on the statistical characteristics, its value being 1 for the temperature-limited diode,

e = electron charge in coulombs,

I_P = average value of the plate current in amperes,

Δf = bandwidth in cps.

When the diode is operating in the space-charge-limited region, the value of Γ is less than unity.

Applying superposition and considering only the noise source of a diode in a circuit with an external load resistor R_L, the equivalent circuit may be drawn as in Fig. 8-12. Combining r_p and R_L in parallel[1], $R = r_p R_L/(r_p + R_L)$. The noise power is $\overline{V_{pk}^2}/R = \overline{I^2}R$, and

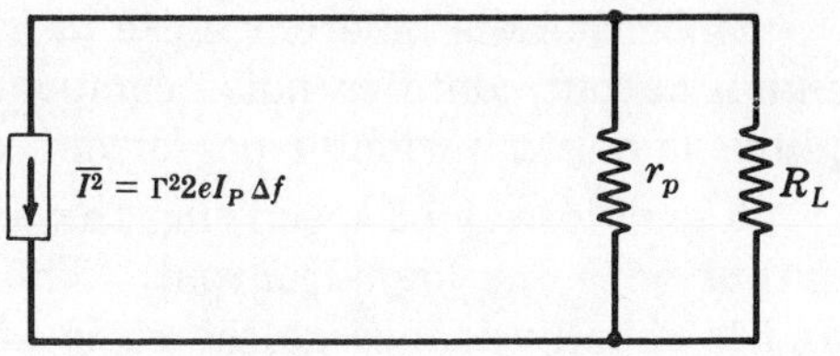

Fig. 8-12. Diode Noise Model

$$\overline{V_{pk}^2} = \overline{I^2}R^2 = \Gamma^2 2eI_P \Delta f R^2 \qquad (8.10)$$

This derivation neglects the thermal noise generated in the resistor R_L since this is usually small compared to the diode noise. If the noise in R_L is included, the noise powers are added because the two sources are statistically independent.

The problem of noise in triodes becomes more complicated. The interest is in the noise voltage appearing from the plate to the cathode but it is convenient to refer everything to the grid circuit. Triode noise is actually specified in terms of an equivalent resistance R_{eq} from the grid to ground.

The voltage gain for a simple triode amplifier with the cathode grounded is

$$|A_v| = V_{pk}/V_{gk} = g_m R \qquad (8.11)$$

where R is the total shunt resistance in the plate circuit. Only the magnitude of the gain is of interest because the average noise power is being calculated and phase shift is of no interest in the present discussion.

Referring the noise source to the grid circuit as an effective equivalent resistance R_{eq}, $\overline{V_{gk}^2} = 4kTR_{eq}\Delta f$. From equation (8.11),

$$\overline{V_{pk}^2} = |A_v|^2 \overline{V_{gk}^2} = g_m^2 R^2 (4kTR_{eq}\Delta f)$$

which is combined with (8.10) to yield

$$R_{eq} = \Gamma^2 eI_P/2kTg_m^2$$

If g_m is assumed proportional to I_P, then R_{eq} is approximated as

$$R_{eq} \cong 2.5/g_m$$

This equation is frequently used in specifying the equivalent noise of a triode. The mean-squared noise voltage from plate to cathode of a triode is the same as would be obtained if a noise free triode amplifier having the same gain had a resistor of magnitude R_{eq} from grid to cathode. Hence R_{eq} is just a convenient way of specifying the approximate noise generated in a triode.

The noise generated in tetrodes and pentodes is higher than for triodes because of the random division of the cathode current between screen grid and plate. This additional noise component is called *partition noise*. The equivalent resistance R_{eq} for pentodes is approximately

$$R_{eq} \cong \frac{I_P}{I_P + I_{S2}}\left(\frac{2.5}{g_m} + \frac{20\,I_{S2}}{g_m^2}\right)$$

There is also a low frequency noise which is inversely proportional to frequency and hence is often called $1/f$ noise. This is significant only in the low frequency range, usually below 1000 cps.

Physically there are more sources of noise in junction transistors than in tubes. In addition to thermal noise in the "ohmic" resistances, there is shot noise, $1/f$ noise, and other sources of noise. Noise generators are sometimes included in the collector, emitter and base of the transistors. The parameters used for calculating the magnitudes of these noise generators are usually not given on data sheets. Instead, a noise figure is given for the transistor which includes all the noise generated within the transistor.

[1]In the temperature-limited region, $g_p \cong 0$ and then $R = R_L$. In space-charge-limited diodes, g_p is not negligible and its effect is often included in the constant Γ.

Since there is inherent noise in any physical system (an amplifier is a special case), the signal output contains noise components. The amount of noise generated within an amplifier is one criterion of performance.

A simple way of specifying the noise generated within a system is to compare the noise output with the signal output. This is called the signal-to-noise ratio (abbreviated S/N) and is usually defined as the ratio of the rms value of the rated signal output to the rms value of the noise output. This is difficult to measure because the noise is present when measurements are made on the signal output. When S/N is high, there is little error if the actual ratio of the signal-plus-noise-to-noise is used instead of the signal-to-noise ratio. It is usually the $(S+N)/N$ ratio which is specified and measured.

The $(S+N)/N$ ratio is not a good indicator of amplifier performance for comparing different amplifiers. An amplifier which has a high output signal rating will have a higher $(S+N)/N$ rating than one which is rated for lower output signal levels, other things being equal. Hence a better indicator of goodness of performance is needed for comparing amplifiers. A criterion used in comparing amplifiers is the noise figure.

A simple definition of noise figure is the signal-to-noise ratio of the source (amplifier input) divided by the signal-to-noise ratio of the amplifier at the output terminals:

$$NF = \frac{(S/N)_{\text{in}}}{(S/N)_{\text{out}}}$$

If a system is ideal and introduces no noise, the noise figure is unity. Frequently the noise figure is given in decibels and is then

$$NF_{\text{db}} = 10 \log_{10} NF$$

Either the $(S+N)/N$ ratio or the noise figure may be used for showing how feedback may be employed to improve amplifier performance. When two different amplifiers are being compared, the noise figure is a better indication of amplifier performance.

8.6 SIMPLE INVERSE VOLTAGE FEEDBACK

Feedback has been used to improve the performance of electronic amplifiers. The subject of feedback is usually introduced by considering a simple case of inverse voltage feedback. The block diagram of Fig. 8-13 shows an amplifier with a feedback network connected from the amplifier output to the input circuit. The amplifier section has a voltage gain $A_v = \mathcal{A}\angle\alpha$ when terminated with the load resistance R_X. The voltage gain A_v is complex and is written in polar form with magnitude denoted by $\mathcal{A}$ and phase angle by α. The transfer function of the feedback network is also complex with magnitude β and phase shift θ.

In general the amplifier and the feedback network are represented by two port models. However, for simple voltage feedback it is usually assumed that the amplifier input impedance is very high and that all the output load resistance is included when calculating the voltage gain. The feedback network usually contains only passive elements, but it could

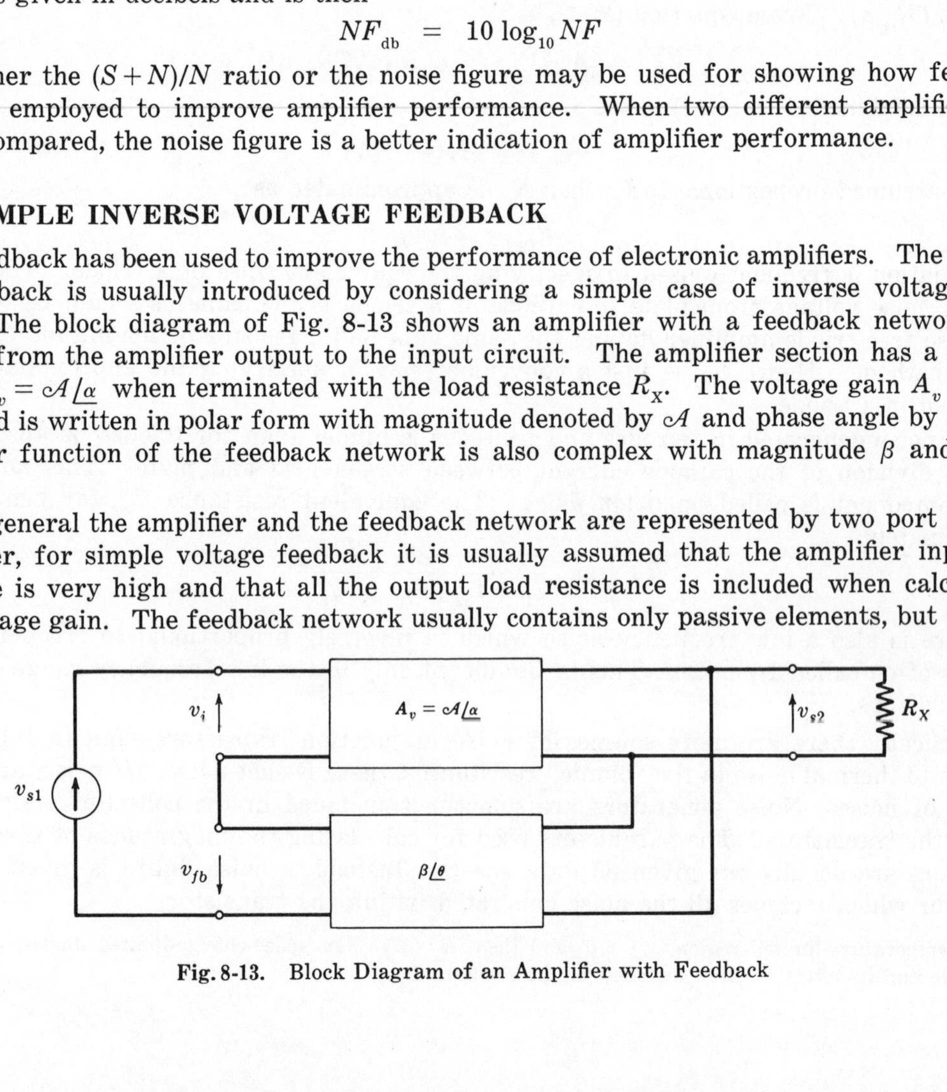

Fig. 8-13. Block Diagram of an Amplifier with Feedback

also include electronic control devices. The feedback network is assumed to have a high input impedance, and it is assumed that the current in the output of the feedback network is zero (amplifier input impedance is very high). The voltage at the output of the feedback network is denoted v_{fb}. The connections are made in the input circuit so that the amplifier input voltage v_i is the sum of the signal voltage v_{s1} and the feedback voltage v_{fb} as shown in Fig. 8-13.

$$v_i = v_{fb} + v_{s1} \tag{8.12}$$

Using the assumptions given above[1], the equivalent circuit for the feedback amplifier of Fig. 8-13 is drawn as shown in Fig. 8-14. The circuit equations are written and solved to find the amplifier gain with feedback. The transfer function for the feedback network is $\beta\underline{/\theta} = v_{fb}/v_{s2}$; then

$$v_{fb} = (\beta\underline{/\theta})v_{s2} \tag{8.13}$$

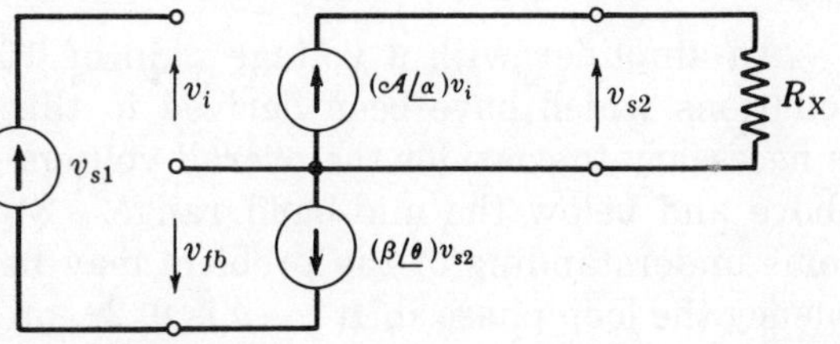

Fig. 8-14. Feedback System Equivalent Circuit

The transfer function for the amplifier without feedback is $A_v = \mathcal{A}\underline{/\alpha} = v_{s2}/v_i$; then

$$v_i = v_{s2}/(\mathcal{A}\underline{/\alpha}) \tag{8.14}$$

Substituting (*8.13*) and (*8.14*) into (*8.12*),

$$v_{s2}/(\mathcal{A}\underline{/\alpha}) = (\beta\underline{/\theta})v_{s2} + v_{s1}$$

which is solved for the transfer function with feedback, A'_v:

$$A'_v = v_{s2}/v_{s1} = (\mathcal{A}\underline{/\alpha})/(1 - \mathcal{A}\beta\underline{/\alpha+\theta}) \tag{8.15}$$

This is a general equation for the circuit of Fig. 8-14. Note that $\mathcal{A}, \beta, \alpha$ and θ are all functions of frequency outside of the mid-band range of frequencies.

Amplifier performance is improved by applying inverse feedback. Consider the amplifier system with feedback for the mid-band range of frequencies. Then if $\alpha + \theta$ is π radians or 180°, the feedback is inverse, assuming the input is connected according to (*8.12*). When $\alpha + \theta = 180°$, $-\mathcal{A}\beta\underline{/\alpha+\theta} = +\mathcal{A}\beta$ and (*8.15*) becomes

$$A'_v = (\mathcal{A}\underline{/\alpha})/(1 + \mathcal{A}\beta)$$

If the magnitude of the amplifier gain is made large so that $\mathcal{A}\beta \gg 1$,

$$A'_v \cong (\mathcal{A}\underline{/\alpha})/\mathcal{A}\beta = (1/\beta)\underline{/\alpha} \tag{8.16}$$

Equation (*8.16*) is very significant and shows the effect of inverse feedback on the voltage gain. When $\alpha + \theta = 180°$ and $\mathcal{A}\beta \gg 1$, then the magnitude of the voltage gain with feedback is a function only of β. If the feedback network contains only passive elements, then the magnitude of the gain is a function only of the passive circuit element parameters. Precision elements may be used so that the parameters of the passive elements are stable. This is the basis for the very stable electronic amplifiers which contain elements that may have unstable parameters and which are powered from lines that may not have constant magnitude terminal voltages. When the magnitude of the amplifier gain remains high and $\alpha + \theta = 180°$, then the overall system gain with feedback is relatively independent of variations in the gain of the amplifier unit. This stability and the independence of the variations in amplifier gain are obtained by sacrificing the overall system gain with feedback.

The amount of voltage feedback in db is defined as the difference between the "db voltage gain" of the amplifier without feedback and that with feedback. This may be written in terms of the magnitude of the voltage gain without feedback $|A_v|$ and with feedback $|A'_v|$. Thus

[1] These are the assumptions usually made even though a block diagram is often used to represent the amplifier system with feedback.

$$\text{amount of feedback in db} = 20\log_{10}|A_v| - 20\log_{10}|A'_v| = 20\log_{10}|A_v|/|A'_v|$$

Insight into some of the practical design problems may be gained by considering the following example. Assume that an amplifier system with an overall gain of 100,000 is desired. Using (*8.16*) to determine the magnitude of β, $|A'_v| = 10^5 = 1/\beta$ or $\beta = 10^{-5}$. Since $\mathcal{A}\beta$ must be large compared to 1, assume that $\mathcal{A}\beta = 100$. Then the magnitude of the amplifier gain is $\mathcal{A} = 100/\beta = 10^7$. Thus the magnitude of the mid-band voltage gain of the amplifier must be in the order of 10^7.

An amplifier with a voltage gain of 10^7 will have many stages of amplification. The equations which have been derived in this section are for mid-band frequencies only. It is necessary to consider the overall voltage gain of the system with feedback at frequencies above and below the mid-band range. Stability will be discussed in a later section, but some understanding of the problem may be obtained from equation (*8.15*). If at some frequency the loop phase shift $\alpha + \theta$ is 0, 2π, or some multiple of 2π radians, then (*8.15*) is

$$A'_v = (\mathcal{A}\underline{/\alpha})/(1 - \mathcal{A}\beta)$$

If $\mathcal{A}\beta = 1$ at this frequency, then A'_v becomes infinite which means that the system is unstable and will oscillate. It is a challenging problem to design a stable amplifier with many stages of amplification and a high overall voltage gain.

The feedback amplifier system discussed in this section is a simple example of voltage feedback. A signal voltage which is proportional to the signal voltage across the amplifier output terminals is applied (fed back) in series with the amplifier input. In this simple case the transfer function of the feedback network is a dimensionless voltage ratio; hence it is called voltage feedback.

Other types of feedback are also used in practical amplifiers. A second type is one where a signal voltage which is proportional to the current in the output load is applied in series with the input of the amplifier. In this case the transfer function of the feedback network is a transfer impedance. The simple common cathode (or emitter) amplifier stage with an unbypassed resistor in the cathode (or emitter) is an example. This type of feedback has often been called current feedback, but this is not true current feedback. The terminology was adopted before transistors, when true current feedback had not yet been realized.

True current feedback is the third type. In true current feedback a signal current which is proportional to the current in the output load is added to the amplifier input current. The feedback network transfer function is a dimensionless current ratio. In vacuum tube amplifiers the input current is usually zero, so true current feedback is seldom realized.

A fourth type of feedback involves adding a signal current which is proportional to the voltage across the output terminals to the amplifier input current. The feedback network transfer function is a transfer admittance. This type of feedback is less frequently used.

In general, the amplifier and the feedback network are each represented by two port equivalent circuits. The circuit equations may then be written and solved without necessarily classifying the feedback network as belonging to one of the four categories. However, some of the feedback amplifier equations are derived by assuming a certain model or type of feedback. Hence it is convenient to have some designation for each of the four categories of feedback systems. The categories are identified by the type of transfer function used to characterize the feedback network. The four types of feedback may be designated as voltage-voltage, voltage-current, current-current, and current-voltage, respectively. This at least serves to distinguish the types of feedback. It will be shown in later sections that equations derived for voltage feedback do not apply for other types of feedback. In some cases different types of feedback give opposite effects. See Problem 8.12.

Only single loop feedback amplifier systems will be studied in this introduction to feedback. As might be expected, a practical feedback amplifier system may contain multiple loops. For example, an amplifier with an inverse voltage feedback loop from the output to the input may include amplifier stages having unbypassed emitter (or cathode) resistors.

8.7 EFFECT OF INVERSE VOLTAGE FEEDBACK ON AMPLIFIER BANDWIDTH

Inverse voltage feedback may be employed to improve the frequency response of an amplifier system. Consider the feedback amplifier system of Fig. 8-13, page 240. Assume $\alpha + \theta = 180°$ in the mid-band range of frequencies and $\beta\underline{/\theta}$ is independent of frequency. Let f_1 and f_2 be the lower and upper half-power frequencies of the amplifier without feedback. Assume that the high frequency voltage gain of the amplifier without feedback is

$$A_v(\text{h.f.}) \quad = \quad \frac{\mathcal{A}\underline{/\alpha}}{1 + jf/f_2} \tag{8.17}$$

where $\mathcal{A}\underline{/\alpha}$ is the mid-band voltage gain. Since $\beta\underline{/\theta}$ is assumed independent of frequency, (*8.17*) may be substituted into (*8.15*) to give

$$A'_v(\text{h.f.}) \quad = \quad \frac{A_v(\text{h.f.})}{1 - [A_v(\text{h.f.})]\,\beta\underline{/\theta}} \quad = \quad \frac{\mathcal{A}\underline{/\alpha}}{1 + jf/f_2 - \mathcal{A}\beta\underline{/\alpha + \theta}}$$

Since $\alpha + \theta = 180°$ in the mid-band frequency range,

$$A_v(\text{h.f.}) \quad = \quad \frac{\mathcal{A}\underline{/\alpha}}{1 + \mathcal{A}\beta + jf/f_2} \tag{8.18}$$

whose numerator is the mid-band voltage gain which is independent of frequency. The transfer function of the feedback network $\beta\underline{/\theta}$ is assumed to be independent of frequency. The upper half-power frequency of the amplifier system with feedback may be found by setting the real part of the denominator of (*8.18*) equal to the magnitude of the imaginary part. If the upper half-power frequency of the amplifier system with feedback is denoted f'_2,

$$1 + \mathcal{A}\beta \;=\; f'_2/f_2 \qquad \text{or} \qquad f'_2 \;=\; (1 + \mathcal{A}\beta)f_2 \tag{8.19}$$

Since $\mathcal{A}\beta$ is usually large compared to 1, the upper half-power frequency is much higher with inverse voltage feedback as shown by (*8.19*). The derivation of this equation involves assumptions which may be only approximately true in practice.

Assume that the low frequency voltage gain of the amplifier without feedback is

$$A_v(\text{l.f.}) \quad = \quad \frac{\mathcal{A}\underline{/\alpha}}{1 - jf_1/f}$$

Using assumptions similar to those made for the high frequency case, the overall low frequency amplifier voltage gain with feedback is

$$A'_v(\text{l.f.}) \quad = \quad \frac{\mathcal{A}\underline{/\alpha}}{1 + \mathcal{A}\beta - jf_1/f}$$

and the lower half-power frequency of the amplifier system with inverse voltage feedback is

$$f'_1 \;=\; f_1/(1 + \mathcal{A}\beta)$$

When $\mathcal{A}\beta$ is large, the lower half-power frequency for the amplifier system with feedback is lower than without feedback. This is significant in terms of reducing the lower half-power frequency. The overall bandwidth is given by $\Delta f = f_2 - f_1$. Since usually $f_1 \ll f_2$, the bandwidth will increase because f_2 is higher when there is a significant amount of feedback. Also, the low frequency response is proportionately improved but this is not apparent in the increased bandwidth.

Inverse voltage feedback is used frequently to improve the frequency response of electronic amplifiers. The factor $1 + \mathcal{A}\beta$ which appears in the equations for the upper and lower half-power frequencies also appears in the denominator of the equation for the mid-band voltage gain. Hence the mid-band voltage gain is decreased by the factor $1/(1 + \mathcal{A}\beta)$. The improvements in performance are achieved at the expense of voltage gain.

8.8 EFFECT OF INVERSE VOLTAGE FEEDBACK ON AMPLIFIER OUTPUT IMPEDANCE

To study the effect of feedback, the equivalent circuit of the simple inverse voltage feedback amplifier system must include the amplifier output impedance. If general two port equivalent circuits are used for the amplifier and feedback networks, the amplifier output impedance is included. However, the simplest equivalent circuit is usually used in order to simplify the computations. The equivalent circuit for the amplifier system with feedback is shown in Fig. 8-15.

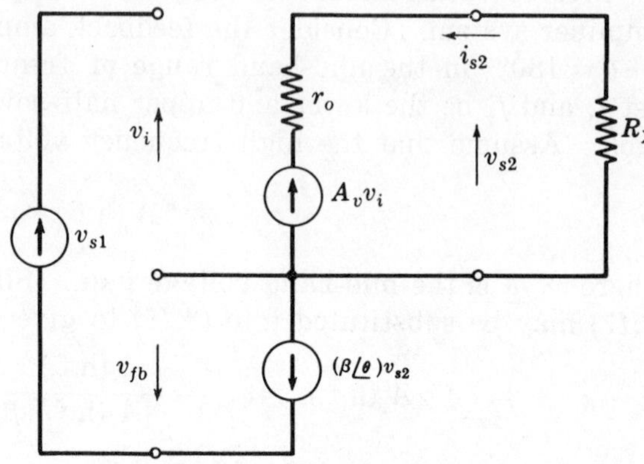

Fig. 8-15. Feedback Amplifier Equivalent Circuit

It is assumed that the amplifier has infinite input impedance and that the mid-band output impedance of the amplifier without feedback is r_o. The amplifier transfer function A_v is the open circuit voltage gain. The terminating load resistance is R_X and is not included in calculating A_v. The assumptions regarding the feedback network are the same as those stated in previous sections.

It is first necessary to derive the equation for the amplifier output voltage v_{s2} without feedback in terms of i_{s2} and v_{s1}. The loop equation for the amplifier output circuit is

$$-v_{s2} + r_o i_{s2} + A_v v_i = 0$$

Substituting (*8.12*) into this equation and then using (*8.13*),

$$-v_{s2} + r_o i_{s2} + A_v(v_{fb} + v_{s1}) = 0$$

$$-v_{s2} + r_o i_{s2} + A_v[(\beta\angle\theta)v_{s2} + v_{s1}] = 0 \qquad (8.20)$$

The open circuit voltage gain for the amplifier is $A_v = \mathcal{A}\angle\alpha$, which is put into (*8.20*) to find

$$v_{s2} = \frac{r_o i_{s2} + (\mathcal{A}\angle\alpha)v_{s1}}{1 - \mathcal{A}\beta\angle\alpha + \theta} \qquad (8.21)$$

When $i_{s2} = 0$, there is no external load resistance ($R_X = \infty$) and (*8.21*) reduces to (*8.15*).

When i_{s2} is not zero and r_o is not zero, v_{s2} is in general a function of both i_{s2} and v_{s1}. The mid-band output impedance of the amplifier system with feedback is found by setting $v_{s1} = 0$ and then solving for

$$r_o' = v_{s2}/i_{s2} = \frac{r_o}{1 - \mathcal{A}\beta\angle\alpha + \theta}$$

The output impedance with feedback r_o' is resistive only in the mid-band range of frequencies where $\mathcal{A}$, β, α and θ are assumed to be independent of frequency. When $\alpha + \theta = 180°$, the mid-band output impedance of the amplifier system with feedback is

$$r_o' = r_o/(1 + \mathcal{A}\beta)$$

which indicates that the mid-band output impedance is reduced by inverse voltage feedback, but voltage gain is sacrificed.

8.9 EFFECT OF INVERSE VOLTAGE FEEDBACK ON AMPLIFIER INPUT IMPEDANCE

In previous sections the amplifier input impedance was assumed infinite. When the input impedance is finite, it must be included in the equivalent circuit. The mid-band equivalent circuit including the input resistance[1] r_i is shown in Fig. 8-16.

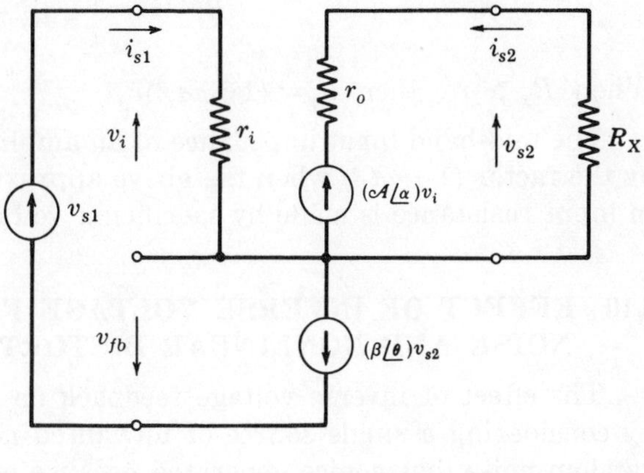

Fig. 8-16. Feedback Amplifier Equivalent Circuit

It is assumed that the feedback network has very high input impedance and very low output impedance. The output impedance of the feedback network is significant because the input loop impedance is no longer infinite and the input current is not zero. Since the amplifier input resistance is usually high, it is only necessary that the output impedance of the feedback network be relatively small. An example of a very simple feedback network is shown in Fig. 8-17. It is assumed that capacitor C is large, resistor R_1 is large, and resistor R_2 is small. Then the output resistance is low and the input resistance (mid-band frequency range) of the feedback network is high.

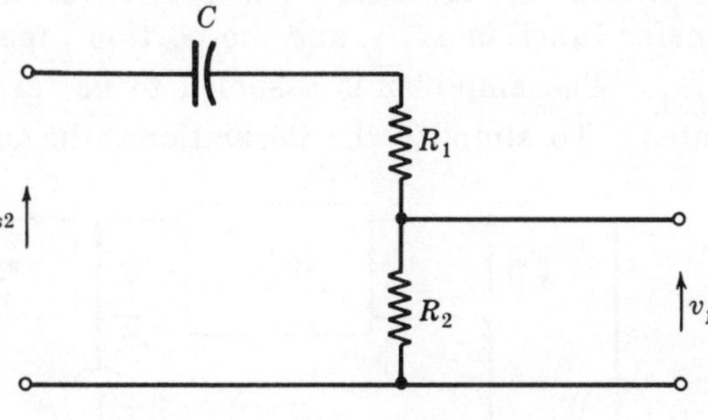

Fig. 8-17. Feedback Network

The input loop equation for the circuit of Fig. 8-16 is

$$-v_{s1} + r_i i_{s1} - v_{fb} = 0 \tag{8.22}$$

The transfer function for the feedback network is

$$v_{fb} = (\beta\angle\theta)v_{s2} \tag{8.23}$$

The output loop equation is $$-v_{s2} + r_o i_{s2} + (\mathcal{A}\angle\alpha)v_i = 0 \tag{8.24}$$

Now put $i_{s2} = -v_{s2}/R_X$ and $v_i = r_i i_{s1}$ into (*8.24*) to find $v_{s2} = \left(\frac{R_X r_i \mathcal{A}\angle\alpha}{r_o + R_X}\right) i_{s1}$ which when substituted into (*8.23*) gives

$$v_{fb} = \left(\frac{R_X r_i}{r_o + R_X}\right)(\mathcal{A}\beta\angle\alpha+\theta)i_{s1}$$

Entering this value of v_{fb} into (*8.22*), we obtain

$$v_{s1} = \left[1 - \left(\frac{R_X}{r_o + R_X}\right)\mathcal{A}\beta\angle\alpha+\theta\right] r_i i_{s1}$$

[1] r_i is a two port equivalent circuit parameter (the reverse transfer parameter is zero). To avoid ambiguity, the two port parameter r_i is denoted with a lower case letter and the input resistance (mid-band one port representation) for the amplifier is denoted with an upper case letter.

and the mid-band input impedance for the amplifier system with feedback is

$$R_i' \;=\; v_{s1}/i_{s1} \;=\; \left[1 - \left(\frac{R_X}{r_o + R_X}\right)\mathcal{A}\beta\underline{/\alpha+\theta}\right] r_i$$

When $\alpha + \theta = 180°$,

$$R_i' \;=\; \left[1 + \left(\frac{R_X}{r_o + R_X}\right)\mathcal{A}\beta\right] r_i$$

When $R_X \gg r_o$, then $R_i' = (1 + \mathcal{A}\beta)r_i$.

The mid-band input impedance of an amplifier with inverse voltage feedback is increased by the factor $(1 + \mathcal{A}\beta)$ when the above approximations are valid. Again, this improvement in input resistance is made by sacrificing voltage gain.

8.10 EFFECT OF INVERSE VOLTAGE FEEDBACK ON AMPLIFIER NOISE AND NONLINEAR DISTORTION

The effect of inverse voltage feedback on amplifier noise and distortion may be shown by considering a single source of unwanted noise signals. The unwanted signals may be a random noise, harmonics generated because of nonlinearity, man-made noise such as hum, or any other unwanted signal. A block diagram of the amplifier with a single noise source n is shown in Fig. 8-18. The portion of the amplifier following the noise source has a transfer function $\mathcal{A}_2\underline{/\alpha_2}$ and the portion preceding the noise source has a transfer function $\mathcal{A}_1\underline{/\alpha_1}$. The amplifier is assumed to be terminated when the transfer functions are calculated. To simplify the derivations, the amplifier input impedance is assumed infinite.

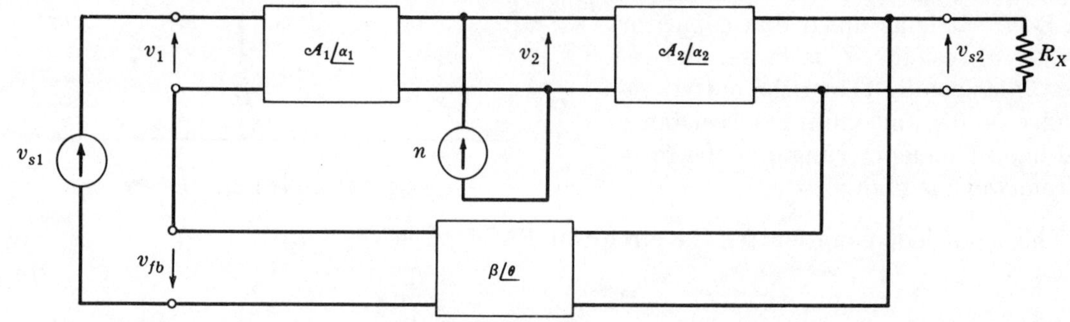

Fig. 8-18. Feedback Amplifier System with Noise Source

The single noise source is located within the amplifier as shown in the block diagram of Fig. 8-18. It is assumed that the instantaneous value of the noise voltage is linearly

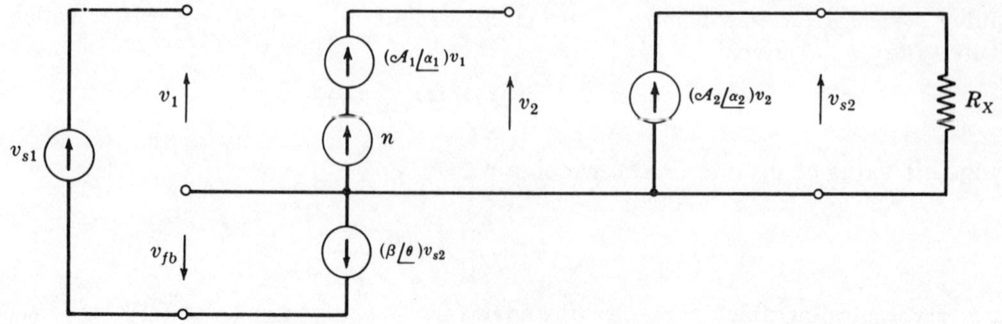

Fig. 8-19. Equivalent Circuit for the Amplifier System with Noise

added to the instantaneous value of the signal voltage at that point. An equivalent circuit for the amplifier system with noise is given in Fig. 8-19. It is assumed that the two amplifier sections are effectively isolated and that the input impedance of each section is infinite. If this is not done, the problem becomes very complicated. The instantaneous noise voltage n is added linearly to the signal output voltage of the first section of the amplifier as shown in Fig. 8-19.

The input loop equation is $-v_{s1} + v_1 - v_{fb} = 0$. Using $v_{fb} = (\beta\angle\theta)v_{s2}$,

$$v_1 = v_{s1} + (\beta\angle\theta)v_{s2} \tag{8.25}$$

The equation for the loop which includes the output of the first section of the amplifier is

$$-v_2 + (\mathcal{A}_1\angle\alpha_1)v_1 + n = 0$$

into which we put (8.25) and find

$$v_2 = n + (\mathcal{A}_1\angle\alpha_1)v_{s1} + (\mathcal{A}_1\beta\angle\alpha_1 + \theta)v_{s2} \tag{8.26}$$

The equation for the output loop is

$$-v_{s2} + (\mathcal{A}_2\angle\alpha_2)v_2 = 0$$

into which we put (8.26) and obtain

$$v_{s2} = (\mathcal{A}_2\angle\alpha_2)n + (\mathcal{A}_1\mathcal{A}_2\angle\alpha_1 + \alpha_2)v_{s1} + (\mathcal{A}_1\mathcal{A}_2\beta\angle\alpha_1 + \alpha_2 + \theta)v_{s2} = 0$$

or

$$v_{s2} = \frac{(\mathcal{A}_2\angle\alpha_2)n + (\mathcal{A}_1\mathcal{A}_2\angle\alpha_1 + \alpha_2)v_{s1}}{1 - \mathcal{A}_1\mathcal{A}_2\beta\angle\alpha_1 + \alpha_2 + \theta} \tag{8.27}$$

The overall amplifier gain without feedback is $\mathcal{A}\angle\alpha = \mathcal{A}_1\mathcal{A}_2\angle\alpha_1 + \alpha_2$. Putting this into (8.27),

$$v_{s2} = \frac{(\mathcal{A}_2\angle\alpha_2)n + (\mathcal{A}\angle\alpha)v_{s1}}{1 - \mathcal{A}\beta\angle\alpha + \theta}$$

When n is very small compared to v_{s1}, the amplifier transfer function with feedback reduces to (8.15).

When the voltage feedback is inverse and $\alpha + \theta = 180°$,

$$v_{s2} = \left(\frac{\mathcal{A}_2\angle\alpha_2}{1 + \mathcal{A}\beta}\right)n + \left(\frac{\mathcal{A}\angle\alpha}{1 + \mathcal{A}\beta}\right)v_{s1} \tag{8.28}$$

which indicates that the noise component of the output voltage v_{s2} is

$$\text{output noise with feedback} = \left(\frac{\mathcal{A}_2\angle\alpha_2}{1 + \mathcal{A}\beta}\right)n \tag{8.29}$$

If there were no voltage feedback, the output noise would be

$$\text{output noise without feedback} = (\mathcal{A}_2\angle\alpha_2)n \tag{8.30}$$

A comparison of equations (8.29) and (8.30) shows that the output noise has been reduced by a factor $1/(1 + \mathcal{A}\beta)$. This is misleading because the overall voltage gain has also been reduced by the same factor.

Consider the signal-plus-noise-to-noise ratio at the output of the amplifier with feedback. The output voltage v_{s2} is the signal plus noise, and the noise is given by (8.29). Writing

$$(S + N)/N = v_{s2}/[(\mathcal{A}_2\angle\alpha_2)n/(1 + \mathcal{A}\beta)]$$

and substituting v_{s2} from (8.28) and simplifying,

$$(S + N)/N = 1 + \left(\frac{\mathcal{A}\angle\alpha}{\mathcal{A}_2\angle\alpha_2}\right)\frac{v_{s1}}{n} \tag{8.31}$$

For given values of the input signal voltage v_{s1} and the noise voltage n, the $(S+N)/N$ ratio is improved only if $(\mathcal{A}\angle\alpha)/(\mathcal{A}_2\angle\alpha_2)$ is increased.

Since the noise generated within an amplifier can never be eliminated, $(S+N)/N$ is always finite. The model assuming only one source is useful even though the sources of noise (and distortion) in an amplifier are numerous. The effect of many sources may be found by applying superposition.

If the gain of the last amplifier section $\mathcal{A}_2$ is small compared to the overall gain $\mathcal{A}$, then the noise source is near the output of the amplifier. When this is true, inverse voltage feedback increases $(S+N)/N$ as indicated by (*8.31*). However, if the noise source is at the input of the amplifier so that $\mathcal{A}_2 = \mathcal{A}$, then

$$(S+N)/N \;=\; 1 + v_{s1}/n \qquad \text{or} \qquad (S+N)/N \;=\; (n+v_{s1})/n$$

which shows that if the amplifier is noise free and there is a noise source in the input, then $(S+N)/N$ at the output equals $(S+N)/N$ at the input; in other words, the noise figure NF is 1. Inverse voltage feedback cannot be used to improve the $(S+N)/N$ ratio in this case. This focuses attention to the real problem of reducing noise in amplifiers. Noise can be reduced by inverse voltage feedback only if there is sufficient voltage gain preceding the noise source. Hence the input stage of the amplifier is the most critical when designing to minimize noise. Noise appearing in the input signal cannot be reduced by the design of the amplifier.

The discussion in this section assumed a single noise source n which may be any source of unwanted signal, including harmonics generated because of nonlinearity. Equations (*8.29*) and (*8.30*) show that the effective output noise is reduced by the factor $1/(1+\mathcal{A}\beta)$; this is the basis for saying that nonlinear distortion is reduced by this factor. Nonlinear distortion is usually a problem in the output stages of an amplifier; inverse feedback will reduce the distortion since most of the voltage gain precedes the source.

8.11 FEEDBACK AMPLIFIER STABILITY

Many problems were ignored in deriving the equations in previous sections on feedback. It was usually assumed that only mid-band frequencies were involved and that $\alpha+\theta = 180°$. Although the derived equations apply only to the simple models shown, they are useful for predicting the effect of inverse voltage feedback on the performance of more complicated amplifier systems. It is now necessary to consider stability, which involves extending the frequency range to the infinite frequency interval.

Equation (*8.15*) is general for the simple voltage feedback system of Fig. 8-13:

$$A'_v \;=\; \frac{\mathcal{A}\angle\alpha}{1-\mathcal{A}\beta\angle\alpha+\theta} \tag{8.15}$$

In general $\mathcal{A}$, α, β and θ are functions of frequency. Most of the interest in previous sections was in the case where $\alpha+\theta = 180°$ and in the mid-band range of frequencies where $\mathcal{A}$ and β are independent of frequency.

A feedback amplifier system will be unstable[1] if the denominator of (*8.15*) becomes zero at any frequency. Hence stability involves investigating the magnitude of the denominator over the infinite frequency interval.

If the magnitude of the denominator of (*8.15*) is less than 1, then the feedback is *regenerative* (positive). The voltage gain of the amplifier is then higher than it would be without

[1]An amplifier is unstable if it maintains self-oscillation or if it is very near the point of self-oscillation.

feedback at the frequencies where the feedback is regenerative. Assuming the magnitude of the denominator does not go to zero at any frequency, the amplifier will not oscillate (engage in self-sustained oscillation). When the feedback is regenerative, the problems of unstable circuit parameters and problems such as poor line voltage regulation are magnified. If the amplifier is regenerative in the mid-band frequency range, the mid-band output impedance increases, the bandwidth decreases, and the effects of amplifier noise and amplifier distortion are magnified.

It might be supposed that regenerative feedback is employed only in oscillator circuits, but this is not the case. Narrow-band low frequency filters using frequency selective regenerative feedback amplifiers have been used in commercial spectrum analyzers. The simple voltage feedback model suggests that this should be possible, but it is not very useful for solving design problems.

If the magnitude of the denominator of (*8.15*) is greater than 1, then the feedback is inverse (degenerative or negative). The voltage gain is then reduced and performance improved as was shown in earlier sections where the special case $\alpha + \theta = 180°$ was considered. In most amplifier design problems, the feedback is inverse in the mid-band frequency range. The entire frequency range is then investigated to determine if the amplifier is stable. Frequencies where an amplifier with feedback will be unstable are usually outside the mid-band range.

Since $\mathcal{A}\beta\underline{/\alpha+\theta}$ in the denominator of (*8.15*) is the only frequency dependent term, amplifier stability may be determined by examining it. This is more convenient than investigating $1 - \mathcal{A}\beta\underline{/\alpha+\theta}$ for the infinite frequency range.

One way of investigating $\mathcal{A}\beta\underline{/\alpha+\theta}$ is to plot this term[1] in the complex plane for all frequencies in the infinite interval. Nyquist extensively studied the problem of regeneration and developed the Nyquist stability criterion based on a plot of $\mathcal{A}\beta\underline{/\alpha+\theta}$. He showed that this plot results in a closed curve in the complex plane and that if this curve encloses the $1 + j0$ point, then the amplifier is unstable and will oscillate.

The Nyquist stability criterion is based on the fact that the amplifier will oscillate if the denominator of (*8.15*) goes to zero. The feedback changes from degenerative to regenerative when

$$|1 - \mathcal{A}\beta\underline{/\alpha+\theta}| = 1$$

Since $\mathcal{A}\beta\underline{/\alpha+\theta}$ may be expressed in rectangular form as $x + jy$,

$$|1 - (x+jy)| = 1 \quad \text{or} \quad (1-x)^2 + y^2 = 1 \tag{8.32}$$

which represents a circle of unit radius and center (1, 0). At frequencies where the plot of $\mathcal{A}\beta\underline{/\alpha+\theta}$ is within the unit circle of (*8.32*), the feedback is regenerative. At frequencies where the plot of $\mathcal{A}\beta\underline{/\alpha+\theta}$ is outside this unit circle, the feedback is degenerative. At the point of oscillation,

$$|1 - \mathcal{A}\beta\underline{/\alpha+\theta}| = 0 \quad \text{or} \quad |1 - (x+jy)| = 0$$

See Problem 8.14.

There are other methods of investigating stability in feedback amplifiers. The Nyquist criterion is convenient to use in a simple voltage feedback amplifier system.

8.12 SIMPLE VOLTAGE-CURRENT FEEDBACK

It was indicated earlier that the voltage feedback (or voltage-voltage feedback) system is not the only feedback model. A simple voltage-current (often called current) feedback

[1]Some prefer to plot the negative of this term, which only changes the statement of the Nyquist stability criterion.

system is briefly discussed in this section. A block diagram of the system is shown in Fig. 8-20. The transfer function of the amplifier is given as the transfer admittance $y_t = \mathcal{Y}\underline{/\alpha}$. The amplifier is represented by using a voltage controlled, current generator as shown in Fig. 8-21. The impedance $Z_{fb} = \mathcal{Z}\underline{/\theta}$ is the feedback impedance connected in series with the output load resistance R_X. The input impedance of the amplifier is assumed infinite as shown in Fig. 8-21.

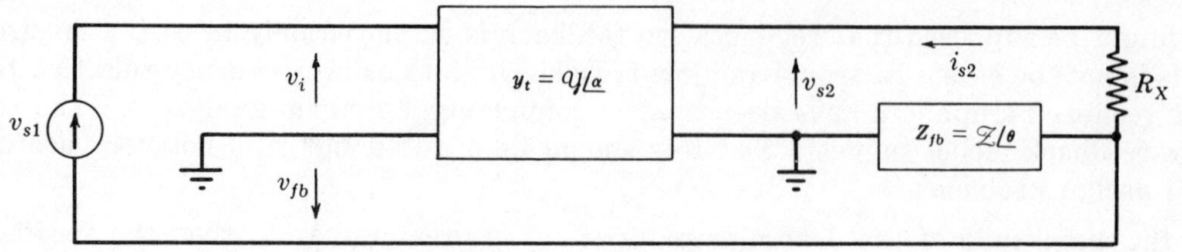

Fig. 8-20. Current Feedback Amplifier

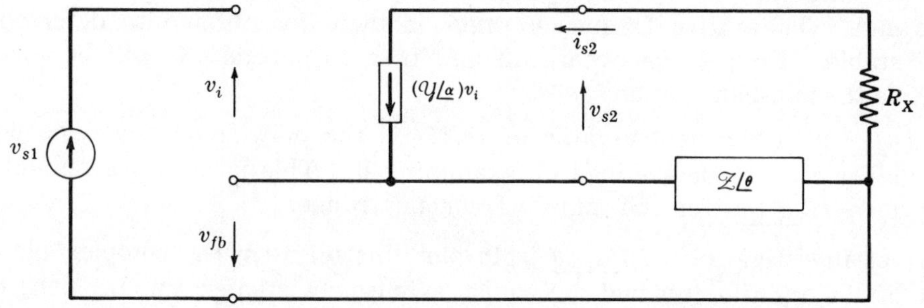

Fig. 8-21. Equivalent Circuit of a Voltage-Current Feedback Amplifier

The transfer function of the amplifier is

$$y_t = \mathcal{Y}\underline{/\alpha} = i_{s2}/v_i \quad \text{or} \quad v_i = i_{s2}/\mathcal{Y}\underline{/\alpha} \tag{8.33}$$

The transfer function of the feedback network is

$$Z_{fb} = \mathcal{Z}\underline{/\theta} = -v_{fb}/i_{s2} \quad \text{or} \quad v_{fb} = -(\mathcal{Z}\underline{/\theta})i_{s2} \tag{8.34}$$

The input loop equation is

$$-v_{s1} + v_i - v_{fb} = 0 \tag{8.35}$$

Substituting (8.33) and (8.34) into (8.35),

$$-v_{s1} + i_{s2}/(\mathcal{Y}\underline{/\alpha}) + (\mathcal{Z}\underline{/\theta})i_{s2} = 0$$

and the overall transfer admittance of the amplifier system with feedback is

$$y_t' = i_{s2}/v_{s1} = \frac{\mathcal{Y}\underline{/\alpha}}{1 + \mathcal{Y}\mathcal{Z}\underline{/\alpha + \theta}} \tag{8.36}$$

which has the same form as equation (8.15), page 241, for the voltage feedback system, except for the plus sign of the denominator term $\mathcal{Y}\mathcal{Z}\underline{/\alpha+\theta}$. If the directions of i_{s2} and the current generator are reversed, then this term has a minus sign. This, however, makes the reference direction of i_{s2} inconsistent with established conventions.

At frequencies where the magnitude of the denominator of (8.36) is less than 1, the system is regenerative; at frequencies where it is greater than 1, the system is degenerative (inverse feedback). Assuming that $\alpha + \theta = 0$ in the mid-band range and that $\mathcal{Y}\mathcal{Z} \gg 1$,

$$y_t' \cong \frac{\mathcal{Y}\underline{/\alpha}}{\mathcal{Y}\mathcal{Z}} = \frac{1}{\mathcal{Z}}\underline{/\alpha}$$

which shows that when $\mathcal{YZ}$ is large, the magnitude of the mid-band transfer admittance with inverse feedback depends only on the magnitude of the feedback impedance. Since the feedback impedance usually involves only passive elements, the magnitude of the mid-band transfer admittance may be made very stable when $\mathcal{YZ}$ is large.

The appropriate equivalent circuits may be drawn and the equations derived which show how voltage-current feedback affects the input impedance, output impedance, upper and lower half-power frequencies, and noise of an amplifier. The general methods are similar to those used in deriving the voltage feedback equations. The representations for the amplifier and feedback networks will be similar to those of Fig. 8-21 but with the appropriate modifications for the amplifier equivalent circuit. See Problem 8.12 and 8.13.

The Nyquist stability criterion may also be applied to the voltage-current feedback amplifier system. In this case it is convenient to investigate $-\mathcal{YZ}\underline{/\alpha+\theta}$ over the infinite frequency interval. The amplifier will be stable if the plot of $-\mathcal{YZ}\underline{/\alpha+\theta}$ does not encircle the point (1, 0). The term $\mathcal{YZ}\underline{/\alpha+\theta}$ is dimensionless, as are the corresponding terms of each of the four types of feedback systems. It is necessary, however, to exercise caution when comparing the different feedback systems because of the established conventions. For example, the term $\mathcal{YZ}\underline{/\alpha+\theta}$ of equation (*8.36*) has a plus sign while the corresponding term of (*8.15*) has a minus sign.

The remaining two types of feedback systems could also be investigated in a similar manner. See Problem 8.15. It should be emphasized that many handbook equations are derived only for the voltage-voltage feedback model. In general, a two port equivalent circuit is used for the amplifier and another for the feedback network. The controlled generators of the two port equivalent circuits may each be either voltage or current generators and each of these may be voltage or current controlled. Since a complete equivalent circuit for a feedback system contains at least two individual two port equivalent circuits, the number of possible variations is rather large.

Simpler models are usually used because the general models may become very complicated when many of the parameters are complex and functions of frequency. Thus the simplest model which is an adequate approximation is usually used.

8.13 OPERATIONAL AMPLIFIERS

Operational amplifiers are specially designed and packaged amplifier units which have the following general characteristics:

1. The input impedance of the amplifier is very high.
2. The output impedance of the amplifier is low.
3. The amplifier is direct coupled[1].
4. The amplifier phase shift is 180°.
5. The voltage gain is high.

The operational amplifier is considered to be a complete unit and the symbol is shown in Fig. 8-22. These amplifiers are designed for use in circuits involving feedback as shown in Fig. 8-23. In this simplified diagram the balancing circuit is designated by a block B. (Since the

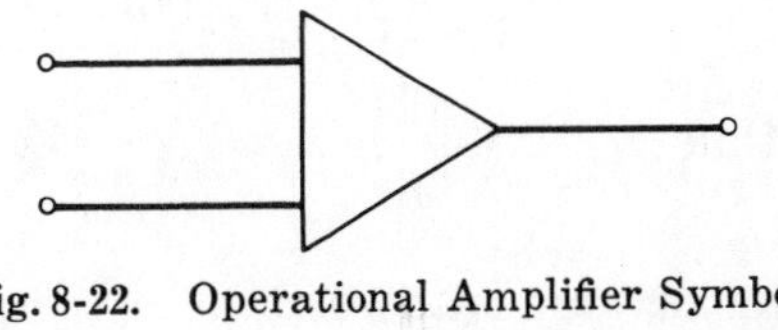

Fig. 8-22. Operational Amplifier Symbol

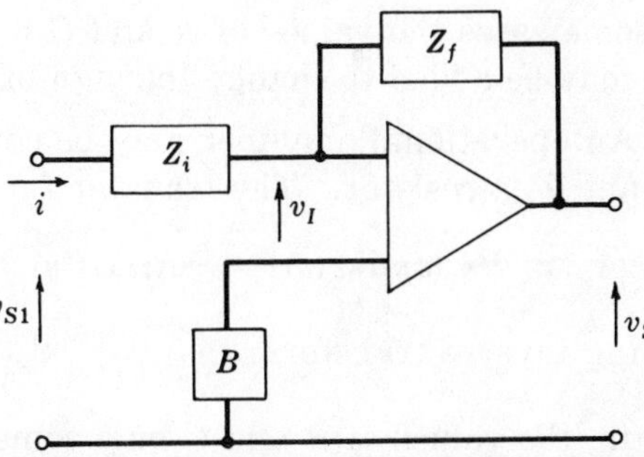

Fig. 8-23. Operational Amplifier with Feedback

[1]Direct coupled amplifiers have not been discussed because only a superficial treatment would be possible in this book. The direct coupled amplifier has no coupling capacitors. A step function applied to the input causes the output voltage to shift and remain shifted (no sag).

amplifier is direct coupled, a balancing circuit is usually provided so that the output voltage v_{S2} may be made zero when the input voltage v_{S1} is zero.) The feedback impedance Z_f is connected from the output of the amplifier to the amplifier input. The input impedance Z_i is connected in series with the amplifier input as shown in Fig. 8-23.

The equation for the transfer function with feedback is to be derived. Since the operational amplifier input impedance is assumed infinite, the input current is zero. The input and output loop equations are respectively

$$-v_{S1} + Z_i i + v_I = 0 \quad \text{and} \quad -v_{S2} - Z_f i + v_I = 0$$

The operational amplifier has 180° phase shift, so the open-loop (no feedback) transfer function is $A_v = v_{S2}/v_I = -\mathcal{A}$. Thus $v_I = -v_{S2}/\mathcal{A}$ is put into the above equations to obtain

$$Z_i i - v_{S2}/\mathcal{A} = v_{S1} \quad \text{and} \quad +Z_f i + (1 + 1/\mathcal{A})v_{S2} = 0 \tag{8.37}$$

from which the transfer function is

$$A_v = v_{S2}/v_{S1} = \frac{-Z_f}{Z_i + (Z_i + Z_f)/\mathcal{A}} \tag{8.38}$$

It was assumed that the magnitude of the operational amplifier gain $\mathcal{A}$ is high. When $(Z_i + Z_f)/\mathcal{A} \ll Z_i$,

$$A_v = -Z_f/Z_i \tag{8.39}$$

which is frequently taken as the transfer function with feedback.

A very stable amplifier may be realized by using two precision resistors as Z_f and Z_i. The voltage gain is determined by the ratio of the resistors, provided the assumptions made in the derivation are valid. The phase shift is 180°.

An inverting amplifier (sign changer) is an amplifier with unity gain and 180° phase shift. Two identical resistors may be used as Z_i and Z_f to realize an inverting amplifier.

An operational amplifier may be converted into an integrator by making Z_i a resistor and Z_f a capacitor. This is most easily shown by writing the Laplace transforms of the voltages and currents. The transfer function is then

$$A_v = V_{S2}(s)/V_{S1}(s) = -Z_f/Z_i = -(1/Cs)/R = -1/RCs$$

and thus
$$V_{S2}(s) = -\frac{1}{RC} \cdot \frac{V_{S1}(s)}{s}$$

having inverse transform
$$v_{S2} = -\frac{1}{RC}\int v_{S1}\,dt$$

In some cases the values of R and C may be chosen so that $RC = 1$. An inverting amplifier may be added to change the sign of the equations if necessary.

An operational amplifier may be converted into a differentiator by making Z_i a capacitor and Z_f a resistor. The transfer function is

$$A_v = V_{S2}(s)/V_{S1}(s) = -R/(1/Cs) = -RCs \quad \text{from which} \quad V_{S2}(s) = -RCs\,V_{S1}(s)$$

having inverse transform
$$v_{S2} = -RC\,dv_{S1}/dt$$

Again, the values of R and C may sometimes be chosen so that their product is unity. An inverting amplifier may also be added as a sign changer.

Operational amplifiers are used widely, including applications where nonlinear elements are taken as Z_i and Z_f. The topic is quite extensive and beyond the scope of this book; however, the derivations of the basic equations are rather simple.

Solved Problems

POWER AMPLIFIERS

8.1. Given the Class-A power amplifier circuit of Fig. 8-24. Cathode bias is not used; there is no cathode resistor power dissipation. A triode tube is employed to eliminate the screen grid circuit power dissipation. An ideal output transformer is assumed, so there are no transformer losses. Show that the efficiency of a Class-A amplifier stage cannot exceed 50%.

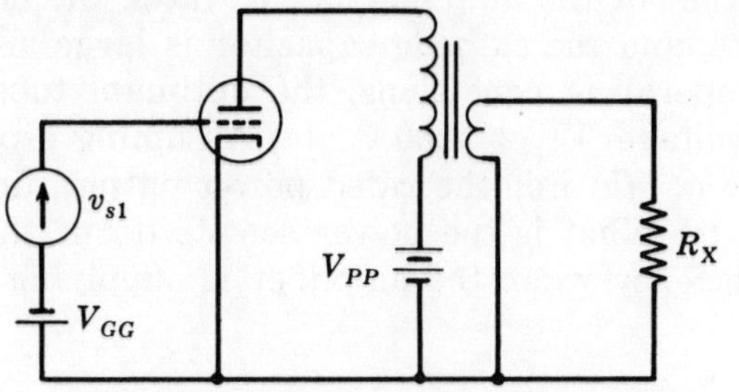

Fig. 8-24. Class-A Power Amplifier

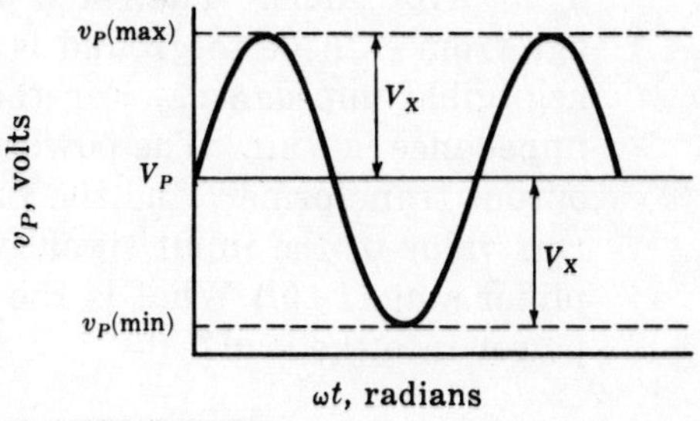

Fig. 8-25. Instantaneous Plate Voltage

Since the output transformer is ideal, there is no power loss in the transformer and the primary dc resistance is zero. The ac resistance looking into the primary is $R_X^* = a^2 R_X$. If operation is within the linear range and a sinusoidal input signal is assumed, then the voltage across the primary of the output transformer is $v_X = V_X \sin \omega t$. (This is written as v_X to indicate that the average value is zero assuming an ideal transformer.) Since the primary dc resistance is zero, $V_P = V_{PP}$. The instantaneous plate voltage is sketched in Fig. 8-25.

It may be seen that (assuming linear operation)

$$V_X \;=\; v_P(\text{max}) - V_P \;=\; V_P - v_P(\text{min}) \;=\; V_P[1 - v_P(\text{min})/V_P]$$

Similarly, let the current in the primary be $i_P = I_P + i_p$ and write the signal current in the primary as $i_X = I_X \sin \omega t$. Then

$$I_X \;=\; i_P(\text{max}) - I_P \;=\; I_P - i_P(\text{min}) \;=\; I_P[1 - i_P(\text{min})/I_P]$$

Since R_X^* is a resistance, v_X and i_X are in phase. The average power into the transformer primary is the product of the rms values of the voltage and current; then $P_{\text{out}} = \frac{1}{2} V_X I_X$. The dc power input is $P_{\text{in}} = V_{PP} I_P = V_P I_P$. The efficiency is

$$(P_{\text{out}}/P_{\text{in}})\,100\% \;=\; \tfrac{1}{2}[1 - v_P(\text{min})/V_P][1 - i_P(\text{min})/I_P]\,100\%$$

and has the maximum value of 50% when the transfer characteristic curve is a straight line and $v_P(\text{min}) \to 0$ and $i_P(\text{min}) \to 0$.

The efficiency of any Class-A amplifier will be much less than 50% because of losses, such as in the transformer and cathode resistor. The efficiency is also lower because the dynamic transfer characteristic curve is not a straight line as $v_P(\text{min}) \to 0$.

8.2. Given the simplified push-pull amplifier circuit of Fig. 8-26 which includes an unbypassed cathode resistor R_K. Show that if the amplifier is perfectly balanced, there is no voltage drop across R_K at the fundamental signal frequency.

It is assumed that the plate currents may each be written in terms of a power series and that the corresponding coefficients are identical (circuits are balanced). The voltage from cathode to ground is

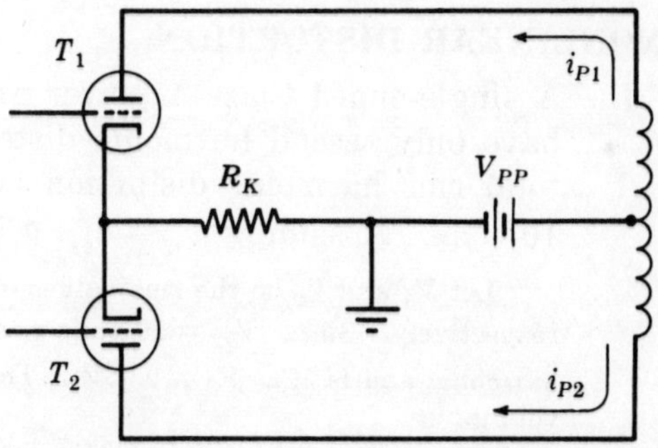

Fig. 8-26. Push-pull Amplifier

$$v_{KO} = R_K(i_{P1} + i_{P2}) = R_K(I_{P1} + a_1 v_{g1} + a_2 v_{g1}^2 + \cdots + I_{P2} + a_1 v_{g2} + a_2 v_{g2}^2 + \cdots)$$

Assuming $v_{g1} = V \sin \omega t$ and $v_{g2} = -V \sin \omega t$, and using $\sin^2 \theta = \frac{1}{2}(1 - \cos 2\theta)$,

$$v_{KO} = R_K[I_{P1} + I_{P2} + a_2 V^2 - a_2 V^2 \cos 2\omega t + \cdots]$$

When the two currents in R_K are added, the odd harmonic and fundamental frequency components of the cathode to ground voltage drop are zero. Hence it is not necessary to bypass the cathode to avoid signal degeneration (inverse feedback).

8.3. Given the Class-A power amplifier of Fig. 8-27 with tube parameters $r_p = 50$ kΩ, $g_m = 4100$ μmho. The rated power output of the amplifier is 2.5 watts, the dc voltage from cathode to ground is 12.5 volts, and the cathode capacitor is large and has negligible impedance. For the given operating conditions, the optimum tube load impedance is 5 kΩ. The power supply voltage $V_{PP} = 250$ v. (*a*) Assuming a perfect output transformer, find the turns ratio *a*. (*b*) For the rated power output, find the rms value of the input signal voltage. (*c*) What is the power sensitivity of the amplifier stage? (*d*) What is the plate efficiency when the amplifier is supplying rated power into the load?

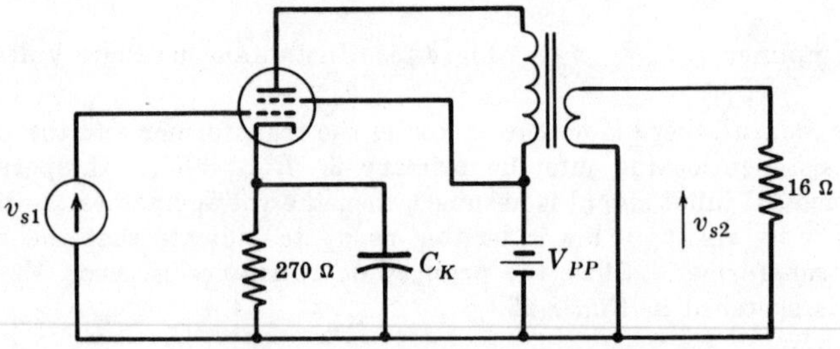

Fig. 8-27. Class-A Power Amplifier

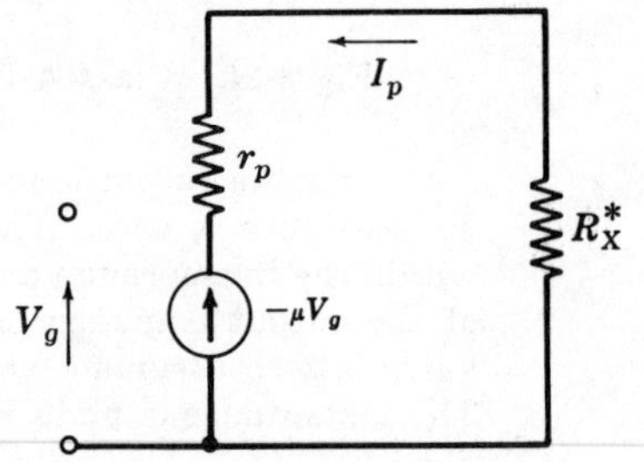

Fig. 8-28. Equivalent Circuit

(*a*) Using equation (*8.1*), page 230: $a = (Z_p/Z_S)^{1/2} = (5000/16)^{1/2} = 17.7$. The actual turns ratio might differ slightly because an integer number of turns would be used for each winding.

(*b*) Assuming the cathode is bypassed, the equivalent circuit is shown in Fig. 8-28. By equation (*8.2*), $R_X^* = a^2 R_X = (17.7)^2(16) = 5000$ Ω. This checks the calculations of part (*a*).

For the tube, $\mu = g_m r_p = 0.0041(50{,}000) = 205$. Since the power output is 2.5 watts, $5000 I_p^2 = 2.5$ or $I_p = 0.0224$ amp rms. Using rms values, the loop equation for Fig. 8-28 is

$$5000 I_p + 50{,}000 I_p - 205 V_g = 0 \quad \text{or} \quad I_p = 0.00373 V_g$$

Then $I_p = 0.00373 V_g = 0.0224$ or $V_g = 6.01$ volts rms. When the cathode is grounded, $V_{s1} = V_g = 6.01$ volts rms.

(*c*) The power sensitivity $= 2.5/(6.01)^2 = 0.0693$ watts/volt².

(*d*) $I_P = 12.5/270 = 0.0463$ amp, dc power input $= 250(0.0463) = 11.58$ watts, and plate efficiency $= 2.5/11.58 = 0.216 = 21.6\%$.

NONLINEAR DISTORTION

8.4. A single-ended Class-A power amplifier of the type shown in Fig. 8-1 is known to have only second harmonic distortion when the input signal is $v_{s1} = 1 \sin \omega t$. The total rms harmonic distortion is 5%. The rms value of the output voltage V_{s2} is 10 volts. Assuming $v_{s2} = a_0 + a_1 v_{s1} + a_2 v_{s1}^2$, find the values of the coefficients a_1 and a_2.

Let V_1 and V_2 be the rms values of the fundamental and second harmonic of the output signal, respectively. Since $V_{s2} = 10$ volts rms, $\sqrt{V_1^2 + V_2^2} = 10$. Since the harmonic distortion is second harmonic and is 5%, $V_2 = 0.05 V_1$. Then $1.0025 V_1^2 = 100$ or $V_1 \cong 10$, and $V_2 \cong 0.5$. The output voltage is

$$v_{s2} = a_0 + a_1 \sin \omega t + a_2 \sin^2 \omega t = a_0 + \tfrac{1}{2} a_2 + a_1 \sin \omega t - \tfrac{1}{2} a_2 \cos 2\omega t$$

from which $V_1 = a_1/\sqrt{2} = 10$ or $a_1 = 14.14$, and $V_2 = \frac{1}{2} a_2/\sqrt{2} = 0.5$ or $a_2 = 1.414$.

NOISE

8.5. Two resistors R_1 and R_2 are in thermal equilibrium at absolute temperatures T_1 and T_2 respectively and are connected in parallel. Derive the equation for the average rms value of the terminal noise voltage.

Referring to Fig. 8-10, page 237, the noise current generators associated with R_1 and R_2 are respectively $\overline{I_1^2} = 4kT_1G_1\,\Delta f$ and $\overline{I_2^2} = 4kT_2G_2\,\Delta f$. The total noise power is

$$\frac{\overline{I_1^2} + \overline{I_2^2}}{G_1 + G_2} = \frac{4k\,\Delta f\,(T_1G_1 + T_2G_2)}{G_1 + G_2} \quad \text{or} \quad \overline{V^2}(G_1 + G_2) = \frac{4k\,\Delta f\,(T_1G_1 + T_2G_2)}{G_1 + G_2}$$

from which $$[\overline{V^2}]^{1/2} = \left[\frac{4k\,\Delta f\,(T_1G_1 + T_2G_2)}{(G_1 + G_2)^2}\right]^{1/2}$$

8.6. A circuit consists of a parallel combination of a capacitor C and a resistor R which are in thermal equilibrium at the standard noise temperature 290°K. What is the mean-squared terminal noise voltage if the effective frequency band is infinite?

The effective impedance of the parallel combination of R and C is

$$Z_C = \frac{R}{1 + \omega^2C^2R^2} - j\frac{\omega CR^2}{1 + \omega^2C^2R^2}$$

In noise calculations the interest is only in the real part of the impedance. Then by equation (*8.8*), page 238, and letting $u = \omega RC = 2\pi RCf$ and $du = 2\pi RC\,df$,

$$\overline{V^2} = 4kT\int_0^\infty \frac{R}{1 + \omega^2R^2C^2}\,df = \frac{4kTR}{2\pi RC}\int_0^\infty \frac{du}{1 + u^2} = \frac{2kT}{\pi C}\Big[\arctan u\Big]_0^\infty = \frac{kT}{C}$$

8.7. A line amplifier is designed so that the input and output impedances are 600 Ω. The effective bandwidth of the amplifier is 20 kilocycles per second. The input of the amplifier is terminated by an ideal 600 Ω resistor which is in thermal equilibrium at the standard noise temperature 290°K. What is the effective input source noise power level in dbm?

The mean-squared input noise voltage is

$$\overline{V^2} = 4kTR\,\Delta f = 4(1.38 \times 10^{-23})(290)(600)(20{,}000) = 1920 \times 10^{-16}$$

and the average rms value is $(\overline{V^2})^{1/2} = 43.9 \times 10^{-8}$ volts rms.

The source noise power $= \overline{V^2}/R = 3.2 \times 10^{-16}$ watts.

In dbm, the source noise power $= 10 \log [3.2 \times 10^{-6}/10^{-3}] = -124.94$ dbm.

FEEDBACK

8.8. An amplifier has an open loop (without feedback) mid-band voltage gain of magnitude $|A_v| = 12{,}480$. When the feedback loop is closed, the amount of feedback is given as 8 db. What is the magnitude of the mid-band voltage gain of the amplifier with feedback and what is the value of β if it is known that $\alpha + \theta = 180°$ for mid-band frequencies?

The db voltage gain without feedback is $20 \log 12{,}480 = 20(4.097) = 82$ db, and with feedback is $82 - 8 = 74$ db. Then $20 \log |A'_v| = 74$ or $|A'_v| = 5000$.

From $$|A'_v| = 5000 = \left|\frac{\mathcal{A}\angle\alpha}{1 - \mathcal{A}\beta\angle\alpha + \theta}\right| = \frac{12{,}480}{1 + 12{,}480\beta}$$ we find $\beta = 1.2 \times 10^{-4}$.

8.9. Given an amplifier with voltage feedback and a single noise source as shown in Fig. 8-18, page 246. The mid-band value of $\mathcal{A}_1$ is 20, of $\mathcal{A}_2$ is 100, and of $\mathcal{A}\beta$ is 4. The rms value of the noise source voltage is 0.01 volts. The output voltage is measured and the rms value is $V_{s2} = 100$ volts. When measurements are made with the feedback loop open, the input signal level is adjusted so that the output signal voltage remains 100 volts rms. Assume the amplifier is linear and free of noise but for the one noise source. (*a*) Calculate the magnitude of the mid-band voltage gain $|A'_v|$ when the feedback loop is closed. Find the $(S+N)/N$ ratio at the amplifier output when (*b*) the feedback loop is open (no feedback), (*c*) the feedback loop is closed.

(*a*) With no feedback, $|A_v| = \mathcal{A} = \mathcal{A}_1\mathcal{A}_2 = 20(100) = 2000$. With $\mathcal{A}\beta = 4$, the magnitude of the voltage gain with feedback is $|A'_v| = 2000/(1+4) = 400$. It is assumed the amplifier phase shift is 180° because a feedback network involving 180° phase shift is unlikely.

(*b*) When the feedback loop is open, the rms value of the output voltage is

$$V_{s2} = [(\mathcal{A}V_{s1})^2 + (\mathcal{A}_2V_n)^2]^{1/2} \quad \text{or} \quad 100 = [(2000V_{s1})^2 + (10^4)(10^{-4})]^{1/2}$$

Since $(2000V_{s1})^2$ must be large compared to 1, $V_{s1} \cong 0.05$ volts rms.

The rms value of the output signal-plus-noise is given as 100 volts and the rms value of the output noise is $\mathcal{A}_2V_n = 100(0.01) = 1$ volt; hence $(S+N)/N = 100/1 = 100$, or $20 \log 100 = 40$ db.

(*c*) With the feedback loop closed, use (*8.29*), page 247, to find

$$\text{output noise voltage} = \frac{\mathcal{A}_2\underline{/\alpha_2}}{1+\mathcal{A}\beta}V_n = (100/5)(0.01) = 0.2 \text{ volts rms}$$

The rms value of the output voltage is $V_{s2} = [(400V_{s1})^2 + (0.2)^2]^{1/2} = 100$. Since $(400V_{s1})^2$ must be large compared to $(0.2)^2$, $V_{s1} \cong 0.25$ volts rms.

The output $(S+N)/N$ ratio with feedback is $100/0.2 = 500$, or $20 \log 500 = 54$ db.

When, as in this example, the $(S+N)/N$ ratio is relatively high, it is numerically the same as the S/N ratio within slide rule accuracy. Hence the $(S+N)/N$ ratio is usually calculated and is often called the signal-to-noise ratio.

This example illustrates that voltage feedback improves the $(S+N)/N$ ratio at the output when there is gain preceding the noise source in the amplifier. It also shows that voltage gain is sacrificed. With feedback, the input voltage $V_{s1} = 0.25$ volts rms; without feedback, $V_{s1} = 0.05$ volts rms. In most practical problems the magnitude of the overall voltage gain with feedback would need to remain the same. The magnitude of the amplifier gain would be increased by an appropriate amount.

8.10. An amplifier is needed for an application which requires that the magnitude of the mid-band voltage gain be 10,000. In order to meet all performance specifications, an amplifier system incorporating voltage feedback is to be used. The mid-band value of $\mathcal{A}\beta\underline{/\alpha+\theta}$ is chosen to be -5. Calculate (*a*) the magnitude of the amplifier gain $\mathcal{A}$, (*b*) the value of β, (*c*) the amount of feedback in db.

(*a*) Using the equation for the system gain with feedback, $10{,}000 = \mathcal{A}/(1+5)$ or $\mathcal{A} = 60{,}000$. Since $\mathcal{A}\beta\underline{/\alpha+\theta} = -5$ and the phase shift of the feedback network is usually zero, the amplifier phase shift is assumed to be 180°. Thus $\alpha = 180°$ and $A_v = -60{,}000$.

(*b*) Using $\mathcal{A}\beta\underline{/\alpha+\theta}$ with $\mathcal{A}\underline{/\alpha} = -60{,}000$, obtain $\beta\underline{/\theta} = -5/(\mathcal{A}\underline{/\alpha}) = 8.33 \times 10^{-5}$.

(*c*) The db voltage gain of the amplifier without feedback $= 20 \log 60{,}000 = 95.6$ db. The db voltage gain with feedback $= 20 \log 10{,}000 = 80$ db. The amount of feedback $= 95.6 - 80 = 15.6$ db.

Or: Amount of feedback $= 20 \log(|A'_v|/|A_v|) = 20 \log 60{,}000/10{,}000 = 15.6$ db.

8.11. An amplifier is represented by the equivalent circuit of Fig. 8-21, page 250. The mid-band amplifier transfer admittance without feedback is $y_t = +1$ mho. The amplifier is terminated by a 500 ohm resistor. (*a*) Assume no feedback $(Z_{fb} = 0)$ and an output signal power of 5 watts into the load R_X. What should be the rms value of the input signal voltage V_{s1}? (b) A 4 ohm resistor is added as the feedback impedance Z_{fb}. If the signal output power is to remain at 5 watts, what is the rms value of the input signal voltage V_{s1} with feedback?

(a) The output signal power $P_o = I_{s2}^2 R_X$; then $5 = I_{s2}^2(500)$ and $I_{s2} = 0.1$ amp.

The amplifier transfer admittance without feedback is $y_t = i_{s2}/v_{s1} = +1$ mho. Then $V_{s1} = (+1)I_{s2} = +0.1$ volts rms. The positive sign indicates v_{s1} is in phase with i_{s2}.

(b) The transfer admittance with feedback is $y_t' = \dfrac{\mathcal{Y}/\alpha}{1+\mathcal{Z}\mathcal{Y}/\alpha+\theta} = \dfrac{+1}{1+4(1)} = +0.2$ mho. Then for the amplifier with feedback, $V_{s1} = I_{s2}/y_t' = 0.1/0.2 = 0.5$ volts rms.

8.12. **Derive the equation for the mid-band output admittance (a conductance) of an amplifier which has voltage-current feedback.**

The equivalent circuit of the amplifier system is given in Fig. 8-29. This is the equivalent circuit of Fig. 8-21, page 250, with the output conductance of the amplifier without feedback (a two port parameter) added as shown in Fig. 8-29.

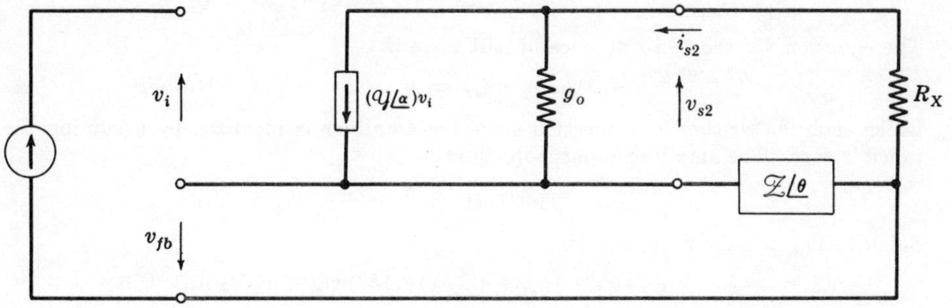

Fig. 8-29

The equation for the non-reference output node is

$$+(\mathcal{Y}/\alpha)v_i + g_o v_{s2} - i_{s2} = 0$$

The feedback voltage is $v_{fb} = -(\mathcal{Z}/\theta)i_{s2}$, and the input loop equation is

$$-v_{s1} + v_i - v_{fb} = 0 \quad \text{or} \quad v_i = v_{s1} - (\mathcal{Z}/\theta)i_{s2}$$

Substituting this value of v_i into the nodal equation,

$$+(\mathcal{Y}/\alpha)[v_{s1} - (\mathcal{Z}/\theta)i_{s2}] + g_o v_{s2} - i_{s2} = 0$$

or

$$i_{s2} = \frac{g_o}{1+\mathcal{Y}\mathcal{Z}/\alpha+\theta}v_{s2} + \frac{\mathcal{Y}/\alpha}{1+\mathcal{Y}\mathcal{Z}/\alpha+\theta}v_{s1}$$

Setting $v_{s1} = 0$,

$$g_o' = i_{s2}/v_{s2} = \frac{g_o}{1+\mathcal{Y}\mathcal{Z}/\alpha+\theta}$$

Setting $v_{s2} = 0$, the short circuit transfer admittance with feedback is

$$y_t' = i_{s2}/v_{s1} = \frac{\mathcal{Y}/\alpha}{1+\mathcal{Y}\mathcal{Z}/\alpha+\theta}$$

which is the same as equation (*8.36*), page 250.

When the feedback is inverse, $\alpha + \theta = 0$ for voltage-current feedback. The mid-band output admittance with feedback is $g_o' = g_o/(1+\mathcal{Y}\mathcal{Z})$. The mid-band output impedance with feedback is $r_o' = 1/g_o' = (1+\mathcal{Y}\mathcal{Z})r_o$. With inverse voltage-current feedback, the output impedance increases. This is opposite to the effect of inverse voltage-voltage feedback.

8.13. Derive the equation for the mid-band input impedance (a resistance) of an amplifier which has voltage-current feedback.

The equivalent circuit of the amplifier system of Fig. 8-21, page 250, is modified by adding the input resistance r_i (a two port parameter) to obtain Fig. 8-30.

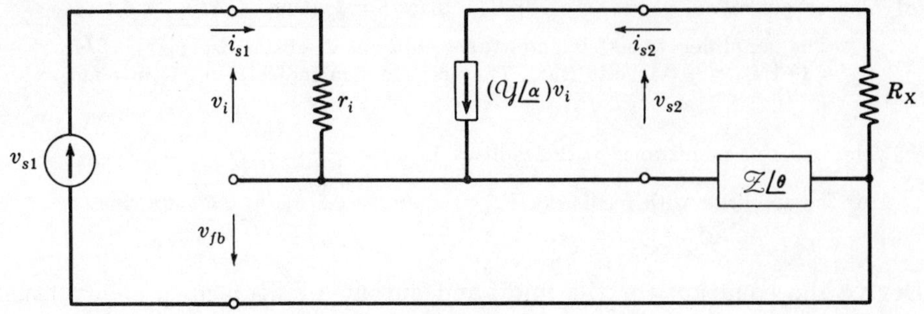

Fig. 8-30

With the feedback impedance included, the input loop equation is

$$-v_{s1} + r_i i_{s1} + (\mathcal{Z}\underline{/\theta})(i_{s1} + i_{s2}) \;=\; 0$$

The equation for the non-reference output node is

$$+(\mathcal{Y}\underline{/\alpha})v_i - i_{s2} \;=\; 0 \qquad \text{or} \qquad i_{s2} \;=\; (\mathcal{Y}\underline{/\alpha})v_i$$

which could be written by inspection since the amplifier is idealized by assuming the output admittance is zero. The amplifier input loop equation is

$$-v_i + r_i i_{s1} \;=\; 0 \qquad \text{or} \qquad v_i \;=\; r_i i_{s1}$$

Substituting into the first equation,

$$-v_{s1} + r_i i_{s1} + (\mathcal{Z}\underline{/\theta})(\mathcal{Y}\underline{/\alpha})r_i i_{s1} + (\mathcal{Z}\underline{/\theta})i_{s1} \;=\; 0$$

from which the input impedance with feedback is

$$Z_i' \;=\; v_{s1}/i_{s1} \;=\; (1 + \mathcal{Z}\mathcal{Y}\underline{/\alpha+\theta})r_i + (\mathcal{Z}\underline{/\theta}) \;=\; 0$$

Since $\mathcal{Z}\underline{/\theta}$ is usually very small compared to $(1+\mathcal{Z}\mathcal{Y}\underline{/\alpha+\theta})r_i$,

$$Z_i' \;=\; (1 + \mathcal{Z}\mathcal{Y}\underline{/\alpha+\theta})r_i$$

When the feedback is inverse and $\alpha+\theta = 0$, the mid-band input impedance $Z_i' = (1+\mathcal{Z}\mathcal{Y})r_i$.

8.14. Given a single stage, RC coupled, common cathode vacuum tube amplifier. The mid-band voltage gain is $A_v = -50$ and the half-power frequencies are $f_1 = 10$ cps and $f_2 = 15{,}000$ cps. A voltage-voltage feedback network is added with $\beta = 0.06$. The value of β is assumed constant and independent of frequency over the infinite frequency range, and the phase angle is zero. Make a polar plot of $\mathcal{A}\beta\underline{/\alpha+\theta}$ and apply the Nyquist criterion to determine if the amplifier is stable.

It is assumed that the low frequency and high frequency gains are respectively

$$A_v(\text{l.f.}) \;=\; \frac{A_{vo}}{1 - jf_1/f}$$

and $$A_v(\text{h.f.}) \;=\; \frac{A_{vo}}{1 + jf/f_2}$$

It is convenient to begin by plotting the mid-band value of $\mathcal{A}\beta\underline{/\alpha+\theta}$ which is $(-50)(0.06) = -3$. See Fig. 8-31.

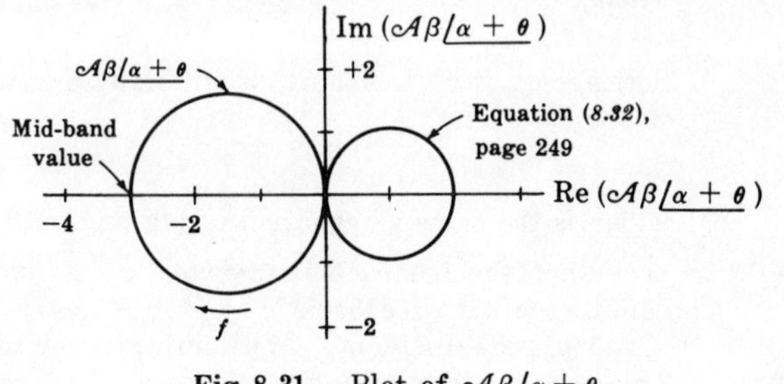

Fig. 8-31. Plot of $\mathcal{A}\beta\underline{/\alpha+\theta}$

The locus of $\mathcal{A}\beta\underline{/\alpha+\theta}$ for frequencies below the mid-band range is obtained by plotting a sufficient number of points to define the curve. Using $f_1 = 10$ and the equation for the low frequency gain,

$$\mathcal{A}\beta\underline{/\alpha+\theta} \;=\; \frac{-50}{1-j10/f}\,(0.06) \;=\; \frac{-3}{1-j10/f} \tag{8.40}$$

and the parametric equations for the locus are

$$x \;=\; \text{Re}\,(\mathcal{A}\beta\underline{/\alpha+\theta}) \;=\; \frac{-3}{1+100/f^2}, \qquad y \;=\; \text{Im}\,(\mathcal{A}\beta\underline{/\alpha+\theta}) \;=\; \frac{-30/f}{1+100/f^2}$$

The low frequency locus of $\mathcal{A}\beta\underline{/\alpha+\theta}$ is a semicircle in the third quadrant beginning at the origin for $f = 0$ and ending at the mid-band value of -3 for f large compared to f_1. In plotting points of the locus, it is convenient to substitute values of f into equation (*8.40*) and make a table as follows.

f, cps	$1 - j10/f$	$\mathcal{A}\beta$	$\alpha+\theta$
0	$-j\infty$	0	
1	$(1-j10) = 10.05\underline{/-84.3}$	0.298	$-264.3°$
2	$(1-j5) = 5.1\underline{/-78.7}$	0.588	$-258.7°$
5	$(1-j2) = 2.24\underline{/-63.4}$	1.34	$-243.4°$
10	$(1-j1) = 1.41\underline{/-45}$	2.12	$-225.0°$
20	$(1-j0.5) = 1.12\underline{/-26.6}$	2.68	$-206.6°$
50	$(1-j0.2) = 1.02\underline{/-11.3}$	2.94	$-191.3°$
100	$(1-j0.1) = 1.005\underline{/-5.7}$	2.99	$-185.7°$

The locus of $\mathcal{A}\beta\underline{/\alpha+\theta}$ for frequencies above the mid-band range may be plotted in a similar manner. Using $f_2 = 15{,}000$ and the equation for the high frequency gain,

$$\mathcal{A}\beta\underline{/\alpha+\theta} \;=\; \frac{-50}{1+jf/15{,}000}\,(0.06) \;=\; \frac{-3}{1+jf/15{,}000} \tag{8.41}$$

The locus of $\mathcal{A}\beta\underline{/\alpha+\theta}$ for high frequencies is a semicircle in the second quadrant beginning at the mid-band value of -3 for f small compared to f_2 and ending at the origin for $f \to \infty$. The tabulated low frequency values may be used to plot the high frequency locus since equations (*8.40*) and (*8.41*) are similar when written in terms of the ratios f_1/f or f/f_2, but for the sign of the angle of the denominator. For example, when $f = 30{,}000$ cps,

$$\mathcal{A}\beta\underline{/\alpha+\theta} \;=\; \frac{-3}{1+j2} \;=\; \frac{-3}{2.24\underline{/+63.4}} \;=\; 1.34\underline{/116.6°}$$

Equation (*8.32*), page 249, is also plotted in Fig. 8-31. Since the locus of $\mathcal{A}\beta\underline{/\alpha+\theta}$ does not pass through the circle representing (*8.32*), there is no frequency at which the feedback is regenerative. Hence this amplifier is stable.

An amplifier with two stages involving a mid-band phase shift of 180° each, will be regenerative at frequencies near the mid-band value because the mid-band value of $\mathcal{A}\beta$ is on the positive real axis. If the mid-band value of $\mathcal{A}\beta$ is equal to or greater than 1, the two stage amplifier will be unstable and oscillate.

The locus of $\mathcal{A}\beta\underline{/\alpha+\theta}$ is a circle only for a simple amplifier of the type assumed in this problem. When the high or low frequency gain ratio equations are different from those assumed, then a circle is only an approximation to the true locus.

8.15. Derive the equation for the transfer function of an amplifier with current-voltage feedback. The transfer function is the transfer impedance of the amplifier system. This system is seldom realized in practice but is included to illustrate the method of deriving the equations.

The equivalent circuit is shown in Fig. 8-32 below. The representation for the amplifier includes the input resistance r_i (a two port parameter) because the input current i_i is not zero.

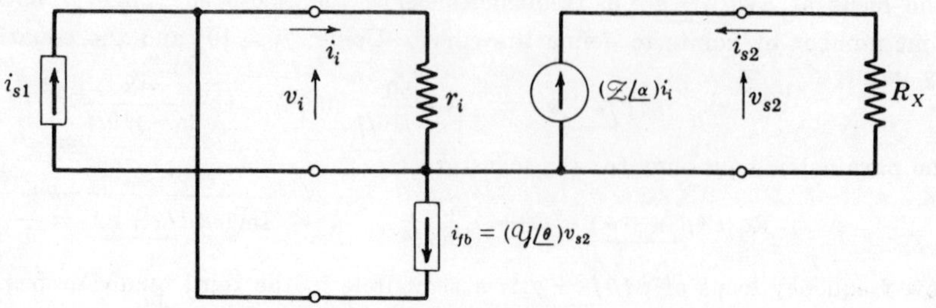

Fig. 8-32

The nodal equation for the non-reference input node is

$$-i_{s1} - i_{fb} + i_i = 0$$

By inspection, $v_{s2} = (\mathcal{Z}\underline{/\alpha})i_i$ or $i_i = v_{s2}/(\mathcal{Z}\underline{/\alpha})$. The feedback transfer function $\mathcal{Y}\underline{/\theta} = i_{fb}/v_{s2}$ or $i_{fb} = (\mathcal{Y}\underline{/\theta})v_{s2}$. Substituting into the first equation,

$$-i_{s1} - (\mathcal{Y}\underline{/\theta})v_{s2} + v_{s2}/(\mathcal{Z}\underline{/\alpha}) = 0$$

from which the transfer impedance with feedback is $z_t' = v_{s2}/i_{s1} = \dfrac{\mathcal{Z}\underline{/\alpha}}{1 - \mathcal{Y}\mathcal{Z}\underline{/\alpha + \theta}}$. The feedback is inverse when the denominator is greater than 1. The mid-band value of $\alpha + \theta$ is 180° for inverse feedback.

8.16. An amplifier has the following specifications: mid-band voltage gain $A_v = -50{,}000$, upper half-power frequency $f_2 = 20{,}000$ cps, lower half-power frequency $f_1 = 30$ cps, and mid-band output impedance $r_o = 2000$ ohms. Voltage feedback is to be added to the amplifier and the value of β is to be chosen so that the mid-band output impedance of the amplifier system with feedback is $r_o' = 600$ ohms. Determine (*a*) the value of β, (*b*) the mid-band voltage gain of the amplifier system with feedback, (*c*) the half-power frequencies of the amplifier system with feedback.

In most voltage-voltage feedback networks, the phase shift is zero. In this case the amplifier phase shift is 180° so that $\alpha + \theta = 180°$.

(*a*) The mid-band output impedance of the amplifier with feedback is

$$r_o' = \frac{r_o}{1 + \mathcal{A}\beta} \quad \text{or} \quad 600 = \frac{2000}{1 + 50{,}000\beta}$$

from which $\beta = 4.67 \times 10^{-5}$. The phase angle θ is zero.

(*b*) The mid-band voltage gain of the amplifier system with feedback is

$$A_v' = \frac{A_v}{1 + \mathcal{A}\beta} = \frac{-50{,}000}{1 + 50{,}000(4.67 \times 10^{-5})} = -15{,}000$$

(*c*) The lower half-power frequency of the amplifier system with feedback is $f_1' = f_1/(1 + \mathcal{A}\beta) =$ 9 cps. The upper half-power frequency of the amplifier system with feedback is $f_2' = (1 + \mathcal{A}\beta)f_2 = 66{,}600$ cps.

8.17. Given the operational amplifier circuit of Fig. 8-23, page 251, derive the equation for the input impedance of the amplifier system with feedback.

Equations (*8.37*), page 252, are solved for the input impedance with feedback,

$$Z_i' = v_{S1}/i = Z_i + [1/(\mathcal{A} + 1)]Z_f$$

When $\mathcal{A}$ is so large that $Z_f/\mathcal{A} \ll Z_i$, then $Z_i' \cong Z_i$.

8.18. An operational amplifier is available which has the following specifications: open loop gain $A_v = -15{,}000$; open loop input impedance of the amplifier is above 100 megohms; open loop output impedance is below 1000 ohms. An amplifier system is to be designed which has a voltage gain of magnitude 33. The input impedance of the amplifier system must be no less than 10,000 ohms. The output impedance is not critical but should not exceed several thousand ohms. Show how the amplifier system may be realized by using the operational amplifier.

Since an operational amplifier is available, this suggests using two precision resistors for the impedances Z_f and Z_i of Fig. 8-23, page 251. The ratio of the impedances is found by equation (*8.39*), page 252, as $33 = |Z_f/Z_i|$; then $Z_f = 33Z_i$.

Since the amplifier system input impedance Z_i' must be no less than 10,000 ohms, let the input impedance $Z_i = 10{,}000$ ohms (see Problem 8.17). The feedback impedance is then $Z_f = 33Z_i = 330{,}000$ ohms. If necessary, an adjustable precision resistor may be included in series with Z_i so that the magnitude of the voltage gain may be set to 33.

Substituting values in (*8.38*), we find $|A_v| = 32.9$. Thus in this case the approximate equation (*8.39*) gives adequate accuracy.

Supplementary Problems

8.19. Assume that the output signal voltage of an amplifier is given by equation (*8.3*), page 232, and that the input signal is the linear sum of two sine waves as in equation (*8.7*), page 236. Assuming the a_i of (*8.3*) are all zero for $i \geqq 4$, write the general equation for the instantaneous output signal voltage v_O. If $\omega_1 = 200\pi$ cps and $\omega_2 = 14{,}000\pi$ cps, what frequencies appear in v_O?

8.20. Refer to the push-pull Class-A amplifier circuit of Fig. 8-3, page 231. A power supply is used which has a small ripple voltage v_r across the terminals and which is represented by the equivalent circuit of Fig. 8-33. Assume that the transformer is perfectly balanced at the ripple frequency and that the tubes are also matched. Show that the output signal voltage does not contain any ripple components (assuming the ideal conditions).

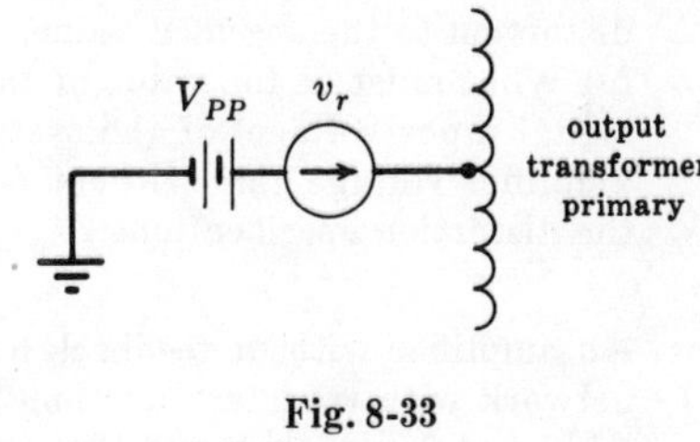

Fig. 8-33

8.21. Given the single-ended Class-A power amplifier circuit of Fig. 8-1, page 230. Assume the 6AQ5 tube parameters are $r_p = 50$ kΩ and $g_m = 4100$ μmho. The effective plate load resistance R_X^* is 5000 Ω when the load resistor R_X is 8 Ω. The rated power output is 4 watts.

(*a*) Calculate the approximate turns ratio of the output transformer.

(*b*) Assuming the efficiency of the output transformer is 75% at the rated output power, find the rms value of the signal input voltage V_{s1} and the power sensitivity of the amplifier.

8.22. Given the push-pull power amplifier circuit of Fig. 8-3, page 231. Assume the 6AQ5 tube parameters are $r_p = 50$ kΩ and $g_m = 4100$ μmho. The total effective load resistance R_X^* is 10,000 Ω (5000 Ω per tube) when the secondary of the output transformer is terminated with a resistor $R_X = 16$ Ω. The rated signal power output (into R_X) is 8 watts.

(*a*) Assuming the output transformer is ideal, find the rms values of the input signal voltages V_{s1a} and V_{s1b} which will give the rated power output.

(*b*) Repeat part (*a*) assuming the efficiency of the output transformer is 70% when delivering the rated output power.

(*c*) What is the approximate turns ratio of the output transformer?

8.23. A medium power transistor is used in a single-ended Class-A power amplifier circuit. The no-signal collector current has an average value $I_C = -200$ ma. A signal in the mid-band frequency range is applied and the signal level adjusted to give the rated power output. The average value of the collector current with signal is $I_{CQ} = -207.5$ ma. Assuming there is only second harmonic distortion, what is the magnitude of the nonlinear distortion if the magnitude of the output signal current at the applied input signal frequency is 150 ma?

8.24. A diode is operating in the temperature-limited region and the average plate current is $I_P = 2$ ma. The diode differential plate conductance is zero. A 10 kΩ load resistor is connected across the diode as shown in Fig. 8-12, page 239, and the resistor R_L is in thermal equilibrium at $T = 290°$K. The bandwidth is assumed to be 20,000 cps. Find the rms noise voltage across the terminals of R_L which is due to (*a*) the thermal noise generated in R_L, (*b*) the diode noise. (*c*) What is the error in the rms voltage across R_L if the thermal noise component is neglected? (The rms noise voltages do not add linearly.)

8.25. Given a triode vacuum tube in an amplifier which has an effective bandwidth of 15,000 cps. Assuming the equivalent noise resistance of the triode is $R_{\text{eff}} = 2.5/g_m$ and that the g_m of the tube is 3000 μmho, find R_{eff}. What is the mean-squared value of the noise voltage across R_{eff} at the standard noise temperature 290°K?

8.26. Given an amplifier identical to that of Problem 8.7, page 255, but for the bandwidth. The bandwidth is 10 kilocycles per second instead of 20 kilocycles per second. Assuming all other conditions identical, what is the effective source noise power in dbm?

8.27. An amplifier has an open loop voltage gain $A_v = -100{,}000$. A voltage-voltage inverse feedback system is to be designed which has 10 db of feedback. Find the voltage gain of the amplifier system with feedback and the value of β.

8.28. A single-ended Class-A power amplifier stage has 5% nonlinear distortion when operating at the rated power output of 4 watts into a 600 ohm load resistor. There are *R-C* coupled amplifier stages preceding the power amplifier stage to provide the necessary signal voltage gain. The amplifier input signal voltage $V_{s1} = 0.1$ volts rms. It is necessary that the nonlinear distortion be reduced to 2% or less. A voltage-voltage inverse feedback network is to be added to reduce the nonlinear distortion to the specified value. It is assumed that all the distortion is in the power amplifier stage. (*a*) What must be the value of the overall voltage gain (signal voltage across the load resistor to the signal input voltage) of the system with inverse feedback? (*b*) What must be the magnitude of the amplifier voltage gain without feedback? (*Hint.* The required value of $\mathcal{A}\beta$ may be determined from the distortion specifications.)

8.29. An amplifier without feedback has a mid-band db voltage gain of 60 db. A voltage-voltage feedback network with transfer function $\beta = 2 \times 10^{-3}$ is added. It may be assumed that $\alpha + \theta = 180°$ for mid-band frequencies and that the amplifier is stable. Determine (*a*) the magnitude of the amplifier mid-band voltage gain with feedback, (*b*) the mid-band db voltage gain of the amplifier with feedback, (*c*) the amount of feedback in db.

8.30. The mid-band voltage gain of an amplifier without feedback is $A_v = -8000$. A voltage-voltage feedback network with transfer function $\beta = 2 \times 10^{-4}$ is added. The half-power frequencies of the amplifier without feedback are $f_1 = 10$ cps and $f_2 = 40{,}000$ cps. The mid-band value of $\alpha + \theta$ is 180° and the amplifier is stable. (*a*) Find the magnitude of the mid-band voltage gain of the amplifier with feedback. (*b*) Find the half-power frequencies of the amplifier system with feedback. (*c*) Discuss some of the assumptions made in the solution to part (*b*). In particular, discuss the assumptions made in the derivations of the equations for the half-power frequencies of an amplifier with voltage-voltage feedback.

8.31. Given the voltage-current feedback amplifier system of Fig. 8-21, page 250. The half-power frequencies of the amplifier without feedback are f_1 and f_2. These are the frequencies at which the transfer admittance of the amplifier is 0.707 times the mid-band value. It may be assumed that the normalized low frequency transfer admittance is $1/(1 - jf_1/f)$ and that the normalized high frequency transfer admittance is $1/(1 + jf/f_2)$. Derive the equations for the half-power frequencies of the amplifier system with inverse feedback.

8.32. The following data apply to the amplifier system of Problem 8.31: $y_t = +1$ mho, $Z_{fb} = 2$ ohms, $f_1 = 30$ cps, $f_2 = 14{,}000$ cps. (*a*) Calculate the transfer admittance of the amplifier system with feedback, y_t'. (*b*) Sketch the locus $\mathcal{YZ}\underline{/\alpha + \theta}$ for the amplifier. Is the amplifier stable?

8.33. Given the amplifier system with current-voltage feedback as in Problem 8.15, page 259. Derive the equations for the mid-band output impedance of the amplifier system with inverse feedback. The circuit diagram of Fig. 8-32 is modified by adding the mid-band output impedance r_o (without feedback) in series with the voltage generator of the output circuit.

8.34. Derive the equation for the output impedance of the operational amplifier circuit of Fig. 8-34. The operational amplifier itself is represented by the simplified two port equivalent circuit which includes r_o (output impedance of the operational amplifier).

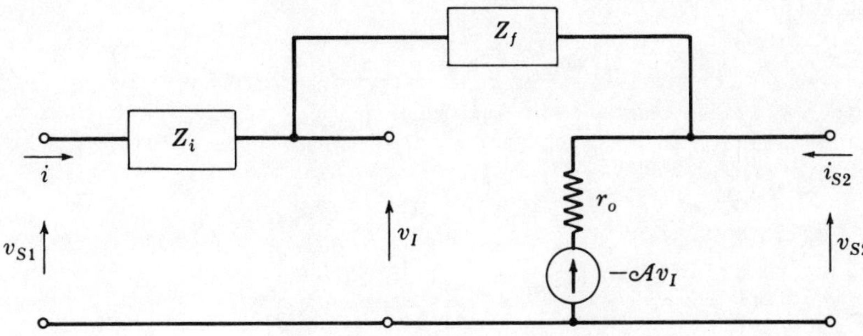

Fig. 8-34

8.35. Given the operational amplifier circuit of Fig. 8-35 with three inputs. Derive the equation for the output voltage v_O as a function of the three input voltages v_1, v_2 and v_3. Assuming $R_1 = R_2 = R_3 = R_f$, is this circuit an adder?

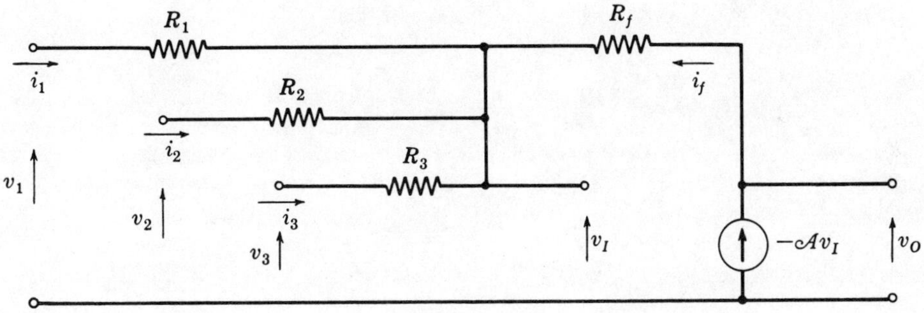

Fig. 8-35

Appendix A

Notation and Symbols for Vacuum Tubes

Notation	Description
v_P	Instantaneous voltage drop, plate to cathode
V_P	Average value of v_P
v_p	Instantaneous value of the varying component of v_P
V_p	rms or effective value of v_p
V_{PP}	Plate supply voltage
i_P	Instantaneous plate current
I_P	Average value of i_P
i_p	Instantaneous value of the varying component of i_P
I_p	rms value of i_p
v_G	Instantaneous voltage drop, grid to cathode
V_G	Average value of v_G
v_g	Instantaneous value of the varying component of v_G
V_g	rms value of v_g
V_{GG}	Grid supply voltage (bias supply)
i_G	Instantaneous grid current
I_G	Average value of i_G
i_g	Instantaneous value of the varying component of i_G
I_g	rms value of i_g

Grids are numbered 1, 2, 3, . . . , beginning with the grid nearest the cathode. Thus the voltages and currents are correspondingly identified. For example: v_{G1}, v_{G2}, i_{G2}, I_{g2}, etc.

V_f	rms value of heater or filament voltage
I_f	rms value of the heater or filament current
v_S	Instantaneous value of the signal voltage
V_S	Average value of v_S
v_s	Instantaneous value of the varying component of v_S
V_s	rms value of v_s

Signal voltages are identified by number depending on the circuit. For example: v_{s1}, v_{s2}, etc.

If a supply voltage has a varying component, the instantaneous value is shown as $v_{PP} = V_{PP} + v_{pp}$.

Appendix B

Notation and Symbols for Transistors

The notation used for the voltages, currents, and powers of all electronic control devices is consistent with the IRE Standards on Letter Symbols for Semiconductor Devices, 1956. This notation scheme was introduced and applied to vacuum tube diodes in Chapter 2.

Each of the three transistor terminals may be used as the reference terminal, hence a complete listing of all letter symbols is long and quite redundant. This would include a listing of the notation for a transistor with the emitter, the base, and the collector as reference elements.

The notation scheme may be summarized as follows:

1. All instantaneous values of voltage, current, and power are denoted using the lower case of the appropriate letter symbol.

2. All maximum, average, or rms values of voltage, current, and power are denoted using the upper case of the appropriate letter symbol.

3. The subscripts of all average, maximum, or instantaneous (total) values of voltage, current, and power are upper case letters. If it is necessary to distinguish the maximum and average values, then the subscript includes a "max" or "av", respectively.

Examples: I_{BE} is the average value of the current in the base terminal with the emitter common (this is the average value of i_{BE}).

v_{CB} is the instantaneous voltage drop from the collector to the base (reference terminal).

4. The subscripts of the instantaneous values and the rms values of varying components are lower case letters.

Examples: i_{be} is the instantaneous value of the varying component of the current in the base with the emitter common.

V_{ce} is the rms value of the varying component of the collector to emitter voltage drop (this is the rms value of v_{ce}).

5. The first subscript designates the non-reference terminal and the second subscript designates the reference terminal. The current references are in the non-reference terminals and the voltage references are the voltage drops from the non-reference terminals to the reference terminal.

6. Power supply voltages are denoted by repeating the non-reference terminal subscript. The reference terminal may be given as the third subscript.

Examples: V_{EEB} is the emitter power supply voltage with the base common.

V_{CCE} is the collector power supply voltage with the emitter common.

Answers to Supplementary Problems

CHAPTER 2

2.6. $i_{P1} = i_{P2} = i_{P3} = 4$ ma, $v_{P1} = 1.8$ v, $v_{P2} = 2.3$ v, $v_{P3} = 2.8$ v

2.7. $i_{P1} = 7.7$ ma, $v_{P1} = 2.6$ v, $i_{P2} = i_{P3} = 3.2$ ma, $v_{P2} = 2$ v, $v_{P3} = 2.3$ v, $v_R = 4.4$ v, $i_R = 4.4$ ma

2.8. $i_{P1} = 4.7$ ma, $v_{P1} = 2.5$ v, $v_{P2} = v_{P3} = 2.5$ v, $i_{P2} = 2.7$ ma, $i_{P3} = 1.9$ ma

2.9. (*a*) $r_{P1} = 532\ \Omega$, $r_{P2} = 927\ \Omega$, $r_{P3} = 1315\ \Omega$
(*b*) $g_{P1} = 1880\ \mu$mho, $g_{P2} = 1080\ \mu$mho, $g_{P3} = 760\ \mu$mho

2.10. $v_P = 2.9$ v, $i_P = 2.1$ ma

2.11. (*a*) $V_P = 3.8$ v, $I_P = 6.2$ ma, $R_P = 613\ \Omega$
(*b*) $r_p = 360\ \Omega$
(*c*) $v_p = 0.8 \sin \omega t$ v, $i_p = 2.3 \sin \omega t$ ma
(*d*) $v_p = 0.8 \sin \omega t$ v, $i_p = 2.3 \sin \omega t$ ma
(*e*) $v_{P(\max)} = 4.7$ v, $v_{P(\min)} = 3.0$ v
$i_{P(\max)} = 8.7$ ma, $i_{P(\min)} = 4.0$ ma
$r_{P(\max)} = 750\ \Omega$, $r_{P(\min)} = 540\ \Omega$

2.12. $r_p = 360\ \Omega$

2.13. $i_{P1} = i_{P2} = 4.1$ ma, $v_{P1} = 3.3$ v, $v_{P2} = 4.7$ v

2.14. (*a*) $V_P = 5.3$ v, $I_P = 4.7$ ma. (*b*) $r_p = 686\ \Omega$. (*c*) $v_p = 0.4 \cos \omega t$ v, $i_p = 0.6 \cos \omega t$ ma

2.15. (*a*) $V_P = 4.2$ v, $I_P = 3.0$ ma. (*b*) $v_{pp} = 2.9 \sin \omega t$ v

2.16. $V_{PP} = 5.1$ v

2.19. For $v_P = 0.8$ v: $g_P = 0.0096$ mho, $g_p = 0.077$ mho. $G_1 = 0.0218$ mho

CHAPTER 3

3.11. $2V/\pi$

3.12. Ripple = 15.5 v

3.17. (*a*) $R = 4000\ \Omega$.
(*b*) $P = 0.355$ watt

3.19. $V_C = 8$ v

3.23. $V_C = 8$ v, $V_C = 13$ v

3.26. $V_C = 12$ v

3.28. $V_O = 1.59$ v

CHAPTER 4

4.10. $h_i = 24.5\ \Omega$, $h_r = 0.123$, $h_f = -0.123$, $h_o = 0.01385$ mho

4.11. $g_i = 0.12$ mho, $\alpha_r = -0.6$, $g_o = 0.0667$ mho, $\alpha_f = -0.333$

CHAPTER 5

5.9. (*a*) $V_G = -4$ v, $V_P = 155$ v, $I_P = 7.2$ ma
(*b*) $G_P = 46\ \mu$mho
(*c*) $g_p = 130\ \mu$mho, $g_m = 2500\ \mu$mho
(*d*) $\mu = 19$
(*g*) $A_v = -13.9$

5.10. (*a*) $V_G = -1.2$ v, $V_P = 140$ v, $I_P = 0.8$ ma. (*b*) $\mu = 90$, $r_p = 75$ kΩ. (*d*) $A_v = -51$

5.11. (*a*) $V_{PP} = 310$ v. (*b*) $V_{GG} = 4$ v

5.12. (*a*) $V_P = 143$ v, $I_P = 1.06$ ma, $V_G = -1$ v. (*b*) $R_P = 135$ kΩ

5.13. $V_P = 184$ v, $V_G = -7.1$ v, $I_P = 3.55$ ma

5.14. $V_P = 167$ v, $I_P = 13.3$ ma, $V_G = -3$ v

5.15. (*a*) $V_P = 137$ v, $I_P = 3$ ma, $V_G = -6$ v. (*b*) $\mu = 16$, $r_p = 12.5$ kΩ. (*d*) $v_{s2} = 16 \sin 2000\pi t$ v

5.16. (*a*) $V_P = 282$ v, $I_P = 1.75$ ma, $V_G = -17.5$ v. (*b*) $g_m = 850\ \mu$mho, $g_p = 35\ \mu$mho

5.17. (*a*) $v_p = 14 \sin(\omega t + \pi)$ v, $i_p = 0.0007 \sin \omega t$ amp. (*b*) $A_v = -14$

5.18. (*a*) $I_P = 3.1$ ma, $V_P = 86$ v, $V_G = -3.1$ v.
(*b*) $g_m = 1700\ \mu$mho, $g_p = 88\ \mu$mho
(*e*) $A_v = -13$
(*f*) $P = 9.6$ mw

5.19. (*a*) $v_{s2} = 1.13 \sin(2000\pi t + \pi)$ v
(*b*) $v_{s2} = 1.74 \sin(2000\pi t + \pi)$ v
(*c*) $P = 3.3 \times 10^{-6}$ watt, $P = 7.6 \times 10^{-6}$ watt

5.20. (*a*) $V_P = 92$ v, $I_P = 3.3$ ma, $V_G = -3.3$ v
(*b*) Power dissipated: $P_{PS} = -0.99$ watt, $P_{R_K} = +0.011$ watt, $P_{R_L} = +0.661$ watt, $P_P = +0.301$ watt
(*c*) $r_{P(\max)} = 41$ kΩ, $r_{P(\min)} = 16$ kΩ

5.21. (*a*) $V_{S1} = 10$ v. (*b*) $V_P = 144$ v, $I_P = 7$ ma. (*c*) $V_2 = 158$ v

5.22. (*a*) $V_P = 160$ v, $I_P = 4.5$ ma, $V_G = -6.5$ v
(*c*) $\mu = 17$, $r_p = 11$ kΩ
(*d*) $v_{P(\max)} = 165$ v, $v_{P(\min)} = 153$ v

5.23. (*a*) $A_v = +9.3$. (*b*) $A_v = +20.9$

5.24. (*a*) $A_v = +0.69$. (*b*) $A_v = +0.87$

5.25. (*a*) $V_P = 187$ v, $I_P = 33$ ma, $V_G = -9.9$ v. (*b*) $A_v = -4.8$

5.26. $g_m = 3500$ μmho, $g_p = 150$ μmho

5.30. (*a*) $V_P = 165$ v, $V_G = -7$ v, $I_P = 35$ ma (*b*) $\mu = 9.5$, $r_p = 2150\ \Omega$ (*c*) $v_p = 27 \sin(\omega t + \pi)$ v, $i_p = 8 \sin \omega t$ ma (*d*) $P_{PS} = -11.2$ w, $P_{R_L} = 4.78$ w, $P_{R_K} = 0.25$ w, $P_P = 5.67$ w

5.32. (*b*) $V_P = 40$ v, $I_P = 2.2$ ma, $V_G = -1$ v. (*d*) $v_{s2} = 14 \sin(\omega t + \pi)$ v

5.33. (*b*) $A_v = -5.36$

5.34. $v_{\text{out}} = 0$

5.36. $r_{P(\max)} = 445\ \text{k}\Omega$, $r_{P(\min)} = 75\ \text{k}\Omega$

5.37. (*a*) $A_v = -14$. (*b*) $A_v = -27$

5.38. (*a*) $V_P = 142$ v, $I_P = 5.3$ ma, $V_{G1} = -2.25$ v, $V_{G2} = 147.7$ v, $I_{G2} = 2.2$ ma (*c*) $g_m = 3500$ μmho (*d*) $A_v = -92.5$

5.40. $R_L^* \to \infty$

5.41. $A_v = 27.5$

5.42. $R_K = 960\ \Omega$, Power rating > 3.3 mw

5.43. (*a*) $v_{S2} = 130 + 59 \sin(2000\pi t + \pi)$ v. (*b*) $v_{s2}/l_k = -295$ volts/lumen

5.45. (*b*) $A_v = +0.875$

5.46. $V_{GG} = +99$ v

5.47. (*a*) $R_K = 1500\ \Omega$. (*b*) $V_P = 86$ v, $I_P = 2$ ma

5.49. $A_v = -12$

5.50. $\mu = 18.4$, $r_p = 3750\ \Omega$, $g_m = 4900$ μmho

5.51. (*a*) $A_v = -364$. (*b*) $\mu = 2550$. (*c*) $A_v = -10.1$

5.53. $A_v = 10.6$

5.54. $h_i = 2475\ \Omega$, $h_r = 0.0476$, $h_o = 12$ μmho, $h_f = -1.0$

CHAPTER 6

6.15. (*a*) $I_{CE} = -2.75$ ma, $V_{BE} = -166$ mv, $V_{CC} = 23.7$ v. (*d*) $R_1 = 1.18\ \text{m}\Omega$ (*c*) $h_{fe} = 100$, $h_{oe} = 106$ μmho.

6.16. (*b*) $r_{C(\max)} = 4650\ \Omega$, $r_{c(\min)} = 780\ \Omega$

6.17. (*a*) $V_{CB} = 8.4$ v, $I_{CB} = 2.9$ ma. (*b*) $R_{CB} = 2900\ \Omega$. (*c*) $h_{fb} = -0.93$, $h_{ob} = 0.5$ μmho

6.18. (*a*) $R_B = 500\ \text{k}\Omega$. (*b*) $V_{CE} = -5.7$ v, $V_{BE} = -180$ mv, $I_{BE} = -30$ μa

6.21. $v_{s2} = 3.45 \sin(\omega t + \pi)$ v

6.22. $h_i = 1640\ \Omega$, $h_r = 1640 \times 10^{-6}$, $h_f = 41$, $h_o = 191$ μmho

6.23. (*b*) $A_i = 22.2$. (*c*) $A_v = -62.5$. (*d*) $A_p = 1388$

6.24. (*b*) $V_{CE} = -7.5$ v, $I_{BE} = -20$ μa, $V_{BE} = -168$ mv. (*c*) $R_2 = 22.7\ \text{k}\Omega$

6.25. (*b*) $A_v = -86.2$. (*c*) $R_i = 1450\ \Omega$

6.26. (*a*) $A_v = 126$. (*b*) $R_i = 31\ \Omega$

6.27. (*a*) $A_v = 0.997$. (*b*) $R_i = 473\ \text{k}\Omega$

6.28. $r_t = 473\ \text{k}\Omega$

6.30. (*a*) $R_i = 47.3\ \Omega$. (*b*) $v_{s2} = 0.23 \sin 2000\pi t$ v

6.31. $A_i = 46.8$

6.32. $v_{\text{gen}} = 0.0017 \sin(2000\pi t + \pi)$ v

6.33. $h_i = 1500\ \Omega$, $h_r = 5 \times 10^{-4}$, $h_f = 50$, $h_o = 320$ μmho

6.34. $r_i = 473\ \text{k}\Omega$, $r_r = 6660\ \Omega$, $r_o = 6670\ \Omega$, $r_F = 473\ \text{k}\Omega$

6.35. (*a*) $A_i = -33.7$. (*b*) $A_p = 33.5$

6.36. $g_t = 0.0125$ mho

6.37. $A_v = -83.3$

6.39. $v_{s2} = 0.144 \sin 2000\pi t$ v

6.41. $A_v = -129$

6.42. $A_v = -145$

6.44. (a) $V_{s2} = 0.978\underline{/148.8^\circ}$ v. (b) $A_v = 139.6\underline{/148.8^\circ}$. (c) $A_i = 46.7\underline{/2.3^\circ}$

CHAPTER 7

7.21. (a) $A_{vo} = -100$. (b) $f_1 = 127$ cps. (c) $f_2 = 795{,}000$ cps. (d) $\Delta f \cong 795{,}000$ cps

7.22. (a) $f_2 = 3.5$ mc/s. (b) $A_{vo} = 0.91$

7.23. (a) $A_{io} = 35.5$. (b) $f_1 = 0.883$ cps. (c) $A_{vo} = -73.4$. (d) $f_1 = 1.66$ cps. (e) $y_{to} = 0.0147$ mho
(f) $f_1 = 1.66$ cps. (g) $r_{to} = -177{,}600\ \Omega$. (h) $f_1 = 0.883$ cps

7.24. (a) $A_{vo} = -34.3$. (b) $f_1 = 34.1$ cps. (c) $f_2 = 55{,}600$ cps

7.25. (a) $C_C = 0.14\ \mu$f. (b) $A_{vo} = 9.26$, db voltage gain = 19.4 db

7.26. (a) $y_{to} = -99.2\ \mu$mho. (b) $f_1 = 3.2$ cps, $f_2 = 65$ mc/s

7.27. (a) $f_1^* = 19.6$ cps, $f_2^* = 12{,}700$ cps. (b) slope magnitude = 18 db per octave

7.28. $f_1^* = 33$ cps, $f_2^* = 9240$ cps

7.30. (a) $A_{vo} = -449$, $f_2 = 73{,}000$ cps. (b) $A_{vo} = -12.3$, $f_2 = 2.65$ mc/s. (c) GBW $= 32.8 \times 10^6$

7.32. $P = 75$ mw

7.33. db = 81.3

7.34. db = 34.3

7.35. 21.9 dbm

7.36. $r_o = 9750\ \Omega$, $A_v = -84.3$

7.37. $r_o = 3160\ \Omega$

7.39. (a) $R_i = 30.1\ \Omega$. (b) $r_o = 2310\ \Omega$, $A_v = 3.5$. (c) Yes

7.41. (a) $v_{\text{gen}} = 0.0017 \sin(\omega t + \pi)$ v. (b) $C_S = 81.7\ \mu\mu$f

7.42. (a) $h_i = 2000\ \Omega$, $h_r = 0.998$, $h_f = -51$, $h_o = 250\ \mu$mho
(b) $A_i = -51$, $g_o = 250\ \mu$mho (c) $G_i = 6.78\ \mu$mho

7.43. (a) $r_o = 451\ \Omega$ (c) $A_v = 0.905$
(b) $f_1 = 0.791$ cps (d) $z_o = (2200 + j12.6f)/(0.022 + j0.0278f)\ \Omega$

7.44. $r_o = 12{,}950\ \Omega$

7.45. (a) $r_o = 96{,}800\ \Omega$. (b) $r_o = 42{,}800\ \Omega$

7.46. (a) $r_o = 9700\ \Omega$. (b) $A_v = 14.5$

7.47. (a) $A_v = -24.4$. (b) $r_o = 454\ \Omega$

7.48. $A_v = 398$

CHAPTER 8

8.19. Frequencies of v_O in cps are as follows:
100, 200, 300, 6800, 6900, 7000, 7100, 7200, 13,900, 14,000, 14,100, 21,000

8.21. (a) $a = 25$. (b) $V_{s1} = 8.76$ v, P.S. = 0.052 ω/v^2

8.22. (a) $V_{s1a} = V_{s1b} = 7.58$ v. (b) $V_{s1a} = V_{s1b} = 9.07$ v. (c) $a = 25$

8.23. $D = 5\%$

8.24. (a) $\overline{V} = 1.79\ \mu$v. (b) $\overline{V}_{pk} = 35.8\ \mu$v. (c) Less than 0.1%

8.25. $R_{\text{eff}} = 833\ \Omega$, $\overline{V} = 0.45\ \mu$v

8.26. $P = -127.95$ dbm

8.27. $A'_v = 31{,}600$, $\beta = 2.17 \times 10^{-5}$

8.28. (a) $A'_v = -490$. (b) $A_v = -1225$

8.29. (a) $A'_v = -333$. (b) 50.5 db. (c) 9.5 db

8.30. (a) $A'_v = -3075$. (b) $f'_1 = 3.85$ cps, $f'_2 = 104{,}000$ cps

8.32. (a) $y'_t = 0.333$ mho

INDEX